North American Cattle-Ranching Frontiers

NORTH AMERICAN CATTLE-RANCHING FRONTIERS

Origins, Diffusion, and Differentiation

Terry G. Jordan

HISTORIES OF THE AMERICAN FRONTIER
RAY ALLEN BILLINGTON, GENERAL EDITOR

HOWARD R. LAMAR, COEDITOR
MARTIN RIDGE, COEDITOR
DAVID J. WEBER, COEDITOR

Published at the quincentennial of the first introduction of cattle into the Americas

University of New Mexico Press: ALBUQUERQUE

*To Madeleine Vera,
born on her grandfather's birthday
in a year of celebration, anniversary, and death;
a hope for the future and a continuity of the family.
Remember and cherish your proud heritage in Texas,
wherever life takes you.*

Library of Congress Cataloging-in-Publication Data

Jordan, Terry G.
　　North American cattle-ranching frontiers : origins, diffusion, and differentiation / Terry G. Jordan. — 1st ed.
　　　　p.　　cm. — (Histories of the American frontier)
　　Companion to: The American backwoods frontier.
　　Includes bibliographical references (p.　　) and index.
　　ISBN 0-8263-1421-X (cl.) — ISBN 0-8263-1422-8 (pa)
　　1. Cattle—North America—History.　2. Beef cattle—North America—History.　3. Cattle—History.　4. Ranchers—North America—History.　5. Human geography—North America.　6. North America—Historical geography.　I. Title. II. Series.
SF196.N7J67　1993
636.2'13'097—dc20　　　　　　　　　　　　　　　　　　　　　　　　92-36428
　　　　　　　　　　　　　　　　　　　　　　　　　　　　　　　　　　　CIP

Design: Susan Gutnik

© 1993 by the University of New Mexico Press.
All rights reserved.
First edition.

Contents

Foreword
ix

Preface
3

ONE
The Nature of Cattle Ranching
7

TWO
Atlantic Fringe Source Regions
18

THREE
Implantments and Adaptations in the West Indies
65

FOUR
From the Indies to the Mainland
86

FIVE
Cattle Frontiers in Northern Mexico
123

SIX
Carolina's Children
170

SEVEN
THE ANGLO-TEXAN RANCHING SYSTEM
208

EIGHT
PASTORAL CALIFORNIA EXTENDED
241

NINE
THE MIDWEST TRIUMPHANT
267

TEN
CONCLUSION
308

NOTES
315

SELECTED ANNOTATED BIBLIOGRAPHY
391

INDEX
429

Figures

1. Spatial model of commercial land-use.
2. Old World range cattle herding areas.
3. Map of traditional livestock herding regions of Spain.
4. Map of the vegetation of Las Marismas of the Guadalquivir.
5. Map of selected stock trails of Las Marismas.
6. A mounted vaquero of Andalucía.
7. The traditional hornless saddle of a province in Costa Rica.
8. Some livestock brands from W. Africa, N. America, and Spain.
9. Andalusian reddish *retinto* longhorn cattle.
10. A metal sign marks the entrance to the Marisma de Hinojos.
11. Map of land tenure and toponyms of Las Marismas.
12. Map of livestock ratios in the Spanish cattle zone, 1865.
13. Map of the highlands and lowlands in the British Isles.
14. Map of Scottish cattle drove roads about A.D. 1600.
15. Map of the cattle-herding belt of tropical West Africa.
16. Diagrams of camps of cattle-raising Fulani in West Africa.
17. Cattle ranching implantments and spread in the W. Indies.
18. Ranches and plantations in British colonial Jamaica, c. 1690.
19. "Feral dog of St. Domingo."
20. A 1579 map of a cattle ranch in Mexico.
21. Map of the Pánuco delta, Mexico.
22. Map of cattle-ranching grants in Mexico to 1620.
23. A prototypical form of lassoing from horseback.
24. Mexican cattle of mixed breed in a rock corral.
25. Herders driving Spanish longhorn cattle along a cañada, Mexico.
26. The colonial South Carolina cowpen implantment and diffusion.
27. Map of the diffusion of cattle ranching in Mexico, 1530–1830.
28. Cattle distribution in Nuevo Santander, 1757.
29. Hispanic influence on American cattle-raising culture.
30. Mexican-type palisado, or picket corral, Comal County, Texas.
31. Map of range resources and climate in California.

32. Spatial evolution of Hispanic Californian cattle ranching.
33. Coastal southern habitats and the cattle: population ratio.
34. Evidence for the diffusion of the Carolina herding complex.
35. Culturoethnic view of cattle herding in the South to 1850.
36. An East Texas "cow dog."
37. A "straight-rail" stock pen in Blanco County, Texas.
38. The cattle complex in the upland South and Midwest.
39. The Texas System of cattle ranching: origins and spread.
40. A drawing of Anglo-Texan cowboys branding cattle.
41. Population: cattle ratios in Texas, 1860 and 1870.
42. The William Perryman Ranch in Frio County, Texas, c. 1870.
43. Migrations of cattle ranchers to and from north-central Texas.
44. Migrations of cattle ranchers to montane West Texas.
45. Texas cowboys "cutting out" cattle on the Great Plains.
46. Origin and diffusion of Anglo-Californian ranching system.
47. Cattle distribution and population ratio in California, 1852.
48. Ranches and other landholdings of Miller and Lux.
49. Californian and Texan tradition equestrian material culture.
50. A "Spanish windlass."
51. The Old Peter French beef wheel at Frenchglen, Oregon.
52. Routes of seasonal cattle transhumance in the western U.S.
53. Transhumant cattle being driven to summer mountain pastures.
54. The diffusion of midwestern cattle ranching through the West.
55. Maj. J. S. Smith's herd wintering on hay in central Kansas.
56. Beef cattle distribution in midwestern-influenced West, 1850.
57. Distribution of large beef cattle herds in Oregon, 1850.
58. Beef cattle distribution in midwestern-influenced West, 1870.
59. The Centerville Ranch on the Camas Prairie, Idaho, 1880.
60. A willow corral near Frenchglen, Oregon.
61. Log horse barn in the ranching country of British Columbia.
62. A traditional "chock-and-log" fence in southwestern Montana.
63. A traditional roofless log haycrib, Wyoming.
64. Three "Beaverslide" hay stackers in a Montana valley.
65. A sagebrush cattle pasture enclosed by a pole "worm" fence.

Foreword

Although the era of open-range cattle ranching on the Great Plains lasted only briefly, from the end of the Civil War through the mid-1880s, that moment in time has bred a stunning number of memoirs, scholarly studies, and fictional accounts. Over and over again, via cinema, story, and song, Americans and aficionados of American culture around the world have taken imaginary journeys on long, heroic cattle drives, pushing doggies north from Texas over dusty, dangerous, unfenced trails to boisterous towns along the railhead in Kansas. At the same time, scholars have studied the cattle business—from its financing to its symbolism, from its Iberian origins to individual ranches, ranchers, and ranching techniques.

Now, in his original and erudite *North American Cattle-Ranching Frontiers*, Terry Jordan has recast this highly familiar and much-studied subject. He has done so, in part, by leaving the North American West behind, crossing the Atlantic, and seeking the origins of western ranching in Europe and Africa. There, in the Old World, Jordan finds that a combination of cultural and environmental influences produced several distinctive ways of raising beef cattle. As the Europeans and Africans expanded to the Americas in the century after Columbus, four regions exported their particular methods of cattle ranching across the Atlantic: the British highlands, Extremadura in western Spain, Andalusia in southern Spain, and tropical West Africa.

Beginning in the late fifteenth and early sixteenth centuries, Africans and Spaniards transplanted their characteristic cattle herding cultures to the Caribbean, then moved westward into central Mexico and northward to what would become the North American South and West. In the seventeenth century, Englishmen transferred their own distinctive ways of herding cattle to the Atlantic coast of North America, and from the Carolinas in particular, pushed into the heart of the continent. Along the way, as Jordan vividly explains,

New World cattle raisers of both English and Spanish descent modified their respective ranching techniques as they adapted to local conditions and as they adopted techniques from one another.

By the mid-nineteenth century, three distinctive "herding cultures"—the Californian, the Texan, and the midwestern—vied, as Jordan puts it, "in a contest for survival of the fittest" (p. 313). In modified form, all three survived, but contrary to conventional wisdom the more capital- and labor-intensive midwestern tradition, with its British antecedents and its emphasis on quality of livestock, came to prevail.

Jordan, then, sees what western historians have long called *the* cattle or ranching frontier as the scene of several competing cattle frontiers—as a place of diversity. He also sees the origins of western ranching as more complex and ethnically diverse than has been supposed. Building on an argument that he advanced in earlier work, Jordan challenges the thesis advanced by Walter Prescott Webb and others that South Texas, with its Hispanic herding tradition, was the sole "cradle" of ranching on the Great Plains.

Terry Jordan brings to this subject the training of a cultural geographer and a wealth of experience in explaining the transplantation of European material culture to American soil. An indefatigable researcher and writer, Terry Jordan is the distinguished Walter Prescott Webb Professor of History and Ideas in the Department of Geography at the University of Texas at Austin. His writing includes five textbooks, six earlier scholarly books, and over sixty scholarly articles. Three of his books treat Texas subjects: nineteenth-century German immigrants, log buildings, and graveyards. Beyond Texas, he is best known for three titles: *Trails to Texas: Southern Roots of Western Cattle Ranching* (1981); *American Log Buildings: An Old World Heritage* (1985); and (with Matti Kaups) *The American Backwoods Frontier: An Ethnic and Ecological Interpretation* (1989).

It is with pleasure and pride that we present Terry Jordan's newest book in the Histories of the Frontier series, for it ranks among those rare works that one might justly describe as magisterial. Asking the biggest of questions about his subject, and marshalling evidence from an astonishing range of sources, Jordan has presented his findings in clearly organized, jargon-free prose, enhanced by a wealth of splendid maps, charts, and illustrations, and a bibliography that will itself be consulted for decades to come. *North American Cattle-Ranching Frontiers* not only summarizes much that is known, but provides fresh understandings and challenges old assumptions, advancing the literature on the cattle frontier to a new level of sophistication.

Like other books in the series, *North American Cattle-Ranching Frontiers* tells a complete story, but it is also intended to be read as part of the broader history of western expansion told in these volumes. Each book has been written by a leading authority, intimately acquainted with his or her subject and skilled in narration and interpretation. Each provides the general reader with a sound, engaging account of one phase of the nation's frontier past, and the specialized student with a narrative that is integrated into the general story of the nation's growth.

The series, conceived by the distinguished historian Ray Allen Billington in 1957 as a multivolume narrative history of the American frontier in eighteen volumes, has expanded substantially in recent years to include topics that Ray had not originally envisioned. Those include titles such as Sandra L. Myres's *Western Women and the Frontier Experience, 1800–1915* (1982), Elliott West's *Growing Up with the Country: Childhood on the Far Western Frontier* (1989), Donald J. Pisani's *To Reclaim a Divided West: Water, Law, and Public Policy, 1848–1902* (1992); volumes now in progress treat a variety of other topics, including photographers, cities, violence, lumbering, and the law.

From the outset, Ray Billington had planned to include a book on "The Frontier of the Cattlemen" as part of the series, but the project languished for three decades until Terry Jordan generously agreed to take up the challenge. In so doing, he has made a major contribution not only to understanding North America's western frontiers, but Euro-African trans-Atlantic frontiers as well. Indeed, *North American Cattle-Ranching Frontiers* represents a more wide-ranging and interdisciplinary work than Ray Billington might have imagined possible in the scholarly milieu of the late-1950s. We trust readers will agree that it has been worth the wait.

HOWARD R. LAMAR
Yale University

MARTIN RIDGE
The Huntington Library

DAVID J. WEBER
Southern Methodist University

North American Cattle-Ranching Frontiers

Preface

North American Cattle-Ranching Frontiers serves as a companion piece to my 1989 book, coauthored with Matti Kaups, *The American Backwoods Frontier.* The earlier volume dealt with the agrarian frontier in the temperate forests of North America, while the present work concerns the pastoral frontier in the Northern Hemisphere grasslands of the New World. Together the colonization systems connected with these two habitats facilitated the rapid European settlement of the far greater part of the continent.

While the appearance of yet another book on cattle ranching will strike some as unnecessary, I believe my work is sufficiently revisionistic, both in scope and content, as to warrant attention and space on the bookshelf. The basis of the revisionism lies in my training as a geographer. By its very nature, geography is an interdisciplinary enterprise, bridging as it does the earth sciences, life sciences, humanities, and social sciences; the present bibliography reflects this interdisciplinary content and approach. I have drawn upon the published works not just of geographers and historians, but also of botanists, range ecologists, zoologists, anthropologists, archaeologists, folklorists, etymologists, economists, sociologists, ethnologists, and climatologists. When so wide a net is cast, the relevant literature assumes a simply staggering quantity. My attempt to synthesize these diverse works, while at the same time adopting a revisionistic stance, led me eventually to refer to this project as the "book from hell."

Nor does the work rest entirely upon such secondary sources. As often as possible and necessary, I resorted to primary materials, ranging from the *Relaciones geográficas* of New Spain to tax lists, manuscript census schedules, and travel accounts. I also relied very heavily upon that most geographical type of primary research—field observation. I personally inspected the relict landscapes and remnant herding cultures of virtually every North American region touched significantly by the cattle-ranching frontier, journeying seven times to Mexico, twice to the West Indies, once to Central America, and numerous times to the western parts of the United States and Canada, ranging from Guanacaste to interior British Columbia. I sought similar traces in the countrysides of the major Old World source regions of the cattle-herding culture complex, including southwestern Iberia, North Africa, and the highlands of Britain and Ireland. If I have detected some little part of the truth about the cattle frontier, my achievement rests more heavily upon those field experiences than any other single element. I am grateful to the Webb Chair endowment fund, administered by Robert D. King, dean of the College of Liberal Arts at The University of Texas at Austin, for financing those diverse travels. I am blessed, indeed, to occupy the Walter Prescott Webb Chair.

Many people have assisted me along the way. I am most indebted to the three editors of the Billington Histories of the Frontier series—David J. Weber, Howard R. Lamar, and Martin Ridge—who honored me by asking that I write the book, allowed me an extraordinary amount of time to complete it, provided many useful suggestions for improving the manuscript, and permitted me, as a geographer, to range rather far afield from their discipline of history. Several other scholars gave generously of their time to read early versions of individual sections and chapters of the book, including Louie W. Attebery of the College of Idaho, Robert C. West of Louisiana State University, William E. Doolittle of The University of Texas, Alfred Siemens of the University of British Columbia, and James J. Parsons of the University of California at Berkeley. Bill Doolittle also accompanied me on several field excursions, most notably to Sonora, his principal area of specialization, sharing insights all the while. These diverse readers, editors, and field companions are not, of course, responsible for any errors or omissions in the book.

Austin cartographer John V. Cotter's gifted hand and creative eye are abundantly revealed throughout this book. His attention to detail, readability, and visual aesthetics make him at once draftsman, designer, and artist. He responded successfully to my most outrageous demands for detail and format, treating each request as a challenge to his ingenuity. In an age when sterile, mechanical maps spew from computer graphics programs, deadening the soul of the humanist, John reminds us that traditional ways are usually better. Long may his eye remain keen and his hand steady.

My research assistant, Webb Fellow Jon T. Kilpinen, put in many hours in diverse tasks, such as reading manuscript census schedules, as well as providing numerous insights derived from his own research on ranching in the high valleys of the Rocky Mountains. He accompanied me in the field several times. Michael Melvin, a graduate student in our department, contributed a useful analysis of nineteenth-century Texas tax lists.

I am also deeply indebted to the staffs at several facilities and institutions, most notably at the Centro de Información of the Parque Nacional de Doñana in Las Marismas of the Guadalquivir, who allowed me unrestricted access to the old grazing grounds, longhorn-cattle herds, and *cañadas* of the remnant marsh; at the splendid Nettie Lee Benson Latin American Collection of The University of Texas, who subscribe to that noblest and rarest principle of librarians—let the people have the books (and even the most precious of manuscripts); at the Edwin J. Foscue Map Library of Southern Methodist University, who possess a far finer collection than they know; and at the Idaho State Historical Society Museum and Archives in Boise, who allowed us to paw through and bring into disorder large sections of their excellent photograph collection.

Beverly Beaty-Benadom, Carol R. Vernon, and other office staff members did a heroic job typing and retyping the manuscript, making sense of my juvenile-arrested handwriting, and occasionally providing helpful stylistic suggestions. I am indebted to them for their professionalism, promptness, and general helpfulness.

T.G.J.
Austin, Texas

ONE

The Nature of Cattle Ranching

A frontier, in the geographical sense, is a zone of contact between "two contrasting types of land use," where bearers of a new and different way of using natural resources advance, displacing older, indigenous modes and peoples. In large parts of the North American continent, cattle ranching provided an innovative land-use strategy that facilitated the advance of the Euroamerican settlement frontier at the expense of the native peoples. In the process, cattle ranching came to be a prestigious and romanticized way of life. Perhaps no other form of economic activity seems so captivating, at least in retrospect, or produced a comparable mythology. On a more prosaic level, open-range or free-grass cattle ranching constituted a business, yielded a society with attendant subculture, reflected an adaptive strategy of land use, and required a favorable institutional-political framework in which to operate. Its prerequisites included an abundance of free range containing plant life suitable for grazing and browsing; sizable, enduring markets and transport access to them; a small, skilled underclass of laborers; an antecedent cultural tradition of cattle herding; modest amounts of risk capital for investment; and colonialistic support institutions.[1]

The traditional raising of range cattle by Euroamericans was at base simply another business, another form of western capitalistic free enterprise, involving the specialized, market-oriented production of livestock, initially on large, unimproved land units at a small expenditure of labor and capital. Ownership of these commercial operations often lay in the hands of wealthy entrepreneurs, whose influence on the industry over the centuries was profound, although the "small rancher" almost invariably also played a role, one too often overlooked in frontier studies.

Grazing large cattle herds upon unfenced natural pastures was the dominant, or occasionally even the sole, activity of the pioneer ranchers, with the aim of producing for sale meat, hides, horns, and tallow. If the goal was meat production, the normal pattern involved shipping the range cattle to other areas nearer market for fattening. These "breeder" and "feeder" districts were usually linked by a network of established trails, over which the range cattle were driven. Often in frontier settings, the cattle business yielded mainly hides and tallow rather than meat, producing leather for clothing, shoes, rope, harnesses, bookbindings, and the like, as well as candles, soap, and lubricants. Population density necessarily was low in the ranching areas, and the level of land use represented the most extensive form of commercial agriculture. The yield per unit of land remained rather minimal, requiring operators to control comparatively large tracts and maintain sizable numbers of stock.[2]

If open-range cattle ranching was a business, it also entailed, to a quite remarkable degree, particular forms of culture and society. A clear distinction normally existed between the ranch owners, often absentee entrepreneurs only weakly linked to the daily operation, and their cowboys, who formed a caste of hired or enslaved laborers. Range cattle raising carried more prestige than most other forms of frontier land use, a prestige that, remarkably, extended down even to the unpropertied, often mixed-blood men who worked the herds or served as overseers. Cowboys, whether in Latin or Anglo cultures, practiced a distinctive way of life, forming a folk community or subculture with its own tools, techniques, jargon, and code of behavior. The prestige and mythology attached to this livelihood are difficult to explain, particularly given the low socioeconomic status of the working herders. Perhaps they derived in part from the use of horses in stock management, for the mounted man has always shared the formidability of warriors. Then too, the prowess and cleverness needed to control large, horned, semiferal beasts were considerable, beyond the athletic capabilities of most, and the ancient, preagricultural prestige males derived from the successful hunting of large game animals continued to reside in cattle herding. The tilling of plants was, by comparison, a safer and ancestrally more female enterprise, while the herding of small animals such as sheep or goats required fewer risks and athletic skills, often being left to children and the elderly. Perhaps the appeal of the cowboy was his imagined freedom, as the carefree child of nature, as Rousseau's "natural man," unburdened by the cumbersome baggage of civilization. Whatever the reasons, owning and working range cattle brought character and prestige, if not generally wealth, to those involved.[3]

Mounted herders of cattle all around the world, whether American cowboy, Argentinian *gaucho*, Australian musterer, Venezuelan *llanero*, Russian Cossack, or South African *veld* Boer shared, to a degree, the same culture and prestige. Equally important, though, were regional differences. Even in North America, the culture of range cattle raising was by no means monolithic, and each ranching region developed its own special character, reflecting the particular cultural heritage of its people and the nature of the local physical environment. To speak of *a* cattle ranching frontier in the singular is to oversimplify and, in the final analysis, to fail to understand or explain.[4]

Physical environments were of crucial importance, and the various types of frontier cattle ranching should be viewed ecologically, as adaptive strategies for human use of the land. The adaptive systems of traditional open-range cattle raising each involved the highly specialized use of a particular floral region, or niche, in its natural state, allowing conversion of forage, mainly grasses and browse plants, into meat, hides, horns, and tallow. Indeed the word "range" implies feeding upon native or uncultivated plants. Although cattle convert at most one-twentieth of what they consume into meat, they utilize cellulose, a food humans cannot directly eat.

Attention to the ecological context of frontier cattle ranching will not lead us to environmental determinism—the belief that the physical setting determines the form of human economy. The historian Walter Prescott Webb, in his classic early work, *The Great Plains,* erroneously viewed ranching in this way, as an inevitable consequence of the treeless, semiarid character of western North America. Cattle ranching, instead, thrived in a great variety of New World physical environments, from tropical savannas to subtropical pine barrens and midlatitude prairies, from fertile lowland plains to rugged mountain ranges, from rainy districts to semideserts. Ranching did not originate in harsh environments; rather, as will be shown, several potent causal forces shunted it to such places, which served as refuges. When necessary, native American feral herd animals, such as bison, were driven away or exterminated to allow cattle to occupy their grassy niche, but in fact there were few formidable native competitors and still fewer predators equipped to challenge the introduced cattle. Amerindians probably came closest to achieving successful predator status. By and large, cattle proved able to defend themselves, compete successfully, and propagate on the open ranges of America.[5]

In spite of the successful introduction of open-range cattle ranching into a variety of New World environments, this extensive herding strategy failed to achieve, in any setting, a sustainable adaptation. Indeed no herding system has ever attained, in any locality, a stable ecological balance, except at a lower productivity level than existed there when pastoralism first began. The open-range cattle-ranching strategy invariably caused habitat modification and damage. As range ecologists William Galbraith and William Anderson observed, "the history of grazing . . . followed a similar pattern throughout the world," in which "an abundance of native forage" underwent "deterioration or loss." At first glance such damage might seem curious, since most botanists acknowledge that grazing is not inherently destructive. Grasslands and herbivore herd animals evolved together, so that grazing is a natural ecological factor, and moderate grazing yields more grass than does complete protection. The animals, by eating grass, stimulate herbage production, create contour terraces or slopes with their trails, fertilize with their manure and carcasses, disperse seeds, and prove beneficial in various other ways. After millennia of grazing under natural conditions, the dominant herbaceous growth remains palatable to grazing animals, suggesting a symbiotic relationship.[6]

Unfortunately, herds of domesticated animals, such as cattle, did not permit this ancient balance of nature to persist. Human pastoral strategy, particularly in commercial ranching, is to maximize herd size and fertility, accomplished through such devices as predator control and the culling of steers, allowing cows to form an unnaturally large part of the herd and leading to overpopulation. Cattle ranching also removes animals from the range before death, depriving the land of carcass fertilization, and promotes specialized grazing by only one herd species. The ecological problem of ranching, then, is not grazing, but specialized overgrazing. Herd size, almost invariably, soon surpassed the carrying capacity of the range in traditional ranching. By trampling, soil compaction, selective foraging, repeated close cropping of the grasses, and seriously overgrazing near sources of water and salt, cattle in excessive numbers typically eliminated the more palatable, accessible perennial floral species and diminished the growth and size of roots, reducing the variety and volume of the native vegetative cover. Overgrazing tends to destroy perennial grasses, causing a deflection of ecological succession toward annual species. Range firing, by contrast, favors perennial grasses and was very widely practiced by pioneer ranchers, but the great reduction of biomass through overgrazing eventually depleted the fuel needed for successful burns. Brush infestation and dominance by annual grasses was all too often the result. Alternatively both soil compaction and the diminished vegetation could result in more arid microclimates, since evaporation increased with the greater exposure of the ground to sunlight, mulch diminishment, decline of water infiltration, and augmentation and evaporation of runoff. On slopes, treading by cattle loosened soil particles, creating the courses of eventual gullies.[7]

In short, cattle herders often left erosion, brushland, and even incipient desertification in their wake. The resultant reduction in range carrying capacity almost always led to a decrease in herd fecundity and health. Ultimately some ranges were rendered unfit for continued commercial cattle ranching. Repeated relocation to new, undamaged habitats became an attractive strategy, and the cattle frontier often took on a mobile character.

Habitat damage was only one force at work encouraging the expansion, shunting, and relocation of ranching frontiers in the Americas. Cattle herding also faced disruptive economic and demographic pressures. For example, the spatial land-rent model proposed over a century and a half ago by the German scholar Heinrich von Thünen is relevant to the cattle-ranching frontiers of North America. According to von Thünen, the dictates of land rent are invariably obeyed in a market-oriented capitalistic setting, with profound geographical, or spatial, results. Intensive forms of commercial agriculture, characterized by high inputs of capital and/or labor, tend to be located nearest to the markets, where land values are highest and access to consumers easiest. Farmers there place their resources of labor and capital into production, needing only minimal outlays for transporting goods to market. Output per unit of land is therefore high. With increasing distance from market, operators spend proportionately more on shipping costs, with the result that less-intensive land utilization necessarily characterizes the hinterlands. The result is a series of concentrically arranged zones of land use around the central market. If applied on a macroscale, western Europe becomes von Thünen's "market," surrounded by zones of dairying, truck farming, livestock feeding, cash grain farming, and herding. Open-range cattle ranching, as the least-intensive form of commercial agriculture, thus becomes relegated to the regions most remote from market, to the outermost zone (figure 1).[8]

These Thünenian rings did not, however, remain fixed in place. The great market complex in western Europe expanded spectacu-

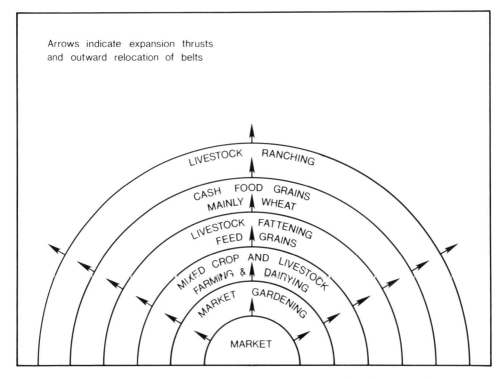

Figure 1. Commercial land-use pattern as dictated by land rent. Based loosely on the seminal work of Heinrich von Thünen, this spatial model proposes that ranching, as an extensive mode of land use, will be relegated to the outermost periphery of market-oriented agriculture. If the market grows and expands, as was true of the European-based industrial economy, then a shunting process will be put in motion, pushing ranching into still more remote regions. Source: Johann Heinrich von Thünen, Von Thünen's Isolated State, *trans.* Carla M. Wartenberg (Elmsford, NY, 1966).

larly, due to the Industrial Revolution, causing an immense growth in demand for agricultural produce while at the same time providing improved transportation technology, lessening shipping costs. The concentric belts of agriculture were set into outward, ripplelike motion, allowing more intensive forms of land use to invade the ranching periphery. Ranching regions often evolved into feeder zones or cash-grain belts in this manner. Looked at another way, land that could produce foods directly edible by humans became economically inefficient for ranching.

At the same time, continued population growth in the ranching zone, both through continued immigration and natural increase, eventually produced densities too great to permit continuation of the free-grass herding adaptive strategy, particularly in light of the diminished carrying capacity of the damaged range. For these economic and demographic reasons, open-range cattle ranching could ultimately survive only in ecological refuge locations, marginal and remote environments where the pressure of land rent became subdued and population remained sparse. Even there ranching eventually had to evolve into a more capital-intensive form, characterized by steel fencing, windmills, scientific range management, and improved breeds. The frontier type necessarily vanished, often rather quickly.

The ring-zone model can be carried too far. Economic determinism is as much to be avoided as environmental determinism. Von Thünen's concepts do not always seem to be obeyed, as will be demonstrated. The same fate befalls another, related spatial model of zonation, proposed by the famous historian Frederick Jackson Turner. He saw the Thünenian belts in motion across the American heartland and developed a frontier stage model to fit. Turner spoke poetically of the frontier as witnessing a succession of economic forms, a "procession of civilization, the fur trader and hunter, the cattle-raiser, the pioneer farmer."[9] But as we will see, not all North American frontiers experienced a cattle-ranching phase, since the environmental and cultural conditions had to be suitable for this adaptive strategy, and certain other areas preserved it long after the frontier had passed. Ranching was not exclusively a frontier enterprise, nor did every frontier witness it. Such simplistic models as the ones proposed by von Thünen and Turner obscure more history than they reveal. In the final analysis, these models often founder because of their acultural nature. They allow no place for human tradition, ingenuity, prejudice, accident, nonconformity, and irrationality. The ranching frontier abounded in these very model-destroying qualities. Nor do the models permit a role for habitat and ecological contrasts. The real world is complex; cattle ranching can only be explained in that context.

The cattle-ranching frontier, then, was for various, intertwined reasons mobile, exploitive, diverse, and frequently ephemeral. Open-range commercial herding often bore the seeds of its own destruction, both ecologically and economically. Before its demise, however, free-grass ranching provided a very effective adaptive strategy for opening vast areas of the New World, especially the grasslands, to rapid European colonization. Ranching, together with the equally mobile and expansive forest colonization of the eastern woodlands of the United States, accommodated the Euroamerican occupation of the far greater part of North America. Open-range herding was adaptively to the grasslands what the Pennsylvania-derived backwoods pioneer system was to the eastern forests.[10] Both changed or disappeared with the passing of the frontier, but not before serving the causes of Europeanization and manifest destiny. To understand the American frontier, one must study these two remarkably efficient, preadapted, and destructive colonization systems.[11]

Ranching was not a product of the American frontier, as Turner would have us believe, or of the semiarid West, as Webb proposed. Westerners in the United States, scholars and laymen alike, are fond of viewing their region and its enterprises, such as cattle ranching, as indigenous and uniquely western.[12] This self-image is largely illusion and myth. Ranching did not originate in the American West, or even in the East. Nor did it arise in Latin America. We must not imagine that cattle ranching developed spontaneously as an adaptive system when Europeans set foot in the Americas. Instead certain immigrant groups were well preadapted for such an activity. The open-range herding of beef cattle, a profoundly old rural profession, originated in the eastern hemisphere, the source of all major animal domestications. Diverse ancient peoples of the eastern Mediterranean raised, venerated, and even worshipped cattle. Along the Nile in Egypt, the great sky goddess of fertility, Hathor, was often depicted as a cow, whose milk gave power to the pharaohs, while Apis reigned as the sacred bull of Memphis. Depictions of these Egyptian gods regularly link the shape of the horns to the sun or moon. In the nearby land of Canaan, in pre-Israelite times, the bull often appeared in art, serving as the symbol of Hadad, the ancient Semitic god of storm and the west, and was associated with Ba'al as well; in Minoan Crete the bull became the feared earthquake god, the destructive shaker of the land. Perhaps the mystique of ranching on the American frontier is, in part, a distant echo of ancient Mediterranean cattle veneration. At the very least, we owe to these ancient places and peoples many of the practices associated with cattle ranching, such as branding and earmarking.[13]

By A.D. 1500, the Old World distribution of range cattle raising had become far-flung, crossing many ethnic and environmental borders and presenting a vivid cultural geography. The most notable surviving concentrations of beef-cattle herding by that time lay in the Atlantic coastal fringe of Europe and Africa, forming a fragmented belt from the British Isles and Scandinavia southward to Angola, a peripheral pattern both relic and refugee in character (figure 2). Occupying the windswept, rainy, rocky "tattered ends of Europe" and the African semideserts bordering the great Sahara, these cattle herders were by 1500 practitioners of a truly ancient and largely vanished way of life.[14] They clung to the westernmost highlands, islands, marshes, moors, and savannas, having been shunted during the course of millennia from an Asiatic cradle almost into the Atlantic, that sinister and feared end of the earth, by tillers of the soil practicing more intensive agricultural systems. The herder folk had been displaced and driven to refuge by the same demographic, economic, and ecological pressures that would later plague many of their American successors.

The Atlantic fringe of Europe, while offering multiple refuges for the pastoralists, also provided from time immemorial a coastwise path of contact and diffusion, allowing exchange among the several herder groups and helping preserve an ancestral kinship and similarity among them.[15] Perhaps all had links to the ancient peoples of the Atlantic fringe, especially the mysterious megalith builders, whose remnant standing stones still startle visitors to the British moors, Scandinavia, France, Iberia, and North Africa. Their roots also surely lie partly among the Celts, whose linguistic retreat to Atlantic refuges paralleled the herders' flight. The African cattle nomads apparently have a very different and equally obscure genealogy.

Collectively the Old World cattle folk throughout history formed a periphery to civilization's core. In Classical times the Greeks and Romans saw the barbarian herders who surrounded them as noble savages, occupying fringe areas beyond the pale. Free of the burdens and demands of civilization, these Arcadian children of Pan lived comfortably from their abundant herds without undue exertion of labor, leading a carefree life in humankind's natural condition. The Greeks and Romans simultaneously admired, envied, and reviled these Scythians, Celts, Goths, Arabs, Berbers, and other peripheral herder folk. Similar mixed feelings characterize the later Anglo-Saxon view of the Irish, Welsh, and Scots or the Castillian image of Extremadura and Andalucía. Still later the same we/they, core/periphery dichotomy would reappear in the Americas, finding a final mythical refuge in Hollywood movies and Marlboro Man advertisements.

The precarious western perches of the range-cattle herders by

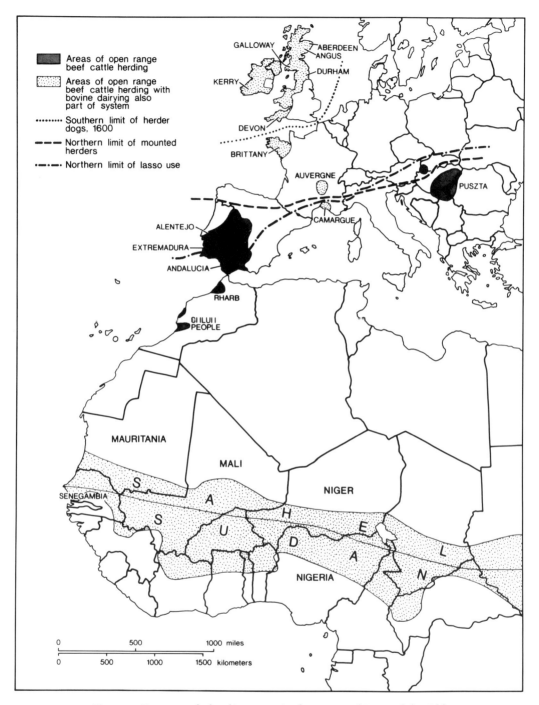

Figure 2. Range cattle-herding areas in the western fringes of the Old World, about A.D. 1600. The border of lasso usage in Iberia is speculative. Sources: Maurice Le Lannou, Géographie de la Bretagne *(Paris,*

France, 1952), vol. II, pp. 14–22; The Atlas of Africa *(New York, NY, 1973), p. 63; Lázló Földes, ed.,* Viehzucht und Hirtenleben in Ostmitteleuropa: Ethnographische Studien *(Budapest, Hungary, 1961), pp. 465–528, 581–645; Xavier de Planhol, "Le chien de berger,"* Bulletin de l'Association de Géographes Français, *370 (1969), p. 360; Andre Meynier, "Le Cantal, premier massif montagneux français pour l'estivage bovin," in* Études géographiques offertes à Louis Papy *(Bordeaux, France, 1978), pp. 309–19; H. Geoffroy Saint-Hilaire,* L'élevage dans l'Afrique du nord: Maroc—Algerie—Tunisie *(Paris, France, 1919), plate XXII; Augustin Bernard,* Afrique septentrionale et occidentale *(Paris, France, 1939), part II, p. 437.*

1500, while peripheral to the Old World and clearly endangered, placed them in an ideal position to profit from the discovery and colonization of the Americas. They would not, after all, be driven into the sea, but rather cross it and find vast new, less-contested domains in which to tend their herds in the old ways. At almost the last moment, a reprieve had come. Thus it happened that the cattle cultures of the Old World's Atlantic fringe were transplanted to both North and South America early in the colonial era.

TWO

ATLANTIC-FRINGE SOURCE REGIONS

ALL THREE MAJOR ATLANTIC-FRINGE HERDING CENTERS IN the Old World—southwestern Iberia, the British highlands, and the sub-Saharan steppes of West Africa—contributed to the development of North American cattle ranching. Each of the three regional complexes possessed a special cultural character and was further subdivided into local variant forms of cattle herding. As a result the New World transplantations displayed substantial diversity, and the individual strands could be detected in the Americas, even amid vigorous blending. The earliest and in some ways most influential diffusion emanated from Iberia.

The Iberian Source

The traditional cattle-raising districts of Iberia lie in the Atlantic periphery, a land of bovine herding since prehistoric times and home to an ancient bullfighting tradition (figure 3). Within the cattle zone, one must further distinguish the "humid crescent" of the north and northwest, where a continental European mixed-farming system prevailed, whose bovine dairying, draft oxen, and grain-fattened beef offered no viable prototype for New World ranching. The remaining, southwestern cattle zone constitutes the traditional range beef-cattle area, where dairying was unimportant and in which the origin of Latin American ranching has correctly been sought. There, as elsewhere in medieval central and southern Spain, many events had conspired over the centuries to encourage and perpetuate pastoralism, from the introduction of new herding traditions by the Goths, Arabs, and Berbers to the long wars of reconquest, which destroyed or diminished farm villages, producing abundant fallow land for pasturing, and lent adaptive advantage to seminomadic herder groups.

It has long been presumed that a single southwestern Iberian herding system provided the basis for subsequent cattle ranching in

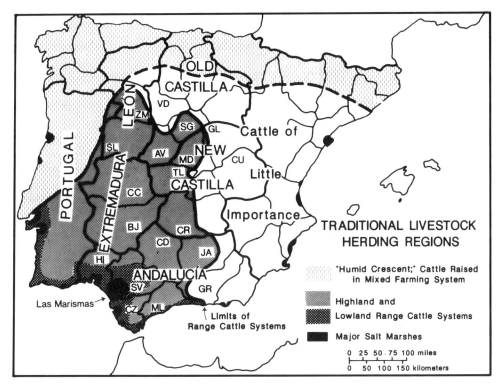

Figure 3. Key to abbreviations: AV = Avila, BJ = Badajoz, CC = Cáceres, CD = Córdoba, CR = Ciudad Real, CZ = Cádiz, GL = Guadalajara, GR = Granada, HL = Huelva, JA = Jaen, MD = Madrid, ML = Málaga, SG = Segovia, SL = Salamanca, SV = Sevilla, TL = Toledo, VD = Valladolid, ZM = Zamora.

Latin America.[1] Sixteenth-century Extremadura and Andalucía, the source provinces, reputedly housed a single cattle-raising system of relatively uniform character, which was bequeathed to the Hispanic colonies overseas.[2] This traditional view is oversimplified and disregards important regional contrasts in Iberian ecosystems and related herding practices. Instead at least two distinct range-cattle complexes existed by 1500, one based in the western part of the interior plateau, or Meseta, of Spain and the other in the Andalusian coastal lowland. Both left abundant remnants in the present-day landscape, and both achieved diffusion to comparable ecological niches in colonial Latin America. Previous efforts to document Iberian influence in Latin American ranching, while correctly emphasizing Andalucía and Extremadura, failed to acknowledge basic differences between the highland and lowland herding systems, lumping together "*meseta* and Andalusian plain."³

The lowland Iberian range-cattle complex, based in an ancient herding tradition, was centered in and near a series of Atlantic coastal estuarine salt marshes, lying at regular intervals between Lisbon and Gibraltar. By far the largest and most important were Las Marismas of the Río Guadalquivir, occupying a huge expanse in three Andalusian provinces (figure 4). The cultural geographer William E. Doolittle was the first to propose that a cattle-ranching complex derived specifically from these Guadalquivir salt marshes was transferred to very similar low-lying wetlands in Latin America.⁴ Other, smaller Andalusian and Portuguese coastal marshes at the mouths of the Guadiana, Piedras, Odiel, Tinto, Guadalete, Barbate, Tajo, and Sado were also traditionally used for cattle raising. One is tempted to add as outliers the marsh-cattle complexes of the Camargue at the mouth of the Rhône in the south of France, the Aveiro Lagoon in northern Portugal, and even the Ebro Delta of Catalonia, where some range cattle were once raised in the outermost fringes.⁵

The region of Las Marismas can be divided into two local ecosystems: the marsh proper and a bordering wreath of open forest and cropland (figure 4). The marsh proper forms an almost completely flat plain of sand and alluvium, causing the Guadalquivir and its tributaries to develop a maze of branches and seasonal channels, called *caños*. Saline intrusions characterize all of Las Marismas, though the southern areas bear the heaviest impregnations, and tides ascend almost to Sevilla. As a result, the vegetation association of the marshes is halophytic, or salt-tolerant, and rather sparse, including bunchgrasses, rushes, and glasswort. The relatively sweet seasonal waters of the caños leave behind the lushest growth, while *lucios,* the lowest and most saline parts of the marsh, offer only a very sparse vegetation. The annual rhythm of Las Marismas is dictated by water rather than temperature. The climate is classified as Mediterranean with Atlantic influence, featuring a prolonged summer drought, accompanied by searing *solano* winds from the southeast. July and August bring less than 5 mm. (0.2 in.) of precipitation, average temperatures rise to 23.5°C (74°F), and Las Marismas become a mirage-rich, dusty alluvial level given up to herds of cattle. By the end of summer, the caños are completely dry, the grasses yellow or brown, and the lucios covered with a whitish crust of dried salt. Fissures 0.5 m. (1.5 ft.) deep crisscross the marsh. Cool-season rains, beginning about the equinox, and subsequent river floods gradually transform

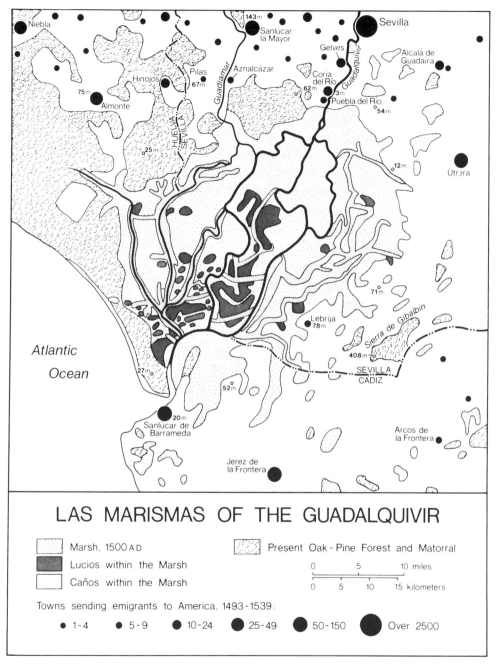

Figure 4. Sources: Spanien 1:50 000 *(see note 19)*; Boyd-Bowman, Indice geobiográfico, *(see note 30)*.

Las Marismas by December into a grassy, watery wilderness. The Guadalquivir experiences two peak flows, the first in December and another in February–March, corresponding to the details of the winter rainfall regime and snowmelt. The inundations, most severe in the second peak, often occur in the manner of flash floods, endangering people and livestock alike. The weather cycle produces two blossomings and growth spurts annually in Las Marismas. One follows the first rains of autumn, but winter floods and the cooler temperatures of December and January, which average 9.3°C (49°F), blunt the growth of vegetation, even though frost is rare. The second flowering comes after the floods recede in spring. Then drought conditions are steadily reestablished, although water remains in the caños well into the summer.[6]

The second major ecosystem in Las Marismas consists of a circummarsh wreath of higher-lying, often hilly land. Surviving forests in this area are dominated by cork oak and stone pine, mixed with two types of *matorral*, or Mediterranean brush, formerly wolf habitat. In lower, moister areas the brush is called *monte negro*, dominated by a growth of juniper, heather, and bramble up to 1.8 m. (6 ft.) in height, while some higher, drier places have *monte blanco*, featuring plants such as thyme, sweet marjoram, rosemary, and rock rose. Dotted here and there in the perimeter are seasonally wet *lagunas*, which can be regarded as small outliers of the marsh proper. The circummarsh ecosystem exhibits some important internal variation, based principally in soil contrasts. In the southwest a sterile spit of dunes lies between the ocean and the marsh. The greater part of that area is stabilized with pine plantations, but where the dunes remain mobile, matorral clings precariously to clefts interestingly called *corrales*. West and north of Las Marismas, the perimeter consists of a low tertiary diluvial plateau, whose dry sands and fine gravels similarly offer few agricultural enticements, helping preserve a large amount of woodland. In the north the plateau is eroded into a fringe of low, sandy hills, also heavily wooded. Only in a few small pockets, as around Pilas, does a more fertile, finer-textured, grayish soil provide the basis for intensive crop farming. Good reddish soils are common farther north, in the fertile belt called the Aljarafe, stretching westward from Sevilla through Sanlúcar la Mayor and similar farm towns. To the east of Las Marismas, beyond the fertile, irrigated, alluvial strip that parallels the Guadalquivir, lie extensive stretches of rolling farmland, based on the marly soils of limestone uplands. Even there, though, the quality of the land varies. Best are the *tierras negras,* or "black lands," excellent for wheat and vines, which provide a startling contrast to the white limestone hills. Quite different are the *tierras rojas,* labeled "poor lands for poor people" and supporting a population of small farmers. Here and there lagunas dot the eastern margin of Las Marismas, as do sandy patches covered with pine and the oak-mantled Sierra de Gibalbín, an outlier of the folded mountains farther east.[7]

The traditional range-cattle system of Las Marismas was adapted to this dual ecosystem. It involved a local shift of the animals between the marsh in summer and the adjacent woodland and brush in winter, using a maze of stock trails (figure 5). Some of these, such as the Cañada Mayor, leading from Hinojos to its nearby marshland, remain intact to the present day. Livestock drifted into the marshes in late spring, after the major flooding peak, advancing as the waters receded. They grazed on the bunchgrasses and, as the dessication intensified, retreated to the caños and the banks of the perennial channels of the Guadalquivir and Guadiamar. The autumnal flowering allowed the stock once again to disperse through the plains. By November the *vaqueros* tensed in anticipation of the inevitable flash flooding, having devised a system of alarm signals to warn of the coming waters. While herders accompanied the cattle in the marsh, living in temporary thatched huts of clay that were annually destroyed by the floods, they exercised little control over the cattle for months at a stretch. The perennial river channels partitioned Las Marismas into separate ranges, and the cattle were allowed to drift to and from the caños as the grazing season progressed. Neglected and largely uncastrated, the marsh cattle became semiferal, apparently to a degree unknown elsewhere in Spain. Cattle wildness in Iberia was not produced by the reputedly frontier conditions of the Reconquest; instead, it was uniquely the result of marsh herding, the only system in which cattle lost daily contact with their keepers. Even as late as the 1920s, the rare summer visitor to the marsh ranges was advised to keep a proper distance from cattle herds.[8]

When the time came for the urgent, flood-threatened autumnal shift into the woodlands, accompanied by the annual culling and marketing, vaqueros were faced with unruly cattle that would not be led along by salt, since the halophytic vegetation provided the animals' needs. The herders' response was to manage the herds from horseback. Spanish culture, and its Andalusian regional variety in particular, endowed horsemanship with a high status, reinforcing the use of equestrian skills in herd management.[9] In the area west of

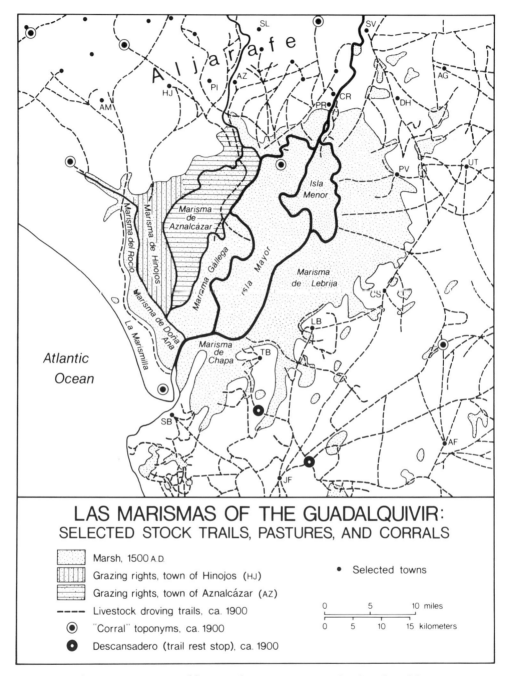

Figure 5. Key: Most abbreviated town names can be deciphered by consulting figure 4. Exceptions are CS = Las Cabezas de San Juan, DH = Dos Hermanas, PV = Los Palacios y Villafranca, TB = Trebujena. Source: Spanien 1:50 000 (see note 19).

the Guadalquivir about 1500, the largest numbers of horses were owned by the residents of Aznalcázar and Coria, both marshside settlements.[10] The raising of horses and the common usage of equestrian skills in herding remained concentrated largely in Sevilla and Cádiz provinces. In the nineteenth century, over half of the proprietors in Spain who owned fifty or more horses lived in those two districts. Mares shared the summer pastures with cattle herds in Las Marismas.[11]

The chief device used by the mounted Andalusian vaqueros was the *garrocha*, or lance (figure 6). Even though the lasso had been present in the Mediterranean lands since ancient Egyptian times, and the word *lazo*, derived from Latin, early achieved general use in Castilian Spain (usually to mean a cord for fastening loads), no Iberian vaqueros cast the lariat from horseback. Instead, in common with the cattle herders of the Camarque salt marshes of Mediterranean coastal France, they used the lasso mainly to pull floundering animals from the mire, after affixing the rope to the tail of their horse, in the absence of a saddlehorn (figure 7). A *taravita*, or spinner, was employed to make rope for lassos from the native esparto grass. Not until about 1640 was a prototypical form of roping from horseback first demonstrated in Spain, in the Madrid bullring by several visiting American vaqueros (see figure 23). Onlookers admired the novel method and bravery of the ropers and clearly were not previously familiar with the practice. Apparently, then, the grass-rope lasso existed in pre-Columbian times in the Andalusian marshes but was not cast from horseback.[12]

At the autumnal drive to the higher-lying ranges bordering Las Marismas, the cattle were collected in corrals made of stakes, branches, and brush, located in the woodland and matorral. Reminders of these now-vanished pens survive in local folklore and toponyms, especially in the woodlands southwest and west of the marsh (figure 5). Farther up the Andalusian coast, in the pine savanna heights above the Río Odiel salt marsh, the town of Corrales perhaps commemorates in its name the former roundups. The animals destined for market were culled and trailed off to Sevilla and Córdoba, while the rest of the herd was set free in the winter pastures, called *montes*. Prior to the emergence of the cattle from the marshes, at the end of the dry season, the montes had been fired, and by the time the herds arrived, equinoctial rains produced a lush growth of tender grass among the trees and brush. Much or most of the winter pasture constituted *monte realengo*—lands open to all livestock owners—

Figure 6. "El ganadero," the mounted vaquero of Andalucía, tending open-range longhorn cattle with a pike. Copied from Paul Gwynne, The Guadalquivir: Its Personality, Its People and Its Associations *(London, UK, Constable & Co., 1912), following p. 240.*

and one of the largest such pastures lay in the municipalities of Hinojos and Aznalcázar. Large herds of swine shared the winter range with the cattle. In spring the cattle were once again rounded up at the corrals, and calves were branded. Sevilla province established one of the earliest compulsory brand and mark registrations in Iberia. Brand designs of the marsh perimeter were generally rather ornate and

Figure 7. A traditional hornless leather saddle of the Dominican Republic in the ranching country near Hato Mayor. This may well represent the stock saddle of the colonial period, implying that roping from horseback was not part of the Antillean Hispanic system. (Photo by William E. Doolittle, 1992, who accompanied the author in the field and did not run out of film; used with permission).

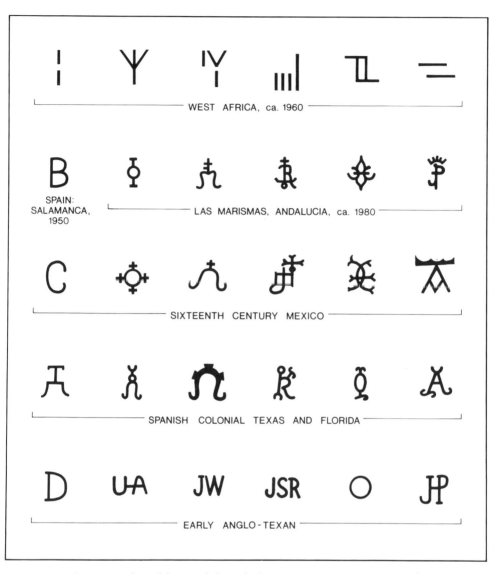

Figure 8. Selected livestock brands from Las Marismas area, North America, and West Africa. The Moorish-influenced brand designs of lowland Andalucía became the prevalent type in Latin America. Sources: Luis L. Cortés y Vázquez, "Ganadería y pastoreo en Berrocil de Huebra (Salamanca)," Revista de Dialectología y Tradiciones Populares 8 *(1952), p. 429;* William H. Dusenberry, The Mexican Mesta: The Administration of Ranching in Colonial Mexico *(Urbana, IL, 1963), pp. 215–28;* Joe A. Akerman, Jr., Florida Cowman, A History of Florida Cattle Raising *(Kissimmee, FL, 1976), p. 14;* Gus L. Ford, ed., Texas Cattle Brands *(Dallas, TX, 1936), pp. 1–9;* Terry G. Jordan, Trails to

Texas *(Lincoln, NE, 1981), p. 94; and* Johannes Nicolaisen, Ecology and Culture of the Pastoral Taureg *(København, Denmark), 1963, p. 141.*

embellished, in contrast to the simpler block letters of the mesetan interior, and it is the Andalusian designs that became most typical in the Spanish colonies in America (figure 8). Earmarks, or *señales*, were apparently also in use.[13]

The breed of cattle raised in Las Marismas was the famous Iberian longhorn, familiar on the American scene (figure 9). Herds still exist in the area, exhibiting reddish-brown as their prevalent coloring. The local dialect also includes words for a variety of other cattle colors, including the spotting and speckling typical of semiferal stock.[14] In the circummarsh area about 1500, the total numbers of such cattle were high, reaching 7,000 in villages of the Aljarafe district, comparable to the 9,500 head enumerated in the same general district, centered on Sanlúcar la Major, in 1865. The very revealing ratio of *ganado mayor* (cattle, horses) to *ganado menor* (sheep, goats, pigs) in the Aljarafe was, by southern Iberian standards, high both in 1500, at 1:1.4, and in 1865, at 1:3.3. In Sevilla province as a whole, small livestock outnumbered large only two to one about 1750, according to tax lists, indicating the great importance of cattle. About 1500 the highest numbers of cattle per owner in the Aljarafe occurred in Coria, Hinojos, Aznalcázar, and Puebla—all villages with direct access to the marshes. The extensive wetland pastures granted to the municipalities of Hinojos and Aznalcázar are indicated on an accompanying map, as are the connecting stock trails (figures 5, 10). East of the river, inhabitants of Alcalá de Guadaira alone owned over 3,000 cattle in 1493, and the citizens of Jerez de la Frontera possessed nearly 18,000 in 1491.[15]

Despite the large numbers of bovines present in the circummarsh area about 1500 and their relative importance in comparison to other livestock, ownership of cattle, except as draft animals, was relatively uncommon. The raising of beef cattle apparently bore little direct relationship to the typical Mediterranean agricultural system of the local farm towns and estates, focused upon wheat, vines, olives, sheep, goats, and pigs. At Aznalcázar, only about 15 percent of the inhabitants owned range cattle, at Hinojos only 11, at Puebla 27, and at Coria 29 percent. Seemingly, then, ownership of range cattle was confined to a small wealthy group, who appear in contemporary records as having herds of, for example, 127 and 190 head. This

Figure 9. Andalusian *reddish* retinto *longhorn cattle, like many in Latin America, at rest on the perimeter of Las Marismas near the pilgrimage village of El Rocío in Huelva Province. The month is June, and the cattle have made their seasonal move into the dried marsh. (Photo by the author, 1986.)*

pattern still prevailed in the nineteenth century, when 415, or 56 percent, of the proprietors in Spain owning over 100 head of cattle lived in the three provinces bordering Las Marismas.[16]

We can gain further insight into the distinctive pattern of cattle ownership by inspecting the land-tenure systems of the period around 1500. These varied quite considerably within the area. West and north of the marshes, estates were rather uncommon, though huge areas there, even entire municipalities such as Almonte, were owned by nobles, to whom tributes had to be paid.[17] To the northeast and east of the marshes, smallholders were also apparently common, but as early as the 1400s, estate building began in earnest there, ultimately producing a major cluster of *haciendas* on the northeastern perimeter of Las Marismas (figure 11). Some of these estates had a continuity dating to Roman times. Vineyards and orchards provided the economic focus for most, although some few developed into the prestigious, brand-flaunting, fighting-bull-raising enterprises that survive today (figure 11).[18]

Southeast of the marsh, along the border between Cádiz and Sevilla provinces, the word and toponym *estancia*, traditionally re-

Figure 10. A venerable metal sign with perforated lettering marks the northern entrance to the Marisma de Hinojos, the section of Las Marismas of the Guadalquivir awarded to the people of Hinojos as a grazing area. The sign stands at the end of the old cañada, or stock trail, leading from the town of Hinojos into the marsh. (Photo by the author, 1986.)

garded as an Americanism, appears, although apparently not to designate a livestock-raising estate, as in the New World (figure 11). An example is the Estancia de los Llanos, located just south of Trebujena. A dictionary of the Andalusian dialect lists *estancia* as a Cádizan term meaning "place for sheltering livestock."[19] Both as a word and

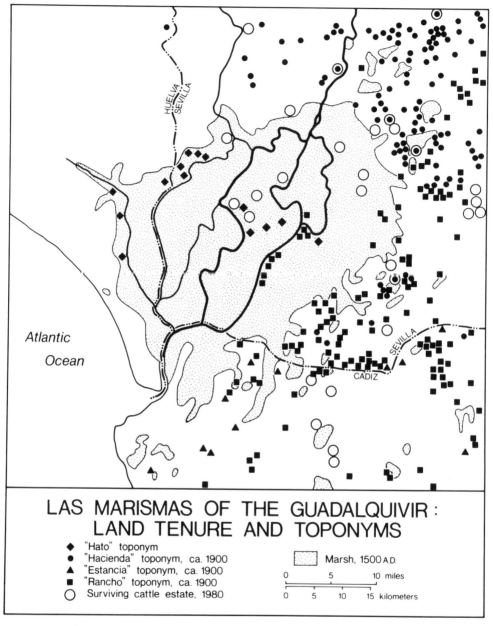

Figure 11. Sources: Spanien 1:50 000 *(see note 19)*; J. Rubiales, et al., El Río: el bajo Guadalquivir *(Madrid, Spain, 1985)*.

concept, *estancia* probably derives from this southeastern perimeter of Las Marismas, where it may have described a private pasture where cattle were *estantes*, or "stationary," and did not have to be driven to and from public domain.[20] The toponym *hato*, which meant "cattle ranch" in some parts of Spanish colonial America, occasionally appeared as a toponym in Las Marismas, merely designating a herder's camp or, in fifteenth-century Sevillan documents, herds of 500 or so cattle. Large livestock owners apparently bore the title *señor de hato*.[21]

Also of interest in the lower Guadalquivir area is the toponym *rancho* (figure 11). It retains there its original Iberian and Mexican meaning of a humble hut or homestead, often only seasonally occupied and usually inhabited by a worker or tenant on an estate.[22] Bearing no particular livestock connotation, it appears among fields, orchards, vineyards, and marsh pastures alike, although the ranchos situated in Las Marismas on the natural levee of the Guadalquivir as late as 1900 were perhaps the seasonal quarters of marsh herders. *Rancho* as a toponym is Andalusian, being confined to that portion of the Guadalquivir plain east and south of the river to the adjacent western part of the Bética Range.[23]

It seems possible, then, that some cattle-based estates existed in the hacienda complex east of Las Marismas by 1500. Clearly, though, another tenure system prevailed north and west of the marsh, where estates were rare. To understand the cattle-herding economy there, it is necessary to realize that a great part of Las Marismas, and in particular the Islas Mayor and Menor, belonged to the town of Sevilla, whose citizens enjoyed grazing rights there. A number of other smaller places had, by petition, also been granted such privileges, including Hinojos, Aznalcázar, Almonte, Coria, and Puebla (figure 10). In turn these towns often were required by law to allow access to their marsh pastures to herders from certain other nearby settlements. Hinojos, for example, shared its marsh range with Almonte, Niebla, Villarossa, and several other places. The nobles who owned such towns could not deny access to citizens. The most likely señores de hato who owned the cattle herds grazing upon Las Marismas were nonagricultural citizens of the town of Sevilla and perhaps the privileged farm towns, including nobles, merchants, and artisans. Often such influential people conspired to gain special private grazing rights on the municipal commons that were denied to ordinary citizens. All answered, however, to a municipal Sevillan *mesta*, or stockraisers' guild, which sought to regulate the industry.

Some of the cattle entrepreneurs owned thousands of head by 1500, forming a viable prototype for the ranching barons of America, but others were small operators.[24]

These town-based, entrepreneurial señores de hato hired or held as slaves the necessary vaqueros and presumably specialized in producing beef and hides for the Sevilla and Córdoba markets. Moors and other African and mixed-blood slaves were particularly common in Sevilla during the 1500s, forming over 7 percent of the total population. Some villages in Huelva province reputedly still have populations displaying negroid traits. Certain grand nobles of lowland Andalucía known to have been engaged in large-scale marsh cattle raising, such as the dukes of Medina-Sidonia, had documented connections to the institution of slavery. Perhaps most typical by about 1500 were hired vaqueros who often received calves instead of wages as payment. These Marismas cowboys formed a rather notorious bunch, whose habitual misbehavior in the seasonal marsh cattle camps caused municipal regulations to be enacted regulating gambling, prostitution, and the bearing of arms there. The itinerant prostitutes could stay only one night in a camp. Each hato in Las Marismas had a chief herdsman, usually a salaried man, although some smaller operators served as their own overseer, and each by law needed four vaqueros for every 500 head of cattle.[25]

It is no exaggeration to say, then, that an entrepreneurial, market-based, specialized cattle-ranching system had developed in Las Marismas by about the year 1500. Beef had become so abundant and cheap as to be commoners' fare in Sevilla, a model for Latin American dietary preference, and cowhide had supplanted sheepskins and goatskins in the Andalusian leather industry.[26]

Many such señores de hato, particularly the smaller operators, would have had good reason to consider emigration to the Americas after 1500. Population growth in the circummarsh area began to restrict the availability of pastures, even as early as the fifteenth century. The town of Aznalcázar, for example, more than doubled in size between 1435 and 1493, Utrera grew from 700 to 1,500 in the same time span, Almonte surpassed 2,000 inhabitants in 1534, and big increases also occurred in Sevilla city, Hinojos, Alcalá de Guadaira, Puebla, Coria, and Jerez de la Frontera. The results were an expansion of cropland at the expense of pasture, and larger herds competing for the diminished range. All around the marsh perimeter, graziers were feeling increasing pressure from tillers by the time of the discovery of America and struggling to retain their privileges and pastures.[27] As early as 1454, certain stock raisers of Sevilla city could no longer use traditional pastures along the Guadalquivir near Puebla due to cropland expansion, and had to relocate to Isla Menor (figure 5).[28] Estate owners east of Las Marismas may have begun about this time to encroach upon the public marsh-grazing domain by extending their properties into the wetlands, further diminishing the range open to urban livestock entrepreneurs. Additional pressure was applied by ever-larger herds of transhumant sheep brought from interior Spain to pasture seasonally in Las Marismas. Eventually a new generation of commercial cattle herders would need to look overseas in search of pastures, a move also motivated by the increasingly precarious financial situation of the petty nobility.[29]

Emigration from settlements around the perimeter of Las Marismas has been well documented. Before 1540 almost 19 percent of the total Iberian migration to the Americas emanated from the marsh perimeter, and the city of Sevilla, together with its suburbs, alone sent 15 percent of the total (figure 4). If one adds emigrants from towns near the other Atlantic estuarine marshes of Andalucía, the proportion rises to 22 percent.[30]

The Andalusian marsh cattle-herding system, then, represented a distinctive adaptation to a unique physical environment. By the time of the Columbian discovery, the lowland system was labor- and capital-extensive, characterized by large herds of semiferal, uncastrated longhorns managed from horseback by a vaquero underclass, by a high ratio of cattle to small livestock, by a regulatory municipal mesta, by only local seasonal stock movement, by an absentee and perhaps mainly nonagricultural owner group resident in the marsh perimeter towns, by a high degree of commercialization, and by increasing competition for pasturage, with significant displacement by farming. Here, truly, was found the embryonic cattle frontier of Latin America.

The highland, or mesetan, range-cattle system is by no means coextensive with Extremadura, although that province provides the core of the region, and its two constituent parts—Badajoz and Cáceres—ranked second and fourth, respectively, as sources of sixteenth-century migration to America (figure 3).[31] The adjacent provinces of Salamanca and Avila also form integral parts of the Iberian range-cattle zone in the western Meseta. The physical environment of the region differs fundamentally from that of the coastal marshes, as does the regional herding system. Well-forested and, by southern Iberian standards, an abundantly watered land, Extremadura defies its popular image as a barren, rocky land that hammered its men into cruel *conquistadores*.³²

Indeed this western area is the most humid part of the Meseta, a trait that helps explain the importance of cattle, which thrive best in lands with adequate moisture. Climatically Atlantic influence is able to move upslope through Alentejo on the west, creating quite humid conditions in the winter half of the year, especially in hilly areas. Everywhere the summer drought of the Mediterranean climate occurs, giving the country a sere appearance by August. In terms of terrain, a distinctly highland character prevails, in spite of the relatively low elevation. The broad plains of the Guadiana and Tajo rivers in Extremadura are fringed by east-west oriented ranges of hills or low mountains, which encompass the greater part of Extremadura, reinforcing its upland character. The higher elevations of the mountains in the north take on a central European or even alpine character. Vegetation reflects both the terrain patterns and the abundance of precipitation. Much of Extremadura and Salamanca, even today, is covered by live-oak forests of varying thickness. The dominant species is holm oak, although the cork oak is represented, as is the chestnut. Often an understory of matorral is present, but usually a parkland or wooded savanna prevails. Mountain areas forming the border between Extremadura and Avila have some pine forests at elevations above 1,200 m. (4,000 ft.). Open grasslands are not uncommon, as on the Trujillo-Cáceres Meseta or in the La Serrena rain-shadow district of eastern Badajoz, but the prevailing image is that of oak parkland.³³

The traditional crop-livestock complex of the western Meseta was also rather different from that of the marshy coastland. In keeping with more purely Mediterranean practice, small livestock far outnumbered ganado mayor (figure 12). Fifteenth-century tribute lists from the area indicate that swine and sheep were more important than cattle, and mid-1700s tax lists reveal that, in Extremadura and Salamanca together, the proportion of small to large livestock stood three times higher, at 6.5:1, than in the Andalusian province of Sevilla.³⁴ Badajoz in 1865 housed more sheep than any other Spanish province, while Cáceres led in goats, followed by Badajoz in the second position.³⁵ Huge flocks of native merinos remain one of the agricultural mainstays of the region still today, and transhumant sheep from other provinces formerly sought winter pasture in Extremadura. Most of the open grasslands, including La Serrena, were sheep pastures belonging to great military orders in the 1400s and 1500s.³⁶

Perhaps the most distinctive feature of traditional herding in Extremadura and Salamanca was the major importance of swine, "one of the greatest treasures of the region."[37] The first three ranking Spanish provinces in number of swine in 1865 were Badajoz, Cáceres, and Salamanca, which collectively dominate the upland herding area under study; a century earlier, swine formed a larger proportion of the total livestock population in Extremadura than in any other region of Spain.[38] Huge droves of native black or red pigs have traditionally been driven through the oak forests from October to January or February, fattening on acorns. Pork sausages, bacon, and well-aged hams are among the greatest prides of the region. A study of the regional dialect of Extremadura contains special sections on pigs and sheep, but none on cattle or horses. The emphasis on pigs is traceable at least to Roman times, and by the middle 1500s up to 100,000 hogs were being fattened annually just in the forests around Jerez de los Caballeros in Badajoz. Both Cortés and Pizarro, the conquerors of Mexico and Peru, had reputedly been *porqueros,* or swineherders, in their native Extremadura, though the term may have been derisively applied simply as a regional stereotype.[39]

Cattle enjoyed regional importance, too, in a tradition dating at least to Celto-Iberian times, two and a half millennia ago. The family name Vaquero appears in the region, and at Guadalupe, the greatest place of pilgrimage in Extremadura, the monastery housing the famous effigy of the Virgin contains a series of wall murals depicting the story of a virtuous old vaquero. Toponyms such as Casa de los Vaqueros occasionally appear, suggesting a cattle-based pasturage.[40] The area most dedicated to range cattle, contrary to Julian Bishko's claims touting the upper Guadiana in Badajoz, is northern Cáceres and adjacent parts of Salamanca and Avila, centered on the flanks of the Sierra de Gredos and the adjacent live-oak parklands of the Campo de Arañuelo and other districts along the Río Tietar, a left-bank tributary of the Tajo.[41] Indeed, cattle to sheep ratios were higher in Avila and Salamanca than in Extremadura in the nineteenth century (figure 12). Herds of longhorns—mainly black in color in Salamanca and Avila, of a reddish-brown hue in the Tietar valley, and belonging to a rare white breed near the town of Cáceres—can still be seen today. Overall, however, cattle were and are less important than sheep and swine. In 1865 only eighty-seven Extremaduran proprietors owned as many as one hundred cattle.[42]

The methods of raising stock were also different in the uplands. Most essentially, the mesetan herding system was labor intensive. In

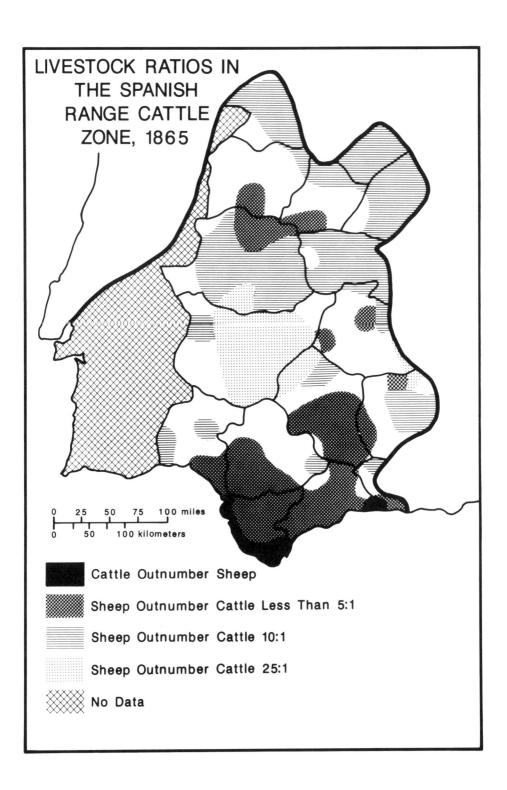

Figure 12. Source: Censo de la ganadería *(see note 11).*

many areas the pastures were enclosed and partitioned with rock walls, rather than forming open range, so that different types and age-groups of stock could be segregated. Cattle grazed the best pastures, especially the breeding stock, and they also enjoyed the ancient Castillian access to grain stubble after the harvest. Seasonal shifts of cattle among pastures occurred, usually local and certainly not so dramatic as in the coastal marshes.[43] Some bovine transhumance along rock-walled *cañadas* occurred in Extremadura and adjacent mesetan provinces, in which herds, following a belled cow, migrated seasonally to pastures distant from the home village. The major area generating such transhumance lay along the flanks of the Sierra de Gredos, forming the border between Extremadura and Avila, in particular the town of Hoyos del Espino in the valley of the upper Tormes, where a guild of carters gained privileges from the king in the 1400s to take their draft animals to winter pastures near the town of Cáceres and along the upper Guadiana, both in Extremadura. More typical in hilly areas was a diurnal, or daily, movement of cattle, as was also traditionally practiced in the Sierra de Gredos. The cattle were driven each morning in the summer up to high pastures above 1,600 meters (5,250 feet) elevation.[44]

Overall, sheep, rather than cattle, remained the dominant migratory livestock of the western plateau region. A mesetan institution, the royal mesta, a powerful national guild of aristocratic sheep raisers, saw to it that the cañadas remained open for their transhumant livestock and that access to seasonal common pastures was not denied. Indeed royal mesta domination and regulation of livestock raising in the highlands to serve the purpose of the wealthy, even to the point of intervention, was far more pervasive than in the Andalusian salt marshes. The royal mesta was a mesetan rather than an Andalusian institution.[45]

Mesetan vaqueros generally remained unmounted while herding cattle. The animals were carefully tended, normally by old men and boys, and calves were castrated, so that the animals did not become unruly or semiferal, making management from horseback unnecessary. Rock-walled corrals still today dot the landscape, suggesting frequent penning, and longhorn herds in mesetan areas often include belled cows, displaying much in common with alpine dairying. Certainly the *ahijadores*—specialized vaqueros who tended only cows—were pedestrians. The herders who accompanied transhumant cattle usually rode on horses, wielding staffs called *varas* to prod the animals along, but the basic function of horses in such situations was to provide transportation. The major exception was the mounted, vara-wielding *novillero*, a type of mesetan vaquero who herded only bulls, including those destined to fight in the ring; but such men, while influential in the later evolution of the Mexican *charro*, were not all that common in the 1500s. A humble vaquero on foot or a peasant riding a donkey provides the persistent image of traditional rural Extremadura. Donkeys and mules outnumbered horses four to one in the province in 1865, and relatively few horse breeders operated in the area. The prestige-bearing horsemanship evident in Extremadura today, reflected in toponyms such as *Dehesa de las Yeguas* ("mares' pasture") and *Dehesa del Caballo* ("horse pasture"), seems to have been confined largely to novilleros and the landed gentry. In one poor village of 229 inhabitants in southern Salamanca province, the entire population in the 1940s owned only five horses.[46]

The typical Mediterranean array of crops is also much in evidence, especially in Extremadura. Vines and grains share dominance, but the olive, though well represented south of the Sierra de Gredos, plays a lesser role than in most of southern Europe. Wheat, described by one native son as "the fundamental base of Extremaduran farming," thrives in the province still today, in places on an agribusiness scale. In traditional farming, wheat was often grown in a system of shifting cultivation on level places among thinned, scattered, pruned trees of the uncleared oak parkland.[47]

Extremeños themselves regard the lands between the Guadiana and Tajo as containing the purest form of their proud subculture. I selected one municipality from that core region—Montánchez in southern Cáceres province—for more detailed study. The district proved to epitomize many of my generalizations about Extremadura and the upland range-cattle area at large. Its tightly clustered hill villages and farming town seat, collections of inward-looking courtyard farmsteads, are surrounded by a maze of dry-rock-walled orchards, vineyards, gardens, and stock pens, beyond which lie wheat fields, some still harvested by sickle. Dotting the grainland, here and there, are haystacks enclosed by rock fences. Areas covered by forest and live-oak parkland, in places reaching near the outskirts of the settlements and now enclosed by barbed-wire fences, served, during my visit in the heart of summer, as range for mixed herds of longhorn cattle, sheep, and goats. Soon, when the grain harvest was completed, the livestock would expand their range to include the stubble, although some farmers choose instead to burn the harvested field. Swine at this season generally remain penned in their rock-walled sties, and local stores proudly offer the *jamones ibéricos* for which Montánchez is justly famous. Menus in the district confirm the fact that Extremeños are partial to pork and mutton, not beef, which they tolerate only in the form of veal. So it has been in Montánchez for at least seven centuries, since well before the migration to America, to which the municipality contributed.[48] Feudal tribute lists for the 1460s mention wine and pigs as the most common commodities, followed closely by lambs, while calves rank a distant fourth. In addition the grain fields of Montánchez yielded even more valuable annual tithes of 2,000 *fanegas* (about 1,050 hl.) each of wheat and barley. By 1865 Montánchez municipality had changed little, reporting 45,210 sheep, 20,030 pigs, 4,609 goats, and 7,169 cattle in a livestock census and also listing five times as many donkeys as horses.[49]

Traditional mesetan cattle-herding practices were also clearly revealed in a 1940s study of another highland community, Berrocal de Huebra, located in Salamanca province. Citizens of the town owned 600 sheep, 500 swine, 250 cattle, and 100 goats. A constantly vigilant vaquero managed the docile cattle on foot, keeping them in designated open-range pastures, and calling each cow by name. He used only a leather sling, ropes for tying the animals by the horns, and a *porra*—a stick about one meter (a yard) in length, with a pear-shaped, solid oaken mass weighing about a kilogram (2.2 lbs.) attached to one end. Rocks were slung or the porra thrown at the animals, as needed, to maintain order. The term *la reata*, which in North America invariably means a lasso, in mesetan usage refers either to a rawhide rope or, more commonly, to the team of oxen used to pull mired cattle out of mudholes. Similarly *rodeo* in the Salamanqués dialect means simply "a place where cattle rest." During November in Berrocal, calves were branded at the festival of San Martín on their right flank with a Latin capital letter B, for Berrocal, and earmarks were also employed. The simple mesetan brands differ strikingly from those of the Andalusian lowlands (figure 8). Most yearling bulls were castrated, assuring controlled breeding and herd docility. No cattle transhumance was practiced, although sheep were moved seasonally. Bovine-derived milk and cheese were unimportant in Berrocal, and veal was the major meat.[50]

Mesetan villages and farm towns such as Montánchez or Berrocal normally belonged to the nobility or to military and ecclesiastical orders in the 1500s. Large estates remain common in the area today, although many of these are of later origin, resulting from private encroachment upon public grazing lands in areas of poor soil and deficient water. This latter type of estate, devoted largely or entirely to stock raising, certainly bears a similarity to New World ranches, but in the crucial, formative sixteenth century, the feudal fief encompassing one or multiple village communities provided the norm. Attempts by the historian Julian Bishko and others to depict "frontier" conditions in which seminomadic pastoral folk raised cattle in Extremadura at the time of the American discovery are erroneous. The depopulated military frontier of the Reconquest, which had in earlier centuries provided convenient and abundant ranges for pastoralists, had by 1500 been replaced by a feudalistic, labor-intensive, village-based agrarian complex, in which cattle were generally tertiary in importance behind crops and small livestock, herdsmen remained unmounted, calves were castrated, and herds stayed docile.[51] The typical mesetan vaquero, a rock-slinging, porra-heaving pedestrian who knew each animal by name hardly offers a viable prototype for the American cowboy.

The British Source

The British Isles, like Iberia, display a lowland/highland dichotomy (figure 13). The essential human geography of these influential islands has long obeyed topography, for upland and lowland housed two fundamentally different rural ways of life. At root the British dualism separated sedentary, village-dwelling lowlanders, for whom crop tillage on the open fields formed the central livelihood, from dispersed, wandering clans of highland cattle folk, among whom grains remained subordinate to livestock.

This ancient partitioning does not duplicate the present English/Celtic divide, for language offers at best a rather poor guide to basic British geography and, in any case, the division is apparently far older than the Anglo-Saxon and Celtic invasions. To be sure, ancestrally Celtic-speaking populations, including the Scots, Picts, Welsh, Cornish, Manx, and Irish, lived on the highland side of the partition; but so did many border English in the hilly northern and western shires. Norse Viking settlers also influenced the herding practices of the highlands, further undermining any Celtic ethnic generalizations.

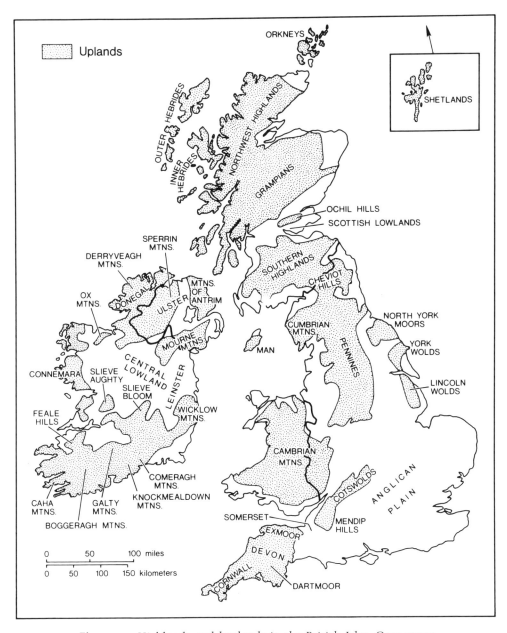

Figure 13. Highlands and lowlands in the British Isles. Open-range cattle herding was traditionally important in all of the highland districts.

North America drew heavily from the uplands of Great Britain and Ireland for its colonial population, causing the highland herding systems to assume a potential importance for New World cattle ranching. Scholars in several disciplines have proposed British antecedents for many aspects of Anglo-American ranching, including open range, overland drives, pasture firing, and western livestock law, among others.[52]

The various British highlands and hilly smaller islands have much in common environmentally: a repeated pattern of very old, hard rocks—granites, gneisses, and quartzites so resistant as to yield little soil. Here and there elevated sandstones and carboniferous limestones offer more generous bedrocks, as in southern Ireland and the Pennines, and basic geological contrasts do exist within the highlands, but the prevailing condition is that of unyielding, ancient rock. The climate is excessively rainy, and the British highlands are well defined by annual precipitation exceeding, 1,000 millimeters (40 inches). In some mountainous places, over 2,500 mm. (100 in.) fall each year. Such surplus moisture causes the thin soils to become leached, waterlogged, acidic, and often peaty. Though mild for the latitude, winters in the British uplands can be so bitter as to endanger livestock, although Ireland tends to be less severe than Great Britain. Most of the upland area originally bore a forest cover, but these woodlands long ago surrendered to human-induced, altitudinally zoned grasslands and, more abundantly, moors of various kinds, covered with heather, coarse seasonal grasses, gorse, and bracken. In some localities, as in the central Pennines, the quality and character of rough grazing vary more with parent bedrock than with altitude. In northern Scotland, higher elevations offer only poor mountain tundra of no value to cattle. Even moderately fertile land is rare in highland areas, usually tucked away in the myriad vales indenting the rainy, rocky masses. Some coastal plains, due to excessive precipitation and poor drainage, are covered by marshes. The comparative absence of good pasture and cropland, coupled with a climate unsuited to the collection of much winter fodder, encouraged the production of lean cattle for export.[53]

While regional differences existed, traditional open-range herding in the British highlands and lesser islands exhibited many common traits, shared by the clans of Cornwall, Wales, Ireland, Scotland, and the English hill shires, as well as the Shetlands, Hebrides, Orkneys, and Isle of Man. The herding system developed most elaborately and consequentially in Ireland, Scotland, and Wales, but all of the hill zones shared it in some measure. Emigrants from most of the participating regions came to colonial North America, but the key contributing areas included Northern Ireland, Wales, the hill shires of England, and the Scottish Highlands.[54]

Almost universally before about 1750, cattle were the most important and venerated livestock in the British uplands, lending prestige and wealth to the owners, passing as currency in most places, and, in the cattle raid, providing the theme for epic Gaelic tales. The Irish possessed "infinite multitudes of cattle" and pastured for "the most part of the yeare . . . upon the mountaine and waste wilde places . . . , driving their cattle continually with them," with the result that "for long stretches of Irish history the cattleman has been king."[55] Among the hill English of the northern shires, "the mainstay of their economy was cattle keeping," while in Scotland the highland black cattle provided "nearly the sole source of cash for laird and cotter" alike.[56] Clearly cattle dominated the traditional British herding system, but from Neolithic times onward diverse other livestock were also present, especially sheep, goats, horses, and pigs, though the latter animal declined with the retreat of the highland forests. Sheep did not rise to dominance until the 1700s and 1800s.[57] At the time of the great emigrations to colonial North America, cattle retained their dominance in the uplands of the British Isles, while swine had become unimportant.[58]

Crops, mainly oats, barley, and rye, also formed an important part of the traditional rural economy, though secondary to herd animals. Wheat, the preferred grain in most of Europe, would not thrive in the cloudy, rainy climate and poor soils. The British hill folk employed an "infield/outfield" system for cultivation. A small, manured infield, tilled every year without a fallow, formed the garden-like nucleus of the cropland and was often worked by spade and hoe rather than plow. Beyond lay the much more extensive outfield, only a small portion of which was cultivated in any particular year, since restoration of its fertility took a decade or more of bush fallowing. Hay meadows occupied the remainder of the land near the infield, often occupying damp, poorly drained places. More remote were the common pastures, in the hills, an area never cropped or mown.[59]

Most of these highland stock farmers lived in small, loose settlement clusters known as *clachans,* consisting of perhaps five to ten farmsteads adjacent to the communal infield, in which each family had a parcel of cropland. The dwellers of these hamlets were normally bondsmen, who owned allegiance and tribute to a local clan chieftain, lord, or other freeman who, by contrast, lived from ancient times in dispersed steads, often surrounded by sturdy protective walls of dry stone. Ruins of the scattered ring forts of these cattle lords still dot the highland areas, including some thirty thousand in Ireland alone, built over a period of fifteen hundred years. In some areas, such as Ulster, dispersal of settlement had extended down even to the commoners by 1700, and hamlets disappeared altogether from many areas. Traditionally the inhabitants of the freestanding steads relied more completely upon herding than did the hamlet people.[60]

The dichotomy of freeman and bondsman, of rich and poor, extended to the ownership of cattle, particularly since these animals constituted the standard measure of wealth and prestige. Individual chiefs and lords often owned hundreds or even thousands of head, rivaling the later cattle barons of the American West. A favored activity was the cattle raid, designed to rustle stock from neighboring herds. By the 1600s the great Celtic cattle kings had largely disappeared, replaced by a Norman-inspired feudal system in which the lords were often absentee. The ordinary folk usually farmed their small fields and owned a few head of cattle, herded communally. They paid a share tribute to the lord for pasture use, or else obtained such rights in exchange for tending the herds of the landed aristocracy. Many hill commoners became tenant "grassmen," tilling no land but acquiring grazing rights for their small herds as a form of payment for laboring on the estate. Some squatter graziers successfully evaded payments to the lords.[61]

The diverse cattle breeds of the British highlands usually bore the names of local shires, counties, districts, and islands, reflecting the importance of cattle raising. Examples are the Devon, Durham, Aberdeen-Angus, Galloway, Anglesey, Hereford, Skye, Teeswater, Craven, Kerry, Highland, and Irish Maol breeds.[62] These exhibited great variations in skull shape, horn length, and color, including longhorns, shorthorns, and polled breeds, with coloration ranging from black to red, dun, brindle, and white.[63] Most common, and probably reflecting an ancient British type, were small, hardy black shorthorn cattle, the best survival of which is the Irish Kerry breed.[64] However, the widely used English term "black cattle" should not be taken literally, since it is merely a generic name for all bovines, just as "horned cattle" was for ovines (sheep) plus bovines.[65] The black cattle of Wales were reputedly the smallest of the highland types, although an observer in Ireland reported that the cattle were "in generall very little, and onely the men and Grey-hounds of great stature."[66]

The range ecology of the British highland herding system was complex. Cattle grazed both on unimproved common pastures and on overgrown portions of the outfield destined shortly to be put into cultivation. Some of the rough grazing involved low-lying marshes, as in Somerset in the southwest of England, but normally these remained hay meadows.[67] By far the most abundant pastures lay above the cultivated zone, in the hills and mountains. Even there conditions varied a great deal. The best pastures, reserved largely for cattle, were grassy, dominated by fescue, often on limestone bedrock. While deforestation accompanied cattle pasturing, preventing, for example, regeneration of the Scots pine, the grazing and trampling by cattle retarded the spread of heather, bracken, and coarse matgrasses. The least desirable hill pastures in the peat moors went in the traditional system to sheep and goats. Indeed perhaps two-thirds of the Scottish Highlands were unsuited for cattle. Fire played an essential role in British highland herding ecology, regardless of pasture type. Heath, grassland, and marsh alike were fired, usually in the late winter, before the middle of March, in order to help maintain plant associations beneficial to grazing and to stimulate a lusher growth of succulent young plants.[68] The Irish, for example, burned the vegetation "lest it should hinder the comming of new grasse."[69]

Seasonal nomadism, or transhumance, formed an almost universal trait of the traditional British cattle-raising system, normally involving a summertime movement of people and livestock to pastures in the hills or, more rarely, coastal marshes. As in Iberia, the summer pasture, called *buaile* in Ireland and *shieling* in Scotland, usually lay close to the settlements, but migrations of up to about 30 km. (20 mi.) sometimes occurred. Entire families once accompanied the herds and flocks, although by the 1600s the task more often fell to the elderly or to boys, girls, and women. Huts or cabins located in the high pastures provided shelter. May Day marked the approximate time of departure for the summer pastures, amid much celebration, and the herders usually returned in time for Michaelmas or, at the latest, Halloween, fleeing colder temperatures in the hills or winter flooding in the coastal marshes of Somerset, a startling reminder of the Marismas of Andalucía. In fact, *somerset* means "summer pastures." Many local variations on this calendar existed, and scrub cattle often remained behind in the hills for most or all of the winter.

A particularly severe winter in 1664–65 caused a die-off of cattle in the Irish mountains. The ecological and adaptive advantages of transhumance included fuller utilization of available range, conservation of scarce fodder, and safeguarding of crops from livestock depredations during the growing season. It also promoted a genetically beneficial mixing, not only for the livestock, but also for the people, since the summer pastures provided a chance to meet the inhabitants of other settlements.[70]

As in the Iberian system, an open range generally characterized British cattle herding, at least insofar as the common pasture was concerned. Even so, most cattle remained far more confined by barriers than was the case in lowland Spain or, later, North America. Those who claim that completely open range prevailed in the British system misrepresent the actual situation.[71] Throughout most of northern England and Scotland, for example, a dry-rock barrier called the *acrewall* or *head dyke* ran along the upper limit of the outfield, separating common pasture from tillable land. Only a third of the land area of Scotland lay above the acrewall, with the result that pasture stood insular in the cropland, hardly the situation that would later prevail in American ranching areas. At planting time, before the migration to the hills, livestock were driven outside the acrewall, so as not to interfere with the crops. After harvest in the autumn, the wall was opened, allowing the cattle to descend. While in the high pastures, the herds often confronted another fence, the *moor wall*, which separated the better grassy pastures intended for cattle from the poor peat moors above, used mainly for smaller livestock and hunting.[72]

During the greater part of the year, while below the acrewall, cattle remained in fenced pens and pastures. Some livestock, confined by earthen or turf barriers, grazed that portion of the outfield intended for planting the following spring. Others were penned in winter on the infield, ringed by a stout stone *inner dyke*, helping to fertilize the cropland. Milch cows, during the winter, stayed in special folds surrounded by constantly maintained fences, or else were actually taken into shelters. Because cattle raids remained common into the 1600s and, in more ancient times, wolves prowled the countryside, even those cattle grazing on the common pasture in summer were often taken at night into pens or the courtyards of ring forts for protection. While at pasture, cattle often wore hobbles to restrict their movement. Rock walls even lined many of the trails over which cattle moved to fairs and markets. Such prolonged and diverse penning and confinement, coupled with calf castration, kept British highland cattle relatively tame, with the exception of some scrub animals not giving milk that were driven beyond the acrewall early, without herders following them, and sometimes left all winter to fend for themselves in the hills. Some of these escaped to become fully feral, but in such cases they ceased to be part of the herding system. In the Somerset marshes, a midseason roundup of cattle, called a "drove" or "drift," was held at least by the 1200s to detect illegal usage of the summer pastures by herders lacking rights there.[73]

Herd docility, in turn, made the task of management and control easier. British herders remained unmounted while in the pastures, although in some districts horses provided transportation during transhumant movements, cattle raids, and drives to market.[74] A highland fondness for horses and racing survives to the present day, and in Ireland "turf accountants" rival pubs in popularity. Transplanted to Australia, the highland British horse complex found a particularly exuberant expression. Even so, the horse never served as an important device for herd control. That role belonged, instead, to the dog. Herding canines, as opposed to larger guard dogs, apparently originated in insular northwestern Europe, spreading from the British Isles to the continent only after about 1600 and not reaching Spain until the present century.[75] An early Welsh reference, from the tenth century, mentioned a "herdman's cur."[76] Because bulls were the most difficult to manage, even within relatively docile herds, British herders developed bulldog breeds. The term *bulldogging* to describe the dragging down of cattle by canines with a grimly fixed toothhold on the lip or loose neck skin is known in Britain from at least the 1700s.[77] The great majority of British cattle-herding dogs were apparently ordinary curs rather than selectively bred.

By and large, the highland herders did not neglect their cattle to the extent found later in North American open-range ranching.[78] The Irish, for example, were said around 1600 to "watchfully keepe their Cowes," and the frequent penning of stock has already been mentioned. In Scotland some winter fodder was usually provided, including hay, dried seaweed, and milled gorse, though usually in small amounts. The more valuable cattle, including the best milk cows, were even brought into human dwellings during winter, occupying one end of elongated single-room houses.[79] Even so the successful wintering of stock remained a problem and, as earlier noted, less valuable scrubs had to fend for themselves in the hills.

Brands and earmarks were apparently both in use in the British herding districts. Edmund Spenser, describing conditions about 1600, mentioned "that good Ordinance, which I remember was once proclaimed throughout all Ireland, that all Men should mark their Cattle with an open several Mark upon their Flanks or Buttocks, so as if they happened to be stoln, they might appear whose they were." Spenser implied that branding had fallen from general usage. Estyn Evans noted a traditional practice in parts of Wales and Ireland "to ear-mark . . . by cutting or notching the ear with proprietary shapes," and A. R. B. Haldane mentioned the Scottish requirement that two witnesses had to verify the owner's brand before cattle could be slaughtered. Branding and marking occurred in spring in the British highlands, before cattle moved above the acrewall, and therefore did not require a roundup. Some drovers also applied trail brands. The prevalent British brand designs consisted of block letters and Arabic numerals.[80]

Against this evidence, we must weigh Ray August's claim that cattle branding, while part of the Anglo-Saxons' Germanic heritage, was "never adopted—either in practice or law—in England" in medieval or modern times. The practice, he claims, enjoyed no basis in English common law.[81] However, the very early appearance of branding and marking laws in the several English colonies of the eastern United States, coupled with the testimony of Spenser, Evans, and others, leads to the conclusion that branding and ear-marking remained informal, perhaps regional and often neglected practices in the British tradition, to be carried across the Atlantic to find renewed vigor and a place in law.

The people who labored as cattle herders formed a lower socioeconomic group in the British Isles. Perhaps most typically, they tended the stock of chieftains and lords as hired or indentured laborers, possessing at best a small herd of their own. Males and females, young and old alike, labored as herders, often as part of an extended family group, presenting a sharp contrast to the cowboys of the New World. Even so the term "cowman" is in general use in most of England to describe a cattle tender, and we may logically assume, given the similarity in word structure and the frequent presence of youthful herders, that "cowboy" also originated in Britain, long before America was colonized.[82]

Cattle played a large dietary role in the British uplands, and many food practices there differed fundamentally from North American ranching fare. One widespread custom involved the bleeding of cattle for food.[83] Scots performed this bloodletting in both spring and autumn in order to make a favored "black pudding," while the Irish, on the summer pastures or in times of hunger, were said to "open a vaine of the Cow and drinke the bloud" or to boil and mix it with meal.[84] Far more important was the highland dependence upon dairy products, especially milk, butter, whey, and curds. The Irish reportedly "drinke milke like nectar," after heating it with fire-warmed stones, while also exhibiting "esteeme for a great daintie sower curds" and swallowing "whole lumps of filthy butter."[85] Dairy work, especially butter making, formed the chief occupation in Irish and Cornish summer pastures, and throughout the British highlands and islands, milk cows were the most privileged animals. They grazed the best grasses, received most of the meager winter fodder, migrated later (in milder weather) to the hill pastures, and enjoyed shelter during severe winter weather.[86] Highlanders also ate beef in quantity during good times, and for the Irish, "meat" meant beef.[87] They "devoure great morsels of beefe unsalted," including the entrails, and also favored "Beefe-broath mingled with milke," though the Irish reportedly ate mainly animals that died of natural causes and "seldome kill a Cow to eate."[88]

While traditional British cattle herding and the farming system to which it belonged remained largely subsistent, substantial export and trade occurred. For centuries, perhaps millennia before about 1500, culled scrub cattle were driven overland along established droving trails to Roman and Saxon markets or fattening districts in the lowlands. Some drove roads on the grassy uplands of Great Britain are reputedly six thousand years old, and a lively trade in Welsh and Scottish cattle on the hoof had begun at least by the 1200s and 1300s. Export of hides, salted beef, and tallow was also of considerable antiquity, ultimately drawing even upon areas as remote as the Shetland and Orkney islands. From Scotland alone in a single year, 1378, came some 45,000 exported hides.[89]

Shortly before the great emigrations of British highlanders to North America, some relevant and fundamental changes occurred in the herding system, all of which hinged upon increased commercialization. In the 1500s and particularly the 1600s, the lowland demand for cattle products and lean animals on the hoof rose dramatically, profoundly altering the traditional way of life.[90] Beef cattle increased greatly in number and importance, at the expense of sheep, goats, pigs, and dairy cows, creating a commercial meat-cattle focus. Ireland annually sent 100,000 live cattle overseas by the 1620s, and forty years later the Irish export of cattle products was worth almost three times as much as sheep and fifty-seven times as much as hog products. So successful did the Irish commercial beef herders become that the English passed a series of punitive Cattle Acts in the 1660s, banning imports from Ireland; but the Irish responded by smuggling and by increasing their already substantial exports of tallow, hides, leather, salted beef, and live cattle to continental Europe. In 1662 alone, almost 320,000 cattle on the hoof passed through the English town of Carlisle (figure 14) on their way south from Scotland and Ireland. Similarly by the 1640s, live beef cattle had become one of the key exports of Wales.[91]

To handle the increased overland trade in live cattle from Scotland and Wales after about 1500, in the commercial era, a class of professional drovers arose. They performed repeated trail drives all through the summer, beginning around June 1 with lean, overwintered stock that had regained sufficient strength for the journey, and ending in the early autumn with culls brought down early, in August, from the high pastures, or shipped over from Ireland. From diverse collection points, stock trails led to markets, fairs, and fattening areas, the most important of which lay in the counties near London (figure 14). These vast droves, "all with their heads directed to London," demonstrated the primacy of the great capital and "the force of its attracting-power."[92] Lesser fattening districts lay in the Scottish lowlands, northwestern England, and, in Ireland, in the northern plains of Leinster, near Dublin. Various major cattle fairs developed, particularly the Falkirk Tryst near Edinburgh, at Thornton-in-Lonsdale, Yorkshire, and at St. Faith in Norfolk.[93]

Armed drovers, usually on foot but frequently mounted on ponies, used dogs to control the trail herds, and one bullwhip-wielding man could manage a hundred cattle, and some say as many as four hundred. At rivers and even at narrow straits separating Welsh and Scottish islands from the mainland, the cattle swam across, in spite of their generally poor condition. In the later 1700s, after road improvement, it became necessary to have the cattle shod for the drive. Herds moved in short daily stages of about 16–20 km. (10 to 12 mi.) between resting places called "stock stances," where rented fallow fields provided overnight penning.[94]

Under such commercial pressure, hill pastures became overstocked, damaging the ranges. Hay shortages developed, leading to the marketing of very lean cattle. By the middle 1700s, pasture deterioration contributed to the widespread replacement of cattle with

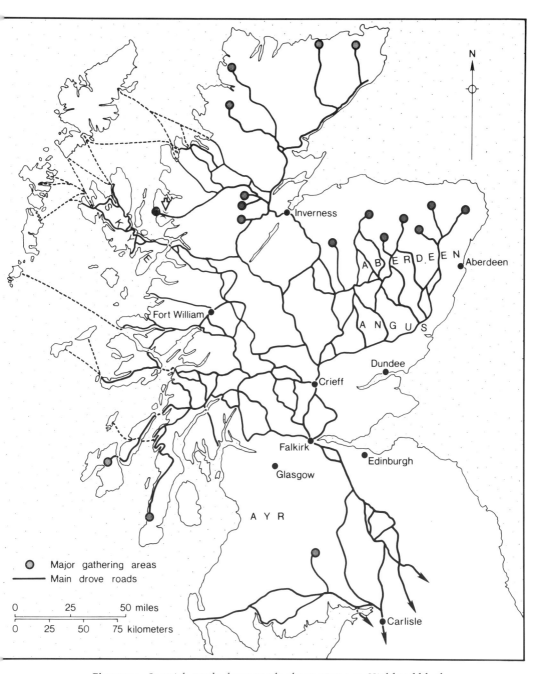

Figure 14. Scottish cattle drove roads about 1600 A.D. Highland black cattle were driven to markets and fairs, both in the Scottish lowlands and in England. The hollow arrow points to the famous Bealach-na Bo, or "Pass of the Cattle," in Ross and Cromarty. Sources: A. R. B. Haldane, The Drove Roads of Scotland *(London, UK, 1952)*; field research 1992.

sheep, which could more effectively utilize the degraded ranges.⁹⁵ The brief era of commercial lean-cattle raising had ended. Today relatively little remains to remind the visitor to the highlands of the beef-cattle episode, although the "Great Glen Cattle Ranch" near Fort William, Scotland, represents a spectacular if atypical and Americanized revival.

The relatively brief era of commercialization, specialization, and professionalism in the British herding areas also weakened the clans and folk culture, forever altering the highland way of life. The result, however, was a far more suitable British herding prototype for American cattle ranching. The altered highland system gave full rein to individualistic, privileged entrepreneurs, who pursued a productive capitalistic system of lean-beef cattle raising using hired labor on large, privately owned estates, with access by stock trails to fattening areas and distant urban markets.⁹⁶ Moreover, the changes in highland herding had occurred before about 1650, prior to the great emigrations of hill Britons. Indeed the shift to specialized cattle raising produced some of the notorious evictions of rural folk in Scotland that fueled the emigration, since fewer laborers were required in the new system.

However, other changes occurring in the same era substantially weakened possible British influence on North American herding. In effect, Ireland was removed from the trans-Atlantic equation. The only substantial group of Irish immigrants to colonial America consisted of the Ulster Protestants, who swarmed to the eastern seaboard of the United States in the 1700s. In Ulster the colonization by Protestants had been accomplished after 1600 largely by lowland Scots and English, who lacked a range-cattle tradition. One of the chief results of their settlement in Ulster and the wars that preceded the plantation was to destroy the cattle-herding system of the local Catholic Irish. Only in the southern upland refuges of Ulster did the decimated natives succeed in perpetuating the older pastoral way of life. The Protestant immigrants, given their lowland background, preferred to settle the plains or valleys and did not practice range-cattle raising.⁹⁷ As a result the Scotch-Irish arrived in colonial America with no substantial open-range herding experience, and the great Catholic Irish immigration came far too late, in post-colonial times, to exert significant influence on American ranching. We must look instead to the highland Scots, Welsh, and hill English for transferal of British influence.

Across the Channel, in France, several hill districts housed cattle-herding systems similar to the highland British type, though lacking the herder dog before about 1600. In both Celtic Brittany and the interior district of Auvergne in the Central Massif, transhumant cattle herding with pronounced but not exclusive dairy emphasis was found, and almost 5 percent of the French emigrants arriving in America before 1700 came from these two regions combined (figure 2).[98] The previously mentioned Mediterranean salt-marsh beef-cattle pastoral system, thriving near the mouth of the Rhône River, in the Camargue, offered a more promising French prototype for New World ranching, but this area was not a notable source of emigration or influence, in spite of vague claims to the contrary.[99]

The African Source

Another great Old World Atlantic zone of traditional open-range cattle herding lay in tropical West Africa (figure 2). Slaves derived from this pastoral complex may also have contributed to the development of ranching in the Americas, although claims that "Africans shaped [the] American cattle industry" and that the Texas cowboy "is indebted to the Negro for his culture" are exaggerated.[100]

The cattle region of West Africa forms a belt, stretching from the Senegambian coast some 5,000 km. (3,000 mi.) eastward, beyond Lake Chad (figures 2, 15). Wedged between the humid, tsetse-infested Guinea coastal region to the south and the arid Sahara to the north, both of which present environments hostile to cattle, the herder belt occupies a broad, favorable ecological niche consisting of the Sudan and Sahel, both also oriented in east-west belts. Of these component parts, the Sudan, not to be confused with the independent state of the same name in northeastern Africa, borders the Guinea district and forms the southern half of the range-cattle region, while the Sahel, bordering the Sudan on the north, adjoins the Sahara.[101]

The environmental basis for the West African pastoral belt lies largely in climate. Precipitation decreases sharply from south to north, and also proves highly erratic both over time and space. The southern margin of the Sudan corresponds fairly well to the 1,000-mm. (40-in.) isohyet, or line of equal rainfall, and it gives way to the Sahel at about the 500-mm. (20-in.) isohyet. The drier Sahel, in turn, yields to Saharan desert conditions at the 200-mm. (8-in.) isohyet. Rainfall occurs largely in a well-defined, high-sun wet season, lasting from May or June until September or October, during which time moist southwest winds from the Atlantic dominate the weather. The

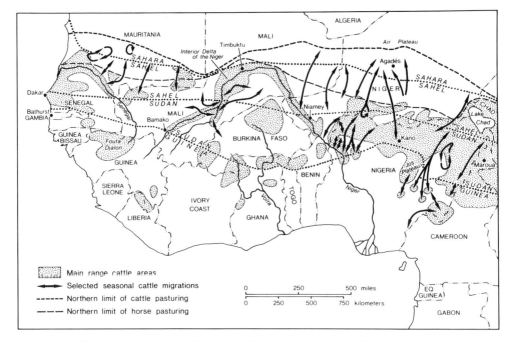

Figure 15. The cattle-herding belt of tropical West Africa. Sources: Bernard, Afrique septentrionale, II, p. 437 (see note 118); Atlas de la République Islamique de Mauritanie *(Paris, France, 1977)*, p. 42; Church, West Africa, pp. 64, 131 (see note 101); Dupire, Peuls nomades, p. 80 (see note 106); Doutressoulle, L'élevage, pp. 42, 78, 255 (see note 107); Gallais, Delta intérieur, II, p. 366 (see note 105); Hopen, Pastoral Fulbe, p. 21 (see note 104); Johnson, Nature of Nomadism, pp. 129, 147 (see note 103); Stenning, Savannah Nomads, p. 230 (see note 106); Udo, Geographical Regions, pp. 146, 171 (see note 109); Fricke, Rinderhaltung, p. 138 (see note 103).

opposite, low-sun part of the year brings drought, causing the Sudan-Sahel rainfall regime to be exactly the opposite of that in Andalucía and Extremadura. During the unpleasant dry months, a parched, dusty northeast wind named the *harmattan* often blows off the Sahara. The impact of the dry season is somewhat mitigated by the presence of two great exotic rivers, the Niger and Senegal, both of which rise in humid Guinean lands to the south and flow through lengthy stretches of the Sudan and Sahel to reach the sea. The Niger, one of the major rivers of the world, is 4,200 km. (2,600 mi.) long (figure 15), and range cattle come to water along most of its length.

Since cattle have less drought resistance than smaller livestock, they are more numerous in the Sudan than in the Sahel.[102]

The range-cattle belt of West Africa lies north of the forests of the Guinea Coast, in tropical grasslands. Vegetation cover in the Sudan is best termed an acacia savanna, a prairie studded with low trees. Grasses, both annuals and perennials, stand about a meter to 1.5 m. (3 to 5 ft.) in height and form a solid carpet well suited to cattle grazing. The trees, small-leafed and pyrophytic, or fire resistant, occur singly and are rarely more than 12 m. (40 ft.) tall. Acacias, the dominant trees, are represented by numerous varieties, some of which are evergreen, and the savanna also contains palm, tamarind, kapok, baobob, and locust bean trees, among others. In places a bush savanna prevails, characterized by shrubs of 2–6 m. (6–20 ft.) in height. Cattle, in addition to grazing, forage on acacia leaves and pods. Herders often lop off acacia branches or cut down entire trees in the dry season to provide more food for livestock. In the Sahel, vegetation reflects greater aridity. Grasses are tufted, short annuals that do not form a solid cover, and the overstory consists of a thorn scrub 5–10 m. (15–30 ft.) high, becoming bushy thorn thickets in some western plateaus. Various deciduous acacias, which remain leafless for about three months beginning in November, form the dominant thorny growth. Herdsmen are remarkably knowledgeable about the range flora of the Sudan and Sahel, having over two hundred names for plants and special pasture qualities.[103]

By midlatitude standards, these West African grasslands provide inferior grazing, averaging only about 5 percent digestible crude protein. From 12 to 14 hectares (30–35 acres) per head are required to support mature cattle. In places overgrazing in time diminished range quality, causing thicket formation, even before the recent catastrophic Sahel drought. Fire is not employed in range management by cattle raisers, but hunters from other, nonpastoral tribes often set dry-season fires, destroying forage precisely when it is in shortest supply and causing distress for the herders.[104]

The terrain of the Sahelo-Sudanic belt belongs to lowland Africa. The western reaches of the continent consist largely of two enormous, shallow basins shaped like saucers. Each ranges from about 180 to 500 m. (600–1,600 ft.) in elevation above the sea and is ringed by higher hills and plateaus. Lake Chad occupies the center of the more eastern of the two saucers, most of which forms a zone of interior drainage. The great inland delta of the Niger, where the river breaks into numerous distributaries and lakes, lies at the middle of the western basin, but the Niger is able to escape and eventually reach the Atlantic (figure 15). During the high-sun wet season, the inland delta becomes inundated, an event and rhythm of great importance to the local cattle herders and similar to conditions in Las Marismas of the Guadalquivir. Collectively the two basins present vast, monotonous, undulating plains resting upon a low platform of ancient rocks. A shallow water table permits numerous wells for watering stock. Still farther west, a broad, flat coastal plain borders the ocean in Senegambia and Mauritania, reaching some 400 km. (250 mi.) inland, where cliffs mark the edge of the low plateau lying on the perimeter of the western saucer. Across these level lands, the West African pastoralists move with ease, driving their herds with them.[105]

While cattle raising occurs through almost all of the Sahelo-Sudanic grasslands, certain areas possess particular concentrations (figure 15). A vivid ethnic patterning underlies the regional and local importance of cattle, and some groups base much of their cultural distinctiveness upon cattle specialization. The Fulani people are the preeminent and most widespread cattle raisers of West Africa, occupying substantial portions of the Sudan and Sahel, from Senegambia eastward to Cameroons, with particular concentrations in northern Nigeria and along the Niger River in Mali. Today the purest form of their ancestral livelihood is practiced by the so-called "Cattle Fulani," who represent the group much as they were at the time of the slave trade. Of mixed Negroid and Caucasoid origin, the Fulani differ both from their more purely black neighbors to the south and the swarthy Mediterranean whites of the Saharan north. The Cattle Fulani are slender, tall, and relatively fair. Until relatively recently they remained pagan, but most are now Islamicized, although the Cattle Fulani display the least Muslim influence within their ethnolinguistic group. Fulani origins lay in the low coastal plain of the far west, in Senegambia; but about A.D. 1000 these people began a remarkable eastward expansion, perhaps prompted by Islamic pressure from the Arabs and Berbers of the coastal area. In the course of half a millennium, the Fulani dispersed through the great shallow basins as far as Cameroons, and the map of cattle concentration today is in the main a depiction of the Fulani diaspora. The movement out of Senegambia was achieved not as a tribal mass migration, but instead by numerous small clans operating independently. If West Africans did, indeed, help shape American cattle ranching, the bearers of the diffusion would almost certainly have been Fulani captured into slavery.[106]

Still, certain other West African ethnic groups also raised range cattle. Among them were the southern Maures, an Arabo-Berber Muslim people of Senegal and Mauritania. The Maures, or Moors, are very light complected and rarely entered the slave trade. While they keep some cattle, they use them principally as beasts of burden and for riding, placing greater importance upon sheep and goats.[107] Similarly the racially mixed but fair southern Taureg, staunchly Islamic Berbers of the Sahara and Sahel in Mali and Niger, are essentially camel nomads, for whom cattle are of marginal importance. By way of comparison, the Fulani own from five to ten cattle per capita, while the southern Taureg own fewer than two.[108] Prior to the eastward expansion of the Fulani, other cattle folk had apparently occupied the West African grasslands. These earlier groups are poorly known, but perhaps the cattle-tending Gamergu of northeastern Nigeria are a remnant. The Arabo-Hamitic Kanuris and Shuwa Arabs, also cattle herders of northern Nigeria, seem to be more recent arrivals, though much mixed with indigenous groups.[109]

These ethnic categories of West African herders extend also to the breeds of cattle. All of the major cattle-tending groups raise zebus, or *Bos indicus,* present for millennia in West Africa and distinguished by a hump on the back from *Bos taurus,* the cattle of Europe. Zebus vary in appearance from one herder ethnic group to another, and the particular breed tells the knowledgeable observer whether the owners are Fulani, Taureg, or Maures.[110]

Traditionally the Fulani were cattle specialists, engaging in no farming and keeping only small numbers of sheep and goats. Some groups of Cattle Fulani, such as the Wodaabe of Niger, remain true to the ancient ethnic livelihood. A symbiotic relationship existed with farmers, in which the Fulani traded for grain and brought their herds to graze on the stubble of harvested fields, fertilizing them in the process. By contrast cattle were far less important to the southern Maures and Taureg, who placed greater emphasis upon sheep, goats, camels, and horses, forming a more diversified adaptive strategy. They, too, traded livestock products for grain and dates.[111]

The Fulani focus upon cattle was based in a veneration of these livestock. For them, cattle raising was less a commercial venture, or even subsistence activity, than a religion. It is incorrect to speak of "ranching" among them, since that word connotes free enterprise and profit. Cattle Fulani still link the myth of their tribal origin to the creation of cattle, and for them ethnic distinctiveness rests in part on the attitude toward these animals. They value cattle far beyond utilitarian or monetary worth, measure their wealth in cattle, and determine personal prestige on the basis of herd size. The Fulani discuss cattle more than any other topic, awarding them a centrality in their lives, and the ceremonial role of bovines exceeds their economic value. Cattle represent a treasured gift to the present generation from the ancestors, a link between the living and the dead, and each father strives to pass more cattle on to his sons than he inherited. By in effect holding livestock in trust for future generations, the Fulani allow unproductive wealth to accumulate in their herds and assure overstocking of the range. Maures and Taureg do not endow cattle with these special attributes.[112]

Perhaps the most essential geographical aspect of West African cattle herding is nomadism, involving seasonal migrations far more profound in character than occur in the local shift of the Andalusian marshes or the transhumance of the British highlands (figure 15). Typically West African cattle folk have no fixed settlements, but instead the whole group continually relocates. The very name of the Wodaabe Fulani means "the wind which blows through the bush." At the root of these movements is the rhythmic seasonality of rainfall, prompting the herders to move southward in the low-sun dry period, seeking out the valleys of dwindling rivers and the shores of shrinking lakes, where water and grass can be found. With the onset of rains as the high-sun season begins, the pastoralists drift northward or into local uplands, following the showers and renewed forage, seeking out as needed zones of saline soils and halophytic plants. Among the Fulani, annual migrations vary in straight-line, one-way distance from about 15 to 200 km. (10–120 mi.). They move their camps seven or eight times a month on the average and never allow cattle to graze the same pasture longer than one week, even in the wet season. Traditionally the Fulani did not own or have formal rights to pastures, trails, water, or campsites, and they often changed migration routes and destinations, reacting to local weather conditions, plant growth, and other environmental cues. Among the southern Maures and Taureg, cattle are separated due to their lesser mobility from herds of sheep, goats, or camels, and the distance and duration of bovine migration are usually short.[113]

Nomadism, so essential to West African pastoralism, was facilitated by social structure. The cattle folk were traditionally based in nuclear, patronymic families, each consisting of the husband, one or more wives, children, the elderly parents of the husband, and perhaps a slave or two. Normally a small number of such families, related by blood, migrated as loose clans, engaging in mutual defense and pastoral cooperation. Before about 1800, the camps of the constituent families were clustered for protection. Assembly into larger groups of multiple clans occurred in the wet season, when territorial friction between herders and farmers peaked. Even in such settings, each family camp remained largely autonomous and self-sufficient.[114]

Frequent movement meant that housing had to be portable. The Maures and Taureg, reflecting their northern origin, dwelled in tents, but the Cattle Fulani built huts of poles, woven mats, and grass thatch. Those in charge of the herd slept on platforms close to the cattle. The typical Fulani cattle camp had as its focus a long rope to which the suckling calves were tethered. On one side of the rope stood the pole huts of the camp, while the thorn-brush-fenced cattle pen and sleeping platforms were located on the other side (figure 16).[115]

Ownership of cattle rested at the level of the individual nuclear family, and in the economic sense the Cattle Fulani family was an independent, herd-owning enterprise. The male head owned the livestock, and his sons, from the age of about seven to the time of marriage, tended the herd. Often Fulani also took care of cattle belonging to sedentary farmers. The Maures and Taureg usually had darker-skinned slaves to tend cattle. Cattle Fulani signified their ownership of zebus by earmarking them on set days in the dry season at the camp pen, amid rites and magic, and some say they also used pelt marks; but apparently branding cattle rarely if ever occurred in West Africa. The Taureg did brand camels and donkeys, using distinctive designs, but not cattle (figure 8). Nor were Taureg cattle earmarked, so that they bore no sign of ownership.[116]

Herd-control techniques offer some of the most revealing and diagnostic aspects of West African cattle raising. The Cattle Fulani exerted more control over their stock than did other herders of the region, providing the animals with meticulous care and protection at all times. Every evening the stock were brought into the camp pen, and a corral fire believed to possess magical qualities burned constantly, taking the chill off the dry-season night. The boys on their sleeping platforms roused at any sign of disquiet, protecting the herd from predators and thieves. Assistance was rendered during calving, salt offered to the animals as needed, ticks removed daily and the bites cauterized, weaning forced upon calves at a certain age, and folk

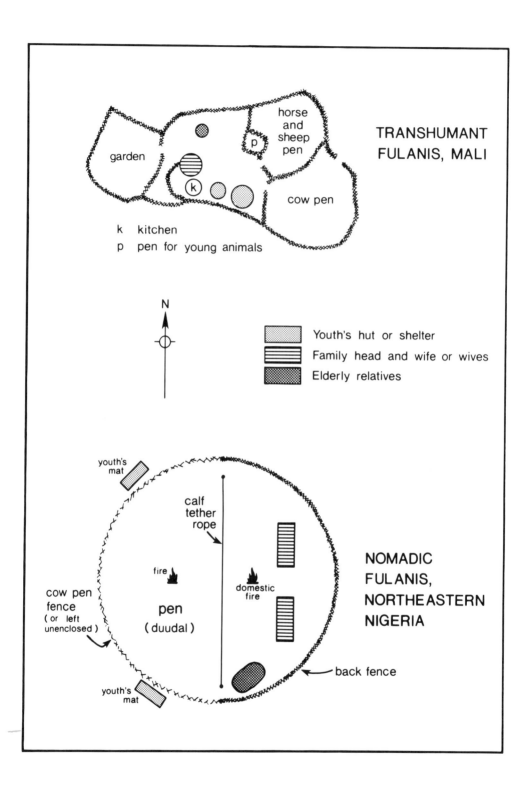

Figure 16. Dwellings and camps of cattle-raising Fulani in West Africa. The Nigerian Fulani prefer to face their camps toward the west. Sources: Stenning, Savannah Nomads, p. 106 (see note 106); Udo, Human Geography, p. 50 (see note 102); Hopen, Pastoral Fulbe, pp. 56–57 (see note 104).

veterinary care administered. The Fulani cattle tenders knew each animal in the herd. During the day, the cattle dispersed through the adjacent pasture, under the careful watch of pedestrian herders employing long staffs, cattle calls, thrown stones, and hand signals to direct the animals and keep them away from crops. Fulani cattle were tame and trained to follow the herder. The modern Fulani conjure up images from ancient Egyptian art of staff-wielding, pedestrian cowherders, suggesting an African continuity of cattle management methods reaching back twenty-six centuries. Selective breeding was practiced by the Fulani, who castrated undesirable males at two or three years old and prevented the remaining bulls from siring by their own offspring. By contrast the Taureg left their cattle pretty much to themselves, often without protection or accompaniment.[117]

The horse is found through nearly all of the West African Sahel and Sudan, among the Maures, southern Taureg, and Fulani alike. Maures exhibit a true passion for horses, and the Sahel Taureg also regard them as prestige animals deserving of special care. Among the Fulani, however, horses made a late arrival, probably as an aspect of their fairly recent Islamization, and they do not venerate them as their northern neighbors do. All three groups used the horse primarily for military, ceremonial, and transport purposes rather than in cattle herd control. Skilled horsemen abound in West Africa, but not mounted herdsmen.[118]

Even so, the present-day Fulani do, on occasion, use both horses and lassos to help manage troublesome bulls, which can be rather fierce. Fulani have been observed to "pursue the bull on horseback and, at a gallop, strike him" on the hamstring with a staff "or throw a lasso over his horns." They also cast lassos while on foot. "A swift, agile herder boy tries to snag the loop on one of the rear feet of the fleeing animal."[119] Given the relatively recent acceptance of the horse by the Cattle Fulani, it seems unlikely that they stalked bulls from horseback with ropes several centuries ago, when the slave trade and forced emigration to America occurred, but they probably did use lassos on foot. In any case, Fulani roping normally does not secure the animal at once; instead a weight tied to the end of the trailing rope causes the bull to tire. The use of ropes to secure cattle, quite aside from lasso casting, seems to be an ancient practice among the Fulani. Herder dogs are not used.[120]

The traditional dietary uses served by cattle in West Africa differed rather fundamentally from those in American ranching. In common with the British highlanders, the Fulani pursued essentially a dairying livelihood, even though zebus are poor milk yielders. Women and girls did all of the dairy work, and fresh milk was the principal product. Butter was also made, but no cheese. Surplus milk and butter were sold or bartered by the females in village and town markets, among farmers and urban folk. Beef appeared only occasionally in the Fulani diet, during festivals or at major social events, and the animals chosen for slaughter were steers or else sick and maimed stock. Dried meat was sometimes prepared. Nowhere in West Africa was beef a staple of the diet, nor were blood or blood-based foods consumed, in contrast to East Africa. The Fulani compulsion to have their herds ever increasing, quite aside from the ecological problems produced, made large-scale or regular beef consumption impossible, and it did not allow the marketing of many live cattle.[121]

A leather industry reminiscent of that at Córdoba in Andalucía developed long ago at places such as Timbuktu and Kano, based partly but by no means entirely upon cattle hides. The Fulani themselves make hides and skins into clothing.[122] Nevertheless, the Fulani devotion to range-cattle raising had rather inconsequential commercial results, aside from milk sales.

Emigrants from the Iberian, British, and West African segments of the Atlantic coastal range-cattle zone in the Old World, bearing distinctive herding systems, reached colonial North America and were in a position to apply their preadapted skills on the cattle-ranching frontiers. Herder folk from the Atlantic fringe first touched North America in the West Indies, an area that would in turn have a profound impact on the mainland, in Spanish, British, and French colonies alike. Our attention now turns to the Antillean implantment.

THREE

IMPLANTMENTS AND ADAPTATIONS IN THE WEST INDIES

THE CATTLE-HERDING CULTURES OF THE OLD WORLD ATLANtic fringe achieved initial American footholds not on the continental mainland, but instead in the West Indies. The implantments that would ultimately help shape the frontier in much of the American West occurred in the Greater Antilles—the four large islands of Española (Hispaniola), Jamaica, Cuba, and Puerto Rico—where an insular confluence of cultures brought together bearers of the Iberian, African, and British cattle-raising traditions, a meeting never achieved east of the Atlantic (figure 17). A diversity of livestock-rearing techniques came to the Indies, from which several related herding systems suited to the American setting coalesced. Complex Old World adaptive strategies could not survive diffusion across the Atlantic unaltered, due to environmental, economic, and social conditions that encouraged cultural change and synthesis.[1] The West Indies witnessed these initial, highly significant introductions and modifications.

The Iberian Implantment

The Iberian cattle folk, "heretofore arrested at the Pillars of Hercules in their advance" westward from the ancient Asiatic hearth of livestock herding, achieved the earliest and most consequential American implantations. Spaniards colonized all four of the Greater Antilles between 1493 and 1512, attaining bases that would facilitate their remarkable seizure of the greater part of the western hemisphere. Española, the first island settled, became the mother colony not just for the Spanish Main, but for Latin America at large. Andalusians led the way, forming by far the largest Iberian provincial element in the West Indian population during the formative period before 1520. Sevilla and its environs were abundantly represented, providing ample opportunity for introduction of the marsh cattle-

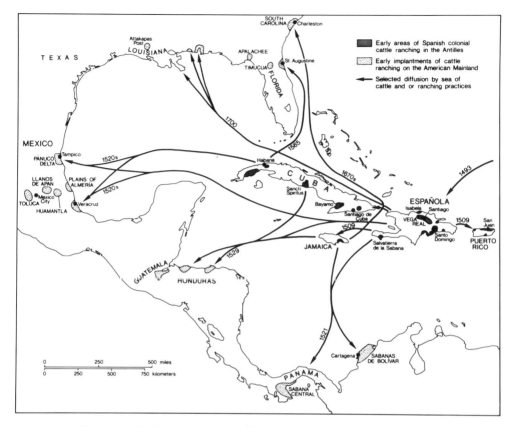

Figure 17. Implantments of cattle ranching in the West Indies and subsequent diffusions to the continental mainland in colonial times. Sources: Harris, Human Ecology, pp. 267–89; Johannessen, Savannas, pp. 27, 37; Dary, Cowboy Culture, pp. 8–10; Street, Historical and Economic, pp. 82–84; Rouse, Criollo, fig. 2; Wright, Early History; and sources by Chevalier, Arnade, and Jordan listed in chapter 4. (See, in chapter 3, notes 1, 5, 7, 11, and 19.)

herding system. For example, of those Spaniards known to have gone to Jamaica before 1520, fully 35 percent came from Sevilla and Huelva provinces, near the great marshes, and more than one of every four early settlers of Cuba was Andalusian. Lesser, but significant, numbers of Extremeños came to the Indies as well.[2]

Faced with a mass die-off of the Antillean Arawak Indian population within the first decade of occupation, the Spaniards began at least by 1501 to introduce acculturated Ibero-African servants from Andalucía, and imports of slaves directly from Africa began soon thereafter. By the early 1520s, Española had some fifteen hundred black slaves and each of the other three large islands about three hundred. The use of people of African birth or ancestry to tend cattle soon became commonplace in the Spanish Indies, and some of these surely possessed Old World experience in raising range cattle.[3]

The hispanicized Christian blacks introduced from Sevilla, Huelva, and other Andalusian districts were descended from a very old slave trade. Beginning in the eleventh century, Muslim traders supplied Sahelo-Sudanic African slaves to the Mediterranean lands, and the pagan character of the cattle-herding Fulani at that time made them attractive to the traders. Many such slaves accumulated in Andalucía under Moorish rule, especially in Sevilla, and others arrived after the Reconquest. Some of them herded cattle in Las Marismas, as was suggested in chapter 2, and possibly brought to the Indies herding skills rooted in both Iberia and the Sahelo-Sudanic belt of West Africa. The slave trade directly from Africa to the Spanish colonies in America began in earnest about 1517, and it, too, tapped the cattle-herding hinterland of West Africa, though not as its principal source. Fulani appeared repeatedly on the slave lists of ships bound for the Indies, as did certain other herding groups. Berbers, conceivably Maures or Taureg, resided in Española as early as 1506, as did "Canene," possibly a garbled reference to the cattle-rich Kanuri of Nigeria. The African ethnic heritage of slaves in the Spanish colonies will always remain imperfectly known, but some certainly had roots in the Sahelo-Sudanic zone.[4]

Accompanying the human immigration, in stalls on the caravel decks, were diverse Iberian livestock, including yearling longhorn calves of many descriptions, from black to roan and even white. Some, the *retintos,* displayed the reddish-brown hide still today evident in herds grazing on Las Marismas of the Guadalquivir (figure 9). Another common type, piebalds called *berrendas,* were white with black marks, especially on the neck and ears. Some Spanish cattle doubtless displayed the spotting and speckling typical of semiferal herds. Longhorn imports to the Indies continued from 1493 to 1512, totalling perhaps five hundred animals, although a third or more perished en route. All of the subsequent huge American herds of *criollo* cattle descended from these few ancestors, derived from both the Canary Islands and Andalusian ports.[5]

As sizable mountainous islands, the Greater Antilles in 1500 offered varied environments and multiple ecosystems. Climatically all four islands possess a distinctly tropical character, experiencing only a modest annual range of temperature. The rainfall regime, generally featuring a low-sun drought, provides the only seasonal rhythm. In these ways, the Antillean climate resembles fairly closely that of the Sudan belt of West Africa, although, with the exception of a few local pockets of semiaridity, the West Indian dry season is shorter and precipitation totals greater. Spaniards, unlike Africans, had to adapt to the absence of a winter as well as to an annual rainfall/drought cycle precisely reversed from that of Iberia. They made the latter adaptation simply by ignoring their calendar and calling the rainy time of year "winter." The vegetation cover at the time of European contact reflected the moister conditions of the Antilles. Far more woodland existed than any of the Old World cattle-rearing folk had known, ranging from montane rain forests to dry tropical scrub forests. Early Spanish accounts remark on the abundance of trees. Approximately two-thirds of the land area of Cuba, for example, was heavily wooded.[6]

Of far greater importance to cattle raising were abundant tropical grasslands in the poorly drained Antillean plains. Perhaps these lush pastures struck the Spaniards as somewhat alien in appearance, for on Española they immediately adopted the Arawak Indian word *sabana* as the generic term for the grasslands, rather than using a Spanish word such as *pradería*. Subsequently the conquistadores carried sabana throughout the Greater Antilles, to many places on the American mainland, and, as *savanna*, eventually to the jargon of botanical science, in which it means a gramineous (grassy) area with widely scattered trees or shrubs. Early Cuban cowboys went by the name *sabaneros*, and to the present day the herdsmen of Guanacaste province in Costa Rica bear that title.[7] No adequate contemporary descriptions of the savannas at the time of Spanish contact exist, and botanists have long debated whether they were induced by human activity or represented a floral climax. The tropical pastures were diverse vegetationally, containing grasses, sedges, forbs (perennial, broad-leafed herbs), other small herbaceous species, canes, and scattered trees. One modern study lists 13 genera, encompassing 110 species, of grasses in the Indies (including some Old World introductions), while another study mentions 24 plant genera just in Cuban pastures. Diverse soil types appear in conjunction with the Antillean savannas, although sands are very common. Eleven varieties of tropical savanna occur in the coastal plains of Puerto Rico alone.[8]

Perhaps the first sort of savanna encountered by Iberians was a lowland aquatic type inundated in the rainy season and including some salt flats near the sea.⁹ Accordingly, the earliest Españolan meaning of sabana was "a marshy treeless floodplain." A luxuriant growth of tall grasses, sedges, and cane occupied these wet savannas, including some halophytic species, as in the Andalusian salt marshes. The Spaniards acknowledged the similarity of these coastal plain savannas to Las Marismas of the Guadalquivir by using the Andalusian generic toponym *caño* in naming deltaic distributaries, as for example the Caño Gallardo in northeastern Puerto Rico. Because the Indians possessed no wetland crops, these marshy savannas lay empty and open to cattle grazing from the very first. Surely the marine marshland herding of southwestern lowland Spain preadapted the conquerors for almost immediate successful cattle raising in the aquatic savannas of coastal Española, giving rise to the American cattle-ranching industry perhaps even before 1500. As the other Greater Antilles were colonized, the same early implantment of cattle raising in the wet littoral savannas occurred. The Sabana la Mar in western Jamaica and initial herding footholds in the Boriquén, Río Cibuco, and Río Coamo areas of Puerto Rico provide examples of early wetland cattle ranching that surely drew upon Andalusian experience.¹⁰

Adjacent to the fully aquatic grasslands lay others, less frequently flooded but still having a high water table and lush, tall grasses. Above the level of periodic inundation was the *sabana seca*, or dry savanna, still in the coastal lowlands. An expanse of tall grass that provided refuge for herds during the time of flooding, the dry savanna was dotted with palms or pines, presenting the stereotyped image of a tropical grassland. In some areas the trees became more numerous, forming a parkland or wooded savanna best described as palm barrens or pine flatwoods. Inland, at higher elevations generally above 500 m. (1,650 ft.), lay the *sabanas de las mesas*, or "savannas of the plateaus," often underlain by a hardpan of iron-impregnated sand that caused waterlogging during the rainy season and interfered with tree-root growth. The Vega Real in interior Española offered such savannas, covered by a hip-high growth of succulent thatch grass and wiregrass (figure 17). In the wet season, these interior grasslands took on more of an Andalusian than mesetan character, and soon they, too, became cattle-ranching centers.

In addition the Indies possessed dry grasslands, not properly classified as savannas. These lay in rain-shadow districts on the leeward side of mountain ridges, sheltered from the moisture-bearing northeast trade winds, and included both dry upland basins and sheltered low valleys such as the elongated Cul de Sac, leading eastward from Port-au-Prince in Haiti. Xerophytic, or drought-tolerant, short grasses, brush, and even cacti grow in such places, but grama grasses offered suitable forage for cattle.

Collectively the different types of grassland occupied a substantial part of the Greater Antilles in 1500. On Cuba, for example, open tropical grasslands covered over one-tenth of the land area, and wooded savanna accounted for an additional 15 percent. Iberian ranchers early encouraged survival of the savannas by perpetuating the Indian practice of firing the grasses in April, at the end of the low-sun drought period. Perhaps, too, these fires represented a continuation of the lowland Andalusian practice of burning wooded montes pastures on the perimeter of Las Marismas and other coastal marshes. The fires suppressed woody sprouts and seedlings, while aiding grasses.[11]

Spanish livestock of all kinds accompanied the initial colonization of the four islands, but with very different results. Sheep and goats did not thrive nearly so well in the new environment, perhaps partly for climatic reasons, but horses prospered and both cattle and swine attained an almost outrageous success. Pigs took to the abundant forests of the Antilles as if to a native habitat, easily crowding out a variety of small indigenous mammals to occupy a largely uncontested ecological niche. Oak-rich Extremadura itself never witnessed such a proliferation of swine. Criollo cattle achieved a parallel preadapted success on the Antillean savannas, attaining almost at once a prodigious rate of reproduction at nearly their biological potential. Longhorn herds reputedly increased tenfold in less than half a decade, quite possible assuming that heifers formed most of the imported stock. In the short run, they encountered no competing large herbivores, suffered no diseases or parasites, and faced no predators in these bovine Elysian Fields. The island Indians had made little use of the savannas, leaving them open to the invading cattle, and the incredible mortality rate among the Antillean tribes quickly created grassy old fields, extending the cattle range. In fact, the depredations of Iberian cattle and hogs in the unfenced fields and gardens contributed to Indian mortality. Eventually European dogs gone feral filled the vacant predator niche in the islands, especially Española, controlling the herds and droves, but the initial proliferation of cattle and swine can only be described as astounding. As a result the Antillean Spaniards became beef eaters to an extent unknown in Spain outside of Sevilla.[12]

So prolific were the cattle, and so intensive the Spanish pastoral occupance, that damage to the West Indian savannas soon occurred. Españolan thatch grass became greatly diminished and some said even extinct as early as 1519. Certain Old World grasses and herbs entered the Antilles through accidental diffusion. Severe ecological disturbance of the islands' pastures had certainly occurred by the 1570s. Overgrazing led to brush invasion in the upland savannas, caused both by sod deterioration, which allowed the seeds of woody plants places to germinate, and by fire retardation, since less fuel remained available at the annual burning season. Españolan cattle spreading guava seeds created thickets. In the long run, such floral alteration opened the way for a botanical Africanization in the islands, when Guinea grass was introduced in the 1700s, to be followed by other African species.[13]

The preadaptive success of cattle and swine in the Antilles was not shared by the typical array of Iberian orchard, vineyard, and field crops. In effect the traditional diversified Mediterranean agricultural trinity of bread grains (wheat and barley), tree and vine crops (olives, grapes, citrus), and herding of sheep and goats collapsed as an adaptive strategy in the West Indies. Climatic factors played a role in this demise, contributing to the sterility or poor performance of wheat and olives, but more importantly the Spaniards failed to transfer to the New World the peasant society necessary to perpetuate the Mediterranean agrarian system. In addition, large-scale olive growers in Spain sought royal restrictions against New World cultivation in order to avoid competition. Even sugarcane, an Andalusian crop preadapted to the humid tropical climate and introduced very early into the Indies, achieved only a very modest success in the 1500s, perhaps because of its image as a Moorish crop. A drastic simplification occurred, and only bits and pieces of the Iberian rural way of life passed through the filter of the new environment and greatly altered socioeconomic conditions in the Antilles. With the demise of the diversified Mediterranean agricultural strategy, Andalusian salt-marsh cattle raising realized an added preadaptive advantage, for it already represented, prior to transferral, a simplified, specialized, commercialized venture operating outside the context of peasant farming, a labor-extensive venture controlled by a town-based elite. The marsh system, not dependent upon a diversified base or numerous peasantry, almost at once acquired a viable American foothold when its two essential types of livestock—cattle and horses—successfully entered the Indies.[14]

Also contributing to the rise and durability of livestock ranching in the Greater Antilles was the small size of the population. The majority of Spaniards, seeking gold, went to the American mainland after 1519, leaving the Indies largely depopulated. Española, where some 9,000 Spaniards resided in 1509, retained only about 700 five years later, after Cuba, Jamaica, and Puerto Rico were colonized. In the 1560s Cuba had only 240 Spaniards in its population, Puerto Rico 200, and Española 1,000. Jamaica housed a mere 120 Spaniards in 1596, all reportedly descended from only three immigrant parentages. The total Jamaican population in 1611 stood at only 1,510, including slaves, free blacks, Indians, and "foreigners." By the 1650s a relative handful of *hidalgos* still owned all of the ranches on Jamaica.[15] In such demographic desolation, a system of extensive land use with minimal labor requirements, such as livestock ranching, could achieve and retain dominance.[16]

Spanish Antillean Cattle Ranching

In the colonial Indies, then, cattle ranches became the most common type of Spanish rural enterprise, a condition that persisted for centuries. Ranching was established very early in certain savanna-rich plains in the islands (figure 17). The Vega Real in Española early developed as the major ranching focus on that island, while on Puerto Rico the largest herds accumulated in the northeast, and in Jamaica the plains on the southern side of the island provided the main herding region. One Jamaican ranch had 1,650 cattle and 60 brood mares as early as the 1520s, while on Puerto Rico a certain grazier was said to own a herd of 12,000 longhorns in the 1590s.

The most immediate model for this New World herding industry can be found in the royal breeder farms of Española, created on choice pastures to provide cattle and horses for the new colony. These enterprises, in turn, may have owed much to the commercial cattle-herding system of the Spanish coastal marshes. In many important respects, cattle raising in the Indies closely resembled the Andalusian lowland type. Large herds of longhorns roamed a seasonally wet, grass-covered, open range under minimal supervision. By contrast the Spaniards took care to pen and feed their swine each day, in a much more labor-intensive system.[17]

The use of brands and earmarks became established practice very early in the Indies, and the semiferal cattle were collected once or more each year at pens during the low-sun dry season for marking, but calf castration, as in Andalucía, was neglected, and only a small number of mature bulls ever felt the cut of the emasculating knife. Vaqueros, mounted like their counterparts in Las Marismas, managed the cattle with Andalusian lances, or *garrochas* (figure 6). While these herders presumably possessed the southern European lasso, they had not yet developed the technique of casting it from horseback, and leather saddles seen even today in ranching areas of the Dominican Republic often lack horns. Many of the Antillean cowboys were African slaves or free blacks, as suggested earlier, but young, unmarried Spaniards also performed such work. Some Indians who survived their demographic holocaust entered service as cowboys, initially through the *encomienda* system, by which Spaniards early obtained legal rights to Indian labor. The native workers became increasingly available to the herding industry after 1516, when laborers were reassigned from the mines of Española. Ultimately, as encomienda rights were extinguished, the Indian cowboys became hired hands. While few Indians remained by 1600 in Jamaica, most of these survivors worked as freemen for small wages "on the ranches and at the hunts," where their services were deemed so essential that a suggestion for removing them to a separate ethnic village led to the complaint that such an action would destroy the island's herding economy. In all probability, many of the Antillean cowboys by that time were triracial mixed-bloods, prototypes of the typical Latin American vaqueros of later centuries. They clearly formed a *casta*, one of the lower socioeconomic classes in the Antilles, although they fared better than plantation workers. Whether Spaniard, African, Indian, or of mixed blood, the cowboys worked under the supervision of a ranch boss, or *mayoral*, who answered to an absentee owner, residing in town. On smaller ranches, apparently always abundant in the Spanish colonies, a resident owner served instead as boss and his sons as cowboys, often assisted by a few blacks or Castilian indentured servants.[18]

Herding in the Spanish colonial West Indies also shared with the Andalusian marsh system a commercial focus. Hides, tanned using the bark of the local mangrove tree, provided much of the economic base of the Greater Antilles in the 1500s, equalling or exceeding in value sugar and other plantation crops. The hides found a market at Sevilla, the same destination as those produced in las Marismas, supplying the venerable Andalusian leather industry. Española alone shipped up to two hundred thousand hides annually in its ranching heyday, between 1530 and 1540, and as late as the 1580s over thirty-five thousand hides were sent each year from the island. The emphasis upon hide production favored cattle among the Iberian livestock in the Indies and perhaps helps explain the lesser role of goats and sheep.[19] Tallow, lard, and to a lesser extent, dried meat provided additional market commodities for the colonial Spanish Indies. These Antillean products, and live cattle as well, most often found their market in American mainland colonies, especially Panama, Colombia, and early Mexico, rather than in Spain. Jamaica alone sent as many as a thousand cows and yearlings to Panama in 1521. Cattle ranches in some areas, as around Habana, had a legal obligation to supply meat for the island towns.[20]

Cattle herds in Antillean colonial cattle ranching remained essentially stationary, performing at most local seasonal shifts to avoid inundations, as in the Old World salt marshes. This model of lowland herd stability would later influence much of Latin America. Mesetan transhumance played no role in the island system, and the seasonal drifting of herds normally occurred within the bounds of individual ranches. Mobility did not extend beyond local pasture shifts and overland drives of livestock to market, a practice rendered less important by the practice of killing cattle in the pastures solely for the production of hides.[21]

The Antillean ranching system also exhibited some fundamental contrasts to lowland Andalusian herding. Not the least of these was its cattle/hog focus. Some swine had foraged in the woodlands bordering las Marismas, but they formed a relatively unimportant aspect of the local livestock economy. In the Spanish Antilles, by contrast, hog ranches, which were called *corrales,* dotted the forested regions, and pork played a significant role in the diet, especially for slaves. Pork fat or even beef lard replaced olive oil in cooking. Perhaps the greatly increased role of pigs in Antillean stock raising reflected Extremaduran influence, but more likely the rise of swine in the West Indies rested mainly upon their preadaptation to the rich and abundant forests of the islands.

The land-tenure system both resembled and differed from Iberian prototypes. Initially cattle herders used crown lands for pasture without permission, a practice that never entirely ceased during the Spanish colonial era in Latin America, and which had Castillan roots. Beginning in 1507, the monarchy allowed all graziers in the Antilles free usage of the royal domain, so that ownership of herds brought the right to pasturage, even though no private land titles had been awarded. The 1507 decree merely legitimized the status quo, for individual herd owners had years earlier begun preempting certain ranges, seizing de facto squatter control over sometimes sizable acreages. While this departed from the ancient Old World custom of common range, it was consistent with the Castillan Reconquest institution of squatter's rights and duplicated the encroachment by private estates upon the pastures of Las Marismas underway in Spain by 1500. Accordingly the local Antillean governors and municipal councils typically approved such pasture appropriations, if only in usufruct and without royal permission. Soon thereafter, probably before 1520, graziers began petitioning the crown for formal usufructuary ownership of these preempted pastures. For example in 1527 a certain Jamaican, Francisco García Bermejo, asked the king for "a grant of the valley of Goayrabo for a pasture ground for his cattle," probably an area García already utilized for his herds. On Puerto Rico, private pastoral holdings obtained royal recognition only in 1545, following local resistance to an earlier decree setting up common pastures, or *hatos públicos*. The evolution of privately owned ranches, then, apparently progressed rapidly from unauthorized herding of privately owned livestock on preempted royal domain, to approved use of the domain, to grants approved by local officials, and finally to formal usufructuary land grants from the monarch. In spite of the development of landed pastoral estates, much common pasture survived, particularly in hilly and mountainous regions. Even some of the choicest grazing grounds in Española's Vega Real long remained common land, open to all. Small-scale, landless operators survived because of access to these pastures, even though the herds of estate owners also drifted onto the commons. The development of grazing estates, then, did not bring an end to public domain in the Spanish Indies, nor did such estates wholly lack Spanish antecedents.[22]

The royal pastoral grants often had a circular shape, based on a radius of two leagues, yielding a huge property in excess of 22,500 hectares (55,700 acres), although most remained vaguely bounded and some were never surveyed. The circular shape probably derived from common pastures of that configuration in sixteenth-century Sevilla province, reflecting still more Andalusian lowland influence. These herding grants also bore typically Andalusian generic names, such as hato and estancia. Hato, which in Andalucía and the early Antilles meant merely a herd or, more commonly, the place chosen by herders to camp while they tended open-range livestock, soon came to designate a privately held property in the Indies devoted to cattle raising, or the areal measure of such a property. Estancia, nothing more than a "place for sheltering livestock" in the dialect of Cádiz and Sevilla provinces, eventually meant a private outlying estate in the Indies. Though hatos and estancias often appeared under separate listings in early Antillean registers, they became virtually synonymous, particularly when bearing the specific designation *de vacas*. Hato eventually was the preferred West Indian term to mean "cattle ranch," attaining a permanent toponymic presence. Similarly *estanciero*, one of the earliest words used in the islands to describe cowboys, eventually lost out to sabanero and the more venerable vaquero. By contrast the word *corral* in the Spanish Indies meant a herding establishment raising pigs.[23] This very process of groping awkwardly for new, altered meanings of Andalusian words suggests that the Antillean system of privately owned herding properties differed to some degree from Iberian pastoral land tenure.

Another contrast between Antillean and Andalusian marsh-herding systems involved degree of specialization. While the entrepreneurial stock raisers of Las Marismas dealt almost exclusively in cattle and often owned no other agricultural enterprises, Antillean graziers had to operate largely self-sufficient estates, not relying upon food imports from the mother country. For this reason many hatos de vacas on the West Indian savannas formed only one part of diversified, spatially fragmented estates that also included hog corrals in the forests, sugar or ginger plantations near the coast, and because of the Antillean failure of Iberian crops, fields devoted to the native American food staples cassava, sweet potatoes, and to a lesser extent, maize, as well as gardens of certain "Castilian vegetables." Cattle raising, then, formed only one element in a diversified and integrated plantation system. Perhaps the crop-oriented haciendas of the Andalusian marsh perimeter provided a viable prototype for this system. As sugar and other plantation cash crops grew in importance, the stock-raising activities gradually became subsidiary and increasingly engaged in providing meat for the field laborers. By the early 1700s, if not earlier, the West Indian plantation demand for meat could no longer be met by island graziers, and imports began.[24]

An even more fundamental contrast between Iberian and Antillean herding systems developed because of the chronic labor shortage in the American island colonies and the remarkable fecundity of the longhorn cattle. Within a very short time after initial settlement, herds simply grew too large to manage, with the result that many cattle, horses, and pigs went completely feral. Sixteenth-century observers often distinguished between tame and wild cattle. On Española several categories were recognized. The tamest cattle, called *corraleros* and forming the least numerous group, never left the vicinity of the ranchstead and could easily be penned, while *mansos,* or "tame ones," drifted somewhat farther afield on the savannas and required considerable effort by vaqueros to round up and drive to the enclosures in a *rodeo de ganado*. The remaining cattle, known as *extravagantes, alzados, orejanos,* and *bravos*, exhibited varying degrees of wildness and did not properly constitute part of the ranch herd. The feral cattle lived in remote areas, usually forested hills and mountains, venturing onto adjacent savannas only nocturnally. Jamaica revealed a similar pattern. Tame herds grazed on the flatter southern side of the island, while feral longhorns inhabited the mountainous north.[25]

The Antillean feral cattle became the focus of a hunting industry, divorced from any pretense of herding. Mounted on specially trained horses, hide hunters belonging to the same underclass as the cowboys employed hunting dogs and a cruelly modified garrocha, possibly derived from the Andalusian marshes, to pursue and kill the bovine game. The main weapon, a *desjarretadera*, or hocksing pole, about 3.5 m. (15 ft.) long, had a crescent-shaped blade, or *luna*, 15 to 16 cm. (6–7 in.) wide affixed to the end. The pole, resting on the horse's head behind the right ear, was thrust forward to sever the hamstring of the pursued animal. The cow hunter, revealingly called a *matador* ("killer") or *lancero* rather than a vaquero, then dismounted and dispatched the prey with a knife attached to a shorter pole. Taking only the hides and tallow, the matadores left the carcasses to the accompanying dogs, probably descendants of the vicious attack canines used in the Conquest. Many, perhaps even most, of the cowhides exported from the Spanish Antilles derived from such hunting rather than from ranching. Dog-assisted hunts for feral swine also occurred. The hunting grounds for wild cattle and pigs, in common with pastures, belonged at least informally to individual Spaniards as private property, although much poaching occurred.[26]

The abundance of feral livestock eventually attracted even foreign poachers, especially to Española in the 1600s. These so-called *buccaneers*, mainly of French Huguenot and British origin, sought both meat provisions to support their piratical activities on the Spanish Main and contraband hides for export to northwestern Europe. Efforts by the Spaniards in the seventeenth century to exterminate the wild herds and to seize the pirate bases so as to rid themselves of the buccaneers failed, and the northern intruders became firmly established in the Indies. They also preyed upon tame herds on the hatos. The buccaneers played an important, if intermediary, role in the development of the North American cattle frontier. Among their more direct gifts, *barbecue*, the cooking of entire cow or pig carcasses elevated over a slow fire, is derived I feel from the French *barbe et queue*, meaning the whole animal, "beard and tail." Their greater contribution, the seizure of Jamaica in 1655, placed that island with its cattle-herding tradition under British rule, allowing it to become a place of experimentation and synthesis, where Iberian, African, and British cattle-raising traditions could meet and blend.[27]

British Jamaica

The Britons inherited a pastoral system in decay on Jamaica, one largely degenerated into hunting of feral stock. However, a small population of Iberian ranchers and racially mixed herdsmen continued to live on the island after the English takeover, permitting a diffusion of their cattle-raising techniques to the newcomers. One town-dwelling Spanish rancher preserved her pastures and herds by establishing a new residence out on the hato. In other cases the new rulers simply seized Spanish ranches and continued extensive open-range livestock raising.[28]

A persistent Ibero-Antillean herding influence in British Jamaica is revealed in various ways, perhaps most convincingly in vocabulary. Abundant Antillean Spanish loanwords entered Jamaican English, including "savanna," "crawl" (from *corral*, hog farm), "jerk" (from *charqui*, to preserve meat by sun-drying), "palisadoe" (picket), "stancha" (farm or small plantation, from *estancia*), "palunk" (from *palenque*, a stockade or small farm), and "paloma." A map drawn about 1690 revealed such Jamaican toponyms as "Spanish Craul" and "Burnt Savana." Some of these borrowings perhaps came indirectly from the buccaneers. Along with "barbecue," they represent the beginnings of the mixed English-Romance vocabulary that would later characterize the cattle-ranching districts of the western United States.[29]

In British Jamaica the Iberian longhorn remained the dominant cattle breed, and additional criollos were imported by the English from Cuba. The Spanish Antillean cattle/hog focus persisted, though the British West Indian settlers had already become heavily reliant upon pork in the diet during their earlier colonization of Barbados and other islands of the Lesser Antilles. As in Spanish times, herders continued to fire the savanna pastures at the end of the low-sun dry season, but perhaps that practice owed as much to British highland precedent as to Iberian Antillean range-management techniques. Confronted with the task of herding unruly longhorns on the open savannas, cowboys in British Jamaica began riding Spanish ponies, although they learned few Andalusian equestrian skills in the process. The horse served them mainly as a transportation device in herding, as well as providing protection from attacks by the longhorns. Apparently the garrocha did not become a British West Indian herding tool, but when animals were "taken with Cords," the southern European lasso may have played a role. More important was the traditional British bullwhip, which could be wielded from horseback.[30]

Large-scale stock raising made the transition to English rule, and some of the new ranches had "immense droves of horned cattle," as many as two or three thousand head on the bigger establishments. One wealthy widow, probably of Spanish origin, reputedly owned forty thousand cattle on the Jamaican savannas by about the year 1700. Smaller-scale operators also engaged in the cattle business, especially before the turn of the century. As in the Spanish system, the ranches often formed one part of self-sufficient, diversified, areally fragmented plantation enterprises, in which cattle raising was only one of several major activities. Hides remained the principal export derived from cattle herding. Jamaica shipped almost forty thousand hides in the nine years between 1671 and 1679, and the total rose to fifty-six thousand for the period 1680–84. British markets, of course, replaced the Andalusian leather towns. Beef, manure, and oxen from the ranches supplied the plantations of British Jamaica, and some barreled beef may have been exported in the late 1600s to Barbados and the other plantation islands of the Lesser Antilles, where no large-scale cattle raising existed.[31]

The traditional hunting of fully feral cattle and swine also persisted, at least briefly, in British Jamaica, among Hispanic guerrillas, renegade blacks, and buccaneers alike. Some hunted on foot, others mounted, using dogs, hocksing poles, and firearms in a mixture of Spanish and English methods. This activity became so widespread after 1655 that the herds of wild longhorns soon dwindled, leaving relatively few by the end of the century.[32]

Cattle raising in English Jamaica did not, however, represent merely a borrowing and resuscitation of Spanish ways. British herding influences entered the island in company with a substantial immigration of English-speaking settlers, including small farmers fleeing estate expansion in Barbados, soldiers, and colonists from the British Isles. Some of these new Jamaicans surely possessed previous cattle-herding experience. Scots and Irish became numerous in the island's population, and at least one influential planter came from pasture-rich Somerset in England.[33]

Some imported British practices, such as range burning or the branding of livestock, served largely to reinforce existing Spanish techniques, testifying to commonalities among the Atlantic-fringe cattle folk, but certain other traits of the new system, such as the use of whips and the preservation of salted beef in barrels, represented changes. Perhaps the most obvious indicator of a distinctively British influence is etymological. The ancient English word *pen*, or less commonly, *cowpen*, replaced *hato* and *estancia* to designate a "cattle ranch." A "pen" included not only the central gathering corrals, but also dwelling, outbuildings, garden for provisions, and outlying stock pens "sprinkled about the pastures." Why English "pen" should have come to mean cattle ranch while the corrupted Romance loanword "crawl" survived to mean a pig farm, with associated sties, gardens, and huts, remains unclear. The Jamaican word *pen* apparently came most immediately from Barbados, where at least as early as 1650 "oxen and kine" were kept in "pends," but the term acquired the altered meaning of "ranch" or "ranchstead" in Jamaica, where the British settlers developed their first trans-Atlantic, large-scale cattle-raising enterprises. Barbados and the other Lesser Antilles provided essentially no prototype for the British Jamaican herding system. Additional etymological evidence of British influence is revealed in the use of standard English *pond* to mean an artificial watering place for cattle, a word used in preference to Spanish *estanque*, as well as *gully*. Both "pond" and "gully" also appear as toponyms in Barbados. The herders in English Jamaica also seem to have been called *cattlemen*, reminiscent of the British "cowman," or else *pen-keepers*. Perhaps most significant change introduced by the Britons was their attention to calf castration, producing herds of steers that were far easier to control.[34]

The British settlers dramatically revived Jamaican cattle ranching. Pens grew rapidly in numbers after about 1665, and twenty years later the island had seventy-three, a total still on the increase, although a map dating from about 1690 showed only forty-seven cattle pens and hog crawls (figure 18). By 1768 the number of pens stood at about five hundred. The population of tame cattle in Jamaica reportedly rose from only sixty to six thousand in the six years preceding 1670, as the settlers domesticated large herds of feral and semiferal longhorns. Additional tame Iberian Antillean cattle were introduced

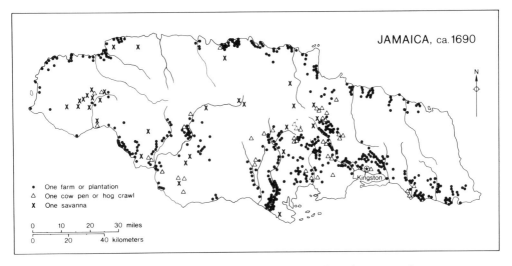

Figure 18. Ranches and plantations in British colonial Jamaica, about 1690. Note that cattle "pens" and hog "crawls" were already being displaced by plantations at this early date. Source: Sloane, A Voyage to the Islands Madera, Barbados, Nieves, S. Christophers and Jamaica (London, UK, 1707), vol. I, following p. cliv.

into Jamaica in the early 1670s. As the numbers of pens and tame cattle increased, ranching spread into the hills and mountains of the interior, including rough karstic areas, former refuges of the dwindling feral herds. Penetration of the areas of rugged terrain by new cowpens also occurred partly as a result of plantation expansion in the deltas, lower river basins, and humid interior valleys of southern Jamaica, where indigo, cotton, cacao, and, after 1670, sugarcane cultivation displaced some herding in the lowland savannas. The British colonists devoted far more attention to agriculture than had the Spaniards. The unequal contest for land between rancher and tiller that would later characterize large parts of Anglo-America, a competition as old as Cain and Abel, finds a prototype in British Jamaica.[35]

Anglo-Celtic Jamaicans developed a more labor-intensive cattle-ranching system than the one practiced by their Hispanic predecessors. Nightly penning of cattle in palisaded enclosures became the rule. In so doing the Britons may simply have been applying the Antillean Spanish methods used in pig raising to cattle husbandry. As during Spanish times, swine in British Jamaica returned to sties at the crawls each evening, responding to the sound of a conch-shell horn, to be rewarded with corn feed. As a result livestock remained tame and did not stray. Jamaican law required at least one white cowboy for each hundred cattle, and they received assistance from servants of other races. Every plantation needed "a good lusty fellow that hath some skill in horses and Cattell." These Jamaican pen keepers, like their successors in the American West, ranked among "the most independent" of rural folk. Nightly confinement of the herds facilitated yet another intensification of land use. The cattlemen periodically relocated the palisaded pens and planted gardens for provisions and feed in the abandoned, manure-rich enclosures, a practice called "penning over" or "cowpenning." This custom almost certainly derived from the highland British practice of penning cattle on the infield during the winter. Because the cattle remained docile and no annual roundups were necessary, the menial jobs of castration, culling, and the application of typically British brands and earmarks could be accomplished with a minimum of difficulty. The elaborate horseback skills of the Andalusians were not needed.[36]

The role of herder dogs in the British Jamaican ranching system remains unclear, but in all probability canines assisted the daily back-and-forth shifting of herds. Buccaneers in Española employed fleet, slim cur dogs (figure 19), likely of Spanish origin, to hunt down wild pigs and cattle, a practice they transferred to Jamaica in 1655. These intelligent, nimble hunting curs probably became herding dogs under British highland influence on Jamaica, when the hunting of feral stock gave way to the tending of tame droves and herds during the 1660s. Indeed the Spaniards themselves apparently had earlier converted continuously domesticated descendants of their attack dogs of the Conquest period into guardians of their livestock herds, and the Britons probably completed this transition by making Antillean guard dogs into true herder dogs.[37]

British Jamaican cattle raisers also instituted a fundamental change in the land-tenure system. Many open-range savannas reverted to the status of commons, rather than remaining private property, as under Spanish rule. Ranchsteads—the pens and crawls—belonged to planters and other individuals, but the adjacent pastures initially formed part of the public domain. Before long, however, estate owners tightened their grip on the pastures and sought to deny smallholders access to them.[38] The pairing of plantations and livestock ranches survived from Spanish times.

African influence in the herding system of English Jamaica remains more problematical. A minority of the black slaves introduced

FERAL DOG OF ST DOMINGO.

Figure 19. "Feral dog of St. Domingo," a cur likely descended from Spanish attack dogs and used later by West Indian buccaneers to hunt wild cattle and pigs. In British Jamaica such curs possibly served as herder dogs for tame livestock after 1660, perpetuating a British highland custom, and the French in Haiti possibly also used them for that purpose. They may be kin to the "Catahoula" hound, the herding canine of the Gulf Coast American south. Source: Charles H. Smith, Mammalia: Dogs *(Edinburgh and London, UK, 1854–56), vol. II, following p. 120.* I thank the geographer Georges Lutz of Strasbourg, France, for bringing this drawing to my attention.

into the island by the Britishers came from Senegambia and the lands of the upper Niger, and presumably possessed cattle-herding skills. African slaves worked alongside white servants on Jamaican pens, and among the holdings of one eighteenth-century sugar plantation were twelve slaves described as "cattlemen." Such herders could have introduced African techniques, in particular nightly penning of the cattle, a custom more closely resembling the ways of the Fulani than the Old World practices of Iberians or Britishers. Etymological suggestions of possible African influence include the Jamaican English *dogi,* reputedly a Bambara word meaning "small" and apparently the antecedent of western American *dogie,* "a motherless calf." The accidental introduction of African guinea grass in 1744 also profoundly affected the Jamaican cattle industry.³⁹

French Saint-Domingue

While English buccaneers facilitated the transfer of Jamaica to British rule, their French counterparts, who formed the majority of the pirate population, played an instrumental role in wresting the western third of Española away from the Spaniards. From a base on the small, barren island of Tortuga, situated off the northern coast of Española, the French pirates progressed from poaching Spanish cattle and plundering ships to colonizing the Españolan coast. In the 1660s Gallic settlements began appearing on the northern and western shores of the island, especially around Cap-Haitien, the first French capital of the colony of Saint-Domingue, later to become Haiti. Very few Spaniards remained by then in the west of the island, due to attacks by the buccaneers, leaving the way open for the new colonists. Spain formally acknowledged French rule of western Española in 1697.⁴⁰

In the process a Franco-American cattle-ranching enterprise came into being. The buccaneers, from about 1610 onwards, had already become familiar with Iberian cattle and swine raising on the island through poaching, and enough of a residual population remained to assist a transfer of herding techniques to the French. Rather than drawing upon the cattle-raising traditions of France itself, as represented by the hill-herding systems of Auvergne and Celtic Brittany and the Mediterranean salt-marsh cattle complex of the Camargue in the Rhône delta, the French settlers of Saint-Domingue simply annexed the Ibero-Antillean ranching system. Hatos, corrales, and sabanas became *hattes, corrails,* and *savanes* without any substantial change in meaning. Under French rule, open-range livestock raising was soon displaced to the hinterland in less desirable, hillier regions by the advent of large sugar plantations, as in Jamaica. By the late 1700s, for example, coastal Gonaives Parish retained only one cattle hatte, while interior Mirebalais Parish, northeast of Port-au-Prince, still had over eleven thousand tame cattle on ninety-two hattes. Some seven thousand cattle and fifteen hundred horses were at pasture then in one portion of the Cul-de-Sac valley.⁴¹

French ethnic contributions to the cattle-raising system in Saint-Domingue remained subtle at best. They may have contributed to its further Africanization through their introduction of more slaves of Sahelo-Sudanic origin, including Fulani, in the 1700s. The French settlers of Saint-Domingue also may have converted the Spanish hunting cur of the buccaneers into a true herding dog, paralleling the British achievement in Jamaica.[42] Indeed, French Saint-Domingue bore more than a passing resemblance to English Jamaica. In both the Spanish Antillean herding system passed, with greater or lesser modifications, to new colonists of different ethnic background. Both, in turn, would quickly transplant island ranching to the North American mainland, though with varying degrees of success.

What, in the final analysis, entered the Indies from the Old World Atlantic fringe? From early-sixteenth-century Las Marismas in Andalucía came preadapted, commercialized, specialized, large-scale, urban-based entrepreneurial ranching of semiferal longhorn cattle on the open range; the rowdy, caste-bound, mounted herdsman; a preference for beef in commoners' fare; the uncast lasso; the concept of a circular pastoral grant; an Andalusian vocabulary; sun-drying of meat; the unimportance of milking; squatters' rights; and the concept of private estate building. From Extremadura, perhaps, came the importance of pig raising. Britain contributed the use of herder dogs and bullwhips, regular milking, the penning of stock on future cropland, and the dominance of public-domain grazing. West Africa sent laborers who, at the very least, introduced some words into the ranching vocabulary and encouraged careful tending of livestock, including nightly penning. Among these diverse gifts, Andalucía's were unquestionably the greatest.

Above all the Greater Antilles played a crucial role in the transfer of Atlantic-fringe cattle herding from the Old World to the Americas and its multiethnic evolution. The islands witnessed the mixing of traditions and the modifications necessary to produce ranching systems suited to the new setting. The role of the Antilles in adaptation and subsequent diffusion to the mainland was essential to the development of the North American cattle frontier, and those scholars who would transport Old World herding directly from Europe or Africa to the American mainland overlook this essential testing ground, diffusionary stepping stone, and arena of ethnic exchange.[43]

FOUR

FROM THE INDIES
TO THE MAINLAND

VARIOUS "ISLAND CATTLEMEN," SPANIARDS AND NON-Spaniards alike, transferred Antillean herding systems onto the North American continent, creating the more immediate hearths of the cattle-ranching frontier of the United States (figure 17). The process started in early New Spain, where cattle introduced from the Antilles allowed the beginnings to be "essentially pastoral."[1]

Gulf Lowlands of Mexico

In Mexico the Spaniards found, truly, a "New Spain," featuring an elevated central plateau laced with mountain ranges that reminded them of the Iberian Meseta and coastal plains in places reminiscent of Andalucía. Given this similarity, the emergence of a lowland/highland, or littoral/interior herding dichotomy in Mexico similar to that in Spain was not surprising (figure 20). The Gulf Coast region of Mexico, largely in the state of Veracruz, early became an area of estate-based cattle ranching. Rather than being merely a passing frontier livelihood there, ranching evolved into a durable regional institution. By 1620, a century after the Conquest, the majority of cattle in Mexico ranged in the Gulf plains, especially in districts closest to the coast, but also at places inland, as around Ciudad Valles and Tamazunchale, at the foot of the mountains. The coastal ranching system, though much modernized through capital intensification, remains vigorous to the present day. Mounted vaqueros still work on the coastal plain, and cattle raising retains high prestige.[2] In the very hearth of cattle ranching on the American mainland, then, we must be warned that Frederick Jackson Turner's midwestern-based model of ranching as merely a frontier occupance stage is not universally or even generally applicable.

Gulf Coast Mexico consists of a variety of terrain types, from table-flat, flood-prone lowlands along the immediate coast and in-

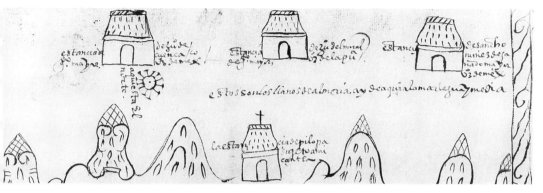

Figure 20. One of the oldest depictions of a cattle ranch in Mexico, this 1579 map, including the plains of Almería, near present Nautla in Gulf Coast Mexico, shows two estancias de ganado mayor at a distance of 1.5 leagues from Misantla. Each ranch is represented symbolically by a thatched Huastecan hut, which may be an accurate depiction of early ranchsteads. Source: Diego Pérez de Orteaga, Gaspar Delgado, and Diego López Bocanegra, "Descripción de Mizantla (o Mazantla) hecha por el corregidor Diego Pérez de Orteaga, 1.° de Octubre 1579," manuscript Relacion Geográfica consisting of 3 leaves and 1 map, in the Benson Latin American Collection, University of Texas at Austin, item no. JGI xxiv–13; see also Howard F. Cline, "The Relaciones Geográficas of the Spanish Indies, 1577–1586," HAHR, 44 (1964), pp. 341–74.

cluding the great Pánuco delta, to a sloping, irregular plain rising gradually but significantly to meet the mountains in the west and cut by stream chasms, or *barrancas*. A low range of volcanic hills inland from the city of Veracruz partitions the interior plains, and an elongated valley north of Ciudad Valles similarly nestles behind an outlying low ridge of the great eastern mountain chain, with the result that Gulf Coast Mexico has a compartmentalized appearance.

Climate and vegetation reflect both the terrain pattern and latitudinal differences. Most of lowland Veracruz receives in excess of 1,500 mm. (60 in.) of rain annually, but precipitation decreases both northward, where the Pánuco delta gets only about 1,000 mm. (40 in.) and, in places, to the interior, where the rainshadow effect of hills and ridges produces moisture-deficient districts such as the Sotavento, sheltered from the rain-bearing northeasterly trade winds. As in the Antilles, precipitation falls mainly in the high-sun, or summer, season, between late May and the middle of October. In the flat coastal floodplains, the Spaniards encountered a tropical rainforest and extensive marshes. Perhaps the first floral complex used for grazing was that of the coastal lagoons and sand dunes, a treeless, halophytic tussock-grass association that survives tidal inundations during hurricanes. Cattle ranching thrives still today in this coastal setting, but even better suited for large-scale herding was the expansive Pánuco delta near Tampico. A seasonally flooded marsh of tall grasses and cane surrounded by a wreath of woodlands including oaks, the delta bears a startling resemblance to Las Marismas of the Guadalquivir, which it equalled in size. Additional smaller marshes, or wet savannas, appear both near the coast and inland, in other alluvial floodplain portions of the lowland. Elsewhere, as in the Antilles, repeated clearing and burning by Indian farmers had created tallgrass wet savannas in the midst of the *tierra caliente* rainforest, as well as extensive stretches of parkland or wooded savanna. These became more expansive to the north, with diminishing rainfall. Much of the elevated, sloping plains above the flood-prone lowlands was covered with a tropical deciduous forest that gave way to stretches of parkland and, especially above about 250 m. (800 ft.) elevation, to extensive dry savannas, particularly in areas of clay soils, where grass grew in great abundance. The rain-shadow districts of the inner coastal plain offered xerophytic scrub and shortgrass vegetation, with scattered thorn thickets. Altogether the tropical Gulf plain of Mexico offered an extension of the insular environments to which Antillean cattle ranching had adapted, as well as abundant reminders of the Andalusian salt marshes.[3]

Also in common with the West Indies, the large Indian population of Gulf lowland Mexico perished in a profound demographic collapse after the Spaniards arrived. In the Pánuco region alone, a native Huastecan population of perhaps one million in 1520 dwindled to about twelve thousand by 1532 as a result of disease, warfare, starvation, and slave export. Cattle ranching is not compatible with dense settlement, and the Indian die-off allowed virtually the entire tropical coastal plain to lie open for large-scale herding within no more than a half-century of the Conquest.[4]

Cattle, introduced from Cuba and Española at both the Pánuco delta and the port of Veracruz in the 1520s, achieved the same impressive multiplication witnessed earlier in the Antilles. From at least three coastal nuclei, each still identifiable in local Spanish dialects, ranching spread through much of the lowland as the livestock multiplied (figure 17).[5] Antillean cattle arrived first at the old site of Veracruz, perhaps as early as 1522, and a concentration of cattle ranches soon developed in the plains hinterland of the port, especially to the south.⁶ A second early focus developed in the depopulated lowland plains of Almería, bordering the Gulf at the narrows of the coastal plain north of Veracruz, in the vicinity of modern Nautla. Tributary to the inland town of Misantla, the Almería plains, occupied by ranchers beginning in the 1550s, housed some thirty Spanish families living on cattle estates a half-century later. Their herds grazed the abandoned fields and pyramid-dotted plains, oblivious to the vanished high Indian civilization of the area. A 1579 Spanish map of the Misantla-Almería area contains perhaps the earliest depiction of cattle ranches in the Mexican coastal plain (figure 20).⁷

The third colonial herding implantment occurred adjacent to the Pánuco delta in the Huasteca, especially south and southeast of the great marshes. An impoverished area in the first years, before cattle were introduced, the delta perimeter soon developed rapidly into a true and thriving ranching frontier. Antillean cattle, acquired in exchange for Huastecan slaves, reputedly doubled in fifteen months on the native pastures. Spaniards attempted little crop farming in the delta region, leaving cattle raising as the dominant local livelihood for centuries. The deltaic marshes, occupying a huge area above the confluence of the Pánuco River and its Tamesi branch, west of Tampico, offered a perfect setting for the implantment of Andalusian/Antillean lowland herding, and the region remains dominated by ranches today (figure 21). Situated on the outermost frontier of the Spanish-ruled area in the middle 1500s, a boundary that coincided with the northern limits of high Indian civilizations at the time of contact, the cattlemen of the Pánuco marshes were ideally positioned to participate in the later northward diffusion of ranching along the coast through Tamaulipas to Texas.⁸ From these three lowland bases—the Veracruz hinterland, plains of Almería, and Pánuco delta—cattle ranches spread rapidly through many parts of the Spanish-ruled coastal plain, in a patchy pattern including a southward expansion into Tabasco and the Isthmus of Tehuantepec (figure 22). Cattle thrived particularly in the portions of the Gulf plains closest to the coast, both in the floodplains and in adjacent portions of the interior sloping plains. To generalize, cattle raising prevailed nearer the coast, while in the higher, drier areas of the interior coastal plains, including the rain-shadow districts, sheep rivalled cattle from the very first (see figure 27).

The origins and ethnicity of the Gulf lowland cattle raisers and their cowboys provide insights concerning the local ranching system. The Spaniards, most of whom arrived after a residential stop-

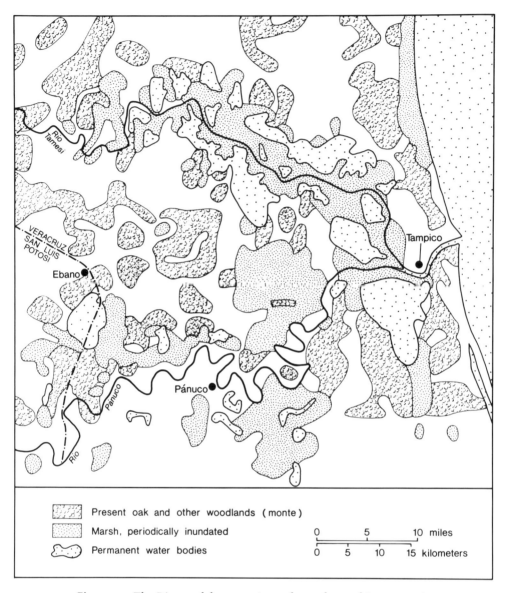

Figure 21. The Pánuco delta, a major early cattle-ranching center in the Gulf lowlands of Mexico. Vegetational data are incomplete. Compare to figure 4. Source: data and sketch provided by William E. Doolittle, University of Texas at Austin.

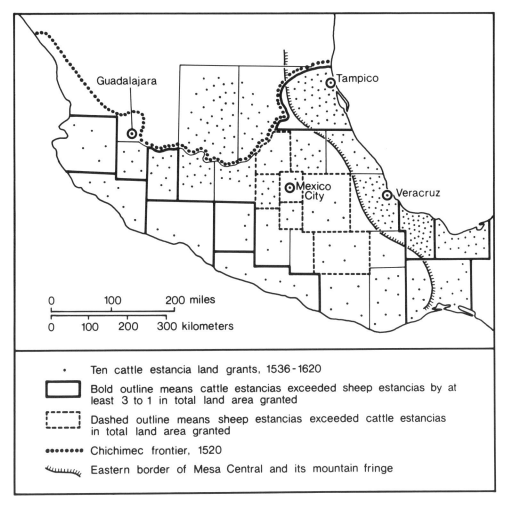

Figure 22. Cattle-ranching grants in Mexico, to 1620. Note the regional variation within Mexico, with particular attention to the lowland/highland dichotomy. Source: Lesley B. Simpson, Exploitation of Land in Central Mexico in the Sixteenth Century *(Berkeley and Los Angeles, CA, 1952),* tables pp. 28–87.

over in the Antilles, were as Andalusian as the islanders. Some 29 percent of the pre-1540 Gulf Coast settlers came from the lowlands of Andalucía, and fully 23 percent from marsh-perimeter towns. Some colonists specifically identified as cattle owners in early Mexico can be linked to origins in the region of Las Marismas, although Mesetan cattle districts were also well represented.[9] In any case, the Antillean experience may have been more influential than the Iberian origins.

Toponymic evidence of the Andalusian cultural character of the Mexican Gulf lowlands abounds. *Mata,* meaning a portion of pasture, or *monte,* covered with trees in the dialect of Andalucía, appears as a generic place name in the fringes of Las Marismas, as for example, Matalascañas and Mata de los Domínguez, to the west of the marsh. These reappear abundantly in the Mexican lowlands, in Tabasco and Veracruz states. Examples include Mata Caña and Mata Espino ("thorn woods") in the Sotavento, Mata Redonda ("round grove") near Tampico, and many others.[10]

The cowboys of sixteenth-century coastal Mexico were most often blacks or mixed-bloods with some measure of African ancestry, although the chief herdsman, or *caporal,* was usually a Spaniard. In early New Spain at large, herding was "done almost universally by mulattoes," working under Castillan bosses or overseers. One Spaniard in the Veracruz area, who owned twenty cattle ranches, employed two hundred blacks as vaqueros by 1571. Since some Mexican blacks at the middle of the sixteenth century were back-country Senegambian Sudanese by origin, knowledge of African herding techniques probably reached the Gulf lowlands. Use of *encomienda* Indian labor ended legally in 1549 but in reality had declined earlier, as the demographic collapse progressed. Some indigenous ancestry apparently survived in the vaquero group, due to racial mixing in the labor force. A cowboy caste or underclass, then, quickly formed, as had happened in the Indies and, still earlier, in Las Marismas. Initially African herders in Mexico probably lived in slavery, although they were reportedly well treated and given considerable freedom of movement. After about mid-century, however, they and the various mixed-bloods evolved into fiercely independent wage laborers, who even dared to threaten a cowboy strike for higher pay in 1576. Such boldness suggests that acculturated Iberian blacks of Andalusian origin may have played a role. Disparaged by the privileged class as vagrants and "low fellows" given to orgies of drink and sex, these early Mexican vaqueros had already become the stuff of which western dime novels could be made. Some of them became feared, mounted, and rebellious vagabonds or squatters who poached or stole livestock from the Spanish ranchers, as in the Pánuco delta area around 1600. Their small, extralegal stock-raising settlements, called ranchos in the proper Andalusian usage of that term, would ultimately provide one prototype for the northward littoral expansion of cattle herding, perhaps ultimately helping give the word *ranch*, in the altered pastoral meaning, to Texas and the American West.¹¹

Cattle-herding techniques employed in the hot, humid Gulf coastal plains of sixteenth-century Mexico exhibited direct linkages to practices in the Antillean savannas and Andalusian salt marshes. A labor-extensive system prevailed, in which a herd of several thousand cattle could be tended by at most three mounted vaqueros including a boss, assisted by several herdsmen on foot, the latter perhaps employing some Sahelo-Sudanic African techniques. They apparently resided out in the pastures. Many or most lowland ranches had insufficient labor. Bulls remained uncastrated as a rule, in the Andalusian manner, making herd control even more difficult, and many cattle became wild, as for example around the coastal lagoons of the Pánuco delta, where fierce feral bulls inhabited the adjacent woodlands and brush, emerging only on moonless nights. As a result, hunting of cattle with hocksing poles, though illegal after 1574, continued, as in the Indies. In part to supply the ranching industry of the Gulf lowland, the Huasteca and especially its Pánuco delta became a major center of horse raising. During the sixteenth century, Mexican cowboys of African, Indian, or mixed blood could not legally own horses, but instead could only use mounts belonging to the ranchers.¹²

In managing tame herds, lowland vaqueros initially used both garrochas and ropes, the latter made of grass with a Cádizan spinner, as in the Andalusian marshes, or, in an adaptation to Mexican raw materials, manufactured from maguey. This Andalusian plant-fiber rope would remain characteristic of Gulf lowland cattle ranching in Mexico and, subsequently, Texas. Sixteenth-century vaqueros of other than pure Castillan ancestry were not permitted personal ownership of either garrochas or ropes, apparently out of a not-unwarranted fear that they would be used after hours to poach hides or steal livestock. The principal early use of the rope was to secure cattle during slaughter. The lazo, described in sixteenth-century Mexican records, was not initially cast from horseback.

A light, small Moorish-inspired combat saddle called a *jineta* was probably used by early mounted vaqueros in the Antilles and lowland Mexico, and it lacked a horn, or *cabeza*, of sufficient size, proper shape, or adequate strength to permit roping from horseback. In addition the jineta stirrups were short, requiring the rider's knees to be bent, while straight legs braced against long stirrups were needed as support against the shock of a suddenly taut rope. Some early Mexican cowboys may, instead, have used some variety of the heavier, larger European *estradiota* saddle, equipped with longer stirrups and a better-developed horn, but even these remained insufficiently strong for roping. Some scholars believe the evolution from jineta or estradiota to the western American stock saddle began on Española or Jamaica, but this development probably occurred instead in Mexico, perhaps as late as the 1700s. No surviving Mexican roping saddle has been securely dated prior to 1830, and the lasso may not have been tied to the saddle horn before about 1750 or even 1780. In the 1640s Mexican lowland vaqueros, probably perpetuating Andalusian salt marsh practice, apparently still tied the lasso to the tail or neck of the horse or to the saddle cinch and, rather than casting it, used the garrocha or a stick to convey the slip-knotted rope over the horns of the target bull or cow (figure 23). The Gulf lowlands may have been the region where both lasso skills and the roping saddle were ultimately perfected, although Mexican tradition favors Jalisco State in the highlands as the scene of this development. It is noteworthy, however, that skilled casting of lassos from horseback is well documented in central Chile and California by 1820, suggesting a rather early date of origin in Latin America, possibly even in the Antilles rather than Mexico.[13]

Roundups occurred ideally on a weekly basis, in order to care for the livestock and keep them docile, but few if any lowland ranchers achieved this frequency, given the short supply of labor. More typical, certainly, was the mass gathering of over three hundred mounted vaqueros near Ciudad Valles in the lowland Huasteca in the late 1500s, to round up the widely scattered herds belonging to all the local ranches. The assembled cowboys formed a huge circle and drove the cattle toward the middle for herd separation and branding. By the nineteenth century, an annual roundup had become the normal method of herd management, supplemented by the use of salting, calf-capture, and tame oxen as control techniques. The annual roundup became associated with the festival called Herradero.

In Mexico branding became the universal legal device for demonstrating ownership of horses and cattle. The traditional Andalusian brand designs prevailed in Mexico, although Indian influence can reputedly be detected in the configuration of certain brands (figure 8). In a departure from Iberian custom, earmarking was outlawed, due to the ease of modification, but at least some Gulf lowland ranchers continued the practice.[14]

Figure 23. A prototypical form of lassoing from horseback, demonstrated in the bullring at Madrid, Spain, by American creole slaves ("unos Esclavos Criollos de las Indias") about 1640. A stick, possibly a garrocha, vara, or caña, is used to extend the loop of the lasso over the bull's head, rather than the vaquero casting it. Note that the lasso is tied to the horse's tail, in the old manner used by Andalusian marsh herders in pulling cattle from mudholes. The bent legs of the rider reveal that the saddle is a light, small, hornless jineta. Source: Gregorio de Tapia y Salzedo, Exercicios de la gineta al príncipe nuestro señor d. Baltasar Carlos (Madrid, Spain, 1643), following p. 72; the original is in color. I thank David Dary for bringing this remarkable book to my attention.

As in Las Marismas and the coastal sections of the Greater Antilles, a seasonal local shift of cattle from wetlands in the drier, low-sun time of year to pastures higher in elevation during the summer occurred in lowland Veracruz State. Indeed, such seasonal shifting is still practiced by cattle ranchers there. In the early colonial years, labor shortages probably dictated that the herds accomplish this locational shift largely on their own, without attendant transhumance. Early vaqueros may have assisted by pulling mired animals from the mud, as they had done for generations in coastal Spain, but the annual loss of cattle to flooding was likely far greater than in Las Marismas. Given the rapid natural increase of the herds, these losses would not have been regarded as much of a problem. Vaqueros probably devoted more attention to range burning, in spite of futile early colonial efforts to ban the practice. Performed at the end of the dry season in the savannas and montes to prevent the grass from aging and coarsening beyond any value to cattle, as well as to control brush and keep the forest at bay, firing remains a basic method of range management in the lowland ranching areas still today. Also reflecting Antillean practice, a cattle/horse/pig livestock complex prevailed in the coastal margins of the Mexican Gulf lowlands, in which sheep and goats were unimportant. In fact pigs came to the mainland before cattle, as around Pánuco, where encomienda Indians worked as swineherders in the early years.[15]

The privately owned cattle estate also made the transition from the West Indies to lowland Mexico. Following an initial period, prior to 1539, in which all vacant lands served as common pastures, land grants for cattle raising began to be awarded to stockmen, reaching a peak in the 1580s and 1590s. These grants bore the designation *estancia de ganado mayor*, had to be situated in vacant lands at least 4.2 km. (2.5 mi.) from the nearest Indian village, and by law measured one square league, or about 1,750 hectares (4,325 acres), far smaller than equivalent cattle grants in the Indies. However, individual grantees often or even usually obtained multiple cattle estancias, as well as other lands designated for crops. Unauthorized land seizures by conquistadores were not uncommon in the lowlands, and squatting occurred by herd owners lacking pasture grants, including former vaqueros who became rustlers, probably as a reflection of traditional Castillan squatters' rights. The Antillean term *hato* did not take permanent root in Mexico, except in the meaning of "pasture," but estancia did, although hacienda eventually became the preferred word to mean "ranch" in the Gulf lowlands. Recently, reflecting influence from the United States, *rancho* has come into general use to designate the prestigious, capital-intensive cattle enterprises of the Gulf region.[16]

Lowland Mexican cattle ranchers in the 1500s marketed their livestock in two principal ways. Hides derived from both tame and feral cattle were shipped to Sevilla, perpetuating the ancient Marismas connection in Andalucía. More than sixty-four thousand cowhides were exported from New Spain in 1587 alone, the majority probably from the lowlands. Cattle were also driven on the hoof into the adjacent highlands, a second market for the Gulf coastal ranchers. The Gulf lowlands became a beef-eating region, in common with the Sevilla area.[17]

Mexican Highlands

The other early colonial implantments of cattle ranching occurred in the temperate interior plateau of Mexico, variously called the Mesa Central or Altiplano. While remarkably like the Meseta of Spain in many ways, the Altiplano also exhibited some fundamental differences. The widely spaced, forest-clad ranges of the plateau, forming an open mountain topography interspersed with basins and valleys, surely convinced the conquerors that they had found a new Meseta. But the Mexican Mesa Central stood far higher above sea level than the Iberian plateau, access to it from the coastal lowland proved more difficult, and numerous volcanic masses intruded, lending alien aspects to the land. The most fundamental contrast, as in the Indies and Gulf lowlands, was the annual precipitation cycle, which featured a pronounced winter drought—the exact opposite of the regime found in the Spanish Meseta—and in the weakness of thermal seasons, due to the tropical location. Moreover, the Spaniards initially occupied the drier, rain-shadowed, eastern part of the Mesa Central, a land more like Castilla than Extremadura and better suited to sheep than cattle.

In spite of these contrasts, Spaniards developed three major early colonial nuclei of open-range cattle ranching on the Mesa Central, selecting favored basin sites freer of rain-shadow effect (figure 17). In these small districts, livestock multiplied rapidly in the 1530s, and ranching enterprises developed, providing the bases from which a cattle frontier would quickly spread northward. One of the implantment nuclei lay in the western and northern part of the Basin of Toluca, west of Mexico City at 2,600 m. (8,500 ft.) elevation (figures 17, 27). The first cattle entered the Toluca area in 1535, and within a decade and a half over sixty estancias owning some 150,000 cattle and horses had been established. A second nucleus of cattle ranching developed in the expansive Llanos de Apan, a basin centered on Tepeapulco in present Hidalgo State, and the third early highland cattle-herding district lay in the large Basin of Huamantla, southwest of Perote. Lesser cattle concentrations developed in the northern reaches of the Puebla Basin (the Val de Cristo) and the Valley of Mexico (figure 27). Each of these areas where cattle ranching gained an initial foothold in the Mexican highlands lay along the axis of the Spanish route of initial invasion and subsequent *camino real*, which ascended from the coast through Jalapa and Perote and ran westward to Mexico City. As late as 1550, most cattle ranching in the Mesa Central remained clustered in these nuclei, along the road of conquest.[18]

The ranching system implanted in these highland localities initially resembled closely that of the Gulf lowlands, Antilles, and Andalucía.[19] For a variety of reasons, however, the lowland herding system proved unsuited to the Mesa Central. It became a short-lived occupance form, a transitory cattle frontier. Well before the end of the sixteenth century, it had largely disappeared from the implantment nuclei, displaced to the northern frontier of Spanish rule.[20]

Numerous factors contributed to the ephemerality of specialized open-range cattle ranching in the Mesa Central. Population density provided one key problem. While profound die-offs occurred among the high native cultures of the eastern and southern Mesa Central, the catastrophe was apparently not as complete as in the Gulf lowlands and West Indies. Indian farming communities remained basically intact, retaining one-tenth to perhaps as much as one-fifth of their precontact population. Friction developed almost at once between the Indian farmers and Iberian stock raisers, usually as a result of crop damage by marauding livestock, and Spanish officials, desiring to avoid a repetition of the Españolan disaster, early adopted a policy of protecting the Indian communities. They never abandoned that policy, with dire consequences for ranching. By the 1550s governmental edicts were effectively displacing most ranchers to the sparsely inhabited northern frontier, away from the centers of high Indian civilization and colonial Spanish urban culture. In the final analysis, cattle ranching proved incompatible with the relatively high density of population in the southern highlands. Not even the granting of cattle haciendas to a few local Indian nobles in Toluca and nearby areas could avert the displacement.[21]

Economic pressures related to the denser population and greater urbanization in the southern Mesa Central also pressed upon livestock ranching. More intensive, crop-oriented systems of land use were favored by the dictates of land rent, and highland Mexico soon reflected rather well a Thünenian spatial structure of agricultural types, in which ranching was allotted a peripheral location. Cattle ranching proved especially vulnerable, since it had taken early root in favored humid districts, where crop farming had a natural advantage.[22]

Ecological problems also plagued cattle ranching in the highlands. The grasslands of the Mesa Central proved more easily altered by grazing than did those of the coastal savannas. Pasture damage quickly became evident and cattle, in effect, "mined" the highland grasses, diminishing the carrying capacity. This problem developed partly because most ranges in the Mesa Central did not offer year-round grazing, and the tendency was to leave the herds on the pastures too late into the low-sun, drought period, causing overgrazing. A dwindling reproduction rate among the cattle, an indicator of environmental degradation, soon became evident in the implantment districts.[23]

In the final analysis, then, the Andalusian/Antillean lowland system of specialized cattle ranching, based in extensive land use, stationary herds, and minimal livestock supervision, proved demographically unsuited, economically unsound, and ecologically maladaptive in the Mexican highlands. It simply could not be sustained there as an adaptive strategy, and a new system was required for long-term occupance of the Mesa Central. Drawing upon another Iberian prototype, the Spaniards soon developed a Mesetan-inspired, village-based strategy closely resembling that of Extremadura and Salamanca, though also reflecting New World cultures and conditions.[24]

The key attribute of the new highland strategy was diversity. Range cattle remained part of the system, especially in the moister districts, but became subordinate to crops and small livestock, and most male calves were castrated, as in Extremadura. Sheep, which thrived from the very first in the Mexican highlands, including the drier areas, quickly became the most important livestock, partly because they could utilize pastures damaged by cattle. Goats, too, became a key to the diversified highland economy, and swine did moderately well, given the limited amount of forestland available. To the present day in the villages of the Mesa Central, one finds pigs confined in rock-walled sties that any Extremaduran would recognize. Wheat, which had failed in the tropical lowlands of Gulf Coast Mexico and the Antilles, became an essential part of the highland system, as in Spain, although irrigation systems were necessary to allow the Spaniards to raise their traditional winter variety of the grain. Maize vigorously survived from preconquest days to become the summer grain of the highland system. Tree crops, in particular the olive, did not achieve a major place in the crop-livestock association, but they had also been absent in much of mesetan Spain and were more closely linked to Andalucía. Vineyards, too, remained uncommon, in spite of many efforts to develop them. In effect, then, the specialized livestock estancia gave way to the diversified highland hacienda, based in sheep, goats, cattle, pigs, and grains.[25]

The local prototypes for the evolving, diversified highland strategy appeared very early. Cortés himself owned an enterprise near Cuernavaca that produced cattle, sheep, horses, vines, and sugar, not greatly unlike estates in Medellín, his home in Extremadura. By 1531 other diversified operations near Puebla, based on Indian labor, yielded corn, wheat, and all the major Iberian livestock.[26]

The highland Mexican hacienda system was also labor intensive. In fact even during the brief cattle-ranching phase, graziers had eventually been obliged by authorities to tend herds carefully and to keep them relatively tame. From June to November, when cattle grazed the summer pastures, authorities recommended weekly roundups and required minimal limits for the number of vaqueros that had to be employed to work each herd. In the evolving highland hacienda system, cattle became so tame and few in number that mounted cowboys were no longer essential. Castration of most young bulls, nightly penning of cattle in rock corrals, and tethering of individual grazing animals was almost universal (figure 24). In the Mesa Central today, one commonly encounters women, boys, and old men, on foot or riding donkeys, tending small herds of open-range longhorn cattle. They wield sticks, cast rocks, and shout verbal commands to maintain order in the herd (figure 25). Any resident of rural Extremadura or Salamanca would find these images and techniques familiar.[27]

Though increasingly irrelevant in the highland hacienda/village system, the mounted vaquero and equestrian skills would survive in the more humid western parts of the Mesa Central, eventually to become a gentlemanly pursuit and national icon of Spanishness, the *charro* cowboy subculture. While now associated with Mexico at large, the charro was traditionally linked, along with *mariachi* music, to the highlands and particularly to the Guadalajara region in Jalisco State, an early cattle frontier in the uplands of the colony of Nueva Galicia during the period following the initial implantments of ranching in the Mesa Central (figure 27). The highland Mexican charro, skilled in handling and taming horses and cattle, wears a high-crowned, broad-brimmed *sombrero*, tight trousers, dec-

Figure 24. Mexican cattle of mixed breed in a rock corral similar to those of the Spanish provinces of Extremadura and Salamanca and located, appropriately, in the hills north of the town of Salamanca in the Mesa Central of Guanajuato State, Mexico. Any salamantino peasant herder would find the scene familiar. (Photo by the author, 1986.)

orated shirt, short jacket, scarf, and spats. Significantly both the word charro and the associated colorful costume derive from the Spanish mesetan province of Salamanca. Though often erroneously applied to all Salamantino peasants, charro more precisely designates rustic rural folk from the Charrería in the southern districts of Gata and Galache in Salamanca, near the border of Extremadura, some of whom were mounted, bull-tending novilleros. The Salamantino charro garb consists of "a broad, flexible-brimmed *sombrero,* a figured shirt, . . . short jacket, medium-length trousers, leather shoes, cloth spats . . . or riding boots, and . . . a sash." The two highland charro traditions are in fact one, linking Meseta and Altiplano. Interestingly, charro in Salamanca remained a term of derision, perhaps equivalent to "bumpkin" or "hick," while in Mexico it became honorific and even aristocratic.[28] The key to the high status of the charro

Figure 25. Two stone-casting, pedestrian herders drive Spanish longhorn cattle home from pasture along a rock-walled cañada, or drove trail, at Jacala, Hidalgo State, Mexico. One animal has bolted through a gap in the deteriorating wall and is being retrieved. The scene epitomizes the highland herding system of both the central Mexican plateau and the western Meseta of Spain. (Photo by the author, 1988.)

in Mexico lies in his equestrian skills, which are viewed as Castilian and therefore non-Indian. It has become a matter of blood and myth.

Similarly, *la reata,* or rawhide rope, apparently also derived from Salamanca. In Mexico it long remained associated with the charro subculture, and eventually designated a lasso rope. Later it became corrupted as the English loanwords "lariat" and "riata." The Mexican charro also retained, both in word and artifact, the Iberian mesetan vara, or staff, used in Mexico to train horses.[29]

Evoking even more strongly the image of mesetan Castilla and Extremadura was the importance of transhumance in highland Mexico, complete with a network of cañadas, or drove trails, often lined with rock walls (figure 25). The seasonal migrations, as in highland Spain, usually involved sheep, but cattle also had to make shifts between summer and winter pastures. In effect, highland pasture ecology demanded transhumance, because in the dry winter season the native grasses could not support the livestock. Stationary herds were incompatible with the local floral environment. At the very least, cattle herds that had dispersed in summer through highland basins such as Toluca and Huamantla had to gather in winter at the residual interior-drainage lakes and marshes toward the center of the basins, where enough moisture remained to provide forage. Some other herds were driven for winter grazing into nearby hill pastures, or monte, an unsatisfactory practice due to the colder temperatures at higher elevations. Wealthier ranchers rented or owned distant winter pastures on the wet, seaward slope of the Sierra Madre Oriental, high above the Gulf coastal lowlands on the eastern margin of the Mesa Central.[30]

Other graziers, seeking to delay the seasonal migration, demanded the ancient Castillan privilege of grazing herds on the stubble of harvested crops, a resource available initially in Mexico only in the fields of Indian maize farmers. Though practiced earlier without authorization, this traditional right became established by decree in highland Mexico in 1551. In the short run, stubble grazing provided yet another cause of dispute with the Indian villagers, but after crop raising became a part of the diversified highland hacienda system, such pasturing became an integral practice. Stubble grazing remains common in the Mesa Central still today. Haystacks enclosed by rock walls in the middle of grain fields, as in Extremadura, assure the retention of additional fodder for the animals.[31]

The mesta, the Castillan stock raisers' regulatory association, came early to the Mexican Mesa Central. Graziers in the Mexico City area founded a regional mesta in 1529. One of its first actions was to require brand registration, and in 1537 its first regulatory code appeared. A far more detailed and restrictive code followed in 1574. For example membership was required of all large stock raisers, earmarking of cattle outlawed, the running of sheep in cattle pastures prohibited, and overland drives of stock controlled.[32]

It is tempting but ultimately incorrect to regard the appearance of the mesta in the Mexican highlands as another reflection of mesetan Castillan culture, as a transferal of the powerful, royally sanctioned, nationwide stock-raising guild of Spain. The Mexican mesta, instead, was patterned after the regional and municipal mestas of Spain, which were as common in Andalucía as in the Meseta. Indeed, the Sevillan mesta seems to offer a more viable prototype for that in highland Mexico, since the New World organization possessed limited power, retained a local or regional focus, was much concerned with cattle, and dealt far less with transhumance than did the sheep-centered royal mesta of Spain. Indeed the most puzzling question is why such a Sevillan-inspired, cattle-oriented mesta failed to appear in the West Indies and came only late and weakly to the Andalusian-oriented Gulf lowlands of Mexico. Regional mestas first appeared in the Pánuco delta after 1600 and in the Veracruz region in 1629.[33] Perhaps the local mestas had hindered the operations of the Sevillan-based entrepreneurs in their commercial beef operations in the early 1500s in Andalucía.

The Iberian origins of the early Spanish colonial population of the Mexican highlands suggests a diffusionary basis for the rise of a mesetan-inspired culture and agricultural system there. An example is provided by the highland colony of Nueva Galicia, bordering New Spain on the west and having its capital at Guadalajara. Andalucía was underrepresented as a source in the population of Nueva Galicia, in comparison to its contribution to Mexico at large, while the cattle-raising provinces of the Spanish Meseta were over-represented. Andalucía sent 35 percent of all migrants to Mexico in the formative 1520–40 period, but provided only 23 percent of the settlers who eventually reached Nueva Galicia. Extremadura, by contrast, supplied more than its share.[34] Of the pre-1540 immigrant Spaniards who ultimately lived in Nueva Galicia and for whom place of origin was recorded—a total of over four hundred persons—about 36 percent came from cattle-raising areas in the Meseta. Some 23 percent emigrated from Badajoz, Cáceres, Salamanca, Zamora, and Avila—the western core of the Iberian highland range-cattle zone (figure 3).[35] Clearly population origins facilitated a more profound mesetan influence and a weaker Andalusian impact in the Mesa Central.

Highland Mexico, then, evoked various images of the western Spanish Meseta, with abundant agrarian reminders of Extremadura and Salamanca. Its diverse, hacienda-based, mesta-regulated system of grains, sheep, goats, pigs, and cattle was labor-intensive, self-sufficient, and transhumant. By almost any measure or definition, this highland system was not "ranching," any more than the herding in sixteenth-century mesetan Iberia had been. Andalusian cattle herding—ranching, indeed—proved in the main fleeting and maladaptive in the Mexican highlands, but the ways of the Meseta abided there.

Implantment in Spanish Florida

The Spaniards also introduced Andalusian/Antillean cattle ranching into northern Florida, establishing the earliest implantment in territory that would ultimately become part of the United States. This second continental foothold remains poorly studied, but Hispanic herding activity in Florida achieved a greater influence than previously supposed, because colonial cattle ranching was not only early but also achieved both durability and continuity, eventually touching the American West.

Transfer from the Indies was facilitated by a similarity of environment. Northern Florida, though not fully tropical, closely resembled the lowland plains of the Greater Antilles.[36] The seasonal contrast is greater, for Florida has a true if mild winter, although the low-sun concentration of precipitation is far less pronounced than in the West Indies. The similarity finds best expression in the vegetation cover. Abundant seasonally flooded savannas once dotted the landscape, especially in the interior, the largest of which, at 8,900 hectares (22,000 acres), was Alachua Savanna, later called Payne's Prairie, near present-day Gainesville. Some varieties of native grasses, including both andropogons and paspalums, appear in both the Indies and Florida. On the Alachua Savanna, the main paspalum, called "blanket grass," provided abundant spring and summer grazing, though deficient in calcium and phosphorus. Sawgrass, a sedge, also provided an important savanna forage plant, as did maiden cane, water lily, bull grass, and water hyacinth. In the somewhat drier winter, after frost had killed the grasses, livestock withdrew to nearby forests and swamps or else to remaining wet, low places in the savannas.

Adjacent to and surrounding the savannas lay the expansive Florida "flatwoods," or pine barrens, an open forest of slash and longleaf pine that was also fired annually. Two subtypes occurred: a wetland pine–wire-grass–palmetto association, or sedge bog, with high species density and a water table that remained near the surface most of the year; and a well-drained upland sand-ridge pine–little-bluestem complex, with an open canopy of pines. Both wetland and upland barrens offered abundant grazing for cattle, the former in winter and the latter in summer. Low-lying oak-hardwood lands, later to be called "hammocks," also provided good winter forage in their understory of canes, sedges, and grasses, and in moss dropping from the trees; but cattle avoided these areas in the warm season, due to insect infestation. Also present were "high hammocks," well-drained upland live-oak groves that also provided winter browse. On these Florida ranges, unlike the early Antilles, predators, mainly wolves and cougars, presented a problem for the Spanish stock raisers. All things considered, however, northern Florida offered fine opportunities for transplantment of the West Indian herding system.

Curiously the Spaniards were slow to develop the grazing industry, even though their efforts at raising crops did not enjoy much success. Various explorers, beginning in 1521, introduced herds of cattle and droves of pigs, without founding any settlements. Few if any descendants of these livestock survived, although the Spaniards who arrived as colonists beginning in 1565 reputedly called the local Indians vaqueros because of the beef and cowhides found among them, presumably acquired through hunting. When colonization began, cattle were once again introduced, from Cuba, but only ten or twelve head survived by 1570. The first ranch did not appear until at least the early 1600s, and no substantial herding activity occurred until the 1650s. A cattle-raising boom of sorts came in the 1680s, and tax rolls listed some thirty-four ranches by 1700, containing a total of perhaps fifteen or twenty thousand head. One ranch based in the Alachua Savanna had a thousand or so cattle.[37]

At least three separate clusters of cattle ranching developed in Spanish Florida (figure 17). The most important, containing twenty-five ranches, lay in the district called Timucua, in and around the Alachua Savanna. A second concentration took shape in the Apalachee district, at the base of the Florida panhandle inland from the small port of St. Marks around present-day Tallahassee. The third-ranking herding cluster developed inland from the fortress settlement of St. Augustine, along the lower St. Johns River. These ranching complexes apparently arose almost solely through the initiative of local, American-born Spaniards, who received no royal encouragement. One early rancher was Cuban. Landed secular estates called haciendas, granted by the local governor at St. Augustine, belonged to the dozen or so families who dominated cattle ranching in Spanish Florida. The Antillean and Andalusian term *hato* also appeared, but apparently in the meaning of a small ranch. At least some of the scattered ranchsteads were fortified.[38]

The role of the Catholic missions in Florida ranching remains unclear. Franciscan missions flourished by 1650 among the Muskogean speakers of the Apalachee and Timucua districts, but apparently cattle did not fit well into the garden-based, largely aboriginal agriculture of the mission Indians. Some competition for land between Spanish ranchers and native farmers occurred, and herding seems to have remained a distinctly Euroamerican institution. Still the native people learned to tend livestock at some of the missions, which collectively owned as many as six thousand cattle. Although the encomienda labor system never existed in Florida, mestizos and Hispanicized Indians probably worked as cowboys in Spanish Florida. Some Indians, drawn from the nobility, even owned cattle and horses.[39] The involvement of indigenous people in ranching likely advanced with acculturation.

The Spanish ranching system in Florida closely resembled that of the West Indies. Based in a cattle-hog-horse livestock complex, it utilized mounted cowboys to manage open-range herds, yielding hides, tallow, and sun-dried beef, both for export and to supply the garrison of St. Augustine. Roundups took place in the spring at palisaded corrals, where calves were branded (figure 8), and again in the fall, for culling. As in early Mexico, the cowboy underclass could not own horses. Lassoing from horseback did not occur in Spanish Florida, and instead both the garrocha and hocksing pole were in use.[40]

Reputedly both the cattle-ranching enterprise and mission culture of Spanish Florida were destroyed by the South Carolina English and their Lower Creek allies in a series of raids between 1702 and 1706, causing the Spaniards to abandon the hinterlands and withdraw to the great fortress at St. Augustine. Supposedly they remained confined there until the transfer of Florida to the British in 1763. In this traditional view, the Iberian implantment of ranching in northern Florida came to naught.[41] Evidence suggests, to the contrary, that the pastoral industry survived to achieve an enduring legacy, both locally and in the West.

The Spaniards modestly revived cattle herding after the military disaster, and by the 1730s ranches were in operation in the St. Augustine district and had, in fact, probably never ceased to function there. English raiders from Georgia near midcentury encountered a fortified ranch north of St. Augustine, and in 1756 Creek Indians attacked Spanish ranches in the province, slaughtering cattle and taking hostages. The regional scope of Spanish ranching seems even to have expanded in the 1700s to include the Pensacola area of far western Florida, for a 1771 document refers to "the Old Spanish Cowpen" up the Escambia River.[42]

An enduring Hispanic influence in Florida was achieved most obviously by the criollo cattle. Longhorn stock abundantly influenced the "Florida cow" of the nineteenth and early twentieth centuries, and stock from the local longhorn herds taken in 1704 raids were driven to South Carolina to be crossbred with the British-derived cattle of that colony, a process repeated in 1740 for Georgia. Florida longhorns later became the basis of English-owned herds in the Pensacola area, and still others were taken north during the War of 1812, to improve Anglo-American stock in Tennessee and Kentucky, on one pathway to the Great Plains.[43]

Far more than Iberian bovine bloodlines survived as a legacy of Spanish Florida. The entire ranching system passed intact, with minor modifications, to Creek and Seminole Indians. While the Lower Creeks never came under Spanish control, they learned from them the management of cattle on horseback, perhaps even as early as 1700, and the Creek Muskogean term for cattle, *waca*, is a loan word. The Creeks, probably including the kindred Proto-Seminoles, acquired Iberian cattle, hogs, and horses in the destructive raids of 1702–6 and later. These Indians continued cattle ranching on the Alachua Savanna after the Spanish retreat and even sold beef to the garrison at St. Augustine. Additional Creeks shunted southward during the English settlement of Georgia also engaged in ranching.[44]

By 1740, if not earlier, the Florida Creeks centered upon the Alachua Savanna acquired a special ethnic identity as the Seminoles, and cattle herding provided a major component of their culture and livelihood. Raising cattle became so important among them that their reputed founder-chief bore the name "Cowkeeper." The Hispanicized character of the Seminoles is revealed both in their tribal names, corrupted from the Spanish *cimarrones*, or "wild ones," and in descriptions left by certain travelers who visited them. One of these, William Bartram, found them in the early 1770s "tinctured with Spanish civilization," owning "innumerable droves of cattle" and "squadrons of the beautiful fleet Seminole horse," which they rode "to collect together several herds of cattle, which they drove along."[45] Creeks also continued to raise cattle, and in the 1790s in the Tallapoosa River Valley of eastern Alabama, herds of one to four hundred head were observed among them. These Indian groups would later, in the middle nineteenth century, reestablish their cattle industry in eastern Oklahoma, allowing the influence of Spanish Florida to reach the Great Plains.[46]

British South Carolina to 1715

Among the thirteen English colonies on the eastern seaboard of the United States, South Carolina achieved a surpassing importance in open-range cattle raising, becoming in the process a major continental herding hearth that would eventually help shape the ranching frontier of the West. Large-scale raising of "black cattle"—the old English term to designate bovine, as opposed to ovine, stock—became a basic element in the distinctive regional culture of colonial South Carolina. This activity and identity largely reflected British West Indian, and more particularly Jamaican, influence. Barbados furnished more settlers to Carolina, but cultural diffusion surprisingly often is achieved by minority elements in a population, and Jamaica wielded such an influence.[47]

Still, the survival to the present day of the toponyms "Jordan's Cowpen" and "Cowpen Rock" at the northern tip of Barbados raises the question of whether an Afro-British cattle-herding system devoid of Spanish influence might once have existed on that small island, a system that could have influenced early South Carolina. I visited the Barbadian site in 1991 and found that it remains an open-range cattle pasture today, grazed by tethered, British-derived stock that actively serve the Afro-Celtic milking-culture complex. The reason why the Jordan's Cowpen area of Barbados had never been drawn into the sugar-based plantation system of this densely populated island was immediately obvious—the table-flat upland has scarcely any soil at all. Countless patches of limestone bedrock are exposed at the surface, and grasses gain a tenuous foothold in the interstices. Cultivated land appears locally only as "penned over" gardens of artificial soil, in the Jamaican manner. As one progresses southward from Jordan's Cowpen, sugarcane fields and dense rural settlement begin abruptly at the point where an adequate depth of native soils is encountered. Nature, in other words, provided a small enclave on the northern end of Barbados where prototypical British-African cattle herding might have developed and survived. However, an inspection of historical documents, especially old maps, revealed that the Barbadian cowpens appeared too late to have influenced either South Carolina or Jamaica. A detailed 1680 map, while showing northernmost Barbados as "champaign land"—that is, open and unfenced—also recorded no settlement in the area and no resident named Jordan. A 1711 map, though less detailed, similarly failed to reveal any such features. The first appearance of "Jordan's Cowpen" was on the 1722 map of William Mayo, based on data gathered in 1717–21. I conclude that Jordan's Barbadian operation probably commenced about 1715, long after the Carolinian cowpen culture had developed, and that his pastoral enterprise most likely reflected reverse migration from South Carolina to Barbados.[48] Jamaica, not Barbados, explains what happened in Carolina.

While the livestock-raising complex introduced into South Carolina was in the main Jamaican, the animals imported to stock the colony came from elsewhere. Virginia cattle—small, black Irish stock in the main, with some English red-browns—were introduced, and others may have come from New York, Bermuda, and New England. The basic bloodlines remained dominantly British throughout the colonial period in Carolina, and the previously mentioned admixture with Spanish Florida stock seems not to have greatly changed the appearance of the herds, although by the 1730s the Carolina stock were "of a middling size."[49]

While one settler noted in the very first year of colonization that cattle thrived in South Carolina, herds did not immediately increase greatly in the colony, perhaps because most went to slaughter to feed the infant settlement. As late as 1675, South Carolina "was almost destitute of Cows" and other livestock, but from that year onward the herds experienced very rapid growth, and "the great encrease of their Cattel is rather to be admired than believed." By 1680 only newly arrived settlers lacked substantial herds, and two years later some individual graziers owned as many as seven or eight hundred head of cattle. Toward the end of the formative early colonial period, in 1712, cattle, hogs, and horses abounded, including one individual herd of at least three thousand that annually produced some three hundred calves. "Many men hath by their stocks of cattle and hogs, in a few years, become rich."[50]

The outer coastal plain and islands around Charleston developed as the earliest focus of cattle raising, and the first notable spread led southwestward, paralleling the coast (figure 26). By 1694 cattle raisers had reached the mouth of the Edisto River, and a decade later they had ascended well inland along that stream. Beyond the Edisto lay the lands of the Yamassee Indians, and the intrusion of herds into that tribal domain precipitated in 1715 a devastating war that conveniently marks the close of the formative period of Carolina herding. Beginning about 1680, then, cattle raising became a principal form of frontier occupance in South Carolina, although a Turnerian stage

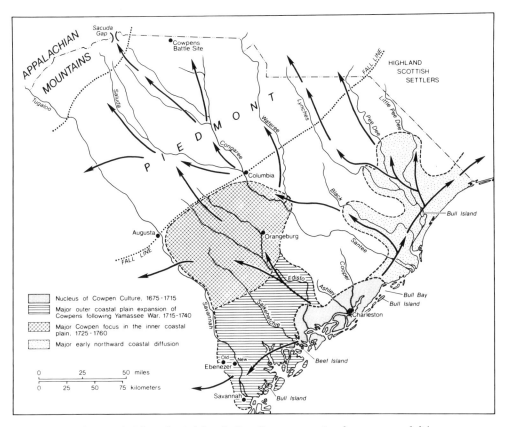

Figure 26. The colonial South Carolina cowpen implantment and diffusion. Sources: Jordan, Trails to Texas, p. 40; Maag, "Cattle Raising in Colonial South Carolina," p. 91; Dunbar, "Colonial Carolina Cowpens," pp. 125–30.

model will not suffice here any more than in lowland Mexico. Open-range livestock herding remained viable in many places long after the frontier had passed, owing mainly to the poor quality of much of the land. Some cattle raising persisted in the Charleston area as late as the 1770s, and in fact cattle raisers still today utilize marshes and canebrakes along the lower Santee and Edisto rivers. This post-frontier persistence, reminiscent of the West Indies, Veracruz, and Pánuco, would remain a basic characteristic of the Carolina cattle complex in its later spread westward to Texas.[51]

Quite compelling cultural evidence links South Carolina cattle raising to a Jamaican origin, and almost every aspect of Carolinian herding is understandable in these terms. Although relatively few

FROM THE INDIES TO THE MAINLAND

immigrants came from Jamaica to South Carolina, some such settlers arrived very early, and trade contacts between the two colonies began almost at once.[52] The most obvious and persuasive evidence of the Jamaican herding connection is etymological and toponymic. Both *pen* (or *cowpen*) and the Spanish-derived *crawl* attained early usage in South Carolina, in their distinctively Jamaican meanings of "cattle ranch" and "pig farm." The place-name Hog Crawl Swamp survives even today in Jasper County, and both generic and specific toponymic use of *cowpen* became very common in the South Carolina coastal plain (see figure 34). Early maps and warrants reveal such names as Salley's Cowpen, Bellinger's Cowpen, and Palachokola Cowpen.[53]

Similarly the use of West Indian *savanna(h)* to mean "a large spot of clear land, where there never was any timber . . . and nothing but grass" occurred from the very first year of settlement in South Carolina.[54] The toponym *savanna*, both as first and last element, became ubiquitous in colonial times, and some of these names survive today in the coastal plain, including Elm Savannah (Richland County), Cow Savanna (Colleton), Crane Savanna (Allendale), Long Pond Savanna (Berkeley), and Pitts Savanna (Sumter) (see figure 33).[55] Other apparently Jamaican lexical transferals to South Carolina English, mostly Spanish loanwords, include *jerked* (sun-dried) beef, *palmetto*—the very nickname of the state—and the Afro-Jamaican *dogey* (a small steer or cull), as well as Franco-Antillean buccaneer *barbecue* and *filibuster*.[56] Additional British-derived West Indian terms, such as *cowhunt, pond* (stock-watering place), and *gully* also entered the Carolina vernacular.[57]

Other parallels between Jamaica and Carolina include the use of enslaved or indentured black, Indian, and white cowboys, the pairing of cowpens and plantations as constituent parts of diversified, spatially fragmented operations, and the cattle/hog focus. Jamaican influence, reaching South Carolina in the 1670s, may have resulted from the initial displacement of some cowpen operations due to the expansion of plantation fields, a process well underway in the 1680s (figure 18). More likely, however, the transplantment occurred because Jamaican cowpens and crawls could no longer fully supply the huge demand for meat in the island's plantation system, requiring an additional source of supply.[58]

South Carolina cattle and pig raising, then, drew very heavily upon "decades of cultural exchange among Hispanics, Britons, and Africans," an exchange most vigorously achieved on Jamaica.[59] However, this hybridization was by no means confined to the Indies. It continued, more modestly to be sure, in early South Carolina. Additional African influence surely entered the colony, since six times as many slaves came to South Carolina directly from Africa as from the West Indies. Moreover many of the new Africans originated in Senegambia, the single most favored source of Carolinian buyers, and some of these surely came from the Fulani hinterlands up the Senegal and Gambia rivers (figure 15). A few early Carolina settlers also sought slaves from the Spanish Antilles, and these, too, may have possessed herding skills.[60]

Additional direct influence from the British Isles also entered early South Carolina, and part of this immigration derived from the highland herding districts. Indeed some scholars maintain, incorrectly I believe, that open-range livestock raising in the American South, including early Carolina, is largely of Celtic origin. English settlers from cattle shires such as Somerset, Devon, and Derby appeared very early in South Carolina, as did some Irish and Welsh and a small contingent of Scots, who settled in 1684 at Port Royal Island, beyond the Edisto River. Between 1680 and 1682, a sizable contingent of dissenter Protestants from Somerset arrived in South Carolina, many bearing the ancient and preadapted marsh-herding tradition of their home shire.[61]

The legal system of South Carolina, including livestock law, reflected this British influence. Carolinian statutes dealing with grazing, open range, fencing, rustling, and straying all came, essentially unaltered, from English common law and had deep roots in Germanic culture. What is more, this body of Carolina livestock laws, in company with the more or less identical statutes of the other eastern seaboard English colonies, was later carried west to become, with a few minor exceptions, the legal tradition governing livestock raising in the Anglo-American ranching West. In lawyer Ray August's words, "the principal rules of western American livestock law trace their heritage to the American East, and even beyond . . . to England and Northern Europe." Germanic rather than Hispanic livestock law would prevail on the United States cattle frontier, a precedent demonstrated in early South Carolina.[62]

The physical environment of South Carolina facilitated a transplanting of the Jamaican cattle-raising system, but was also compatible to British ways. While not tropical, the climate offered winters sufficiently mild that no seasonal fodder was required. West Indian immigrants consistently underestimated the severity of the winters, and stock die-offs did occasionally occur in the colder months, but by and large the Jamaican system worked splendidly in the new habitat, particularly since the troublesome low-sun drought of the Antilles did not occur in Carolina.[63]

A convenient diversity of juxtaposed habitats awaited pastoral colonists in the outer coastal plain of South Carolina, including savannas, marshes, swamps, pine barrens, and hardwood forests. Collectively this ecosystem could support and fatten one cow on about each 6 hectares (15 acres) year-round, with no supplementary feed. The first environmental niche likely occupied by cattle in the new colony, as earlier in the Antilles, were brackish tidal marshes rich in cordgrass and salt grass, for South Carolina reputedly has a quarter of all such marine wetlands on the entire East Coast of the United States, providing yet another reason for the colony's pastoral preeminence. The salt marshes proved particularly valuable in winter, when growth was tender and nutritious. At that season, too, insects presented less of a problem in the wetlands. References to "firm" feeding marshes suggest the importance for grazing.[64] Freshwater, waist-high brakes of switch cane (*Arundinaria tecta*), also called winter or green cane, served as a useful adjunct to the tidewater marshes, particularly since this "sort of short cane growing plentifully on the lower moist land" bore "a long green leaf in winter, on which cattle delight much to feed."[65] Brakes of giant cane (*A. giganta*) served the same function. While plantation-based rice cultivation spread through some of the wetlands beginning in the 1690s, much remained available as winter pasture.[66]

The upland pine barrens, a pyrophytic, or fire-resistant, complex similar to and continuous with the Florida flatwoods, grew on higher-lying, better-drained sands, the best of which displayed the splendid longleaf pine. Some inland barrens took on a sandhill character, supporting only thin stands of slash pine and scrub oak. An understory of shrubs and grasses offered abundant browsing and grazing, and cattle herds foraged in the pine barrens from spring until the first killing frost in autumn, while hogs lived on roots and pine sprouts. Riparian belts of hardwood forest, called "oak and hickory lands," cut through the pine barrens at frequent intervals, and while these streambank woods offered little grass, they provided abundant mast for hogs and, nearer the coast, Spanish moss, a "long sort of green moss which the winds shake off the trees," providing winter browsing for cattle. Wetland pine barrens like those of Florida also occurred in South Carolina and supplemented the winter range.[67]

Above all, cattle raising depended upon the local savannas, "exceeding good for a stock of cattle," of which two types existed, both in South Carolina and in the southern coastal plain at large.[68] In the low, flat outer coastal plain, and near some streamcourses, where the water table remained at or near the surface most of the year, appeared wet savannas, a periodically inundated wire-grass–pine vegetative community, sometimes referred to as a grass/sedge bog and characterized by high species density. In better-drained sandy uplands above the wetlands, dry savannas "of several magnitudes" appeared in the pine barrens, dominated by little bluestem (*Andropogon scoparius*) and often thinly interspersed with pines, both longleaf and slash. These offered not only grazing for much of the year, but also favored sites for pastoral homesteads, and the colonists "frequently settled their cow-pens upon them."[69] The potential of these grasslands as pastures was perceived at once by the immigrants, and a 1674 account, before the cattle boom, mentioned "many large pastorable Savannas."[70] West Indian settlers could easily evaluate the Carolina savannas, since they were similar to those of the Antilles and even shared some of the same herbaceous species. British immigrants directly from Somerset surely found the wet savannas similar to the great Levels, though the grasses were easily overgrazed and nutritionally inferior to those east of the Atlantic.[71] "Our beef is grass fed," reported a Carolinian in 1712, "and in the latter end of August and September is very fat."[72] After the first killing frost, the cattle that had escaped slaughter drifted into the nearby swamps, marshes, and hardwood forests, leaving both the savannas and pine barrens open for burning in the late winter, an Indian and Antillean practice perpetuated by the Carolinians to enhance grass growth, retard shrub and hardwood intrusion, and improve the nutritional value of the forage.[73]

Predators, though present, posed no substantial problems for cattle raisers. In fact, by discouraging sheep raising, a British predilection, the wolves, wildcats, and cougars gave an added preadaptive advantage to cattle and pigs, both of which could more adequately defend themselves. Nor did any major cattle diseases appear during the formative period of Carolina pastoralism.[74] All things considered, large-scale cattle and hog raising with minimal human supervision could be carried out in early South Carolina, and the particular Jamaican prototypes worked very well.

The center of operations for each Carolina cattle raiser was the cowpen, or ranchstead. Generally situated in a savanna or small clearing, the cowpen consisted of a hamlet of small structures, including the overseer's dwelling, quarters for the herders, cook shed, smokehouse, and stock enclosures, as in a Jamaican "pen." These little settlements ranged from permanent, well-developed steads on private property to the temporary camps of drifting squatters. The constituent structures, probably of picket walls, were often positioned to form a funnel, through which range cattle could more easily be forced into the spacious adjacent holding pens during roundups. Other, smaller pens for saddle horses, calves, milk cows, and slaughtering also formed part of the cowpen complex. Periodically, one of the smaller enclosures was relocated and a garden planted in the manure-rich soil, perpetuating the Jamaican and highland British infield system of "penning over." This garden, containing little besides corn and vegetables, usually represented the only crop production performed at the cowpen, for stock raising remained the dominant enterprise.[75]

Archaeological studies reveal even the type of fence used at the Carolina cowpens. Stakes less than 5 cm. (2 in.) in diameter were driven into the ground at close intervals, and vines and branches tightly woven between them, forming a wattle fence that perhaps owed morphological debts to the Fulani brush pens of West Africa as well as to medieval British wattle construction. In the long run, this fence type proved insufficiently strong, leading to its replacement in the late eighteenth and nineteenth centuries, but it had to suffice during early colonial times.[76]

The cattle-herding system centered at these cowpens was also strongly reminiscent of British Jamaica, as were the people who labored as herders. As earlier suggested and as is implied in the term "cowpen," the operation specialized in raising range cattle. That was particularly true when the cowpen was paired with a plantation as its beef supplier, under absentee ownership, but from the very first many operators were independent and as a result ran a somewhat more diversified enterprise, including pigs and a fuller array of subsistence crops. In either case the hunting of wild game was an important subsidiary activity to supplement the diet. Even when the land upon which the cowpen was situated belonged as private property to the planter or herder, the bulk of the range remained in the public domain, or "king's land," in keeping with Jamaican and ancient British custom. A great many cowmen never acquired title even to the stead.[77]

Open range prevailed, for South Carolina early adopted "fence-out" laws governing at-large livestock. In the formative years before about 1690, coastal islands often provided natural enclosures for pastures, and some cattle raisers seated on peninsulas erected wattle fences across the neck to control herd drifting. On the mainland proper, however, pastures remained effectively unbounded, preparing Carolina cattlemen for their later westward expansion across the continent.[78]

The white overseers or owners and their herders—whites, blacks, or Indians—were variously called "cowpenkeepers," "cowkeepers," "pinders," "hands," "rangers," and, in the case of African slaves, perhaps even "cowboys." Eventually the term "cracker" would become attached to the whites engaged in pine-barren herding. Graziers quickly evolved into an independent, disrespectful lot, as in Jamaica, including not a few who could be described as "mean and inconsiderant." The cowpen folk formed "a very peculiar class," viewed by outsiders as nonindustrious, unchurched, and decadent. Perhaps the most typical Carolina cowhands were black slaves, although the frontier situation of many cowpens and the ready availability of horses provided good opportunities to run away. Even as late as the first decade of the 1700s, many South Carolina slaves were Indians, and much mixing occurred. After several generations of miscegenation, the typical cowboy may have had triracial ancestry, repeating the pattern of the Indies and Mexico. Their legacy lives on in such lowland Carolina triracial groups or "little races" as the "Brass Ankles," "Red Legs," "Red Bones," and "Marlboro Blues," "Lumbees" (Croatans), and "Buckheads." Even the overseers sometimes had mixed blood. Yet another cowherding caste had appeared, obeying a prototype at least as venerable as the Andalusian marsh vaqueros and once again belying the misleading WASP image of the American cowboy.[79]

In the earliest years of cowpenning in South Carolina, the herders attempted to establish the British and African-inspired practices of careful tending of the livestock. In 1680 the workers called the cattle to pen "every night that wee may have the milk and keep them from running wild." Similarly crawls included shelters for the pigs, and they assembled each evening at the sound of a horn, as in Jamaica, to be fed with corn and slop at dark as well as in the morning before release.[80] Onomatopoeic echoes of these horn blasts reverberated for centuries through the piney-woods South in verbal pig calls such as "woochie" and "sooey." In the long run, the intensive herding methods could not survive the rapid multiplication of the stock. As early as 1682, some cattle had become semiferal, and nightly penning for entire herds ceased. Some years later, fully feral, unbranded bulls became a problem by luring cows away from the cowpens, and edicts were passed to rid the range of such animals. One law in the first decade of the eighteenth century required one cowboy or cow hunter per hundred head, a precise restatement of the earlier Jamaican law. The less labor-extensive system that soon developed in South Carolina depended heavily upon calf capture and castration. Each spring the newborn calves were gathered during roundups and confined in special paddocks at the cowpen, thereby controlling the milch cows, who gathered faithfully each evening at the cowpen after grazing all day, in order to suckle their calves. Let into an enclosure by themselves, the cows were milked in the evening and again the next morning before being allowed to tend the calves. Daily milk gathering reflected both Fulani and highland British dietary preferences. After feeding the calves, cows were released again to graze. All through the summer and up to the first killing frost, the calves remained penned, but in winter, soon to be weaned, they were released to drift with the cows and steers into the marshes, canebrakes, and hardwood forests.[81]

Most of the herd remained at large all through the year, subject only to semiannual "cowhunts," also called "rallies." The first of these occurred in spring, when the cowboys gathered one small group after another and drove them to the large holding pens until as much of the herd as possible was assembled and counted. The new calves, together with any captured mavericks, were marked and castrated before beginning their summer of captivity in the calfpens. In such open range conditions, ownership marks were essential, and as early as 1694 South Carolina law required registration of brands and earmarks. Most brands consisted of simple block initials, such as JB and H, or Arabic numerals, an enduring British and Anglo-American type that was carried west. Branding remained more common for horses than cattle in the early years in South Carolina, a reaffirmation of the weaker British devotion to bovine branding. In the late fall, when the cattle had fattened on the grass of the savannas and pine barrens, a second, less-inclusive cowhunt took place, largely for the purpose of culling and marketing.[82]

To achieve these roundups, Carolina cowboys relied upon a number of herd-control techniques. They rode horses, as in Jamaica, and by 1712 the colony had "horses plenty," which "we ride out in search for [stock] in the woods." They did not use the lasso, but instead British bullwhips such as those still seen in Australia. In the branding pens, calf wrestling sufficed. While the early Carolina documents, like those of Jamaica, remain curiously silent on the subject of herder dogs, the cowpen and crawl folk almost certainly employed curs, possibly of Jamaican origin, to help gather cattle and hogs, a method that would remain a basic part of piney-woods open-range herding in the South. When black Texan Bill Pickett, of South Carolina ancestry, centuries later introduced the technique appropriately called "bulldogging" to competitive rodeo, he seized the targeted animal with not just his arms, but also with teeth tightly clamped on the cow's lip, faithfully duplicating the canine method. In addition, the nimble horses used in piney woods roundups were taught dog-like back-and-forth movements in controlling cattle, demonstrations of which are still given in cutting-horse competitions. One might say that the Carolina cow pony, a "quarter horse," was, in the behavioral sense, three-quarters dog. Salt, an ancient British herd-control device, was also employed by the Carolina cowpenkeepers. Strewn on the ground or placed atop a tree stump, salt kept much of the herd in the vicinity of the pens, and scattered wild cattle could be lured with salt toward the gathering paddocks.[83]

The early Carolina herding complex rested ultimately, as does all ranching, upon viable market demand. Cowpens and crawls supplied beef and pork, tallow, candles, raw hides, and tanned leather for a variety of markets. The most substantial trade was in preserved meat, although very little slaughtering and butchering took place at the cowpens—just enough to provide smoked or jerked meat for subsistence purposes. Cattle and hogs destined for market were often sold to drovers at the pens and crawls in autumn, then trailed overland to Charleston, the only noteworthy early meat-packing center.[84] At the slaughterhouses there, the stock were converted into salted, barrelled meat for winter export, perpetuating an ancient British preservation technique. Reinforcing the genetic link of Carolina to the Antilles, shipments of barreled meat found an almost exclusive market in the British West Indies, especially Barbados, Jamaica, the Leewards, and the Bahamas, although some went to New England. Carolina pork was already entering the West Indian trade by 1680, but cattle remained in short supply in the Charleston market as late as 1682, and only small quantities of beef reached the Antilles in that year. By the 1690s both pork and beef in abundance moved from Charleston to the Indies, a trade that continued through most of the eighteenth century. In return South Carolina received the salt needed for meat preservation, sugar, slaves, rum, molasses, and cotton.[85]

A local market for livestock and meat products also developed. As late as 1682, a substantial part of the annual cattle-herd increase was sold to newly arrived settlers as breeder stock. Beginning in the 1690s, with the rise of slave plantations devoted to rice and other crops, an increasing share of meat produce went to that market, perhaps mainly through the system of paired cowpens and plantations.[86] Tallow and candles, by-products of the meat-packing industry, also became major Carolina exports, finding a market principally in the West Indies. Unprocessed cow hides went both to the New England colonies and to the British mother country, and a modestly developed Carolina tanning industry sent some leather to the Indies and New England.[87]

While livestock raising in colonial South Carolina has been described as "a rough and speculative business," the cowpens and crawls, at least during the first half-century in the colony, offered a means of accumulating wealth with a minimal expenditure of labor or capital and attended by minimal risk. The profits, invested in land and slaves, assisted the rise of a Carolina plantation system in the 1690s, and the livestock industry survived throughout the colonial era as a major component of the local economy.[88] More than that, the Carolina cowpens would ultimately contribute substantially to the cattle frontier of the West.

French Louisiana, 1700–1760

Antillean herding influences also reached early Louisiana and adjacent Gulf Coast areas by way of the French from Saint-Domingue, establishing yet another continental ranching nucleus. Iberville transported at least 27 Spanish longhorns, as well as some wild swine, from the Saint-Domingue colony to Mobile, Biloxi, and Louisiana proper around 1700. The pigs multiplied very rapidly in the infant Gulf Coast French colonies, but cattle had increased to only 102 by 1708. In the eastern colonies, including the Alabama barrier island of Dauphine and Fort St. Louis on the mainland, some 300 cattle ranged by 1713, while in Louisiana an observer in 1710 reported that "horned cattle have multiplied considerably." In general, though, herd growth remained sufficiently slow in the French Gulf Coast colonies that additional stock were obtained in Spanish Cuba, Veracruz, Florida, and Texas, as well as from the French settlements in the Illinois country. These several introductions provided the basis of modest but enduring cattle- and hog-raising enterprises in Mobile and Biloxi, where some few individuals developed large herds by the 1720s. One French settler north of New Orleans owned 300 cattle by 1724, but around Natchez in Mississippi, French herding was destroyed by Indians later in the decade.[89]

The key area of early French cattle ranching developed in the west, in Louisiana proper around the old outpost of Attakapas, or St. Martinville, on the eastern edge of the great coastal prairies, beyond the Atchafalaya distributary of the Mississippi River. There a system of *vacheries*, meaning "cattle ranches" in Louisiana French, existed at least by the 1730s, when the local brand-registration book began to be compiled. The word *vacherie* may have been a borrowing, at least in meaning, of the Spanish *vaquería*, since in France it meant a "cowshed" or transhumant dairy herd. A census taken by the Spaniards in 1766, shortly after they took control of Louisiana, revealed over sixteen thousand cattle in the Attakapas district, including fifteen thousand on "five old *vaquerías*" alone, well over half of the total cattle listed for Spanish Louisiana at that time and about seventy-five times as numerous as the human inhabitants of the district.[90]

The Attakapas vacheries utilized the subtropical coastal prairies of southwestern Louisiana, a splendid natural setting for open-range cattle ranching. Herds grazed on the prairie from spring until the frosts of autumn, then retired into cane brakes and bottom forests for winter foraging, a seasonal rhythm similar to that in South Carolina. Roundups must have been necessary, and branding clearly was the custom, but in fact we know very little about this prototypical, pre-Cajun Louisiana French ranching system. The failure to use the Haitian-Spanish word *hatte* suggests that West Indian influence might not have been pervasive, and the early Louisiana ranchers did not possess Spanish equestrian skills, although horses became locally abundant by the 1730s. Still Creole ranching probably did not reflect much Old World French influence, and the herding system of early Louisiana likely reflected what had been learned in Saint-Domingue by the 1690s. It belonged to the syncretic Caribbean culture, as did the other early continental ranching enclaves. In that context we should not be surprised that all fifty of the inhabitants of the five Attakapas vacheries in 1766 were slaves. The immigrants from Haiti to Louisiana may also have brought the Spanish herding cur to help manage cattle, and a place named Catahoula is located near the old Attakapas post, possibly explaining the southern term "Catahoula hound" to describe the herder dog.[91]

Multiple diffusions beginning in the 1520s carried the West Indian cattle-herding complex, in its several creolized variant forms, to the mainland, gaining footholds along the North American coastal perimeter from the Carolina low country to tropical Mexico. Spaniards, Britons, French, Africans, and Native Americans all took part in the establishment of Antillean cattle herding in the diverse continental hearths. Collectively these mainland implantments would, in time, spread to form the cattle-ranching frontiers of the American West.

FIVE

CATTLE FRONTIERS IN NORTHERN MEXICO

THE VARIOUS EARLY NUCLEI OF OPEN-RANGE CATTLE herding in Mexico, both along the route of conquest in the interior highlands and in the Gulf coastal lowlands, proved enormously influential in the development of the North American ranching frontier. In Mexico, including most of the southwestern United States before 1836 and 1848, cattle ranchers provided in many districts a cutting edge of Spanish civilization in its surge northward, ultimately allowing substantial Iberian inputs into the herding culture of the American West.

El Bajío

As we have seen, the earliest colonial cattle-ranching centers in the Mesa Central of Mexico, particularly in the basins of Toluca, Apan, and Huamantla, proved ephemeral (figure 17). Displaced due to population pressures, range deterioration, and the dictates of land rent, and succeeded by a mesetan hacienda system, the cattle ranchers of early highland Mexico soon found similar, more remote tropical mountain basins to the northwest in which to continue their type of pastoralism. Collectively these interlaced highland valleys to the north formed El Bajío, a temperate, fertile, and reasonably well-watered plain littered with hilly volcanic masses. Lying somewhat lower in elevation than the herding hearth basins, at 1,525 to 2,000 m. (5,000 to 6,500 ft.) above sea level, El Bajío stretched from Querétaro State westward some 450 km. (280 mi.) to the region of Guadalajara, straddling the colonial border between New Spain and Nueva Galicia (figure 27).

In the short run, El Bajío, in many ways reminiscent of interior Spain, offered abundant advantages for cattle raising and became the great staging ground for the continued northward diffusion of ranching in interior and Pacific coastal Mexico. In spite of its location on

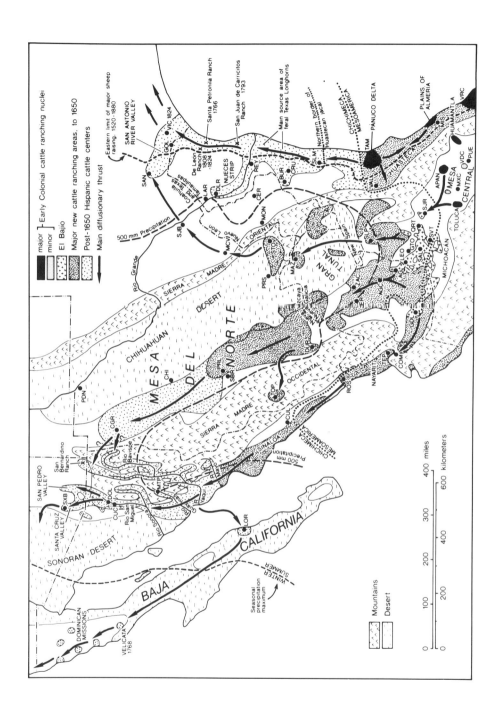

Figure 27. Diffusion of cattle ranching in Mexico, 1530–1830. Key to abbreviations: AGC = Aguascalientes, BUR = Burgos, CER = Cerralvo, CGR = Casas Grandes, CHI = Chihuahua, COM = Compostela, CRU = Cruillas, CUC = Cucurpe, CUL = Culiacán, DLR = Dolores (Texas), DOL = Dolores (Sonora), DUR = Durango, GDL = Guadalajara, GOL = Goliad, GTO = Guanajuato, JER = Jerez, LAG = Lagos, LAR = Laredo, LEO = León, LOR = Loreto, MAZ = Mazapil, MCV = Monclova, MIS = Misantla, MON = Monterrey, MTP = Mátape, MXC = Mexico City, PDN = El Paso del Norte, PRL = Parral, PRS = Parras, PUE = Puebla, QRT = Querétaro, REY = Reynosa, ROS = Rosario (Sinaloa), SAL = Salamanca, SAN = San Antonio, SBR = Santa Bárbara, SJB = San Juan Bautista, SJR = San Juan del Río, SLM = Soto la Marina, SLP = San Luis Potosí, STO = Saltillo, SVT = Salvatierra, SXB = San Xavier del Bac, TAM = Tampico, TEP = Tepic, TOP = Topia, VAL = Valles, VDC = Val de Cristo, VIC = Victoria, VRC = Veracruz, ZAC = Zacatecas. Sources: Chevalier, Land and Society *(see note 1)*, endpaper; Brand, "Early History" *(see note 3)*, p. 137; Doolittle, "Marismas" *(see note 19)*, p. 9; West, Mining Community *(see note 9)*, following p. 58; Lehmann, Forgotten Legions *(see note 21)*, pp. 44, 122; Robert C. West and John P. Augelli, Middle America: Its Lands and Peoples *(Englewood Cliffs, NJ, 2d ed., 1976), pp. 26, 269, 277, 318;* Trautmann, "Geographical Aspects" *(see note 11)*, pp. 243, 247; Dary, Cowboy Culture *(see note 3), p. 26;* Meigs, Dominican Mission *(see note 38), following p. 192;* Serrera, Guadalajara *(see note 8), precedes p. 89;* Dunbier, Sonoran Desert *(see note 27), p. ii;* West, "Flat-Roofed" *(see note 10), p. 123.*

the margins of the dry lands and classified as subhumid, the plain receives adequate rainfall, particularly in the southwest. No part of El Bajío has less than 500 mm. (20 in.) of precipitation on the average annually, and some districts, especially in the southwest, get in excess of 700 mm. (28 in.). Four-fifths of this rainfall comes in the period of June through September, repeating the awkward summer maximum that had troubled the Spaniards since their first entry into the Americas in the West Indies. El Bajío lured the graziers with lush prairies covered by a dense growth of largely perennial seasonal grasses lightly interspersed with oak, mesquite, acacia, and cacti, offering browse. Abundant salt pans further served the needs of the cattle. Deficient low-sun forage and surface water presented the main difficulties for stationary herds, but the cattle could generally survive along the stream courses in the dry season.[1]

Demographically El Bajío offered advantages to cattle raisers and to Spanish culture at large that had not been available in the earlier nuclei of ranching farther southeast. Even though climatically blessed, El Bajío lay, at the time of the Spanish entry, largely beyond the border of high Mesoamerican civilization, in the sparsely settled domain of hunter-gatherer Chichimecs, the "wild" Indians of the north. As a result the Iberians did not usually have to compete with sedentary populations of farming Indians for possession of the land in El Bajío, and once the warlike Chichimecs gave way, the region acquired a very pronounced Iberian cultural imprint, becoming the Spanish heartland of Mexico. Culturally challenged from the very first in the Aztec stronghold farther south, the Spaniards took firmer root in these good lands along the Chichimecan frontier of Mesoamerica. A major early component of the Castilian presence in El Bajío was a vigorous cattle-ranching industry.[2]

Indeed the cattle graziers fairly exploded through the cultural vacuum of El Bajío. In the 1530s ranchers occupied pastures in southern Querétaro State around San Juan del Río, north of the herding hearth basin of Toluca, and within a few years livestock also grazed the vicinity of the present city of Querétaro, to the northeast, a major cattle-raising center by 1550. San Juan remained the cattle portal of El Bajío, initially commanding the route of Spanish pastoral entry into the region and later serving as the official registration and brand-inspection point on the drove trail leading southeast to the markets. Simultaneously, in the 1540s, other ranchers spilled westward through the grasslands of El Bajío, into southern Guanajuato and northern Michoacán states, around the new towns of Salvatierra, Salamanca, Guanajuato, and León (figure 27). Crossing into Nueva Galicia, cattle graziers quickly became established at Lagos and Guadalajara, also in the 1540s. On the Chichimecan frontier of northern Michoacán and Nueva Galicia alone, some 450 ranches existed by the late 1550s. The Chichimecan pastures, from San Juan to Guadalajara, by midcentury surpassed the earlier Aztec basins in cattle population, and by 1580 perhaps a million longhorns grazed the pastures of El Bajío.[3]

The spread of ranching through El Bajío has often been attributed to the discovery of silver at places such as Guanajuato, where mining began in 1554. Ranching and mining frontiers did enjoy a symbiosis in much of interior Mexico, with cattle graziers providing meat for the laborers and tallow for mine candles. In return the mining entrepreneurs invested some of their wealth in cattle and land. Even so the explosive expansion of ranching was well under way in El Bajío before the mineral bonanzas, and cattle raisers occupied the grassy plains below Guanajuato prior to the discovery of silver. Even after mineral exploitation began in the region, ranchers often pressed well beyond the miners. Meat for the Aztec heartland and hides for distant Sevilla or Córdoba remained as important as the mining market, if not more so.[4]

The cattle-ranching system in El Bajío began as a preemption of the common pastures in the royal domain by private graziers, but after about 1550 usufruct land grants became the rule (figure 22). Richard Morrisey even claimed, incorrectly I feel, that "the concept of the *estancia* as a privately owned livestock ranch acquired by government grant," ending the era of common range, originated on the Chichimecan frontier in the San Juan area. In any case such holdings soon became common in El Bajío. Most ranches initially operated on a modest scale, producing a class of graziers that would be called "small ranchers" in the American West today. A few cattle estancias in El Bajío were even granted to acculturated Mesoamerican Indian nobles, some of whom accompanied the Spaniards northward into the region. In time the necessity to fortify estancias against attack by the Chichimecs and to maintain a sizable fighting force of vaqueros encouraged larger holdings, and many of the lesser ranches were destroyed by hostile native tribes. Low cattle prices, a recurring problem from the 1540s onward, also gave an advantage to the large operators. As a result a cattle-herder aristocracy arose in the last half of the 1500s in El Bajío, particularly in its western reaches, in Nueva Galicia.[5]

An increasingly independent, mixed-blood vaquero class came north with the Spaniards into El Bajío. Because of the need for protection from Indians, local ranchers ignored Mesta regulations and armed these vaqueros. The progeny of this unruly and frequently outlaw caste would, in the long run, help give northern Mexico its reputation as a center of banditry and revolution. Nothing like feudal order and labor docility prevailed on the cattle-ranching frontier from El Bajío northward.[6]

El Bajío also witnessed the tentative beginnings of Catholic church involvement in cattle ranching, a venture that would, later and farther north, lead to major participation by the missions in the pastoral industry. Church holdings began in the 1540s with crown grants of unclaimed maverick cattle. In 1550 a convent or monastery on the Chichimecan frontier received a cattle-estancia grant, and about that same time the church began investing some of its wealth in herds. Descendants of these livestock later provided the basis for herds tended by Chichimec neophytes. Missionization of the frontier tribes began in the 1590s, well to the north of El Bajío, when peace finally came to the northern peripheries of the Spanish domain. Even earlier, however, some Chichimecs had become preconditioned for mission ranching by stealing cattle from frontier estancias and, in the late 1500s, beginning on their own to herd them in more or less the Spanish manner.[7]

While El Bajío served as a staging ground, where the Mexican highland cattle frontier took final form and achieved notable vigor, the industry could not endure there, with local exceptions. Along the southern margins of El Bajío, the same conflicts with Indian farmers that had helped evict pastoralists from the basins farther southeast arose again, even as late as 1600. More important in the displacement of cattle ranching from El Bajío, however, was the great fertility of the region. Underlain by deep black soils, El Bajío early attracted agricultural settlers. Before the end of the sixteenth century, the process of converting cattle estancias to the more diversified, crop-based, mesetan-inspired haciendas was well under way. Range deterioration caused by a million head of grazing cattle on a subhumid range also contributed to the retreat of the ranchers. The dwindling of cattle herds and the simultaneous rapid rise of sheep raising presented the most obvious clue that range damage had occurred. By 1580 twice as many sheep as cattle grazed the pastures north of San Juan del Río, and by that time some two hundred thousand transhumant sheep from the southern Querétaro region went west each year to winter pastures near Guadalajara. Cattle ranching survived best in the more humid western part of El Bajío, in Nueva Galicia, especially around Guadalajara, where some powerful estancias held out and the cattle culture became institutionalized. Even there it is the charro culture, more mesetan in derivation, that persisted.[8] The Andalusian-inspired cattle ranchers would have to continue their retreat northward, following the frontier.

The Mesa del Norte

Northward from El Bajío, the Spaniards encountered the Mesa del Norte, an extension of the Mexican interior plateau, a complex of basins and ranges walled in on east and west by the two great Sierras Madre. The core of the Mesa del Norte consisted of drier lands, beyond the 500-mm. (20-in.) isohyet. Semiaridity already became evident in the Gran Tunal, not far north of San Luis Potosí, where grasslands gave way to an expanse of prickly pear studded with tall, woody deciduous shrubs, a visual warning to potential cattle raisers of moisture deficiency. Beyond the Gran Tunal stretched the even more forbidding Chihuahuan Desert, dominated by mesquite, sotol, creosote bush, and lechuguilla agave (figure 27). Reaching northward beyond the present international boundary, the Chihuahuan Desert includes huge areas of internal drainage.

Cattle, of course, browse as well as graze. Some desert plants offer excellent browsing, as for example the mesquite, which provides palatable seed pods rich in fat and protein. The desert fringe can, for this reason, be utilized by cattle raisers, as in West Africa, but the driest areas will normally be avoided and left to smaller livestock. Faced, then, with the barrier of the Chihuahuan Desert, the diffusion of cattle raisers on the Mesa del Norte bifurcated, avoiding the worst of the desert. The main ranching vanguard moved northward along the western side of the Mesa, clinging to a corridor of higher, moister steppe grasslands spread out in the foothills and high plains bordering the mountain wall of the Sierra Madre Occidental, along the axis of the 500-mm. (20-in.) precipitation line (figure 27). In colonial times, most streams there remained perennial, and small springs, or *ojos*, occurred widely, as did marshy areas, or *ciénegas*, where cattle could congregate. The adjacent mountains, easy of access from the east, offered abundant high meadows for seasonal use.[9]

Even with these advantages, pastoralists had to adapt to the drier conditions in the north. Ranches grew larger in size, sheep increasingly shared dominion with cattle, and environmental damage could more easily occur. Far to the north, in Chihuahua, after the initial years of grass "harvesting," 6 hectares (15 acres) of pasture would be required to support a single head of cattle. Perhaps the most visible aspect of the adaptation to drier conditions is evident in ranch architecture. Throughout the Mesa del Norte, the flat-roofed dwelling became the dominant style, including both the humble adobe huts of the vaqueros and the massive fortified stone dwellings of the ranch owners.[10]

In spite of these climatic readaptations, cattle raisers spread up the grassy western corridor of the Mesa del Norte with astounding rapidity, continuing the pace of expansion begun in El Bajío and achieving "one of the most biologically extravagant events" of the sixteenth century. Their expansion was stimulated in part by major mineral discoveries, as at Zacatecas, Durango, and Parral. Ranchers began occupying the region around Zacatecas, in far northern Nueva

Galicia, even before the discovery of silver in 1546. Crossing into Nueva Vizcaya, cattle raisers established themselves around the new town of Durango in the 1560s (figure 27). One rancher, Francisco de Ibarra, operating in the borderlands between Zacatecas and Durango, reputedly owned 130,000 cattle by 1578, and some men were said to own as many as 150,000 head. By the middle 1580s, Ibarra branded 42,000 calves annually, and one of his neighbors 33,000. Owners of fewer than 20,000 cattle qualified only as small operators. These huge numbers made Nueva Vizcaya the cattle center of the Mexican highlands by the turn of the seventeenth century, an achievement facilitated by the end of the Chichimec wars in 1590. But overgrazing soon damaged the pastures of the frontier, resulting in herd diminution around Durango as early as 1600.[11]

Beyond Durango, another climatic readaptation had to be made, for the migrating ranchers not only faced drier conditions but also far more pronounced thermal seasons. They departed the tropics to enter the middle latitudes, and winter became increasingly a season not just of drought, but of cold. North of Durango, through Nueva Vizcaya, the belt of foothills and elevated plains along the western margin of the Mesa del Norte reflected this transition to the middle latitudes, while still offering continued advantages to the graziers. Black grama grass, green in summer and curing to a palatable forage in winter, grew abundantly in better-drained areas, providing a key resource for the cattle industry. In the poorly drained lower parts of the basins and along the streams, matty *tobosa* grass, palatable only during a brief growing season, provided summer grazing, and mesquite thickets supplied browse.[12]

The cattle ranchers did not pause at the Tropic of Cancer, but perhaps dimly recalling the middle-latitudinal character of ancestral Iberia, took quick advantage of the northern pastures. By the 1590s they had pushed into the country around Parral, in present-day southern Chihuahua, where a major mining bonanza subsequently occurred. Another episode of Chichimec warfare between 1644 and 1652 blunted the expansion of ranching, but soon the cattle raisers pushed forward again, reaching the high, cool plains around Casas Grandes in far northwestern Chihuahua by the late 1660s. At that period, one rancher in the Parral region owned 42,000 cattle, as well as several thousand sheep. Many of these advances were lost in the century of bitter warfare with the Chichimecs that began about 1680, but by the late 1700s ranchers had reclaimed the remote wide plains of Casas Grandes. Meanwhile the more humid southern reaches of the western Mesa del Norte, as around Aguascalientes and Jerez, were lost to the crop-based hacienda system.[13]

The northward expansion of cattle raising along the eastern margin of the Mesa del Norte proceeded more slowly and produced less spectacular results, in large part because the Chihuahuan Desert nestles right up against the Sierra Madre Oriental, in its tradewind rainshadow, leaving no substantial grassland corridor. The ranchers moving northward from El Bajío on this eastern fork of diffusion quickly claimed the rich pastures around San Luis Potosí, but beyond that town water became scarce, mineral discoveries few, and competition from Querétaro-derived transhumant sheep more intense. The eastern margin of the Mesa del Norte would, in fact, belong preponderantly to the sheep raisers. To the north of San Luis Potosí, cattle ranching developed in the late 1500s mainly in small oasis clusters where dependable surface water and grass occurred, as around Mazapil and in the Parras Basin of southern Coahuila (figure 27).[14]

In their continued northward spread, the stock raisers of the eastern Mesa del Norte eventually, in the 1600s and 1700s, passed through several easy gaps in the Sierra Madre Oriental to emerge onto the inner Gulf Coastal Plain of Coahuila and Nuevo León. These transmontane plains remained sufficiently elevated and dry as to take on a plateau character, having more in common with the Mesa del Norte than the Gulf lowlands. Monterrey, situated beyond Saltillo at the eastern entrance to one of these gaps, became one major center for the implantation of plateau-derived herding with sheep emphasis, as did the similarly situated mining center of Monclova.

Relatively few parts of the Mesa del Norte retained a specialized cattle-ranching economy in colonial times. Most of the plateau evolved very early toward a mixed system, in which sheep and, to a lesser extent, goats played important roles. Transhumant sheep followed cattle into most parts of the Mesa del Norte, particularly the eastern reaches, following well-established cañadas. In some areas the migrant sheep even preceded cattle. Eventually the seasonal appearance of sheep in the north gave way to a year-round presence, as paths of transhumance became routes of migration. With sheep came a more labor-intensive, mixed-herding system that placed equal value upon pedestrian *pastor* and mounted vaquero. In addition, virtually every ranch in the interior plateau possessed some cropland and sought self-sufficiency in staple-food production. In all of these respects, Extremadura or even Castilla prevailed over Andalucía through most of the Mesa del Norte. The creation of a specialized cattle-ranching economy in much of the far north owes more to twentieth-century influences emanating from the United States than to the Andalusian-inspired colonial cattle frontier.[15]

Reflecting these diverse cultural influences, the land-tenure system on the Mesa del Norte pastoral frontier exhibited marked contrasts. Mission-based ranching became increasingly more important as the frontier moved northward, particularly after 1590, when the Chichimecs agreed to receive missionaries and become settled. Military posts to protect the missions led to the development of royal presidial cattle herds as well. Small-scale secular ranchers, also abundant after 1590, shared the early Mesa del Norte with the mission and presidio system. Aristocratic, estate-building ranchers possessing royal pasture grants became more typical behind the frontier, although some scholars date the rise of pastoral estates in Nueva Vizcaya to the earliest settlement phase in the 1560s. Estate development in the ranching districts intensified in the 1700s and continued well into the nineteenth century. In Aguascalientes and most other southern areas, the aristocratic era yielded crop-based haciendas rather than ranching estates, although the Jalisco-derived Nueva Galician charro tradition remained strong there. To the north, from Zacatecas to Chihuahua, and in Coahuila as well, one perceives today a far less overt aristocratic ranching culture, displaying only faint charro influences, even though the northern ranches are far larger in size than those of the south.[16]

The vaqueros of the Mesa del Norte, in common with their Antillean and early Mexican prototypes, had complex bloodlines. Perhaps representative of the early period was the population of the Aguascalientes region in 1610, where a traveler found over a hundred mulattos, twenty black slaves, fifty mestizos, and ten Native Americans under the control of about twenty-five Spaniards, largely on rural estates. Today one looks largely in vain for racial or cultural evidence of the substantial early African contribution to the vaquero population of the Mesa del Norte, but surely these bloodlines survive. As the frontier advanced northward, the mission system injected many acculturated Chichimecs into the ranch work force, bringing the image of Pancho Villa to mind. In the late 1840s a mounted vaquero encountered near Saltillo, "a rather light-colored Indian," displayed such characteristic items of material culture as a lasso, a garrocha (described as a long, iron-pointed "aspen goad"), a heavy, conical felt hat, a red sash at the waist, sandals, and large spurs.[17]

The Northern Gulf Coastal Plain

In the Gulf coastal plain north of Tampico and the Pánuco delta, cattle ranchers failed to duplicate the rapid northward expansion achieved in the Mesa del Norte. Indeed the delta, which also marked the border of New Spain and the Chichimec frontier in the coastal plain, remained the northernmost outpost of major ranching activity in the lowlands for nearly a quarter of a millennium. Oddly, a major early colonial ranching nucleus remained adjacent to the largely unoccupied, yet suitable pasturelands of Tamaulipas until the middle of the eighteenth century.

It is true that the coastal zone north of Tampico, classified climatically as subhumid, is drier than the lowlands to the south, in Veracruz State, but almost all parts of the outer northern plain receive in excess of 500 mm. (20 in.) of precipitation annually (figure 27). Savannas and even some midlatitude prairies lay along the Tamaulipan coast (figure 28). A seasonal drought prevails from November to May, but the Spaniards had already grown accustomed to that rainfall regime as early as the Antilles, and in any case, abundant cloudiness mitigates the harshness of the dry season in Tamaulipas. Numerous small bolsons, or structural sinks, preserve greenery into the heart of the drought, and while surface water is scarce, seeps called ojos occur fairly widely. Nor were abundant sea marshes lacking along the northern coast to attract the Andalusian eye. Inland from the marshes, in a narrow belt paralleling the coast, scattered grasslands beckoned the cattle raisers, as in the Loreto caliche plain and the huge, sandy Wild Horse Prairie, offering midlatitude bluestem bunchgrasses. In a few areas, the Tamaulipan coastal lands even contained oak-covered hills and moist palm forests that could support droves of swine, so important in the colonial herding system south of Tampico. Hostile Chichimecs, holed up in low volcanic outlier ridges paralleling the coast, such as the Sierra de Tamaulipas, troubled the region as late as 1800, but surely they presented a no more formidable human barrier than that encountered by ranchers on the Mesa del Norte. Perhaps the Veracruz cattle raisers hesitated to leave the watery abundance of the Pánuco delta, which provided such splendid opportunities for cattle raising. Possibly, as children of the constantly warm *tierra caliente*, they did not wish to encounter the cold blasts of the *nortes* in winter, for Tamaulipas belongs as much to the middle latitudes as to the tropics. Nor were they compelled to go north by displacement, for cattle ranching was no transi-

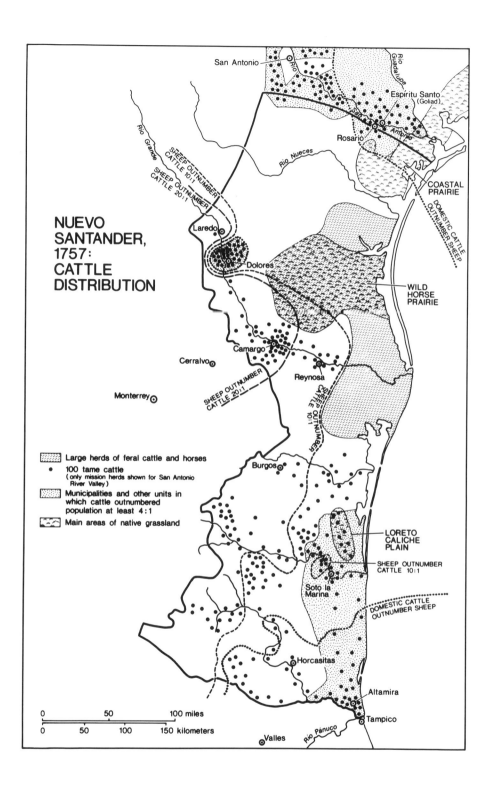

Figure 28. *Cattle in Nuevo Santander and its borderlands, 1757. The figures for Nuevo Santander proper are for 1757, and for the San Antonio River valley missions of Texas: San Juan 1756, Valero 1762, Concepción 1762, San José 1758, Espada 1762, Espíritu Santo 1758, Rosario 1758. Note that at this time cattle were still more numerous in the upper San Antonio Valley than in the lower part, a situation that changed in the 1760s. Note also the coast/interior contrasts in ratios within Nuevo Santander. Sources: census by José Tienda de Cuervo, as published in Prieto,* Historia *(see note 20), pp. 170, 174–75, 195; Jackson,* Mesteños *(see note 44), pp. 24, 36–39, 47; Ramsdell, "Spanish Goliad" (see note 46), p. 38; Castañeda,* Our Catholic Heritage *(see note 44), vol. 4, pp. 6–15, 23; Gerhard,* North Frontier *(see note 11), p. 359; Johnston, "Past and Present" (see note 18), pp. 456–65.*

tory frontier occupance form in the Veracruz littoral, but instead an enduring system. Contributing to the failure to achieve early northward expansion along the Gulf Coast was the absence of precious mineral discoveries that might have generated a demand for meat and tallow, as on the Mesa del Norte.[18]

Perhaps, too, the northern plain was not entirely ignored. Peter Gerhard speaks of "a handful of intrepid shepherds" entering the north as early as the middle 1600s. Small-scale ranchers of mixed blood, operating extralegally, may have utilized southern Tamaulipan pastures from an early time. Lacking the protection of military posts and, due to their small scale of operation, less able to defend themselves against the Chichimecs, they no doubt suffered setbacks, but their ranchos likely appeared in the region from the late 1500s onward. Nor should we overlook the organized colonization effort by Luis de Carvajal, a rancher from the Pánuco delta who in 1584 founded Cerralvo, situated in the inner coastal plain a scant 55 km. (35 mi.) south of the Río Grande (figure 27). His colonists introduced Pánuco cattle and presumably began ranching in the lowland manner. Though Cerralvo failed within twelve years, some of Carvajal's colonists remained in the region.[19]

Nevertheless, successful colonization did not occur until much later, beginning in the 1740s, under the leadership of José de Escandón, founder of Nuevo Santander, the precursor of modern Tamaulipas. Escandón, with apparent ease, founded numerous enduring towns and several large haciendas throughout the northern coastal plain, such as Burgos, Soto la Marina, and Reynosa (figure 27). Many of his colonists eventually received pastoral grants, and livestock raising thrived almost from the first (figure 28).[20]

While Escandón drew heavily upon the Mesa del Norte and even El Bajío for his settlers, he also recruited people from the Huasteca, including the Valles and Tampico areas. For example Spaniards from Valles and Huastecan mestizos and mulattos founded Horcasitas (modern Magascatzín). Some of these lowlanders may have been herders already familiar with Tamaulipas through extralegal seasonal use. The cultural stamps of both of these source regions— littoral and interior—were imprinted upon colonial Tamaulipas. The moister lands near to the coast received the strongest impress of the old Gulf coastal cattle culture, of the Antilles and the Andalusian marshes, while the drier, more elevated interior part of the plains, bordering the Sierra Madre Oriental and including nearly all of the previously settled Nuevo León as well, was shaped mainly by colonists from the plateau and inner coastal plain. The immigrants from interior districts can best be detected by their attention to sheep, which had little place in the culture of the Gulf littoral (figure 27). Accordingly an inspection in the middle 1750s revealed that Escandón's interior settlements normally had ten times or more as many sheep and goats as cattle. The colonies upstream on the Río Grande, from Reynosa to Laredo, had over twenty times as many and accounted for fully 62 percent of all the flocks in Tamaulipas, while possessing only a third of the province's cattle, mainly at one hacienda, Dolores (figure 28). This preponderance of sheep in the interior settlements occurred even in the earliest years of colonization, when undegraded pastures and a labor shortage would seem to have offered an advantage to cattle raising. Only two years after its founding, Laredo reported over 9,000 sheep and goats, but only 101 cattle. Even Dolores, the cattle hacienda, had acquired sizable sheep and goat herds by 1767.[21]

By contrast, coastal cultural influences prevailed in the belt along the Gulf in colonial Tamaulipas. At the 1757 inspection, the coastal municipality of Altamira, just north of Tampico, had five times as many cattle as sheep and goats (figure 28). In addition the mission system played a minor role along the littoral in Tamaulipas, as earlier in Veracruz, a fundamental difference from the Mesa del Norte. Similarly royal presidial herds were far less common than in the interior of Mexico, reflecting the minimal protection provided in the coastal districts. In short the littoral cattle frontier remained overwhelmingly secular.

Near the coast in Tamaulipas, crops received little attention, even as late as the 1930s, and ranch hands often lived in Huastecan straw huts, or *jacales*, with hipped thatched roofs, just as in the sixteenth century on the Plain of Almería, far to the south (figures 20, 27). Today this type of jacal still appears as far north as Soto la Marina, where it gives way to a gabled type. As in Veracruz, horse raising and mule breeding shared importance with cattle raising. Tamaulipan cattle, raised mainly for hides on stationary pastures, exhibited the characteristic coastal semiferality, due in large part to the Andalusian-inspired neglect of calf castration. To the extent that steers existed in the Mexican herds, they resulted from the castration of mature bulls, and consequently not all that many were actually neutered. Even as late as the 1880s, some roads near Soto la Marina became impassable at times, "because bulls attack anything which comes in their way." Such an account brings to mind early twentieth-century descriptions of wild bulls in Las Marismas of the Guadalquivir. This Andalusian impression is heightened by the occurrence in coastal Tamaulipas of the generic toponym *mata* to describe a grove or thicket. Examples include Mata la Monteada near Tampico and Mata de Sandía near the Río Grande in far northeastern Tamaulipas. At roundups cattle were driven through funnel-shaped *mangas*, flanked by crude fences made of brush and branches, which led to corrals of similar construction.[22]

The littoral of Tamaulipas, beginning in Escandón's time, long remained dominated by small-scale, mixed-blood stock raisers devoted to cattle and horses, although some pastoral estates did exist. By 1795, according to a government inspector, Tamaulipas had 437 ranchos—probably mainly small cattle/horse enterprises—but only 17 haciendas, as the local pastoral estates were called. Even the large, legally sanctioned estates lacked fixed boundaries and often shifted location, in spite of the fact that their sites had originally been specified. The disdainful inspector attributed such disorder to the derivation of the inhabitants, who were reputedly "formed from among the vagabonds and malefactors" of other provinces. Much concerning racial ancestry can be read between these lines. By the early twentieth century, pastoral estates had overwhelmed the ranchos to dominate the outer coastal plain.[23]

By 1795 nearly 112,000 domestic cattle reportedly grazed the ranges of Tamaulipas, up from 25,000 in 1757, together with an astounding 130,000 horses. The northern coastal-plain ranching industry had become sufficiently large by then to inflict major environmental damage, particularly given the additional impact of large herds of fully feral cattle and horses in the region. Huisache had become more widely dispersed, its seed spread by the browsing animals, and the coastal grasslands had become strewn with *bosques* of "spiny and harmful bushes." By the 1930s floral modification had reached profound levels, and over 15 hectares (40 acres) were required for each cow. Range burning is still practiced in the coastal parts of Tamaulipas, following earlier Veracruz custom.[24]

From the huge cattle herds of the late eighteenth century Tamaulipan plains, supposedly only about two thousand head reached market annually at Tampico, although others may have been driven to towns in the interior plateau. In part the explanation lies in the local tendency to market only hides. Perhaps, too, some cattle from Nuevo Santander were being driven overland to New Orleans. Certainly that held true for some ranchers in northern Tamaulipas, now part of Texas.[25]

The Gulf lowland cattle frontier, then, advanced very late through coastal Nuevo Santander, ceding the interior parts of the plain to highlanders. Belated though it came, this northward diffusion along a natural, long-neglected pathway would have a profound impact upon the still more remote coastal lands beyond.

Pacific Coastal Mexico

If the advance of pastoralists along the Gulf Coast proceeded slowly, the opposite was true on the Pacific shore, even though access proved difficult and the coastal lowland there did not appear, superficially at least, too inviting. Pacific Mexico, walled off from the interior plateau by a true barrier range, the Sierra Madre Occidental, might well have remained isolated, had it not been for a single relatively easy descent—the legendary "road to Cíbola"—leading from Guadalajara in El Bajío down through Compostela and Tepic to the lowlands of Nayarit (figure 27). To the north, only the difficult trail through Topia and routes skirting the northern end of the Sierra proper, especially by way of Púlpito Pass, into Sonora, allowed access to the Pacific, with the result that the coastal zone developed largely in isolation.[26]

Once descended to the lowlands, the Spaniards found a less than paradisical region. An oppressively hot, rainy summer awaited the conquerors in the tropical portions of the coast, yet another land where the precipitation regime was precisely the opposite of the accustomed Iberian annual rhythm. The low-sun season, while mercifully mild, provided very little rain, producing a dry tropical deciduous thorn forest, called a *montaña*, covering most of the inner lowland and foothills. Acacias dominated the thorn forest, with abundant shrub undergrowth but a scarcity of grasses. In some places, particularly on floodplains, the forest thinned out to present a more open appearance. At least the thorny, leguminous trees, together with the hot, wet summers, discouraged sheep raising, allowing the Pacific to become, in the main, a cattle coast. Indeed all along the Pacific, from Sonora to distant Guanacaste province in Costa Rica, the dry forest eventually came to support an enduring cattle-ranching culture. The cattle fared well in many areas along the Mexican coast, particularly after annual pasture firing diminished the thorn thickets and encouraged grasses. Burning also served to control the cattle-induced guava thickets that, as in the Indies, resulted from livestock spreading the seed. But many thickets, both natural and induced, survived in the coastal plain, providing hiding places for cattle and allowing many to go feral. By the 1630s pastoralists complained that the great majority of the stock, fully wild, had "scattered through the many thickets" and could scarcely be gathered. In response, Pacific coast vaqueros developed *armas*, a slab of cowhide hung over the pommel to shield the horse and rider from brush and thorns. More appealing to the ranchers was a chain of Andalusian-like coastal salt marshes, some rather sizable, as well as palm forests. The Marisma la Tigra lies at the very foot of the descent from the interior plateau. Place names such as Cañada Grande in these marshes perhaps derive from a local seasonal shift of herds to and from the wetlands, in the Andalusian manner. To the north along the coast, about where the tropics gave way to the middle latitudes, the great Sonoran Desert blocked the way, offering little to cattle raisers and prompting them to begin vertical transhumance into the oak savannas of the Sierra Madre Occidental in the summer season, a practice that persisted into the present century.[27]

All things considered, Spanish occupation of the tropical Pacific coast proceeded remarkably quickly. Towns and military posts were founded very early in the Mesoamerican lowland cultural zone, including Compostela and even distant Culiacán, in Sinaloa, both settled in 1530–31, a mere decade after the conquest of the Aztecs. Due to a rapid and almost complete demographic collapse of the native population in the Mesoamerican sector of the Pacific coastal plain, duplicating the Indian holocaust experienced in most other Latin American lowlands, the way lay open for cattle. Descending by way of Compostela, the ranchers turned north along the coast into Sinaloa. By about 1600 cattle ranchers had reached the plains below Culiacán, and some individual Sinaloan herds numbered as large as 33,000 and 42,000 head.[28]

We know rather little about this early Pacific Coast ranching frontier in Mexico. Initially, at least as far north as the Mesoamerican border, it seems to have remained secular, apparently largely dependent upon labor imported from El Bajío. It proved durable, for a ranching economy long dominated the Pacific lowlands. As late as the 1760s, the lowland district of Tepic ranked first as a cattle exporter in the extended Guadalajara administrative region, followed by El Rosario, a marsh-side coastal municipality farther north. These two districts together accounted for 84 percent of the cattle export of the Guadalajara region at that time, eclipsing the old highland *charro* center. Many large cattle estates survive in the Pacific tropical lowlands today.[29]

The entrepreneurial pioneer ranchers soon crossed the old Mesoamerican cultural frontier near Culiacán (figure 27). Some reputedly even reached the plains south of the Río Yaqui in southern Sonora by 1610, at the margins of the great coastal desert. However, they soon had to yield dominance to the Jesuit fathers. The northern tribes survived in far greater numbers than the Mesoamericans to the south, avoiding a demographic collapse and prompting the Jesuits to begin mission work. From the 1590s on, the ranching frontier of the Pacific north was dominated by the mission system, all the way into California two centuries later. One official report, dated 1637, said the Jesuits owned one hundred thousand cattle on their coastal-plain mission ranches in Sinaloa and up to the Río Mayo in southern Sonora, although the Jesuits quickly and with justification denied such affluence, saying they possessed fewer than eight thousand tame head, confined to a single ranch, from which they supplied some ninety thousand neophytes with jerked beef and animals to be slaughtered on feast days. The Jesuits relied upon meat derived from their cattle herds to help attract Indians to the missions and keep them there. They soon realized that meat distribution could more efficiently be achieved when herds were dispersed to the various missions, instead of concentrated at the central ranch in Sinaloa. To that end small seed herds of thirty or more cattle were sent out about 1640, and soon thereafter the neophytes became accustomed to a weekly ration of beef. In this manner the Jesuit cattle ranching of the Pacific Coast frontier originated.[30]

In Sonora the Jesuit-led ranching frontier, still following the old explorers' road to Cíbola, turned inland after 1620, seeking the sedentary population clusters among the Pimas and Opatas in scattered mountain valleys, simultaneously avoiding the inhospitable coastal desert. Pacific cattle raising ceased, at that point, to be a coastal-plain ranching system and would never be one again, even in Alta California. In Sonora the Sierra Madre Occidental breaks down into a series of parallel low ranges, oriented on a north-south axis, presenting a corrugated terrain. The rivers follow the troughs and occasionally cut through ridges, producing a trellis pattern of major streams, along which valleys of widely varying size occur at intervals. Grass proved to be abundant, particularly in the upper reaches of the elongated valleys, and cattle thrived as they never had along the coast. Jesuits, beginning in the 1620s, ascended through the riverine trellis, founding missions initially among the Pima Indians of the Pimería Baja country of central Sonora, in the valleys of the Río Yaqui and its tributaries, the Tecoripa and Chico, as well as along the Mátape, San Miguel, and Sonora rivers (figure 27). Mission Mátape, well-situated in a broad, undulating, grass-rich valley, reportedly possessed forty or fifty thousand cattle by the early 1730s, and every other mission in Pimería Baja had herds of at least moderate size, varying with local range conditions. Continuing their rapid ascent, the Jesuits entered the lands of the Opatas and other related tribes in northeastern Sonora, planting missions and establishing herds in the valleys of the upper Río Sonora; along the upper Yaqui, fed by streams such as the Moctezuma and Bavispe; and beside the upper San Miguel, as at Cucurpe, where a mission was founded in 1647. Some one hundred thousand cattle reportedly ranged near the headwater of the Bavispe in the Opatería by 1694. Clearly the impetus for these riverine diffusionary thrusts came largely from Pacific Mexico (figure 27).[31]

In the 1680s the remarkable Jesuit father Eusebio Francisco Kino began his work still farther north, in Pimería Alta—the northern part of Sonora, including present-day southern Arizona. Headquartered at his first mission colony, Dolores on the upper San Miguel River, Kino played a dominant role in the continued diffusion of mission-based cattle ranching on the Pacific frontier. Working among the resident Pima Indians, Kino by 1701 had established five missions, including Dolores, with a collective cattle herd of 4,200. His activity extended into the Santa Cruz Valley of Arizona, the adjacent San Pedro Valley, and the Altar Valley to the west (figure 27). San Xavier del Bac, where the first cattle arrived in 1697, and Tumacácori missions in Arizona represent his work. Training Pima vaqueros, he left a ranching legacy at nearly twenty different settlements in modern Sonora and Arizona before his death in 1711.[32]

In the following half-century, the Sonoran Jesuit ranching industry largely held its own, aided by a market among local miners, although the period of rapid expansion had ended with Kino's death. Cattle raising took on an increasingly secular character, based in individually owned herds. Private estancias appeared as early as 1649 in the Bacanuchi Valley near the headwaters of the Río Sonora and in 1658 on the Río Yaqui. From that time on, the Sonoran highland frontier pastoral industry began gradually to be secularized. By the 1730s some acculturated Pimas had become small ranchers, and soon even "the most humble Indian" possessed a brand. Some Spaniards owned private herds numbering from five to six thousand head by the 1760s in Sonora. These ranchers may have acquired their cattle from the Mesa del Norte even in the 1600s, since Púlpito and nearby Carretas passes provided reasonably easy access to the Casas Grandes area of Chihuahua. Using the same route, some Sonoran ranchers drove cattle to the mining markets of Parral beginning in the 1670s.[33]

The Sonoran ranching system, both ecclesiastical and secular, became rather diverse, although cattle continued to play the major role. Sheep, which had been rare in the hot coastal thorn forests, acquired at least local importance in Sonora, although in the 1760s they had reportedly become rare, both because they required too much labor and because their fleeces fared poorly among the dense thorn thickets of the desert margins. Sheep haciendas survived in the Santa Cruz Valley of Arizona as late as the 1830s. Crops, raised under irrigation, also received attention, since Sonora had to be largely self-sufficient in food production.[34]

Several descriptions of Sonoran herding techniques in the middle 1700s survive, and they depict a largely Andalusian system.[35] These accounts also reveal an internal diversity of practices, perhaps differing even from one ranch or mission to the next. Some describe two roundups each year, including one in spring for branding and another in autumn for slaughter. Others mention a large fall *rodeo* for branding, with smaller bimonthly gatherings to keep the stock reasonably tame and localized. In some cases semiferal herds were lured into *palisado*, or picket, corrals by tame livestock, then deprived of feed and water until they became sufficiently weak and dispirited that the vaqueros could go among them and remove the calves for branding. Another technique was to erect a post at the place of rodeo, covered with tallow and salt to serve as a lick around which the cattle habitually gathered. During roundups the stock, if called by whistles or frightened by noise makers, would flee to the salting posts, facilitating the cow hunt and making corrals unnecessary. The Indian and mestizo cowboys of Sonora employed both lasso and garrocha in their work with tame cattle. For largely feral animals, they still used the terrible hocksing pole with its sharp steel luna, fully two centuries after the feeble Mexican mesta first sought its abolishment. Rather little had changed since early Española in the Antilles.

By the middle 1700s, the Sonoran cattle industry faced difficulties and decline. Disturbances and revolts by the Pimas in 1737 and again in 1751 proved disruptive and destructive of cattle. Even more serious, a lengthy episode of Apache raiding lasted from about midcentury until 1790. Some 330 Sonoran ranches stood deserted as early as 1763, as the Apaches, seeking leather for their shields and meat to eat, took a heavy toll among the herds. Depredations by wolves and mountain lions added to the losses. Then, in 1767, a royal edict banished the Jesuits, a blow from which the Sonoran missions never fully recovered. Nevertheless, the Franciscans and Dominicans, who replaced the Jesuits, drew upon Sonoran herds to help supply their new missions in California in the 1770s.[36]

Sonoran-Arizonan cattle ranching rebounded impressively after 1790, when a temporary Apache peace was achieved; but even though certain missions thrived again, the revived herding industry lay largely in private hands. Spanish officials issued a number of large grazing grants, and even more were awarded by the government of independent Mexico. Known both as estancias, a name that still today appears repeatedly on maps of Sonora, or haciendas, some of these estates possessed enormous herds in the early 1800s. San Bernardino Ranch, situated in far southeastern Arizona and adjacent Sonora, repeatedly had a herd of one hundred thousand cattle at its peak, while another hacienda, located on a left-bank tributary of the San Pedro in southern Arizona, possessed forty thousand head. In the main this expansion of secular ranching achieved merely a reoccupation of the Jesuits' earlier herding valleys, such as the Santa Cruz and San Pedro, but some new areas were also settled, as at San Bernardino (figure 27). Hispanic cattle ranching had attained its secular golden age in the Sonora-Arizona borderlands by the 1820s, and requests for new pastoral land grants continued vigorously. Then an old menace reappeared, when the delicate peace the Spaniards had achieved with the Apaches collapsed. The transfer of power to independent Mexico in 1821 signalled a renewal of Apache raids. While the 1820s remained more peaceful than generally depicted, the following decades proved catastrophic for ranching in Arizona and also caused major retreat in Sonora. The great San Bernardino Ranch was abandoned in the 1830s, and by the time southern Arizona came under United States rule, with the Gadsden Purchase in 1852, only Tucson and Tubac retained any population at all.[37]

From Sonora, during its Jesuit prime in the 1690s, missionaries began colonizing the southern parts of the sere, largely sterile peninsula of Baja California, across the Gulf. The beginnings were at Loreto, to which hundreds of Sonoran cattle went by sea in 1700, finding adequate grazing and forage on the small plain adjacent to the settlement. Eventually the Jesuits built a chain of missions, maintaining small subsistence herds, flocks, fields, and gardens. From the standpoint of cattle ranching, the legacy of the Baja California missionization following the Jesuit expulsion in the 1760s remained small, indeed. The chain's herds, though limited by range resources and ravaged by predators, supplied enough cattle to stock a new mission, San Fernando de Velicatá, founded by the Franciscans in 1768, from which, in turn, livestock went almost at once to help begin the herds of Alta California. A 1772 census revealed that the mission San Francisco Borja, with 500 head, possessed the largest cattle herd. Borja is positioned about the middle of the enormous peninsula, at the foot of the mountains. Dominican fathers, who gained royal permission to establish missions in the previously neglected, far more humid and grass-rich northwestern part of Baja California, eventually developed cattle herds of noteworthy size in a few places by 1800, and a few private ranches were founded by discharged presidio soldiers. Descendants of the peninsula's colonial longhorn cattle can still be seen today. In no sense, however, did Baja California function as a major cattle frontier.[38]

The northward diffusion of ranching in Mexico, then, proceeded along three major pathways, each of which would lead into the region that today forms the southwestern United States. Texas, New Mexico, Arizona, and California would all receive these impulses, with profound results for the later Anglo-American ranching frontier of the West. Each of the prongs of Spanish ranching had acquired a special regional character by the time the present border was crossed, and we should not think in terms of a monolithic Spanish colonial pastoral system entering the American Southwest. These regional types would help differentiate the American ranching frontier in noteworthy ways.

The Gulf Coast ranching complex, spreading north through Tamaulipas into coastal Texas, emphasized cattle and horse raising; produced principally hides, saddle horses, and mules; remained highly labor-extensive, allowing even the domestic cattle herds to become semiferal; and allowed a major place for the small, landless, extralegal herder. Many or even most northern coastal ranchers were of mixed blood, likely descended from the early colonial vaquero class of the Veracruz lowlands. The mission system played virtually no role in this largely secular Gulf coastal ranching. The Tamaulipan ranching culture also bore the strongest Antillean and Andalusian imprint, representing an undiluted littoral herding system.

The Mesa del Norte pastoralists, with roots in El Bajío and the early colonial highland basins, represented a subculture quite distinct from that of Tamaulipas. Their system was more labor intensive, allowed a much greater role for sheep in the livestock holdings, relied substantially upon transhumance, enjoyed governmental sanction and protection, depended upon the missions and presidios to break the path north and supply acculturated vaqueros, and evolved fairly quickly into dominance by an estate-based aristocracy. In the final analysis, the Mesa del Norte ranchers became more Castilian than Andalusian, more a people of the high plateau than of the coast.

The Mexican Pacific ranching complex is not so easily characterized and distinguished, for even though coastal, it had roots in Guadalajara and El Bajío at large, bearing as a result the mark of both highlands and lowlands. These diverse influences became even more significant when, in the northward diffusion, the Pacific herders diverted away from the coast to enter interior Sonora and achieved contact with the Mesa del Norte. From that point it became a system based neither in coastal lowland nor high plateau, but rather in grassy hills and valleys. Perhaps the most outstanding feature of Pacific ranching was the overwhelming initial dominance of the mission system in its expansion beyond Culiacán, a dominance that would pass northward from Sonora.

New Mexico

Not surprisingly, given the rapid advance of the ranching frontier in the high grasslands along the western side of the Mesa del Norte, the first Spanish settlement of the American Southwest occurred in New Mexico. Departing from the southern Chihuahua mining town of Santa Bárbara in 1598, Juan de Oñate, son of one of the founders of Zacatecas, led a party of colonists, mainly from the Mexican interior plateau, northward. They passed present El Paso, and ascending through the volcanic badlands that border the Río Grande, reached the highlands of northern New Mexico. Oñate reputedly started his expedition with 7,000 cattle, although the correct figure was probably 700, particularly since he had promised in his application for undertaking the project to bring only 1,100 head, in addition to 5,000 sheep and goats. By 1610, as a result of Oñate's efforts, Santa Fe, destined to be the Spanish cultural capital of New Mexico, had been founded, less than a century after Cortés' conquest.[39]

While the 250-odd Spaniards residing at Santa Fe a generation later, in 1630, possessed "ample cattle," Hispanic New Mexico in fact never became a center of cattle ranching. Perhaps the single greatest retarding factor was the presence of a substantial established population of Pueblo Indian irrigation farmers. Protecting the fields and crops of these neophytes, the mission fathers apparently blocked the development of a large-scale cattle industry. The same competition with Indian farmers that had banished the cattle raisers from the highland basins of the early colonial basins of the Mesa Central perhaps prevented their exploitation of New Mexico. Sheep, easier to control, would rule the fine pastures of the new colony, and Oñate had laid the basis by introducing breeder flocks. The pedestrian pastor and guard dog would take the place of the mounted vaquero in New Mexico.[40]

Initially at least, cattle raisers fared better to the south, around El Paso (present-day Ciudad Juárez), founded about 1670 on the Río Grande. Nine thousand cattle ranged along the river there by 1680, the year of the great Pueblo Indian revolt that led to the temporary abandonment of all Spanish settlements above El Paso. Groups of loyal or enslaved Pueblos who retreated southward with the Spaniards began almost at once tilling the fertile valley just below El Paso, forever ending the brief hegemony of cattle ranchers there. Even the arrival of two thousand cattle from Casas Grandes in Chihuahua, sent to relieve the refugees, failed to revive the local ranching industry.[41]

After the Spanish reconquest and resettlement of New Mexico, which began in the 1690s, the Pueblo agricultural presence apparently continued to retard cattle raising. Perhaps, too, the remote ancestry of the Spanish settlers played some role in the ascendant position of sheep in the livestock economy. A map of Hispano Iberian origins, reflecting the recolonization of the 1590s, hangs in the museum on the town square in Taos, revealing many mesetan Castilian family roots in the Spanish sheep provinces but surprisingly few Andalusian and Extremaduran sources. In any case, sheep reassumed dominance on the ranges of the colony. Perhaps representative of the renewed Spanish herding system in eighteenth-century New Mexico were the holdings of a certain Diego Padilla, an estate owner south of Albuquerque who, about 1740, owned 1,700 sheep but only 141 cattle. By 1757 all of the Spaniards in the province together owned fewer than 8,000 cattle and less than 2,500 horses. Three-quarters of a century later, in 1832, not long before the American takeover of the province, New Mexico reportedly had 240,000 sheep and goats but only 5,000 cattle and 850 horses. Given this focus, it is not surprising that ovine transhumance, a basic aspect of sheep raising in the Mesa del Norte, became part of the New Mexico system. Old maps of Santa Fe reveal a Camino de la Cañada leading northward, and one town on this route bore the name Santa Cruz de la Cañada (in the present-day Española Valley).[42]

By the time the great century of settlement expansion by New Mexico highland Hispanos began, about 1790, sheep had become the economic hallmark of the regional Euroamerican culture and had even acquired a mystical, quasi-religious status among some neighboring acculturated Indian groups, in particular the Navajos and Utes. The Hispano expansion was accomplished by sheep pastores who founded scores of small *plazas,* or herder villages. As a result, New Mexico, which at the close of the Spanish colonial period had a larger population than Texas, California, and Arizona combined, never became a cattle-ranching frontier.[43]

Hispanic Texas

In contrast to New Mexico, Texas under Spanish and Mexican rule acquired a viable, if modest, frontier cattle industry. Since so much has been made of the supposedly crucial role of Hispanic Texas in the later evolution of western Anglo-American ranching, pastoral developments in this remote and otherwise unimportant hinterland must be carefully considered.

Aside from several inconsequential and mostly abortive early *entradas,* the Spanish ranching presence in Texas began in the San Antonio River valley in the middle eighteenth century. In and near the present city of San Antonio, a presidio, civilian town, and chain of five major Franciscan missions were in place by 1735. Downstream, not far from the coast, several more missions and a presidio appeared, concentrated in the area around present Goliad (figures 27, 28). Between these two nuclei, a riverine cluster of Spanish colonial cattle ranching eventually developed, based in the rich local tallgrass prairies and oak savannas.

The way, as in Sonora, was led by the mission system, which in Texas also emphasized cattle, perhaps because no indigenous Indian agriculture stood to be harmed and because wolfpacks took a heavy toll among the sheep and goats. Following the first noteworthy livestock introductions in 1721, small herds became attached to each Franciscan mission, although development proceeded very slowly in the San Antonio area. Initially, small numbers of cattle were kept near the individual missions, but the threat they posed to cropland soon caused displacement of the herds to more remote pastures, where the cattle became vulnerable to depredations by "wild" Indians. As a result, cattle multiplied slowly. By 1745 the five largely self-sufficient missions of the San Antonio area possessed some 5,100 cattle, but a generation later these had increased only to 5,500 head, somewhat less than half the size of the sheep and goat herds. Greater cattle-raising success was obtained downstream, in pastures along the lower San Antonio River and its sister stream, the Guadalupe (figure 28). Mission Espíritu Santo, founded on the Guadalupe in 1726, had plentiful cattle herds within a decade. Moved to the present site of Goliad on the San Antonio River in 1749, Espíritu Santo and its neighbor mission, Rosario, surpassed the upper valley in herd size by about 1760 and owned 20,000 cattle by the middle of that decade, roughly four times as many as the five San Antonio missions. At Espíritu Santo, by far the largest and most important ranching mission in Texas, the padres undertook few activities other than cattle herding, planting very little in the way of field crops and developing no irrigation system. They used the profit derived from the cattle business to purchase food supplies. Rosario operated in a like manner. These herding successes did not long endure, for the heyday of mission ranching throughout the San Antonio River valley ended in the 1780s. Defections by neophyte herders, Indian raids, the drawing away of cattle to supply other areas, government taxation, and growing anticlericalism all contributed to the decline. Rosario, which had possessed 5,000 cattle at its peak, retained only 600 by 1803.[44]

As the missions declined, the cattle-ranching industry of the valley shifted steadily into private hands. Competition for grazing land between the civilian town of San Fernando and the San Antonio missions had begun in the 1730s, when the common pastures of the citizens proved too small. By the 1740s presidio soldiers at both San Antonio and Goliad began acquiring ranching grants from the local governor; as the decades passed, these ranches came to dominate most of the valley, spilling over along several adjacent tributaries. The number of private ranches in the valley rose from twenty-three about 1780 to forty-three in 1795, at which time some sixty-nine heads of families worked on these establishments. Two estancias, in addition to much more numerous ranchos, were mentioned in the census of 1792. A partial enumeration in 1810, listing fourteen ranches near San Antonio, counted about four thousand head of cattle, together with slightly fewer sheep and goats. Clearly the valley became a frontier of small-scale ranches, largely lacking substantial estates. Considerable areas of pasture remained common land. This modest cluster of Hispanic ranches stayed fairly intact when the Anglo-Americans overran Texas in the 1830s, and the San Antonio River valley remains the Anglo/Hispanic cultural divide in the state still today.[45]

The methods employed in San Antonio River valley ranching are not clearly described in the surviving records. At the missions, small weekly roundups occurred for the purpose of gathering a half-dozen or so culls to be slaughtered before meat distribution on Sunday. Each Thursday a foreman, or caporal, led four to six mounted neophyte vaqueros, equipped with lassos, out into the pastures to get the desired animals. A major roundup occurred only once each year, in October and November, when branding was done. The failure to castrate calves made the management of cattle herds more difficult, although some small-scale private ranchers sometimes carried out bimonthly roundups in an effort to keep the stock from running wild. In many cases, however, the roundup ethic failed. At Mission Espíritu Santo, cattle were gathered and branded only three times in the 1771–78 period, while in the San Antonio area no systematic roundup occurred between 1773 and 1787. Pens were apparently not always used in roundups, although the typical Mexican palisado, or picket, corral did appear in the valley. Most of the missions maintained outlying ranchsteads, sometimes at a considerable distance from the main settlement, and the ruins of one of these, Rancho de las Cabras near Floresville, Texas, reveals stone-walled structures where the herders could have lived in relative comfort.[46]

Implantments of Spanish cattle ranching also occurred elsewhere in Texas. Escandón's Río Grande colonies were mentioned earlier, although they lay, of course, outside of Spanish Texas, in Nuevo Santander. Beginning in the late eighteenth century, additional developments occurred in the "Nueces Strip," the northern coastal part of Nuevo Santander, between the Río Grande and the Nueces, long a gathering ground of feral herds of cattle and horses (figure 28). The Spanish and Mexican governments issued numerous pastoral grants there, 134 of which were subsequently accepted as valid by Anglo-Texan authorities, and a few even remain partly controlled by descendants of the original grantees. The largest such holding, awarded to José Narcisso Cavazos in 1793, was San Juan de Carricitos Ranch, encompassing 243,600 hectares (over 600,000 acres) near the mouth of the Río Grande (figure 27). Two adjacent ranching grants of 115,150 and 127,690 hectares (285,000 and 315,000 acres, respectively) were also awarded by Spanish officials. This more humid outer part of the Río Grande plain offered some splendid tall bluestem prairie and also included huge adjacent grassy salt marshes, dotted with higher-lying *lomas* for refuge during hurricanes. The Loma de Vacas remains a local landmark in the marshes near Port Isabel near the southernmost tip of Texas still today. Cattle ranching in the Nueces Strip actually began farther to the north, near the Nueces River, where Santa Petronela Ranch was established in the 1760s (figure 27). The coastal part of the Nueces Strip remains cattle-ranching country still today and houses the world-famous King Ranch on the Wild Horse Prairie, among others (figure 28). Here, then, were the pastoral estates lacking in the San Antonio River valley. In spite of a decline in ranching activity in the Strip between 1810 and 1870, caused by Indian attacks and, later, the long period of Anglo-Mexican skirmish and warfare, which rendered the area a no-man's-land, the aristocratic cattle estate survives there to the present day.[47]

This coastal, secular development of Hispanic cattle ranching eventually extended north of the Nueces, into Texas proper. At Victoria, on the tallgrass coastal prairie not far from the earlier mission center at Goliad, a colony of ranchers was founded in 1824 by Martín de León, a native of Tamaulipas (figure 27). The newly independent government of Mexico enlisted him as an empresario to create a pastoral colony in the path of the westward-rushing Anglos.[48]

Assessing these Spanish-Texan ranching implantments in the light of the previously discussed littoral/interior cultural division in Mexico becomes essential, given the subsequent importance of

Texas to the ranching industry. Surprisingly no writer has previously attempted to weigh the imprint of Gulf coastal Mexico against that of the Mesa del Norte, of Andalucía versus Castilla in Texas. Perhaps this oversight occurred because scholars of Spanish Texas have assumed, incorrectly, that a monolithic Hispanic pastoral frontier descended upon the state. Moreover the general and unchallenged assumption has been that Texas was the child of the Mexican plateau. After all, did not the Franciscans, who missionized Texas, operate out of Querétaro and Zacatecas, and cannot the very dominance of the mission system in early Texas be said to reflect plateau frontier influences? Was not Texas "principally settled from Coahuila and Nuevo León," north Mexican states shaped culturally by the interior plateau? Did not the initial livestock introductions come from those same sources? Do we not see these ties revealed in the overland drives of cattle from San Antonio to Coahuila as late as the 1770s? Was Texas not married politically to Coahuila and did it not find its capital city in distant Saltillo on the Mesa del Norte? All of this is true, but the resultant cultural impress of the highlands in Texas should be sought in the relatively *modest* scale of cattle ranching around Spanish San Antonio and in the enduring importance of sheep in interior South Texas, including Escandón's colonies, as late as the 1880s (figures 27, 28).⁴⁹

Surviving material evidence of the Mexican plateau abounds in Texas, usually in conjunction with sheep and goat raising rather than cattle ranching. In Escandón's upstream Río Grande colonies, the oldest dwelling type, represented as a relic in the modern landscape, is the venerable flat-roofed, fortified ranchstead of the Mesa del Norte. More abundant and plebian is the palisado livestock corral, whose present geographical distribution strongly suggests diffusion from the San Antonio mission complex (figure 29). Concentrated in the Texas Hill Country, a dissected plateau north of San Antonio, this traditional picket pen, built of closely spaced small posts set in the ground, is more closely linked to sheep and goat raising than to cattle ranching in Texas, suggesting that the herding legacy of the San Antonio missions may lie in the wool and mohair districts of the Hill Country instead of in the cattle regions (figure 30).⁵⁰

Coastal Texas, by contrast, belonged to the distinctive, dominantly secular littoral subculture, derived from the Gulf Coast of Mexico, from Tamaulipas, the Huasteca, and Veracruz. Indeed Tamaulipas extended to the Nueces in Hispanic times, with the result that the coastal region has been largely ignored by most Texas histo-

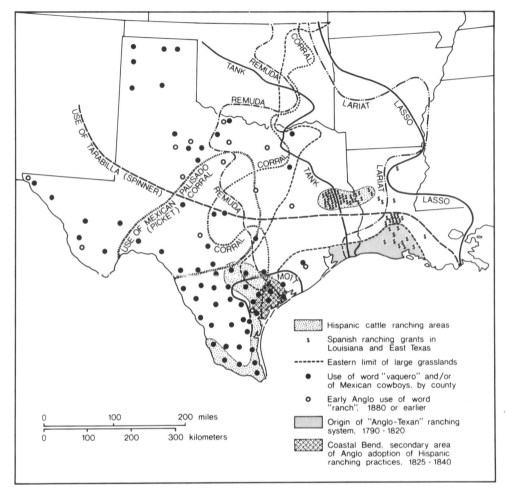

Figure 29. Historic and modern patterns of Hispanic influence in the cattle-raising culture of the south-central United States. Hispanic influences extended well into Louisiana by the late 1700s, and it was there, rather than in Texas, that Anglo-Americans adopted roping techniques, the horned saddle, and equestrian skills in the period 1780–1820. Isoglosses, labelled in quotation marks, show the northern and eastern limits of selected Spanish loanwords as of about 1960, and the "mott" isogloss refers only to its use as a generic toponym. The Andalusian tarabilla, or tarabi, is a spinning device, or whirligig. Sources: Jordan, Trails (see note 53), pp. 145, 148, 151, 189–90; Jordan, "Origin of Mott" (see note 52), p. 164; Atwood, Regional Vocabulary (see note 52), pp. 132, 149, 152, 156, 161, 247, 248, 251; Kniffen, "Spanish Spinner" (see note 58), p. 199; Kniffen, "Western Cattle" (see note 58), pp. 179–81; Jackson, Mesteños (see note 44), p. 492; Milton B. Newton, Jr., Atlas of Louisiana (Baton Rouge, LA, 1972), p. 72; Fred Tarpley, From Blinky to

Blue John: A Word Atlas of Northeast Texas *(Wolfe City, TX, 1970),* p. 304; Lillie Mae Hunter, The Book of Years: A History of Dallam and Hartley Counties *(Hereford, TX, 1969),* p. 119; Jordan, "Origin of Motte and Island," (see note 52), p. 122; United States Census, 1880, manuscript schedules for Texas. Distribution of the picket corral was ascertained through field observations by the author over a period of twenty-five years.

rians, who remained fixated on the Coahuila linkage and gray-robed Franciscans. Even Goliad lay on the margins of Tamaulipas, and in fact it was Escandón who requested the relocation of Espíritu Santo to the Goliad site and prompted the creation of Rosario. Coastal Texas, not the upper San Antonio River valley, formed the major hearth of Spanish cattle ranching, and coastal Texas owed more to Tamaulipas than to Coahuila. I propose that while the secular cultural impulse spreading north along the Gulf Coast reached Texas later and, due to a paucity of records, rather obscurely, it constituted a more important and durable diffusion for cattle ranching. Geographer William E. Doolittle has previously suggested as much, and historian Dan E. Kilgore, a resident of the Nueces Strip, also glimpsed this truth when he recently concluded that the distinctive, hardy Texas longhorn variety of the Andalusian cattle breed developed under feral conditions on the Gulf coastal plains of present northeastern Tamaulipas, not on the Mexican plateau.[51]

Evidence of the durable intrusion of Mexican Gulf coastal culture into the Texas littoral abounds, both in material culture and vocabulary. The vegetable fiber rope and related Andalusian spinner are both found in South Texas, suggesting a coastal route of diffusion. And the Andalusian-derived word *mata* reached the Texas coast by way of Tamaulipas. Corrupted after 1830 by Irish colonists near Goliad and Victoria into *mott,* it remains the dominant generic placename to describe a grove of trees in a prairie in the Coastal Bend region of Texas, as far east as the lower Colorado River (figure 29). Indeed *mott* serves as a useful geographical index to bound the extent of Tamaulipan cultural influence in Texas.[52] Moreover the diffusionary evidence concerning *mott* raises the distinct possibility that certain other Spanish loanwords used in the Anglo-Texan ranching culture, such as *tank, lasso, lariat, remuda, mustang, corral,* and even *ranch* itself may have been acquired by way of the coastal route (figure 29). For example, *tanque,* rather than the more standard Spanish *estanque,* is the preferred word for a stock-watering place in

Figure 30. Mexican-type palisado, or picket corral, in the Texas Hill Country of Comal County, north of San Antonio. Built by setting small juniper posts in a trench and then filling in the dirt, these pens are most common on sheep and goat rather than cattle ranches and probably reflect a material cultural influence of the San Antonio mission complex and of the Mexican interior plateau. (Photo by the author, 1984.)

Tamaulipas, as in most of Mexico. It appears in placenames such as Tanque del Grullo ("tank of the ash-colored horse"). The corruption of *tanque* to *tank* requires little change, and *tank* occurs in an unbroken belt along the Texas coast almost to Louisiana (figure 29). The use of *tank* on the King Ranch in the Nueces Strip dates at least to the 1880s, and that area may be where the word entered Texas English. Similarly the earliest documented uses of *ranch* in the Anglo-Texan cultural context occurred in the Coastal Bend country, centered on Victoria, in the late 1840s and early 1850s (figure 29).[53]

The case for Tamaulipan influence in Hispanic Texan ranching can most forcefully be presented by reviewing briefly the life story of the previously mentioned Martín de León, founder of Victoria and arguably the single most influential person in the transfer of Spanish ranching culture to Anglos in Texas. Born in 1765 in Escandón's recently established town of Burgos in Nuevo Santander, de León married a woman from Soto la Marina in 1795 and began ranching at Cruillas, a place founded in the very year of his birth (figure 27). In common with many ranchers in coastal Tamaulipas, de León raised horses and bred mules, in addition to herding cattle, often drawing upon the abundant feral mustangs found near the lower Río Grande. Seeking more such horses, in 1805 de León explored the Nueces Strip in Texas, going as far as Goliad. Impressed by the native pastures, which did not suffer the troublesome winter drought of Tamaulipas, and equally attracted by the mustangs available in the Texas Coastal Bend, de León migrated in 1806 to establish a new ranch south of Goliad, near the mouth of the Aransas River. He requested, but did not receive, a land grant there, all the while rounding up wild horses, for which he dutifully paid the royal tax. Displaying the mobility typical of coastal stock raisers, de León in 1808 relocated his squatter ranch to the nearby lower Nueces, close to present-day San Patricio (figure 27). Although hostile Indians periodically drove him from this site, it remained his headquarters for some fifteen years, housing thousands of horses and cattle as well as his mule-breeding operation. While driving stock overland to market at New Orleans, de León came to know the lands still farther up the coast. Attracted by the fine tallgrass prairies and salt marshes beyond the Guadalupe River, Don Martín, by then approaching sixty years of age, in the middle 1820s gained permission from the Mexican government to establish a colony of cattle and horse ranchers on the humid subtropical grasslands around Victoria, in the Texas Coastal Bend, obtaining in the process the right to gather wild horses free of duty for ten years (figure 27). The Mexican ranchers thrived at Victoria, building "large stocks of cattle," although the local Karankawa Indians, the last of the Chichimecs in coastal Texas, killed some of their livestock. Established in the new home, de León managed his ranch and colony for almost a decade before perishing in the cholera pandemic of 1833. He had succeeded in establishing the easternmost and best-documented settlement of Tamaulipan ranchers along the Gulf of Mexico coast. Many of his papers survive, including a formula for curing cowhides, and his legacy, perhaps more than that of any other individual Hispanic rancher in Texas, endured on the North American pastoral frontier.[54]

De León's ranching methods at Victoria remain somewhat obscure, in spite of the available documents. The labor on his estates was performed by "peons," and he sought at one point to hire an Anglo caporal, or "straw boss." De León owned separate horse and cattle ranches in the vicinity, and his holdings also included small numbers of sheep. Anglo observers were struck by the fact that Mexicans like de León "had no steers," implying that the traditional Andalusian inattention to castration prevailed in coastal Texas. As a result de León and the other Victoria ranchers, in order to control their stock, had to practice what the Anglos called the "Mexican plan" of herding, which involved a rodeo "once a week or oftener," in addition to the major roundups in spring and fall. All through the Victoria years, de León maintained his marketing link to New Orleans, although he was also able to sell stock to the developing Anglo-Texan colonies to the east.[55]

Expansion into Spanish Louisiana

Martín de León's ties to the New Orleans market reflected an overland trading pattern that had begun much earlier, linking the two Spanish colonies of Texas and Louisiana after 1763. In fact both legal and contraband trade between the two began even earlier, when the French still ruled the lower Mississippi, and had become vigorous by the 1750s. Mid-eighteenth-century Spanish Texas, in many ways, had closer ties to French Louisiana than to the rest of Mexico.

The cattle-ranching frontier was much influenced by the Texas-Louisiana connection. One result was the development of a Spanish cattle-ranching enclave in the grass-rich Piney Woods of deep East Texas and adjacent Louisiana. Following several abortive attempts, the permanent Spanish settlement of that region began in the 1770s, initially around Nacogdoches in Texas. Pastoral land grants by Spanish officials on both sides of the Sabine provided the basis of a modest local cattle industry (figure 29). By the early nineteenth century, many or most of the Piney Woods settlers engaged in herding pursuits, firing the forest annually to enhance grass growth.[56]

A more important result of the trade between the two colonies was the flow of Texas cattle and ranching techniques into the prairies of southwestern Louisiana. After issuance of a 1778 edict permitting Texas ranchers to "round up herds of stock and drive them to other regions," a huge export of cattle and horses to Louisiana began, sufficient to cause herd depletion in the lower San Antonio River valley. The 1780s were a particularly active decade for this trade, and through sovereignty shifts and ethnic changes the Texas-to-New Orleans overland cattle trade persisted into the 1860s.[57]

From the very first clandestine trade in the middle 1700s, Spanish-Texan herding techniques diffused into Louisiana along with the cattle, allowing a significant hispanicization of the Louisiana French vacheries to begin. Most notable among these influences were equestrian skills, in particular use of the lasso, a word which entered the Cajun vernacular speech and the coastal herding lingua franca in Louisiana. By 1797 observers reported the lasso in use as far east as Lake Pontchartrain, not far from New Orleans. Additional evidence of Spanish/Mexican cultural influence in prairie southwestern Louisiana is seen in the presence there of a marsh-side Andalusian device for spinning horsehair, grass, Spanish moss, or wool into yarn or cord. Originally known as a *tarabita* in Cádiz province, Spain, adjoining Las Marismas, the spinner attained a place in Mexican Spanish as *tarabilla* and became corrupted in Louisiana into *tarabi*. Its geographical distribution links Spanish Louisiana culturally to South Texas and coastal Mexico (figure 29). Also reflecting their herding influence in Louisiana, the Spaniards issued numerous large grazing grants. Some of these lay in the Piney Woods of the north, adjacent to the East Texas cluster, but the more important and enduring ones, usually awarded to people with French surnames, were situated in the coastal prairie of southwestern Louisiana, where a hispanicized cattle industry blossomed in the 1780s.[58]

What, in the final analysis, was the legacy of Hispanic cattle ranching in Texas and Louisiana? First, Texas apparently proved less important than Louisiana in the transfer of Spanish techniques to

Anglo-Americans. The chauvinistic defenders of South Texas as the nursery bed of the western cattle industry and of *tejanos* as the agents of diffusion have greatly overstated their case.[59] Nearly every center of Hispanic ranching in Texas, save only the de León colony in the Coastal Bend, suffered severe decline between 1800 and 1825, the period that witnessed the initial Anglo intrusion. Indian attacks contributed greatly to the herding decline, as did secularization of the missions, range deterioration, and the termination of Spanish rule. Even the isolated ranching complex in the forested fastness of East Texas did not escape the decline. By 1840 no Spanish-surnamed person around Nacogdoches reported as many as one hundred cattle to the tax collector, and the remnant East Texas Hispanic population, numbering only about three hundred by 1850, lapsed into subsistence farming or took menial positions in the towns of the region, overwhelmed by the numerically superior Anglos.[60]

Instead, the most important Hispanic ranching legacy was likely that seated in the prairies of Louisiana, for it was there that Anglos and their African slaves adopted the equestrian herding techniques that would play so great a role in the West, as will be discussed in chapter 6 (figure 29). Louisiana, not Texas, witnessed the crucial transferral of Spanish skills.

Only the Coastal Bend section of Texas can correctly be said to have participated in this diffusion of traits to the Anglos, and its role was clearly secondary to that of Louisiana. It is true that the Coastal Bend had been home to the greatest cattle-raising complex of Spanish Texas, but the region had suffered severe decline with the collapse of the mission ranches. Even de León's colony did not escape the deterioration. By 1850 the Coastal Bend retained but ten Spanish-surnamed people owning one hundred or more cattle and only 181 Mexican vaqueros. However, the de León colony ranches had continued to thrive into the middle 1830s and functioned as a secondary center of hispanicization. Some Anglos were present from the earliest years of the colony, and several nearby settlements of Catholic Irish in the early 1830s also became cattle-ranching colonies, based in large part upon de León's Tamaulipan model. When Mexican rule ended at San Jacinto in 1836, a secondary wave of Anglo plunderers, called "cowboys," stole the cattle and lands of most of the de León tejanos, although some remained. Most if not all of these opportunistic Anglos had already learned Spanish techniques farther east. In the Coastal Bend they merely stole Hispanic livestock and land, not skills.[61]

Alta California

Belatedly, in the year 1769, the Spanish settlement frontier finally reached Alta California, bringing the introduction of cattle ranching. The irony in this tardiness lies in the fact that California, perhaps more than any other part of North America, resembled the Iberian homeland. The typical Mediterranean climate regime of winter rain and summer drought prevailed in California, inviting the whole array of Spanish grains, vines, orchards, and livestock to thrive (figures 27, 31). Actually California more closely resembled Portugal than Andalucía, Extremadura, or Castilla, since the coastal valleys and hills, where the colonial activity was focused, experienced cool summers, like those of Lisbon rather than Sevilla, where the hot breath of the Sahara is often felt. At last the Spaniards had escaped the inverted rainfall regime that had plagued them in the Antilles and Mexico, although as criollos of many generations depth, they had no doubt forgotten all about the climate of ancestral Iberia. As in southern Europe, the California climate, while fine, proved far from idyllic. The summer drought yielded a sensitive vegetation cover that demanded relatively light rates of stocking and attention to seasonal herd relocations for conservational reasons. Sometimes, too, the winter rains failed. Cyclically, about once a decade, these prolonged dry spells occurred, as in 1809–10, 1820–21, 1828–30, and 1840–41, causing havoc among livestock raisers, who had to slaughter herds to save the pastures. In certain other years, floods drowned livestock in low places. All things considered, however, Alta California's climate could be called arcadian, particularly in comparison to the thorn forests of Sinaloa or the deserts of Baja California.[62]

The terrain also invited colonization. Paralleling the California coast lay a series of low mountain ranges and hill belts, interspersed with elongated, fruitful valleys and a few low basins. Beyond, in the interior, the huge Central Valley stretched from northwest to southeast, as fertile and attractive as any in lowland Andalucía. Nevertheless the Spanish settlement of California never progressed beyond the coastal ranges and valleys, perhaps understandably, given the small population of the colony and the preexisting concentration of Native Americans, who attracted the missionaries, near the shore.[63]

California's greatest resource for cattle ranching was floral. It offered, in effect, an undegraded Mediterranean habitat such as had not been known in Iberia for millennia. The local Indians, who practiced no agriculture and kept no herd animals, had only lightly touched the

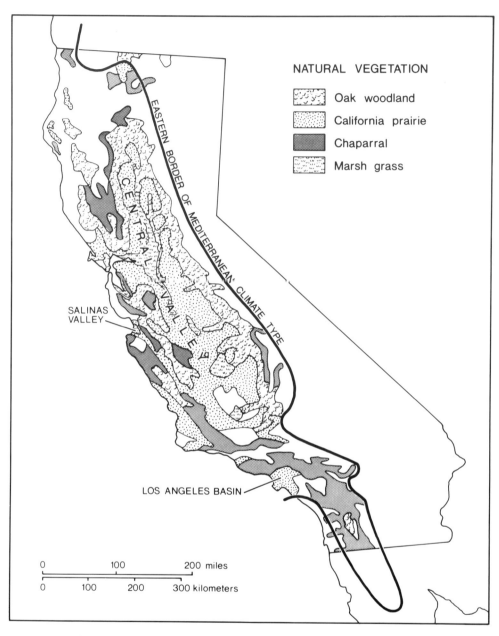

Figure 31. Range resources and climate in California. The four vegetational regions of greatest value to cattle raising are shown, as is the border of the "Mediterranean" climate type. Hispanic Californians fell far short of fully occupying this pastoral habitat in the 1769–1846 period. Sources: Burcham, **California Range** *(see note 38), p. 80;* Kesseli, "Climates of California" *(see note 62), pp. 476–80.*

land, leaving it nearly pristine for the Spaniards. Most promising for cattle were the famous "California prairies," treeless meadows dominated by perennial bunchgrasses, chiefly needlegrasses (*Stipa* species) in the south, with a few sod-forming types such as beardless wild rye, California oat grass, and numerous annuals (figure 31). Both annual and perennial forbs—broad-leafed, low-growing plants—and herbs appeared in association with the grasses, providing additional forage and browse. While concentrated in the Central Valley, these prairies, totalling about 9 million hectares (22 million acres) also occurred in the low, warm coastal valleys and basins, from the Los Angeles area northward beyond San Francisco. In poorly drained, lower parts of these grassy valleys appeared the much-prized "tulelands," marsh-grass associations covering about 200,000 hectares (500,000 acres). These wet areas, too, proved to be a wonderful resource for cattle, particularly in the dry season. Another prime grazing domain, over 3 million hectares (8 million acres) in extent, lay in the live-oak woodlands of the foothills and lower mountain slopes adjacent to the prairies, up to about 750 m. (2,500 ft.) elevation. These open, parklike forests contained abundant grasses that could be utilized by Spanish herds. Above the woods, in the higher, steeper parts of the coastal ranges, extensive tracts of dwarf evergreen chaparral appeared. While less desirable for cattle than the prairies, marshes, and oak woodlands, the chaparral offered both palatable browse and, in numerous openings, reasonably good forage grasses and forbs. In sum the floral resources available for livestock in Alta California can only be described as splendid.[64]

The Spanish-speaking colonists recognized at once the value of the land. In the year of initial settlement, one of the Spaniards described California as "covered with the finest pastures," a forage resource of the first quality, permitting year-round grazing because the grasses were palatable both green and dry. Not just cattle could thrive there, for the country was "well-suited for sheep and goats," since "it is very clean, without anything which might injure the wool." Also acorns aplenty awaited the Iberian hogs. The tropical thorn forests of Pacific Mexico lay far behind and largely forgotten.[65]

The first Spaniards to enter this fat land were Franciscan missionaries, who continued the ecclesiastical dominance of Mexico's Pacific frontier begun long before by the now-banished Jesuits. Beginning at San Diego in 1769, the Franciscans established five missions by 1773, the beginnings of a chain that would eventually contain twenty-one churches, stretching northward to Sonoma, beyond San

Francisco, and house at its zenith in the 1820s some twenty-one thousand Indians. Due to California's remoteness, these outposts of Christendom needed to become largely self-sufficient, prompting the early introductions of the whole array of Iberian crops and livestock, including cattle. Two hundred longhorns went north from Velicatá in Baja California to San Diego in 1769, 164 of which reached their destination alive, and 300 more arrived from the lower peninsula in the following year. Having for the moment exhausted that source, the Franciscan fathers turned to ranches in Sonora and Arizona, which sent 65 cattle in 1774. These introductions barely sufficed for local subsistence needs, since the *padres* once again used beef rations to lure Indians to the church. In a later century China would have its "rice Christians," but the Spanish mission frontiers of Mexico attracted "beef Christians." Because of this essential dietary demand, all of the Alta California missions combined possessed only 204 cattle in 1773 and 427 by 1775. Ranching, properly speaking, had not yet begun.[66]

The crucial entrada, which led to the great pastoral era in California, did not occur until 1776, when herders assembled some 350 cattle in the Santa Cruz Valley of Arizona and drove them overland to California, experiencing only minor losses on the way. Additional, smaller numbers came from Baja California in that same year. Never again would beef be in short supply, and by 1780 the California herds had become "ample." A spectacular natural increase of cattle began by the middle 1780s, reaching an especially rapid multiplication in the 1795–1805 period. For example Santa Clara's herd grew from 15 animals in 1778, a year after the mission's founding, to 1,500 by 1793. At the turn of the nineteenth century, the total California mission herd reputedly numbered about 67,000 cattle, and by the early 1830s widely varying estimates placed the herds at from 200,000 to 400,000 cattle. Such rapid expansion was facilitated by the exceptional quality of the range, the nonagricultural character of the local Indians, and the absence of mounted Indian raiders.[67]

In the California mission herding system, specialized cattle ranches were established, often in locations rather remote from the main church compound. These ranchos, as they were called, provided the prototype for the California cattle ranch. The word rancho became the preferred local Spanish term to describe a livestock-raising enterprise. Each mission operated multiple ranchos, scattered through large hinterlands, and some of these tributary herding settlements became, in effect, small missions, complete with a chapel and permanent resident population. San Luis Rey, north of San Diego, had twenty ranchos in a range hinterland of 3,100 square km. (1,200 square mi.), including, for example, the cattle-raising Rancho San Juan, located in the chaparral about 13 km. (8 mi.) east of the parent mission. Mission San Gabriel, in the prairies of the Los Angeles Basin, operated thirty-two ranchos in a tributary area of about 5,450 square km. (2,100 square mi.), including seventeen devoted exclusively to cattle and horses.[68]

In the Alta California mission agrarian economy, cattle raising represented the single most important activity. Iberian grains, vines, orchards, sheep, goats, and swine, though abundantly present and well-suited to the local environment, never seriously rivaled cattle in significance. Perhaps the bovine emphasis reflected the Pacific coastal tradition of Mexico, in which sheep had always been secondary in importance to cattle, a legacy of the hot tropical thorn forests. California's Spanish criollo inhabitants, mainly of mixed blood, came very largely from the Pacific Coast, with Sonora and Sinaloa the most important contributors, followed by Nueva Galicia, including Nayarit, and Baja California. The great mountain barrier of the Sierra Madre Occidental had dictated that Spanish California should be linked culturally to those Pacific parent provinces.[69]

Although a cattle focus existed, the mission livestock industry of Alta California displayed a vivid internal geography. The greatest attention to cattle raising occurred in the southern missions, excluding only San Diego. The three largest herds in 1834, accounting for almost two-thirds of all the cattle in the twenty-one missions, belonged to San Gabriel, San Luis Rey, and San Juan Capistrano, all situated from the Los Angeles Basin southward. The fourth, fifth, and sixth largest herds at that time were owned by San Fernando, Purísima, and Santa Inés, also in the southern half of the chain, and the six leading missions of the south controlled three-quarters of the church-owned cattle in California. Likewise the three missions having the greatest numerical excess of cattle over sheep—San Fernando, nearby San Buenaventura, and Santa Inés—were all southern in location (figure 32). Mission cattle herds in the north remained much smaller, led by San José, and sheep outnumbered cattle there. Accordingly the core area of California cattle ranching should be placed in the southern coastal district, centered on the splendid prairies of the Los Angeles Basin. The reasons for cattle preeminence there remain unclear, but perhaps sheep fared better in the cooler north.[70]

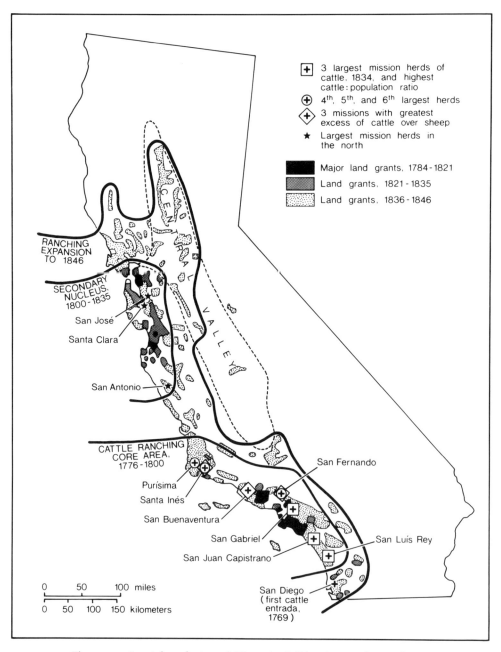

Figure 32. Spatial evolution of Hispanic Californian cattle ranching, 1776–1846. From a southern nucleus, where both mission-based and private ranching developed before 1800, the California herding complex spread to a second nucleus in the north in the early nineteenth

century. By the end of Mexican rule, cattle ranching had entered parts of the Central Valley. Sources: Hornbeck, "Land Tenure" (see note 72), pp. 377, 386, 387; Bancroft, California Pastoral, (see note 62), p. 339; Gentilcore, "Missions," (see note 65), pp. 59, 67–69.

The mission system of Alta California continued to thrive while that of Texas decayed. A new mission was founded as late as 1823, and prosperity lasted even for a decade after that. However, when secularization did finally come, in the middle 1830s, the plunder and ruination of the California missions were sudden and very nearly complete. By 1842 they retained fewer than 4,500 resident Indians and only 29,000 cattle.[71]

The destruction of the mission system did not bring decline to the California cattle-ranching industry, although many cattle were slaughtered. Instead pastoralism simply passed into private hands, completing a transition begun much earlier. Secular herds dated at least to the Sonora-Arizona entrada of 1776, when 30 head of privately owned cattle accompanied a herd of 325 destined for the missions. A few civilian criollo colonists came at that time, and soon, in 1781, the *villa* of Los Angeles was founded. In time some citizens of that town developed sizable herds. Even more important in the evolution of the private ranching sector, as in the Mesa del Norte and San Antonio River valley, were presidio soldiers who drew upon early royal presidial herds of the local *ranchos del rey*. As the private herds grew, the owners soon began requesting pastoral land grants. Beginning in 1784 in the Los Angeles area, Spanish officials issued thirty or so usufruct grants, mainly to soldiers and veterans, before Mexico won its independence in 1821 (figure 32). The early Spanish grants lay largely in the evolving southern nucleus of California ranching, reinforcing the mission-based pattern.[72]

In the development of private cattle ranching, the life story of one José María Verdugo is revealing. Born in the Jesuit community of Loreto in Baja California in the year 1751, Verdugo came north as a soldier with the initial settlement of San Diego in 1769. Two years later he was assigned as a guard at newly founded San Gabriel mission in the Los Angeles prairies. Soon he began running a small cattle herd along the San Gabriel River, and in 1784–86 Verdugo received a local grazing grant, the Rancho San Rafael, of about 14,600 hectares (36,000 acres) in the present Burbank-Glendale area. Attaining his own registered brand in 1788, Verdugo built his private cattle herd to two thousand head by 1801 and possessed vineyards, irrigated fields, orchards, horses, sheep, and mules as well. By the time of his death in 1831, he had acquired a considerable personal fortune.[73]

If under Spain the pastoral grant had served to reward veterans and influential civilians, Mexico eventually developed the privately held rancho into a formidable and pervasive colonization institution. The private sector had never seriously challenged the missions under Spain. As late as 1823 mission Indians outnumbered criollos six to one in Alta California, and the church herds, similarly, dwarfed those of the secular rancheros. In fact mission domination of California ranching lasted into the middle 1830s, for independent Mexico had issued only twenty land grants as late as 1832, almost exclusively in the Los Angeles and Monterrey areas.[74]

The big change came between 1834 and 1846, when coastal California was converted into a loose network of private ranchos. Many church ranches simply passed to individual persons, some of whom were "friends of the mission." Simultaneously various new areas were opened to pastoral settlement through grants of unoccupied land, including the northern part of the heretofore neglected Central Valley of California and the coastal valleys north of San Francisco (figure 32). By 1840 at the latest, the private ranch had become the dominant social and economic institution in Mexican California, a dominance further strengthened by additional grants during the last half-decade before the Anglo takeover. Of the 473 ranchos awarded by Mexico, fully 182, or 38 percent, were made between 1843 and 1846. Some 4 million hectares (10 million acres), encompassing 10 percent of the surface area of California and much of its best pastureland, had passed into private hands by the time Mexican rule ended. Hispano-Californian cattle ranching, in contrast to that in Texas, thrived vigorously at the time of takeover.[75]

To quote the florid words of Hubert Bancroft, Hispanic California's "pastoral days swept by, half-way between savagism and civilization."[76] But what, more precisely, did this herding culture entail? What sort of ranching system had evolved at the northern extremity of the Mexican Pacific coastal diffusion? In what ways did it remain distinctive from the herding frontiers of the Gulf Coast and Mesa del Norte? These basic questions bear upon the subsequent development of Anglo-American cattle ranching in the western states.

First, the California Hispanic system shared with the opposite, eastern coast a cattle-horse focus, and the emphasis upon cattle became even more profound as the private ranchos rose to dominance. The local stock raisers held no bias against sheep or goats and usually kept small flocks, but California, clearly, was a cattle frontier. Crops also played an inconsequential role in Hispanic California, particularly after the collapse of the missions, although a few haciendas existed. As late as 1870 the entire state had only about 36,500 hectares (90,000 acres) under cultivation. In fact, cattle raising, for both the missions and the rancheros, represented the only notable commercial enterprise, providing the raw materials for local industry and the basis for overseas trade. The remoteness of the province dictated that only hides and tallow possessed a market value. Hides, often treated in brine and dried on the beaches, found a market by sea trade beginning about 1800. The purchasers, tanneries and leather factories in the northeastern United States, particularly Boston and New York City, instigated a profitable trade, which peaked between 1826 and 1848. Some tanning vats also operated in California, using the abundant native oak bark. Tallow was shipped to destinations such as Lima in Peru, and California-made candles also entered the sea trade.[77]

The local vaqueros were derived largely from the indigenous Indian population, although California and Sonora bloodlines, as elsewhere in frontier Mexico, included a mixture of Spaniard, Indian, and perhaps African. Even into modern times, they referred to themselves as *sonoreños*, perhaps a way of denying California Indian ancestry. These herdsmen possessed the full array of highly developed Mexican Pacific equestrian skills, including use of the garrocha and the rawhide lasso, the latter made fast in the Californian technique by winding the *reata* around the saddle horn and letting it slip off loop by loop when taut, a technique called *dar la vuelta*. Mission vaqueros received their own saddle and soft leather shoes, but other equipment, as well as their mounts, remained church property. Most California herders used the Spanish great-rowel spur, distinguished by four or five rather sharp, long points, each measuring about 2.5 cm. (1 in.) in length. Spanish Californian vaquero equipment and dress retained much more charro influence than appeared in Texas or even the Mesa del Norte. Given the diffusionary connections back to Sonora and Jalisco, this transfer of material culture from the Guadalajara area, and more remotely from Extremadura and Salamanca, was not unexpected. The charro imprint in California appeared in the embellishment of the riding gear, or rig, and in such items of clothing as the short jacket; medium-length dark pants; red sash at the waist; and low-crowned, flat-brimmed hat equipped with a neck string. At the same time, the Californians tampered with and modified the Sonoran equipment they had inherited. The cowhide armas of the Pacific thorn thickets of northwestern Mexico, which protected the rider's legs and horse's breast from injury, gave way in the less offensive vegetation of Alta California to smaller, lighter deerskin *armitas*, which were fastened to the rider's belt and reached below the knee, held in place by tie-strings. The main purpose of armitas was to prevent chafing by the rawhide reata during roping. Below, covering the calves of the legs, the vaquero wore goatskin *gamuzas*. Similarly the design of the bit underwent modification in California when, in 1814, the "Santa Bárbara" hinged snaffle was invented (see figure 49).[78]

Equipped in this manner, mission vaqueros working under the direction of a *mayordomo*, or ranch foreman, performed the roundups. Each week a small rodeo of twenty or thirty head was held to supply the Saturday beef distribution, although in jerky-making season during the dry summer months as many as one hundred cattle might be gathered and slaughtered with lances each week. A much larger and more complete mission rodeo occurred at the end of the dry season, in September and October, for the purpose of counting, branding, and marking the stock. At Mission San José, fifty vaqueros worked in this annual roundup. As in Sonora the same grounds were repeatedly used for roundups, and the larger ranches had multiple places of rodeo, all generally lacking corrals. Both the embellished Andalusian-style brands, including the "cross-over-crescent" design, and cut earmarks identified ownership. Eventually, on the private ranches, only the annual fall roundup was required by law, but most rancheros also carried out a late spring rodeo for mass slaughter, hide gathering, and tallow rendering. Calf castration seems to have been unusual in California practice, as elsewhere on the Mexican frontier.[79]

The cowboys in the California system also had the task of shifting cattle between ranges seasonally, in an effort to reduce overgrazing and prevent damage to the fragile valley pastures in summer. The details of this range rotation remain unclear, but almost certainly a local-scale vertical transhumance was involved, as in Sonora. In what is now Santa Bárbara County, mission cattle from near the coast were driven seasonally over Refugio Pass to a grazing hinterland in the Santa Ynez Valley. Perhaps such shifting of pastures began as a spontaneous upslope drift of free-ranging cattle in the hot summers. Cattle were also herded onto grain stubble in the mission fields after the harvest in May and June, as in Extremadura and Nueva Galicia. In unusually dry years, older cattle were killed with lances out in the pastures to retard overstocking. In spite of these precautions, extensive damage to the California ranges occurred in Hispanic times.[80]

The spread of open-range cattle raising through northern Mexico, then, yielded two major implantments of ranching in what is now the United States. Both lay at the end of coastal paths of diffusion, one on the Gulf of Mexico and the other in Alta California. The Mesa del Norte expansion, while spectacular and rapid within Mexico proper, played only a minor role in the development of cattle ranching in the American Southwest, but the two littoral complexes, centered in the Gulf coastal prairie and the grassy western valleys of California, would contribute greatly to the ranching frontier of the West in the latter half of the nineteenth century.

SIX

Carolina's Children

SOUTH CAROLINA'S COWPEN CRACKERS DID NOT REMAIN confined to their colonial foothold on the eastern Atlantic seaboard, but instead achieved several diffusions, which together formed the Anglo-American herding frontier of the East.[1] Their most successful and enduring route westward followed the coastal plain corridor of pine barrens, a path that led them eventually beyond the Mississippi River to the great western prairies.

Through the Pine Barrens

Much of the coastal plain in the South invited the spread of the cracker cattle folk by offering a wooded, humid subtropical habitat similar to that of South Carolina all the way into East Texas. Herders moving along this coastal corridor could continue to utilize the grassy pine barrens, marshes, swamps, canebrakes, low hardwood lands, and savannas they had known back East, collectively capable of supporting one cow on about every 6 to 8 hectares (15 to 20 acres). This familiar environment can most readily be recognized on vegetation maps as a fragmented belt of longleaf pines (figure 33). Early Carolina traders and surveyors probing westward made such notations on their maps as "high pine land good for cattle range," "barren land cypress savannas good only for cattle range," and "pine land full of caneruns fit for cow-pens." Coastal expansion of the Carolina herding system, then, filled a familiar ecological niche. While not ideal for cattle, due mainly to protein and calcium deficiencies in the winter, as well as phosphorus shortages in all but the springtime, the coastal pine-barrens range could support essentially stationary herds all year long, with no supplemental feeding needed, except for milk cows. Various types of wire grass and bluestems, in addition to wild oats, sedges, rushes, and cane, produced animals that, though poor at the end of winter, became passably marketable by the time of first frost.[2]

The initial pine-barrens cowpen expansion occurred within the Carolina low country after about 1715, following a devastating war with the Yamassee Indians centered in the outer coastal plain southwest of Charleston. The Yamassee uprising had been caused in part by herder-induced environmental changes that diminished the deer herds upon which the Indians relied, and many cattle and cowpens were destroyed during the fighting. Once hostilities ended, the herders spread vigorously through the former Yamassee homeland, crossing the Edisto River to reach the banks of the Savannah (figure 26). An even more important new focus of the cowpen culture developed beginning in the 1720s in the inner coastal plain, below the Fall Line and centered upon Orangeburg, where abundant upland pine barrens, sand hills, small savannas, and riverine cane swamps and marshes could be found. The cane-rich wetlands between the forks of the Edisto and along the upper Salkehachie River became particularly noteworthy cowpen concentrations. Historical geographer Gary Dunbar found this Orangeburg focus so important that he mistook it for the nucleus of the entire Carolina cowpen complex, confusing child with parent.[3]

Early expansion also led northeastward from the Charleston area, along the coast, where herders occupied the immediate littoral and then spread inland along the valleys of the Black and Pee Dee rivers, reaching the inner coastal plain. In fact by 1750 few if any areas of lowland South Carolina remained untouched by the cowpen culture. In spite of a severe epidemic of bovine distemper, which together with range deterioration precipitated a gradual and irreversible decline of stock raising in the low country, the colony at midcentury housed eighty to one hundred thousand cattle.[4] Expansion into comparable environments in the coastal plain of adjacent colonies soon became necessary.

Diffusion of open-range cattle raising very early reached southern coastal portions of the sister colony of North Carolina, where records dating from as early as 1715 reveal that "cowpen" was being used in the South Carolinian meaning of "ranch" (figure 34). A law passed in that year mentioned "persons that shall have at . . . their respective Cowpens or Plantations any stray cattle," a clear indication the two were regarded as different types of enterprise.[5] In 1735 mention was made of "the Lands called the cowpen" on the Northeast Cape Fear River, north of Wilmington near the Holly Shelter Swamp. Acts passed in 1741 and 1766 in North Carolina contain similar references. "Savanna" achieved a parallel diffusion. This early

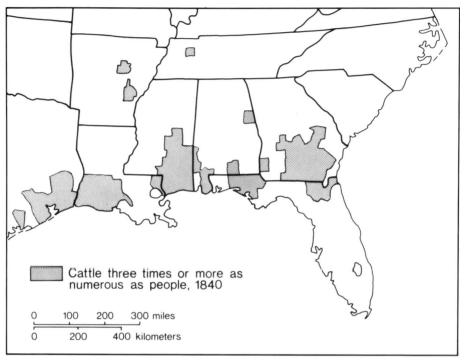

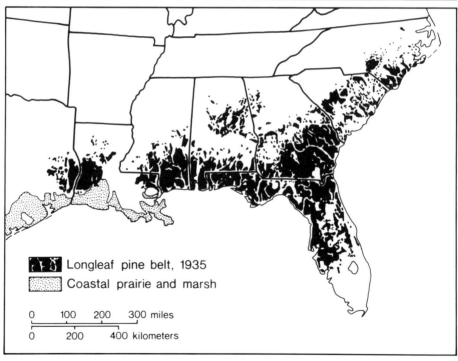

Figure 33. *Coastal southern habitats and the cattle: population ratio. The coastal route of Carolinian expansion followed the longleaf-pine belt, then entered the humid coastal prairies. The cowpen culture did not pass away with the frontier, but instead was an enduring occupance form. Sources: W. G. Wahlenberg,* Longleaf Pine: Its Use, Ecology, Regeneration, Protection, Growth, and Management *(Washington, DC, 1946), p. 46;* Compendium of the Enumeration of the Inhabitants and Statistics of the United States, as Obtained at the Department of State, from the Returns of the Sixth Census *(Washington, DC, 1841); Jordan,* Trails, *pp. 44, 46 (see note 15).*

coastal expansion into North Carolina produced a significant cluster of cattle-ranching activity in the marshlands and pine barrens between Albemarle Sound and the Neuse River. The Pamlico River port of Bath became the major export depot (figure 33). The absentee owner of a ranching operation in this area wrote in 1750 that he owned "two places . . . with tennants on Shares with what we cal Cowe Penns, in which way Cattle and Hogs are easily raised under careful Industrious People." Nearby, across Pamlico Sound, the Outer Banks of North Carolina shared the cattle culture of the coastal mainland. The importance of herding in coastal North Carolina, involving "prodigious numbers" of stock, apparently misled one eighteenth-century observer, who assumed that the entire colony resembled its southern coastal fringe and concluded that "such herds of cattle and swine are to be found in no other colonies."[6] In the long run, this northeastward thrust along the coast of North Carolina proved to be a diffusionary cul-de-sac. Boxed in by the sea and the fertile plantation areas, the coastal cattlemen could not continue the spread northward. Their herding enterprise would, however, endure for well over a century in coastal North Carolina.

The more important spread of the cowpen culture complex lay westward and southward along the coastal plain (figures 33–35). There, too, the raising of open-range cattle proved far more durable than merely a frontier occupance form. In various districts scattered along the pine-barrens belt, even into East Texas, open-range cattle herding remained a viable enterprise into the early twentieth century.

The colony of Georgia, as was dictated by its location, became the first stepping stone on the route of Carolina's crackers to the Great Plains.[7] Only several years after the founding of Georgia, the colony's trustees purchased herds from South Carolina cowpen owners, including 150 cattle driven to the trust's Savannah cowpen in

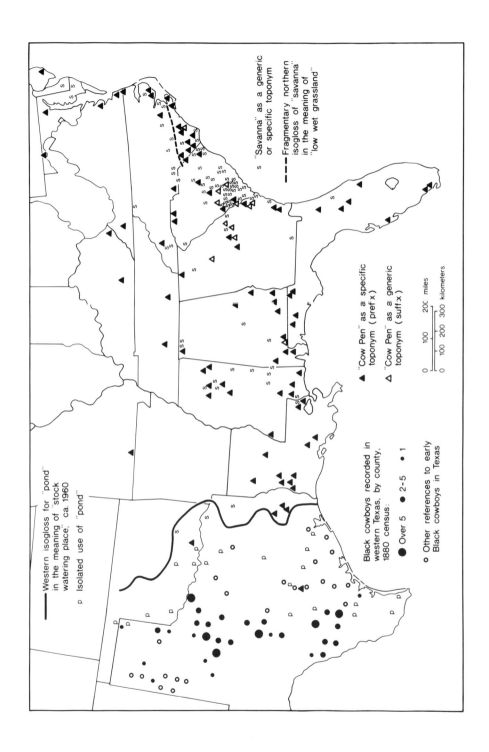

Figure 34. Black cowboys and vernacular distribution of "cowpen," "savanna," and "pond" as evidence of diffusion of the Carolina herding complex. Some of the toponyms are archaic. Sources: United States Geological Survey, National Mapping Division, Office of Geographic and Cartographic Research, Branch of Geographic Names, "Geographic Names Information System" Reston, VA) (alphabetized indices of place names appearing on U.S.G.S. topographic sheets, available on microcards or computer print-outs, by state units); Hans Kurath, Word Geography of the Eastern United States *(Ann Arbor, MI, 1949)*, pp. 47, 61 and fig. 91; John H. Goff, "Short Studies of Georgia Place Names," Georgia Mineral Newsletter, 16 (1963), pp. 88–92, 97; Fred Tarpley, From Blinky to Blue John: A Word Atlas of Northeast Texas *(Wolfe City, TX, 1970)*, p. 304; Names in South Carolina, vols. 1–20 (1954–73); Saunders, Colonial Records, vol. 1, p. 385, vol. 3, pp. 431–32, vol. 4, p. 53 (see note 6); Goff, "Cow Punching," p. 348 (see note 7); Atwood, Regional Vocabulary, p. 156 (see note 20); DeVorsey, Indian Boundary, p. 154 (see note 2); William DeBrahm, "South Carolina and a Part of Georgia," (map) (n.p.: Thomas Jeffries, 1757) (copy at University of Georgia Archives, Athens); Baxter, "Cattle Raising," p. 13 (see note 15); Brooks, "Cattle Ranching," p. 67 (see note 3); Gallay, Formation, p. 85 (see note 19); Virginia Browder, Donley County, Land o'Promise *(Quanah, TX, 1975)*, p. 91; Alma W. Hamrick, The Call of the San Saba *(Austin, TX, 1969)*, p. 40; Philip C. Durham and Everett L. Jones, The Negro Cowboys *(New York, NY, 1965)*; Jordan, Trails, p. 144 (see note 15); U.S. Census, manuscript population schedules, 1880, for western Texas.

1734. By 1738 the trust also owned a "Cow-Pen" at Old Ebenezer, including "a great Stock" of cattle (figure 26). The *pinder*, or head herder, there, together with his counterpart at the trust's Savannah cowpen, was ordered to round up the "near a Thousand" cattle estimated to be running loose in the "North Part of the Province." The Georgia trustees continued for years to maintain a large cowpen. In 1747, for example, they instructed "that the Cowpen Keeper be ordered, that as the Season for Hunting and Driving up Cattle is near at Hand, He immediately puts all his Fences in good Repair." Purchases from the government cowpen were mentioned as late as 1752. Another cowpen, privately owned, had been established by 1734 on the Georgia side of the border "about six miles from Savannah."[8]

Because of the trustees' policy of maintaining Georgia as a slaveless smallholder haven based in compact settlements, cattle raising progressed slowly at first. Only after 1749, when antislavery restrictions were dropped, did South Carolinian immigration begin in earnest, with cattle raisers in the vanguard. "To make room for the yearly immense Increase" of cattle in South Carolina, wrote contemporary

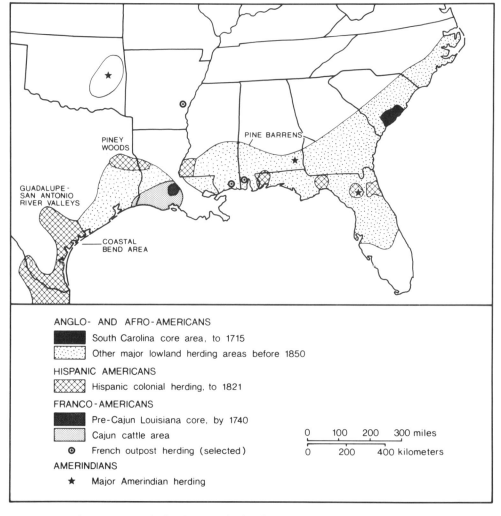

Figure 35. Cattle herding in the lowland South, to 1850: A culturo-ethnic view. Carolinian diffusion and cultural mixing occurred in numerous coastal regions and districts. Source: Jordan, Trails, p. 40 (see note 15).

observer William DeBrahm, "great Herds ... have been driven ... into the Neighbouring Province of Georgia ... since 1757," particularly into the frontier interior between the Savannah and Ogeechee rivers. There, he continued, the cattle were "kept in Gangs under the Auspicie of Cowpen Keepers which move ... from Forrest to Forrest in a measure as the Grass wears them out, or the Planters approach them." The planters' smaller herds, reported DeBrahm, were "prejudicial to the great Stocks, from among which the former draw the Bulls, and sometimes the Calves . . . and as the Cows follow the bulls also, great Gangs are apt to be misled to the Pasturage near the Plantations, which not affording sufficient Range for a great stock, the Cattle are in Danger to grow poor and sick, for which reason the Cowpen Keepers prefer their Solitude to the Neighbourhood of Planters." In 1755 Georgia adopted herding laws similar to those of South Carolina, requiring for example that cropland rather than pasture be enclosed with fences. Mart Stewart's self-contradicted claim that Georgian open-range cattle raising was in considerable measure indigenous in origin and distinct from that of Carolina should be disregarded, in favor of his statement that "upcountry cattle-raising followed the South Carolina model."[9]

By the time of the Revolutionary War, the cowpens of Carolina's children were scattered through much of the inner Georgia coastal plain and pressed against the Indian frontier in the west. An anonymous observer in the 1770s described the scattering of rural population in Georgia, noting that "they generally plant themselves at a distance, for the sake of having an uncultivated country around them for their cattle to range in."[10] William Bartram, a learned traveler, left a rather detailed description of one of these Georgia cowpens on the lower Savannah River in the 1770s. The owner had some 1,500 head of cattle, of which about 40 were milked daily, and he used calf capture as a device to facilitate the milking process and herd control. All dairy products, including butter and cheese in addition to milk, were consumed at the cowpen by his family and laborers, which included black slaves. The fenced enclosures that formed the cowpen were approximately one hectare (2 or 3 acres) in extent. In the long run, perhaps due to environmental damage, such enormous herds disappeared from the Georgia pine barrens, or "wire-grass" country. One large operator in south Georgia in the middle nineteenth century owned 130 cattle and controlled some 2,000 hectares (5,000 acres) of range. His more typical neighbors owned about 50 cattle and a comparable number of pigs.[11] The expansion into Georgia began a long series of marketing shifts. For a time, at least, Georgia's ascendance as cattle-herding country allowed Savannah to challenge Charleston in the export of barreled beef, hides, and tallow. Later, south Georgia cattle products were often exported through Florida ports.[12]

From Georgia the Carolina cowmen quickly expanded into the

old Spanish and Seminole herding districts of East Florida, where the advent of British rule in 1763 facilitated the diffusion. The placename "cowpen" and the profession of "cow keeper" appeared in Florida coincident with the British takeover, and a later observer, in the 1820s, noted that "the greater number of the ancient inhabitants lead a kind of pastoral life, and subsist by rearing cattle."[13] Although repeated changes of sovereignty, by which Florida passed back to Spain in 1783 and to the United States in 1821, fostered disruption, conflict, and persistent outlawry, the Carolina cowpen culture thrived and grew. The Florida flatwoods, cane marshes, hammocks, and savannas—renamed "prairies"—provided an abundant setting, and after 1821 a new surge of Georgian and Carolinian herders spread rapidly down the peninsula, driving the retreating Seminoles before them. By the latter part of the century, the crackers had reached Lake Okeechobee and the Everglades. The traditional West Indian markets remained viable, served through the port of Tampa and, later, Punta Gorda and Punta Rassa. Representative of the peninsular cracker frontier, Manatee County in 1860 had thirty-seven times as many cattle as people.[14] Vital and enduring as the south Florida herding industry was, it formed another diffusionary cul-de-sac, a useful place to study relict cracker and Hispanic traits, but of little relevance to the western frontier.

The more significant Carolinian path led west through the pine barrens of southwest Georgia, the Florida panhandle, southern Mississippi, and the "Florida" parishes of eastern Louisiana (figures 33–35). Along this diffusionary route in the Old Southwest, the Carolinians repeatedly overran the preexisting herding nuclei of the Spaniards, French, and Creek Indians. Contemporary travelers reported individual herds of up to a thousand head of cattle around such places as Pensacola and Mobile, as well as sizable droves of pigs. Along the Tombigbee River in Alabama, cattle outnumbered people almost five to one by the late 1790s. A decade earlier the Natchez area of Spanish Mississippi began its impressive rise as a center of cracker cattle ranching, and a 1794 census of the Natchez District enumerated some 18,000 cattle, three times the size of the human population. Gulf seaports and river towns eclipsed Charleston and Savannah as major terminals of cattle trails, allowing Pensacola, Mobile, Natchez, New Orleans, and some lesser places to become "cow towns." The West Indies remained a major market for pinebarrens cattle, even after American independence.[15]

Beginning about 1810, cracker cattle herders leapfrogged the

Mississippi River floodplain and moved into the extension of the pine barrens in western Louisiana and East Texas, again overrunning smaller Spanish ranching enterprises. The entry into the Piney Woods of Texas began in earnest in the 1820s, and about 1835 Anglo cattle raising in these woodlands began a rapid rise. Texas tax lists for 1840 revealed twenty-seven Piney Woods Anglos owning 100 or more cattle, and a decade later almost three hundred Anglos reported herds of that size in the East Texas woodlands. The focus of the Texas Piney Woods cattle industry lay in the westernmost reaches of the longleaf belt in the lower Trinity River valley, south of the earlier Spanish colonial ranching district there (figures 29, 33). Cattle outnumbered people five or six to one by 1850 in counties such as Polk, Liberty, Houston, Jasper, and Montgomery. The largest Texas Piney Woods herd in 1850, totaling over 1,000 head, belonged to former Louisianan Raleigh Rogers of Montgomery County. A decade earlier Rogers had owned 200 cattle. In 1850 he was closely followed by Sherrod Wright of Jasper County, with 902 cattle, and William D. Smith of Sabine, with 820. Even a decade earlier, the census of 1840 had shown the cracker pine-barrens cattle-herding belt to be intact all the way from the Atlantic coastal plain of Georgia to the forests of East Texas. The thrust westward from the low country of South Carolina to the margins of the western prairies had taken only a century (figures 33, 35).[16] So rapid was the diffusion that a cowpen keeper who had learned the business in colonial South Carolina could have ended his days raising cattle in the Piney Woods of western Louisiana or even Texas.

The Carolinian presence throughout the longleaf pine belt is abundantly documented. By 1850 South Carolinians lived in large numbers all along the chain of coastal southern states. Over 30 percent of the natives of the Palmetto State resided by midcentury in Georgia, Alabama, Florida, Mississippi, and Louisiana, and one contemporary account named Carolina and Georgia as the principal sources of herders in the pine barrens of southern Mississippi. At the same census, South Carolina was the leading state of birth of owners of large cattle herds in the East Texas Piney Woods, and fully 60 percent of all such herders had come to Texas directly from Louisiana, Mississippi, and Alabama, the three pine-barren states immediately to the east.[17]

Several specific migrations are revealing. For example a certain James Alderman, born in coastal North Carolina in 1801, moved to Thomas County in southern Georgia by 1815 and again to Hillsborough County in south Florida by 1851, where he owned nearly eighteen hundred cattle by 1860. The Baxter family began as herders in the Pee Dee Valley of lowland Carolina in the 1750s, relocated to southern South Carolina by 1790, to adjacent coastal Georgia by 1805, to Feliciana in the Florida parishes of Louisiana, and finally, by 1816, to Greene County in the pine barrens of southern Mississippi. The previously mentioned Texan, Sherrod Wright, a 1795 native of South Carolina or Georgia, earned his first four cows and calves back East by serving as a substitute in the War of 1812. After being discharged, he married and moved west, driving his small herd to the shores of Lake Pontchartrain in eastern Louisiana, a district he had visited during his military service. In 1832 he migrated again, bringing sixty cattle to the Piney Woods of Jasper County, in deep East Texas. By 1840 Wright's herd had increased to one hundred. Eventually his cattle, ranging over 80,000 acres in Jasper and Newton counties, numbered fifteen hundred.[18]

The woodland cattle-herding system diffused in this manner continued to bear Carolinian traits. All the way to East Texas along the pine-barrens belt, one finds herder dogs (figure 36), stock salting, free range, calf capture and castration, milking, cow whips, gardening in old pens, range burning, stationary pasturing, slave or hired cowhands of African, European, Native American, or mixed ancestry, paired plantation/cowpen operations, simple British-derived brand designs consisting of block letters and Arabic numerals, a cattle/pig focus, a lack of supplementary feeding, one major spring roundup followed by a less-inclusive gathering of culls in fall, the preemption of large acreages of public domain from small, privately-owned cowpen steads, livestock statutes based in English common law, overland droving to market along established trails, and the use of horses merely for transportation. As in lowland South Carolina, the herders established an enduring land-use system rather than a transitory frontier occupance form. Indeed the open range would persist into postbellum times.[19] In all these ways, cattle herding in the pine forests of East Texas in the 1840s replicated the cowpens of early coastal Carolina.

Etymological and toponymic evidence abundantly reveals the Carolinian diffusion (figure 34). Such distinctive usages as "savanna," "cowpen," "pond," "cowhunt," "dogie," and "cowman" went westward from South Carolina. A volunteer company in the Confederate army from the heart of the Georgia pine barrens dubbed themselves the "Irwin County Cow Hunters" and the "Cowboys." Strewn all

Figure 36. An East Texas "cow dog," bred by V. T. "Cowboy" Williams, Jr., of rural Navasota, Texas, near the westernmost tip of the coastal piney woods in Texas. This breed, linked to the pine-barrens herding system of the Gulf South, has been used for more than a century and a half in eastern Texas for managing range cattle and hogs. (Photo by Jim Hurst, 1976, courtesy of V. T. Williams.)

through the coastal pine belt where such place names as Cow Pen Pond (Jackson County, Florida), Galphin's Cowpen (Jefferson County, Georgia), Jones' Cowpen (Washington County, Georgia), Cowpen Reed Brake (Lamar County, Mississippi), Cowpen Gully (Vernon Parish, Louisiana), Cowpen Branch (Tyler County, Texas), Cowpens Creek (Henry County, Alabama), Savanna Swamp (Autauga County, Alabama), Savannah Bay (Echols County, Georgia), and Savannah Branch (St. Tammany Parish, Louisiana). Many of these survive to the present day. However, *savanna* eventually gave way to *prairie* westward from Carolina, perhaps reflecting Gulf Coast French influence, and the word has essentially disappeared from the vernacular. Similarly "cowpen" in the Carolinian-Jamaican meaning of "ranchstead" apparently did not progress much beyond eastern Alabama, although the word, in its more traditional English meaning, diffused as far as

Texas. The term *cowpen herd*, probably derived from the coastal South, reached parts of the West, where it designated the holdings of a small-scale rancher. Perhaps the single best vocabulary indicator of Carolinian herding influence is the word *pond*, the dominant term for a stock-watering place all the way into eastern Texas and Oklahoma (figure 34).[20]

Although a Carolinian imprint can be detected far to the west, the coastal southern cattle complex also abundantly absorbed other cultural influences during its diffusion. Some of these came from the highly successful, Pennsylvania-derived backwoods colonization system of the eastern forests, a system that challenged the crackers for domination of the southern forests. Borrowings from the Pennsylvanian culture included notched-log construction and several highly distinctive fence types. The wattle stock pens of colonial South Carolina yielded either to the sturdy Pennsylvania zig-zag or "worm" rail fences, strengthened by X-shaped braces or "stakes," upon which rested a locking "rider" rail, or to the even stronger "straight-rail" fence, also of Pennsylvanian origin and consisting of pairs of posts placed at regular intervals and rails stacked horizontally between them (figure 37).[21]

Conceivably the Carolina herding system also absorbed additional highland British influence in the coastal paths of diffusion. Many Scottish highlanders fled to the American South after their crushing defeat in the 1740s, and a notable concentration developed in the inner coastal plain of southern North Carolina, centered in the Pee Dee drainage (figure 26). The county named Scotland remains as a reminder of the highlander settlement. However, the Scots arrived late and positioned themselves poorly to have much influence on cowpen culture. In British Florida a Devonshire immigrant named Denys Rolle helped redevelop cattle raising in the St. Augustine area and personally owned a herd of one thousand head, while other Britons pursued herding interests in the Pensacola region. Even under renewed Spanish rule after 1783, a British company operated cattle ranches in Florida.[22] At the very least, these commercial operations offer a potential prototype for later British capitalist investment in Great Plains ranching. Most likely all of these British efforts simply drew upon the indigenous American herding system, rather than injecting renewed influence from Great Britain and Ireland.

Adding further ethnic complexity to the westward diffusion of the coastal cattle-herding complex were indigenous people of the "Five Civilized Tribes," who by 1780 had experienced abundant

Figure 37. A "straight-rail" stock pen in Blanco County, Texas. These sturdy enclosures, also called "stake and rail," "post and rider," and "post and rail" fences, are derived from the Midland (or Pennsylvanian) backwoods culture and entered the Carolinian cowpen complex in Georgia and the upland South, replacing earlier wattle fences of British origin. (Photo by the author, 1990.)

contact and mixing with Spaniards, Anglo-Americans, French, and Africans. The Seminole and Creek efforts at cattle ranching in the Southeast have already been mentioned in chapter 4. Aside from facilitating the transfer of Hispanic influence to the Anglo-Americans, the Indians also participated in the westward diffusion of cattle raising. Creeks, Choctaws, and Cherokees all herded longhorn cattle in the Piney Woods of East Texas well before 1840. They also introduced cattle ranching into eastern Oklahoma, where "cowpen," "pond," and other evidence of Carolinian influence can still be detected (figure 34).[23]

Far more consequential in the westward coastal diffusion of the Carolina cowpen culture were renewed and almost continual contacts with Spanish cattle ranching, both indirectly, by way of Native Americans and Hispanicized Gulf Coast French, and directly, when crackers repeatedly intruded upon Spanish-ruled provinces in Florida, Mississippi, Louisiana, and Texas. In addition, as we have seen, repeated impulses from Mexico penetrated the western coastal area of the South (figure 29). Creolization, with additional Hispanic influence, was the result for the herding culture of the Gulf. Representative of the Latinized way of life were the Floridians of the 1820s, who exhibited "a compound of Spanish, French, and American manners."[24]

Perhaps the most obvious index of renewed Hispanic influence was an increase in Iberian bloodlines in the cattle population, a result of the previously established longhorn herds in the Gulf coastal area and of imports from Texas and Coahuila in the late 1700s (figure 17). Early Georgia cowpenkeepers disliked Florida longhorn stock, but sought in vain to prevent crossbreeding with their British-derived herds. Indeed, feral English cattle from South Carolina and strayed Spanish stock from Florida had colonized Georgia before European settlement began in that colony, initiating the hybridization process. In southern Mississippi the longhorn breed remained dominant until about 1840. Spanish-derived Andalusian ponies also achieved dominance, to the extent that the local horses of the Mobile area in the 1820s were called "Spanish tackies." The English-derived quarter horse of the South, ancestor of the "cutting" horse of the Great Plains, displayed thereafter a trace of Spanish blood.[25]

With increased Spanish influence also came fuller development of the Indo-Anglo-Afro-Romance lingua franca that had begun forming in the West Indies. By about 1790, for example, Anglo-Americans in the Mobile and Natchez areas had added to their vocabulary additional corrupted Spanish loanwords such as *calaboose* (jail), *caviard* (or cavvyyard, a group of saddle horses used as rotated mounts), and *cabras* (halter, from *cabestro*). Some of the Spanish terms later used in ranching districts of the American West entered the English vernacular in the Gulf coastal South, well east of Texas.[26]

A Prairie Readaptation

The continual infusion of Hispanic influence in the cattle complex of the Gulf Coast took on heightened vigor and significance in the littoral belt west of the Mississippi River, in the southwestern part of Louisiana and in the southeastern corner of Texas, the latter area known in the 1820s as the Atascosita District. There, south of the great pine belt, the migrating cattle raisers entered a fundamentally different floral environment, leaving behind the wooded lands and entering the humid tallgrass coastal prairie. They had departed the eastern forests and encountered the grasslands of the West, passing across the most basic ecotone in North America (figure 29). The Carolinian cattle herders, in making that transition, readapted their system to the new setting, a readaptation that helped prepare them for the West.

The coastal prairie, improbably wedged between the pine barrens and the sea, consists of two major subdivisions. Nearest to the coast lies a narrow strip of salt marsh, a zone inundated periodically by the hurricane-aroused Gulf. Surrounded by the marshes and parallel to the coast stand elongated, low, sandy ridges called *cheniers*, covered by an open forest of live oaks. Adjacent to the marshes on the north is the second, broader, and better-drained subdivision of the coastal prairies, dominated by tall, coarse grasses, mainly species of *Andropogon, Panicum*, and *Paspalum*, particularly bluestems. Both marshes and prairies were fired in February, perpetuating West Indian and pine-barrens custom. Cutting across the coastal prairie at regular intervals are numerous rivers, creeks, and bayous, each paralleled by a ribbon of galeria forest and canebrakes. These interruptions serve to partition the grassland into constituent prairies, which often take the name of the adjacent stream. Thus the Calcasieu and Sabine prairies of southwestern Louisiana lie adjacent to their namesake rivers.[27]

Cattle utilized all of these local floral ecosystems. In the short subtropical winters they clung to the galeria forests, canebrakes, and cheniers, emerging onto the prairies in March, after the annual burning. Many drifted into the marshes, obtaining all the salt they needed simply by grazing, thereby rendering useless the Carolinian herd-control device of salting. These "swampers" became almost fully feral. All things considered, the subtropical, moist coastal prairies offered a splendid natural setting for cattle, one in which they could have thrived without any human interference.[28]

The herders, by contrast, had to make substantial changes. Among other adjustments, they decreased their reliance upon pig raising, a forest-based enterprise, and focused upon cattle. Far more significantly, they adopted the management of cattle from horseback, including roping, a practice absent in the Carolinian and West Indian complexes. Both Anglo- and Afro-Americans, roughly at the longitude of the Mississippi River, began using the Mexican-introduced lasso and acquiring the related equestrian skills. Soon the lasso prevailed among all cattlemen who came to the coastal prairies, regardless of ethnic origin. By the 1830s at the latest, the use of the lasso by Anglo-Americans, and then their African slaves, had spread westward through the prairies to the shores of Galveston Bay.[28] Coincidentally the word *lasso* entered the coastal herding lingua franca, although Anglo-Texans would later come to prefer the alternative loanword, *lariat* (figure 29). For reasons that remain unclear, the garrocha, unlike the lasso, apparently did not gain acceptance by non-Hispanic herders in Louisiana, or elsewhere for that matter, perhaps because it was not needed for steers. On the other hand, one must explain the curious Anglo terms "cowpoke" and "cowpuncher," which imply the use of a goad. The modern cattle prod could be interpreted as a latter-day revival of the Andalusian garrocha. The hocksing pole, similarly, did not pass to the Anglos, having been rendered unnecessary by the castration knife and the shift to beef rather than hide production.[29]

This hybrid Carolinian-Tamaulipan or "Texas system" of ranching, as practiced by the "old-established herdsmen of the coast-prairies" in southeastern Texas and adjacent Louisiana, was revealingly described in 1854 by the noted traveler Frederick Law Olmsted.[30] His account provides a clear picture of the nature of this system as it had developed in the area of origin (see figure 39). Much about the larger-scale coastal prairie "stock-farms" and their resident "first-class Texas graziers" remained Carolinian and Jamaican. The ground adjacent to one typical "large but rude" house "was nearly all fenced and divided into large and small pens," two of which enclosed a garden and cornfield, and "the only system of manuring in use is that of plowing up occasionally the cow-pens." Calf-capture and the British milking culture complex remained vital, for "in a small pen are kept a number of calves," to which the cows "come in morning and evening," allowing a dozen or so to be milked. Calves were sometimes weaned by splitting their tongues with a knife.

Olmsted visited the region in April, "the season for the annual gathering and branding of the calves," some five hundred of which represented "the year's increase of their herd." The cowhunt, called a "drive," was a collective venture by neighboring ranchers and lasted two to three weeks. "They first drive the outer part of the circuit, within which their cattle are supposed to range, the radius of which is here about forty miles." The hunters used two retaining pens "at different points upon the circuit" of the range. "It may be conceived that the labor of gathering and confining in pens so many wild cattle . . . is not slight." In early summer, "after the cattle are in good condition with the spring pasture," drovers purchased fattened four-year-old steers and took them overland to market at New Orleans. At other times of the year the herds were largely left alone, to fend for themselves, suffering much from flies and ticks. The local vocabulary, including words such as "hummock" to describe patches of higher-lying oak land, also reflected the coastal southern connection.

But if much Carolinian was evident from Olmsted's description, Mexico's imprint could also be seen. For supper one grazier "proceeded to lasso and slaughter a bullock," and in the milking pen a lasso served to pull the nursing calf away from the udder. When wild horses were broken for riding, usually at four years of age, "a lasso was first thrown over the neck" and snubbed to a tree, to facilitate bridling, saddling, and mounting. The saddle type in use locally was "Spanish" and had "attached to the pommel a strong wooden bar, rolled in sheepskin" to help the rider withstand the bucking of the horse. In cowhunts each man brought "two or three extra horses" as a cavvyyard, to be "driven before the company" as extra mounts. "Horsemanship" was an essential ingredient of the annual roundup. Aside from castration, "no pains were here taken to improve the breed," by which Olmsted meant to say that Iberian longhorn blood prevailed.

The prairie environment had prompted other changes in the Carolinian system as well. Line riding, later to be a principal Texan device of herd control on the Great Plains, had become established in southeastern Texas. Olmsted noted that "once every month or two they rode through the range, driving in the cattle that were ranging wide." In addition, pigs had all but disappeared from the coastal-prairie system, for "hogs do not flourish upon the grass." Fresh or "jerked" beef had replaced pork on the local dinner tables.

This, then, was the nascent Texas cattle-ranching system in its coastal prairie cradle. Mexican equestrian skills, equipment, and new loanwords, as well as Iberian bovine bloodlines, had been joined to the lowland Carolinian cowpen culture. The coastal dual monarchy of cows and pigs had yielded to the grassland cattle kingdom. If in the final analysis the distinctively Anglo-Texan ranching subculture of the Great Plains arose, as Walter Prescott Webb suggested, at the place where Anglo-Americans began tending cattle from horseback, then that hearth lay in late eighteenth-century southern Louisiana, at the eastern margin of the great prairies (figure 29). The traditional preference for mid-nineteenth-century South Texas as the source area of the Anglo-Texan horseback system is seventy-five years too late and 700 km. (400 mi.) too far west.³¹ Anglo-Americans bearing a mixed Carolinian-Mexican herding culture crossed the Sabine River from Louisiana mounted on horned saddles and swinging their lassos above their heads, accompanied by black and mixed-blood slaves possessing the same skills. "Anglo-Texan" ranching, for want of a better term, was born on the humid prairies of Louisiana.

Midwives to this birth were the Cajun French, whose role in the evolution of western ranching has never been evaluated. Cajuns began settling the coastal prairies of southwest Louisiana about 1765, just when Mexican influence began modifying the older Creole French vacherie culture on the Attakapas prairie around St. Martinville. Quickly the Cajuns became "a pastoral people," and while they never became as proficient as the Mexicans at horsemanship and roping, they perpetuated the important Francophone cattle industry on the coastal prairie. All the while the Cajuns interacted and intermarried with Anglos, French-speaking blacks, and the local Indians, displaying the genius for mixing and blending that gives Cajun culture such vitality and durability. The cattle-brand registers of the Louisiana prairies after 1765 reveal a mixture of Cajuns, Anglos, free blacks, and Native Americans. Jerked and salted beef, tallow, lard, and tanned hides became major Louisiana exports by the 1770s, and the Spanish rulers began awarding large ranching grants (figure 29).³²

Initial cattle-ranching expansion in the Louisiana coastal prairie occurred around the pre-Cajun, pre-Spanish nucleus in the Attakapas district, and in the 1780s the industry, drawing upon Hispanic techniques, spread through the Opelousas prairies to the north, where Cajuns joined in force. By the time of the American Revolution, the Cajun parishes of the Louisiana coastal prairie housed five to seven times as many cattle as people, a ratio that increased to fifteen to one by the early nineteenth century. Well into the 1800s, the Attakapas prairie remained "the centre of the land of shepherds and the paradise of those who deal in cattle," while in the Opelousas area some graziers owned "above 15,000 head of horned cattle and 2,000 horses and mules."³³

At the same time, westward expansion along the coastal prairie occurred. About 1820 Cajun, Anglo, African, and assorted mixed-blood cattle raisers and cowboys began crossing the Sabine into the prairies of southeastern Texas, bearing a herding system well preadapted for the western grasslands. The Cajuns, more tied to place and kin, would not accompany the Anglos beyond the lower Trinity River in any substantial numbers, but they nevertheless assisted greatly in the prairie readaptation. Southeast Texas would, accordingly, belong mainly to the Carolina-derived Anglos. Perhaps representative was Micajah Munson, a stock raiser born in South Carolina about 1789, who came as a child to Mississippi, resided in Louisiana in the early 1820s, and entered the Atascosita District on the Texas coastal prairie about 1824. Fully 86 percent of the cattle raisers of the prairies west of the Sabine immigrated directly from Louisiana, Mississippi, or Alabama. Concealed in these totals were some thoroughly creolized genealogies and diverse ethnicities. For example the largest cattle raiser in early Jefferson County, Texas, bordering the Sabine, was a "redbone" of mixed white, black, and Indian ancestry.[34]

Clearly the Louisiana grassland readaptation of the pine-barrens cattle herders yielded success. By 1850 the Texas coastal prairie between the Sabine and Trinity rivers was home to thirty-eight persons owning five hundred or more longhorn cattle, including five Cajuns and four "redbones." Travelers even as early as the 1830s reported "immense herds of cattle," including five thousand head owned by Taylor White, a rancher on the eastern shore of Galveston Bay. To the present day, cattle still rule the salt marshes east of the Bay. Many of the early southeast Texas graziers continued to use slave cowboys, just as they did back East, but they had been converted into mounted herdsmen skilled in the use of the rope.[35] The prairie coastal Carolinian herders, by now diverse in genealogy and even more so in techniques, were in many important ways ready for the West, a readiness acquired east of the Sabine.

Through the Upland South

While these enormously important events transpired in the Gulf coastal plain, other Carolinians carried the cowpen culture complex into upland, interior portions of the South. Beyond the fall line, which marked the inner border of the plain, lay the Piedmont, a hilly, plateaulike region lying atop ancient rocks. It presented a marked contrast to the geologically youthful sands, sedimentary rocks, tidal marshes, and barrier islands of the low, flat coastal plain. The Piedmont became the first upland area of the South to be colonized by the cattle folk and became so important for herding that some have mistaken it for the hearth area of Carolinian ranching.[36]

One major inward thrust of the cattle raisers that reached the Piedmont came from the coastal district northeast of Charleston, beyond the Santee River (figures 26, 38). Herders followed the rivers

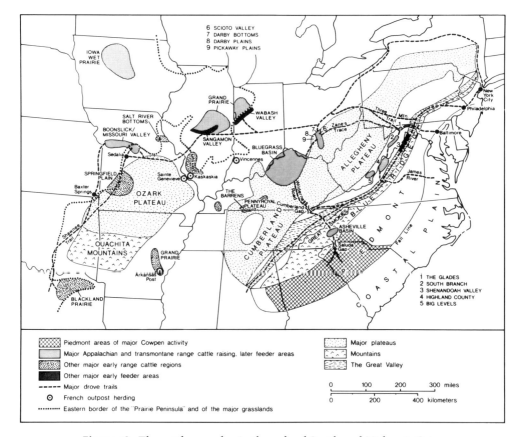

Figure 38. The cattle complex in the upland South and Midwest. Cattle raising was a transitory frontier occupance phase in this region. The feeder areas became prototype nuclei of the Corn Belt. In the Highland County area of Virginia are found the headwaters of the Calf Pasture, Bull Pasture, and Cow Pasture rivers, as well as the South Branch of the Potomac. The Arkansas Grand Prairie is perhaps best regarded as a lowland southern herding area. Sources: Henlein Cattle Kingdom; Wheeler, "Beef Cattle Industry," p. 32 (see note 56); Jordan, Trails, p. 40 (see note 15); Drago, Great American, p. 11 (see note 61); Hilliard, Hog Meat, p. 194 (see note 15); Hewes, "Northern Wet Prairie," p. 315 (see note 65).

inland, along the Pee Dee, Lynch, and Black after about 1730. Some crossed the low divide to the Wateree, Congaree, and Saluda—all belonging to the upper Santee drainage. Ascending above the Fall Line, they entered the Piedmont and pressed against the retreating Cherokees, who were slow to adopt cattle raising. Other cowmen from the inner coastal plain around Orangeburg joined them in an expansion along the axis of the Saluda River. By the 1750s cowpens were sprinkled through the outer South Carolina Piedmont, and a generation later the continued Cherokee retreat allowed cattle herders to approach the mountains. One cluster of ranches, just below the mountains near present Spartanburg, lent its name to the well-known revolutionary Battle of the Cowpens.[37]

The Piedmont of adjacent North Carolina also experienced this intrusion by lowland cowpenkeepers. In 1755 an official in that colony observed that "the great increase of settlements in this Province of late in the Countries at a distance from the sea, has increased the Breeding of Cattle and Hogs," and a decade later North Carolinians had become alarmed by the magnitude of the bovine invasion. In 1766 they passed a law "to prevent the Inhabitants of South Carolina [from] driving their Stocks of Cattle from thence to range and feed in this Province." "Whereas of late years," read the act, "many of the Inhabitants of South Carolina have made it a Practice to fix Cowpens, and settle People with Large Stocks of Cattle (though they are not Owners of any Lands) in this Province, which destroys the Range, and greatly injures the poor inhabitants of several of the Counties bordering South Carolina," landownership henceforth would be a prerequisite for using North Carolina's public range. The Georgia and Alabama Piedmont also eventually received South Carolinian herders, as is reflected in place-names such as the town of Cowpens in Tallapoosa County, Alabama. South Carolina, however, remained the cattle-raising focus in the Piedmont.[38]

While important, the Piedmont never rivaled the coastal plain as a center of cattle raising, nor did the cowmen long abide there. In the interior South, the production of lean cattle remained a transitory, Turnerian-frontier occupance type. Numerous factors combined to lessen the importance and durability of the Piedmont as a grazing region. Outlawry played some role, as is revealed in a 1767 petition by cattle raisers in the backcountry of South Carolina, complaining that "our large Stocks of Cattel are either stolen and destroy'd—our Cow Pens are broken up—and all our valuable Horses are carried off . . . by these Rogues."[39] Bovine diseases also took a heavy toll around midcentury. Further contributing to the difficulties of the Piedmont was its more remote access to the ports serving the West Indian trade in barreled beef and tallow. Coastal-plain graziers had a locational advantage in this respect. In response Piedmont cowmen as early as the 1740s began driving lean steers far northward, to Philadelphia and even New York City, using a trail that passed through Saluda Gap, crossed through the Great Smoky Mountains, and then turned northeastward through the remarkable structural trough known as the Great Valley of the Appalachians to the immediate hinterland of Philadelphia (figure 38).[40]

The main reason for the lesser importance of the Piedmont in open-range cattle raising was environmental. A mesothermal, midlatitude climate prevailed above the fall line, a condition clearly reflected in the deciduous hardwood forest of oak and hickory that dominates most of the upland South. The cowpen system had Jamaican and Carolinian roots in tropical or subtropical lowlands, and a major readaptation of the culture complex had to occur in the colder uplands. Adaptive systems simply do not cross major environmental boundaries unaltered. The most significant problem in the Piedmont was the lesser winter carrying capacity of the range for cattle, a reflection of the longer, often bitter winters. As a result, ratios of cattle to population in the Piedmont never approached the magnitude of those in the coastal plain. Moreover the heavier, clay-rich soils of the Piedmont proved more attractive to farmers than did the sandy pine barrens of the low plains, causing population densities quickly to become too great to allow a large-scale open-range grazing presence.[41]

Even so the Piedmont ranges had much to offer the herds of cattle. Abundant "glades"—a distinctive piedmont/mountain term—provided canebrakes similar to those in the lowlands. Grass and clover grew among the hardwood trees, and small dry prairies abounded, still often called "savannas," a term that would eventually lose out to "barrens" in the uplands. Peavine, or "partridge weed," provided abundant and nutritious forage for cattle.[42]

In spite of this abundance, a major readaptation occurred. Pigs increased in importance, since they thrived better in the mast-rich woodlands. This shift in the cattle:pig ratio, though to a degree spontaneous and Darwinian, prompted as it was by the floral environment, also reflected increased Carolinian interaction with the backwoods pioneers of Pennsylvanian origin, who spread southward into the Piedmont bearing their highly successful forest-colonization culture, in which cattle stood decidedly secondary in importance to pigs. Herders who would have owned hundreds or even thousands of cattle in the pine barrens usually kept mere scores on the Piedmont, compensating for the loss with increased droves of swine. The smaller herds, as a rule, remained tamer. Piedmont cowpenkeepers managed these relatively docile animals quite effectively with salting, whips, herder dogs, and calf-capture. Some even reverted to the older, more intensive methods of belling dominant cows and gathering the herd at night.[43]

Another readaptation involved the rise of vertical transhumance, a practice designed to maximize the amount of forage available by abandoning stationary pasturing. This began by the 1760s at the latest, when cattle raisers along the upper Saluda River sent their herds, communally grouped, in the summer to the mountains bordering the Tugeloo River in westernmost South Carolina, in the Cherokee lands (figure 26). This mountain range, part of the larger Appalachian Blue Ridge complex, measured about 50 by 100 km. (30 × 64 mi.) and reputedly housed twenty thousand transhumant cattle each summer, under the care of hired herders. In part southern Appalachian transhumance exploited the well-known "balds" of the Blue Ridge and Great Smokies, open places of mountain oat grass and heath that lay at higher elevations.[44]

Transhumant Piedmont cowboys departed for the mountains in April, luring the bell cows, each with a following of several hundred head, along the trail with salt. Earlier, at the end of March, the mountain pastures had been fired to enhance new growth, and the steers and cows, while still on their winter range, were deprived of salt. Once in the high pastures, each herder, resident in a log cabin, could control up to fifteen hundred head of cattle, largely by providing salt every eight to ten days. In autumn the herds descended again to the Saluda Piedmont, at which time culling and marketing took place. During the Revolutionary War, some Saluda graziers adopted a more completely nomadic existence in order to escape British depredations, migrating for years through the mountains with their herds and covering great distances.[45]

The hired men who tended the transhumant herds of the Saluda were often of European birth. Many probably came from highland herding districts of Great Britain and Ireland. Almost certainly the emerging system of vertical cattle transhumance along the Piedmont-Appalachian border represented a renewed infusion of British influence. The image of docile animals filing peacefully under the leadership of belled cows reveals a radical change from the coastal-plain system, based in essentially stationary and semiferal stock.[46]

Another fundamental readaptation to the lesser amounts of forage available on the Piedmont involved sending some lean cattle northward to separate fattening farms, where they could be upgraded for market. The Piedmont graziers, instead of producing fattened cattle, trailed stockers to be finished elsewhere. The early feeder areas, as would be expected, lay along the Appalachian Great Valley trail to the northern markets. Some lean cattle went all the way to the wet pastures of Moyamensing, on the outskirts of Philadelphia, for fattening. More important as a feeder area, beginning at least by the 1760s, was the western part of the Shenandoah Valley, a segment of the Great Valley lying in northern Virginia. Prominent from the very first as feeders were farmers of Pennsylvania "Dutch" origin, who had settled in the Shenandoah region. The separation of cattle breeding and fattening into two spatially and ethnically segregated enterprises, connected by drove trails, thus became an essential trait of the upland southern cattle economy.[47]

If the Carolina herders spreading into the upland South modified their system in response to a different physical environment, accepting additional British influences and finding new "Saxon" market towns to serve, they also drew more remote from Spanish and African influences, in contrast to their kinsmen in the Gulf coastal plain. To generalize and risk oversimplification, we might conclude that the cattle-herding system spreading through the southern uplands and, later, the Ohio Valley and Midwest, emerged beyond the Mississippi as a largely British highland system, while its counterpart in the Gulf coastal plain became largely Hispanicized. The Old World Atlantic fringe herding complexes, after lengthy contact in the tropics, reassembled according to their proper latitudes and topographies in the eastern United States. What Carolina's children carried successfully into the southern mountains were elements originally highland British—salting, herder dogs, small black highland breeds, whips, calf-capture, the milking complex, penning over, calf castration, and overland drives to Saxon markets and feeder areas. What they left behind in the lowlands was generally Hispanic or African. In this respect a comparison of upland southern and Australian cattle ranching is instructive. Both had dominantly highland British roots, featured mounted, whip-wielding herdsmen, utilized cow dogs, depended upon castration, and employed overland marketing along established drove trails.[48]

Into the Mountains

That the upland cattle herders of the Carolina Piedmont should have entered and settled parts of the southern Appalachians was inevitable. Adjacent to the Piedmont, the mountains lay astride the westward path. Furthermore the readaptation of the Carolina herding system on the Piedmont had prepared it well for occupance of the Appalachians. In fact transhumance had very early linked the Piedmont functionally to the mountains, and routes of seasonal stock movement generally became later paths of migration. Equally important in guiding the herder penetration of the mountains was the drove trail leading through Saluda Gap and into the Great Valley.

Lured into the Appalachians along these routes of trade and transhumance, the Carolinians initially entered the eastern part of the mountain range, known as the ridge-and-valley section (figure 38). As its name implies, this sector of the Appalachians consists of a succession of strikingly parallel ridges, oriented in a northeast-southwest direction, with elongated valleys between. The easternmost heights, marking the western border of the Piedmont, is the Blue Ridge. In the south the Blue Ridge broadens to form a mountain complex, including the Great Smokies. Immediately west of the Blue Ridge lies the Great Valley, beyond which an orderly succession of lesser ridges and valleys lends a geometric corrugation to the land. Along the western side of the ridge-and-valley sector towers the formidable Appalachian Front, locally bearing names such as Cumberland and Allegheny, marking the transition into the fundamentally different plateau section of the mountains, where a dendritic pattern of dissection prevails and all hint of parallelism vanishes.

The Carolinian occupance of the mountains began remarkably early. As early as 1755, an English soldier passing along Braddock's Road, west of the "settlements" in the Maryland sector of the ridge-and-valley country, noted that "we now got into the Cow-pens," explaining that "a Cow-pen generally consists of a ... Cottage ... in the Woods, with about fore-score or one hundred Acres inclosed with high Rails," divided into calfpens and garden patches for corn or vegetables. This particular mountain cowpen cluster probably was situated in The Glades, a cane-rich area near Cumberland in present Allegany County, Maryland. Continuing his description, the soldier reported that a single cowpen "may perhaps have a Stock of four or five hundred to a thousand Head of Cattle." These ran "as they please in the great Woods, where there are no Inclosures to stop them." In early spring at the time of calving, "the Cow-Pen Master, with all his Men, rides out to see and drive up the Cows with all their new fallen Calves," placing them in pens, following the Carolinian custom of calf-capture. When "the Calf grows Strong, they mark them" and, "if they are Males they cut them and let them go into the Wood." In a second cow hunt, "in September and October, they drive up the Market Steers that are fat and of a proper Age." As a rule, "they reckon that out of 100 Head of Cattle they can kill about 10 or 12 Steers and four or five Cows a Year." The cowpenkeepers "live chiefly upon Milk," including also "Whey, Curds, Cheese and Butter," and they "also have Flesh in Abundance, such as it is, for they eat the old Cows and lean Calves that are like to die." Though disdainful, the soldier concluded that "the Cow-Pen Men are hardy People ... almost continually on Horseback, being obliged to know the Haunts of their Cattle."[49]

In addition to The Glades, certain other districts in the Appalachian ridge-and-valley country became major cattle-raising centers. Just beyond Saluda Gap, near the Carolinian point of entry into the mountains, lay the Asheville Basin, embedded in the Blue Ridge, and it became a very early montane focus of cattle raising. Buncombe County, of which Asheville is the seat, initiated brand registration rather early. From the Asheville area, the old drove trail followed the French Broad River through the Great Smokies into the Great Valley. Moving against the prevailing downvalley migration of Pennsylvania's Scotch-Irish, the Carolinians, still following this market trail northward, became established in some of the lesser glens west of the Great Valley. The Glades was such a place. So was the South Branch country, lying along the Potomac fork of that name and centered in present Hardy County, West Virginia, where another major cattle-raising center developed. The headwaters of the South Branch lie in Highland County, Virginia, near the sources of the south-flowing Bull Pasture and Cow Pasture rivers, and this watershed district also became an important producer of stocker cattle, as did the Big Levels, along the Green Brier River around Lewisburg, in southern West Virginia.[50]

Most producers of lean Appalachian cattle, even in these major districts, operated on a modest scale, usually owning less than a hundred head of basically tame stock. Some, essentially farmers, had fewer than twenty. Many of the smaller producers did not reside in the major cattle districts, but instead in less-endowed back valleys. It was mainly in reference to such small-scale operators that mountain travelers late in the century wrote of Kentuckians "who have ... long been in the custom of removing ... farther back as the country becomes settled, for the sake of ... what they call range for their cattle" and of Tennesseans who used "wild range," where their herds "support themselves among the reeds, pea-vines, rye-grass, and clover."[51] All through the ridge and valley sector of Virginia, Maryland, and West Virginia, one finds vegetational place-names suggestive of early cattle herding, such as Meadows of Dan, Barren Springs, Bald Mountain, Glade Springs, and the trio of James River headwaters named the Cow Pasture, Bull Pasture, and Calf Pasture rivers (figures 34, 38). To a degree, then, lean-cattle production occurred all through the mountains. If, however, we limit our attention to a "ranching frontier," then only the major montane cowpen centers would qualify for inclusion.

Great or small, the Appalachian producers of lean cattle remained linked to the drove trails, feeding districts, and northern urban markets. Other, supplementary marketing patterns arose as the mountain stocker-cattle industry grew, including a James River Valley route to Richmond and the Chesapeake Tidewater plantation districts. To a degree the original Piedmont marketing flow northward eventually reversed, sending mountain hogs and cattle southward along the Great Valley and through Saluda Gap, to help supply the growing plantation economy of the Piedmont and inner coastal plain of the South in the early 1800s. In the process the Asheville Basin shifted from stocker production to livestock fattening.[52]

Carolinian influence penetrated the mountain South in several quite different ways. Perhaps most commonly, Appalachian backwoods settlers, largely of Scotch-Irish origin by way of Pennsylvania, began selling a few head of lean cattle to Carolina drovers headed north, stimulating what had been a minor aspect of their frontier farming-hunting-herding economy. In many other cases, the Carolinians themselves settled promising glady valleys and implanted the Piedmont type of cattle raising. Their etymological spoor tells us as much and is revealed in such terms as "savanna" and "cowpen." A related mountain term, "milk gap"—the pen where milking occurs—is still heard from upland North Carolina and the Virginia Appalachians westward into the hills of Kentucky and is clearly kin to the "cow gap" used in the Georgia pine barrens. Both likely derive from a Carolina lowland prototype.[53]

In this manner both Pennsylvanian backwoodsmen and Carolinian cowpenkeepers became the frontier producers of lean cattle, which went to "Dutch" and "Saxon" farmers back in the "settlements" for fattening. A clear core/periphery pattern had evolved. In the herding periphery of the upland South, the acculturation of Carolinians into a Midland or Pennsylvanian backwoods culture continued. Eventually intermarriage yielded an assimilation, though the more substantial cowmen of districts such as the South Branch and Big Levels reputedly remained "by outlook and culture lowland southerners."[54] Through it all, much of the Carolina Piedmont cattle complex survived, regionally at least, to become part of the mountain frontier culture.

Carolinian or not, the mountain herders would have to play by Pennsylvanian, Turnerian rules of ephemerality. Core always pursued and displaced periphery in this Midland American pattern, causing cowmen to give way to Germanic farmers in the lushest pastoral valleys. In part this displacement was prompted by range deterioration, a consequence of the sort of overstocking reported at The Glades in 1755. Another example is the enormous expanse of canebrake along the Ohio River, just west of the Kentucky–West Virginia border, which was pastured into extinction within a few years.[55] But the main reason the cowmen moved on west was the continual influx of secondary settlers, who converted the grazing valleys into farms, disrupting the system of vertical transhumance by depriving the livestock of winter ranges, and who raised population densities beyond levels consistent with cattle ranching.

The process of herder displacement took on a very regular and predictable pattern in the upland South. Areas that began as major centers of stocker-cattle production evolved into feeder areas. The Asheville Basin, South Branch, The Glades, and Big Levels all followed the Germanic Shenandoah Valley model and became centers of cattle fattening by the 1790s or early nineteenth century. The breeders normally opted to move on west to new ranges, rather than become feeders. Drovers often became feeders, but cowmen rarely did.[56]

Meanwhile the feeding operations themselves became more intensified. In the early years, nearly all fattening of the mountain cattle was achieved on grass, in the lush and increasingly artificial valley pastures. Toward the end of the 1700s, corn had begun to play a role as feed in the valley fattening districts. The Corn Belt, a surpassingly Germanic institution, was being born in those fertile Appalachian valleys, of solid "Dutch" and Anglo-Saxon parents. It would be the destiny of the larger-scale pastoralists continually to flee westward before this feeder system, all the while providing its bovine stocker raw material and regionally vindicating Frederick Jackson

Turner. The pattern of Mexico had been repeated: enduring pastoralism in the Gulf coastal lowlands and transitory frontier cattle ranching in the interior highlands.

Entering the Heartland

The flight of the Appalachian cowmen led them, after about 1780, to enter the eastern reaches of the continental plains of interior North America, one of the largest structural basins in the world. These plains, stretching all the way from the Appalachians to the Rockies, provided a major arena for the ranching frontier. The upland southern cattle raisers reached the interior plains after passing through and largely rejecting the plateau sector of the Appalachians, following the Wilderness Road through Cumberland Gap, the route by way of the forks of the Ohio at Pittsburgh, or one of the other, more difficult paths. Kentucky, their first destination, lay at the end of an older route of transhumance, for some overflow South Branch cattle, among others, had earlier been sent to the Kentucky Glades for summer pasture.[57]

The Bluegrass Basin, centered on present Lexington, became the first great transmontane center of lean-cattle production. Though largely surrounded by the Appalachian plateau, the Bluegrass is best regarded as an embayment of the interior plains. The basin was ideally positioned to facilitate transhumance into the Cumberland Plateau to the east and south or the Pennyroyal uplands in the southwest. If cattle remained off the Bluegrass Basin pastures during the spring spurt of growth, the grass would last most of the remainder of the year, and seasonal movement of stock occurred from the first. Soon Bluegrass stockers began reaching the South Branch, Asheville Basin, and other Appalachian feeder districts. Before long, however, much of the Basin itself became a fattening area, perpetuating the displacement process. While parts of the inner Bluegrass continued for some decades to produce lean cattle, the portions closer to the Ohio River, as early as the 1790s, were heavily engaged in feeding operations. Stocker breeding retreated to the Pennyroyal, including The Barrens, an Indian-induced grassland that very early attracted cattle raisers. In later years stockers were drawn to the Bluegrass feeders from barrens and flatwoods west of the Tennessee River, in the Jackson Purchase of western Kentucky, and even from the Chickasaw lands along the Mississippi in Tennessee.[58]

Other Kentucky cowmen crossed into Ohio by about 1790 and

began raising stockers in the middle Scioto River Valley, around present Chillicothe and Circleville, using the adjacent uplands on either side of the river as summer pastures. The Darby Plains, on the uplands above Circleville, became a particularly favored summer range, attracting transhumant herders even from Kentucky. Scioto Valley stockers generally went east to the South Branch for fattening, revealing the southern links and origins of the fledgling Midwestern cattle industry. Even before 1800, however, portions of the Scioto Valley began shifting to feeding, a transition perhaps begun in the Darby Creek Bottoms outside Circleville. The number of fattened cattle exported annually from the middle part of the valley rose to seven hundred by 1810, to fifteen hundred by 1815, and to as many as seven thousand by the late 1820s. Although some stockers still departed the valley in the 1830s, its conversion to a fattening area was by that time essentially complete. Both grass and grain feeding were employed initially along the Scioto, but as the decades passed corn gained the advantage, because it yielded more feed per acre and because the corn raised in transmontane districts could most efficiently be marketed as fattened animals. Stockers for the Scioto Valley came from as far away as western Tennessee by 1810, Missouri by 1816, and Texas by 1846.[59]

Raisers of lean cattle, once again displaced, temporarily sought refuge in the former summer pastures west of the Scioto, parts of which contained considerable amounts of prairie, as in the Pickaway Plains. So completely did feeders eclipse breeders in the Scioto Valley and, ultimately, elsewhere in the Ohio Valley, that the term "cattle king" there is reserved for the well-to-do, large-scale fatteners, who bore chunky Pennsylvania Dutch surnames like Funk. The herdsmen recalled in local lore were the "boys on horse back" who worked for the feeders, supervising the grass fattening of imported stockers, carefully watching the cattle by day and penning them at night. Largely forgotten are those who bred and rode herd on the lean frontier cattle. Carolina's children hurried almost unnoticed and little remembered through the Ohio Valley; Pennsylvania's Teutons abided and wrote the local histories. To be sure, part of the reason for the obscurity of the real Midwestern cowmen, in addition to the brevity of their reign, lay in the fact that much of the breeding of stockers was done by small farmers, who lacked the glamour of true ranchers. Similarly, part of the fame of the local feeder-cattle kings rests in their practice of breeding some of their own stockers, in a diversified operation. But insofar as durability was concerned, the "complex and changing skein of feeding areas and ranges" worked inevitably to the advantage and fame of the feeders.[60]

Largely under the influence of these feeder-cattle "kings," improved bloodlines began to enter the herds in the Ohio Valley as early as the South Branch and Bluegrass Basin episodes in the 1780s. Carolinians and backwoods Midland pioneers had introduced scrub stock largely of British, and more specifically degraded Devon or Durham, origin into the Appalachians, and such animals continued to spread westward with the stocker-cattle frontier. Gradually, however, the feeders began to demand higher-quality cattle to send to market, including shorthorns and even a few Herefords. The earliest improvements came in the stockers bred by the feeder kings, but in time upgrading affected even the hinterland herds. Eventually improved breeds became a hallmark of the evolving "Midwestern system" of cattle ranching, helping explain the bias many or most feeders had against Texas cattle. Almost simultaneously, these progressive graziers established the first stockraisers' associations, as in 1832 in Ross County, Ohio, creating in the process another Midwestern influence that would spread westward to help shape the ranching industry.[61]

Fattened cattle from the middle Scioto Valley usually found their way to market over trails leading to Philadelphia, Baltimore, and New York City, often passing right through older feeder areas such as The Glades. Drovers used the "Centreville whip" as a principal control device. The main cattle road, well-supplied with drove stands and raunchy taverns, was called the Three Mountain Trail, leading east through Pittsburgh, Bedford, McConnellsburg, Chambersburg, and Harrisburg, toward Philadelphia. Kentucky Bluegrass–fattened cattle either connected with these routes by way of Zane's Trace or, more commonly, went east on the Wilderness Road through Cumberland Gap to the Great Valley.[62]

Onto the Midwestern Prairies

The grasslands of the Kentucky Barrens and Pickaway Plains first announced the crossing of another important ecotone in the westward migration of upland southern cattle raisers. Just like their counterparts entering the Gulf coastal prairies in Louisiana, the cowmen of the Ohio Valley encountered the eastern margins of the expansive grasslands of the West, represented in the Midwest by the so-called "prairie peninsula." While these tall bluestem prairies, in many places sprinkled with oaks, did offer a fine setting for cattle raising, certain readaptations had to be made.[63]

The major changes included increased herd size and a lesser role for open-range hogs, which were in part replaced by sheep—an animal subsequently tolerated in the Midwestern herding system. Vertical transhumance, a practice still possible along the margins of the Appalachian Plateau in Kentucky and Ohio, diminished in importance, although horizontal pasture shifting took its place. Hay production for winter feed became essential. Herds remained small enough that the animals did not become semiferal, and the ancient British devices of calf-capture, herder dogs, salting, calf castration, and whips sufficed to maintain order. The horse retained its Carolinian and British function of range transportation. About the only Hispanic influence to reach the Midwestern herders, and a stigmatized one at that, was increasing numbers of longhorn cattle, although most such imports went directly to the Corn Belt fatteners and played little role in regional stocker breeding.[64]

The first major grassland range to be used for breeding cattle, the Grand Prairie, lay west of the Middle Wabash River, in northwestern Indiana. French colonists had made the beginnings in that general area, and almost a thousand cattle were being kept at Vincennes, to the south, by the 1760s. Apparently, however, little French input into the Anglo system occurred here, and in any case, the upland southern herders established themselves farther up the valley, northwest of Lafayette, centered in counties such as Newton. By the middle 1820s, the Indiana Grand Prairie had become a cattle range supplying stockers to feeding districts as distant as the Scioto Valley and even southeastern Pennsylvania. The adjacent Wabash Valley also began developing a feeding industry, coincident with the rise of the Grand Prairie as a breeding area. Only very slowly did the adjacent prairie become converted to farming, displacing lean-cattle producers, and some areas continued to serve as stock range late in the century. The farmers' readaptation to the prairies proved far more difficult than that of the herders. The slow advance of the tillers was caused mainly by the poor drainage of the Grand Prairie. Over a third of this grassland, bearing the marks of ancient glaciation, remained too wet for cultivation until capital-intensive drainage projects began in the latter part of the nineteenth century. Spartina, or slough grass, the "signpost of the wet prairie," shared dominance with sedges, rushes, tall panic grasses, and big bluestem, so that extensive wet areas displayed no sod formation. Some of the wet prairie served as wild hay meadows for the stocker-cattle raisers, producing winter feed to supplement the rather meager forage found in the Wabash bottoms. Hay became an integral part of the Midwestern herding system in the wet prairies, probably beginning in Indiana, representing another break with Carolinian tradition.[65]

A short distance to the southwest, across the state line in Illinois, another Grand Prairie stocker area developed, centered in the flat countryside of McLean and Champaign counties. There, too, stocker-cattle production began very early, about 1816 or 1818, when shipments to the Scioto Valley commenced. Its complementary feeder area developed in the adjacent Sangamon Valley, where fattening operations began after about 1830. These Illinois prairies, too, were slow to convert to farming, due to widespread wetness.[66]

An even earlier focus of cattle raising lay in southwest Illinois and adjacent Missouri, along the Mississippi River, where the beginnings, as early as 1711, were also French. Cahokia, Kaskaskia, St. Louis, and Ste. Genevieve possessed "large stocks of black cattle," possibly longhorns derived from Louisiana, by the 1770s. A 1766 census of the Missouri side of the river revealed over 1,100 cattle, somewhat more numerous than the inhabitants, as well as 223 horses. At Ste. Genevieve, the abundant local cattle wintered on the stubble wheat fields of the Mississippi floodplain, or *terre basse,* and in the karstic prairie parkland above the plain, in the *terre haute.* During the summer, in a system of vertical transhumance, hired or enslaved herders took the cattle to vacheries, grazing grants issued by the Spaniards and located in the hardwood and pine forests to the west of Ste. Genevieve, in the hilly fringes of the Ozark Plateau. By 1840 upland southern Anglos had overrun these French outposts, appending to them an important new center of lean-cattle production in the counties of Illinois east and southeast of St. Louis.[67]

Farther to the south, another French cattle focus developed by the late 1700s in the Grand Prairie of eastern Arkansas, near the confluence of the Mississippi and Arkansas rivers, probably as an offshoot from Louisiana. At Arkansas Post, the local settlement, the settlers' riches consisted "only in the number of their cattle," which grazed in summer on the prairie open range and in winter on river-bottom canebrakes. By the late 1700s, the French at Arkansas Post were using salt as a herd-management device, a technique generally indicative of Anglo influence. When the area around Arkansas Post began acquiring Anglo-American settlers after about 1820, the newcomers enlarged upon the French cattle economy, and by 1840, the prairies from Arkansas Post north into White County were the principal focus of the state's cattle herding. Arkansas County, which included the old Post, had over four times as many cattle as people, and the ratio in White County was three to one. A tie to French Louisiana is suggested by the practice of driving cattle to market at New Orleans. The role of these various French outpost herding clusters in the evolution of American cattle ranching remains poorly understood, but probably their importance was minimal. Still, many early settlers of Texas who later became important cattle ranchers had made a residential stopover in the Ste. Genevieve area of Missouri.[68]

Simultaneously with the establishment of upland southern herders in the prairies of Indiana and Illinois came the rise of open-range cattle raising in Missouri. Graziers displaced from Kentucky were largely responsible for this leap westward. The Salt River bottoms, tributary to the Mississippi near Hannibal, witnessed some early herding activity, but the major center, before 1820, lay in the Boonslick country, situated along the Missouri River from Boonville upstream to Lexington. Within the Boonslick, the prime herding district was the Petitesas, or 'Tit Saw, Plains in Saline County, lying above the bottomlands. Salt licks dotted the Boonslick region, and some streams even ran brackish, so that it was unnecessary to salt the herds. The cattle clustered naturally at the licks and could be fairly easily controlled. Rich tallgrass bluestem prairies bordered the Missouri River on both sides, offering abundant summer range. Eventually this herding region spread from the Boonslick upstream along the Missouri Valley to the Kansas City area, where Clay County became a cattle-raising center by 1850. To the south the Ozark Plateau, represented by the Springfield Plain, provided additional abundant prairies suitable for seasonal transhumance, restoring the vertical shifts of livestock absent in the prairies of Illinois and Indiana (figure 38).[69]

As early as 1832, cattle raisers also crowded southward out of the Boonslick country and began taking up permanent residence in the Springfield Plain. While in strictly topographical terms this diffusion represented a reentry into the mountain South, the largely undissected nature of the Ozark Plateau in the Springfield sector caused it to function more as a part of the interior plains, a linkage reinforced by its tallgrass bluestem prairies. Cattle raisers in the Springfield Plain, seeking new winter ranges, drove herds southward to canebrakes in the mountain glades of northern Arkansas, continuing to use the Plain only as a summer pasture.[70]

Most Missouri stockers went east to the Scioto Valley, Central Illinois, and other developing feeder districts in the Midwest, continuing the traditional upland southern pattern. Other herds were driven north to Manitoba or west to the new settlements in Oregon and to the goldfields of California, sometimes directly out of the Arkansas winter ranges. Missouri also helped stock the cattle ranges of the Great Plains. Early attention to improved cattle breeds in Missouri, largely a Kentuckian influence, meant that the stockings derived from its prairie ranges were far superior in quality to those of Texan origin.[71]

Another antebellum Midwestern expansion of the upland southern cattle herders occurred in Iowa, beginning in the 1830s. Based initially in the river valleys of the southeastern portion of the state, in particular the Des Moines, Skunk, Iowa, and Wapsipinicon, the herders sent their cattle west onto the wet prairies of central Iowa for summer grazing. Later, when the cattle raisers were shunted onto the grasslands, wild-hay production soared to accommodate the need for winter feed in this colder setting. As one local cowman put it, "cattle do not thrive in Iowa without especial feeding, owing to the long winters." The production of hay was now firmly enmeshed in the Midwestern range-cattle culture. By the 1890s, the Iowa wet prairies, following the pattern of those in Indiana and Illinois, became fully incorporated into the growing Corn Belt.[72]

The southernmost expansion of upland southern-prairie cattle herders into a setting largely uninfluenced by other ranching traditions occurred in the blacklands of northeastern Texas, a tall, lush, humid grassland developed on alkaline materials unsuitable for extensive tree growth. Big and little bluestem, wire grass, Texas winter grass, and buffalo grass, among other varieties, coupled with cane-rich glades in the nearby forests and accessible salt licks or saline springs, made the region ideal for cattle raising. Most preferred by graziers was the local wire grass, "a narrow leaf growing three or four feet in height."

Cattle could graze on these northeast Texas prairies for most of the year, then retreat in the short winter to the river and creek bottoms, rich in cane, ryegrass, and winter grass. Some winter forage could even be found in the "buffalo-wallows," small depressions in the prairie where grass reportedly remained green all year. The mild climate made unnecessary the transhumance typical of upper southern herders in the Appalachians and Ozarks. In some winters the cattle did become "poor," but there were apparently no major die-offs. The only major winter hazard was river flooding, which occasionally caught cattle by surprise in the cany bottoms and drowned entire herds. In February 1843, for example, one stockraiser in Red River County lost 75 or 80 percent of his rather substantial cattle herd in this manner. The grasslands were fired once or twice each year, in late winter and sometimes again in July, causing new green shoots to come forth.[73]

Cattle entered northeastern Texas about 1815 with the earliest Anglo-American settlers, most of whom were upland southerners. As early as 1825, cattle outnumbered people by twenty-two to one in the area, and tax lists for the decade and a half before midcentury include some 172 people who owned a hundred or more cattle. Notable among them was the Kentuckian James E. Hopkins, who had 800 cattle bearing his "EH" brand ranging on the Blackland prairies by 1850, herds that increased to 1,600 by 1855. Nearby, Meredith Hart ran 1,000 head of cattle with his "H" brand by 1850. Hart, the Missouri-born son of Ohioan John Hart, came to northeast Texas as a boy in 1832 and already owned a herd of 20 cattle while still a lad of fifteen. Fully 22 percent of all the large cattle operators in the prairies of northeast Texas had lived in Missouri immediately prior to coming to the Lone Star state. So abundant had cattle herds become by midcentury that a traveler in the prairies of northeast Texas was moved to remark that "for miles in every direction there was nothing to be seen but the immense number of cattle feeding."[74]

Carolina's upland children, or grandchildren, arrived in northeast Texas with much of their largely British traditional system intact. Accounts and documents refer to "a cow pin enclosed with a rail fence," to "cow hunts," British breeds, herder dogs, block-letter brands, and salting. Livestock were not allowed to become too wild and could still be controlled without need for the Mexican lasso or equestrian skills. "Cattle are penned once a fortnight and salted, which is necessary to keep them gentle," noted an observer in 1850. The cattle were "descendants of Durham stock" and exhibited little or no intermixing with longhorns. The indigenous feral Spanish cattle were hunted into extinction. Connections to the Missouri roots persisted in the Shawnee Trail, linking northeastern Texas to the feeder areas and markets of the Midwest (figure 38).[75]

Hispanic influence of any sort remained very weak in northeastern Texas until after the Mexican War of 1846–48. In effect the upland southerners achieved the implantation of an essentially unaltered Ohio Valley prairie-ranching complex on Texas soil, occupying a corner of the state that still reveals strong cultural ties to the interior South.⁷⁶ Some distance to the south, their pine-barren cousins had established their final foothold in the westernmost piney niche, and beyond them, in far southeastern Texas, the heavily Hispanicized Anglo-Cajun coastal-prairie cattle complex had made an entry. Carolina's children had thus reached the pivotal state of Texas by two environmental paths, bearing markedly different herding systems that spoke of their cultural and ecological divergence.

Equally important, the diffusion of upland southern herding across the major prairies of the Midwest had implanted, before 1850, on the eastern margins of the Great Plains in the Springfield Plain, Missouri Valley, Iowa wet prairies, and northeastern Texas, a basically British-derived system involving a higher level of labor intensity than Anglo-Texan herding on the coastal prairie. The stage had been set for a contest between different cattle-ranching systems for dominion in the West.

SEVEN

THE ANGLO-TEXAN RANCHING SYSTEM

THREE AND ONE-HALF CENTURIES HAD BEEN REQUIRED FOR the Euroamerican cattle-herding frontiers to spread from the Greater Antilles hearth, to footholds on the continental mainland, and finally to three major outposts around the perimeter of the American West, the region most closely associated in the popular mind with ranching. Dispersing from these outposts—coastal Texas and Louisiana, the midwestern frontier in the Missouri and Iowa prairies, and the valleys of Mediterranean California—cattle raisers would require only a scant additional third of a century after 1850 to occupy the remainder of the West and bring the final passing of the free-grass ranching frontier.

The role of Anglo-Texans on the western cattle frontier during this final period of expansion has received such abundant scholarly and popular attention that it might seem impossible to add meaningfully to an often-told story. Cattle ranching and its associated culture in the West have usually been depicted in monolithic Anglo-Texas hues, in no small part because famous Anglo-Texan historians and folklorists wrote the most compelling studies of the pastoral frontier. According to this widely held view, hispanicized Anglo-Texan neophytes dispersed the colorful, volatile cattle industry through the West, filling a semiarid environmental niche in a remarkably short span of decades and implanting an enduring pastoral culture.[1]

Several major elements of this standardized version of western ranching history need revision. In earlier chapters the birth of Anglo-Texan cattle ranching was presented as a syncretism of Carolina southern littoral and Mexican coastal herding traditions, rather than a simplistic borrowing. The principal arena where this mixing occurred was relocated eastward to Louisiana, and the time period when it took place pushed back to 1780–1820 (figure 39). In the present chapter, I further suggest that the shaping influence of the resultant Anglo-Texan ranching system, whatever its time and place

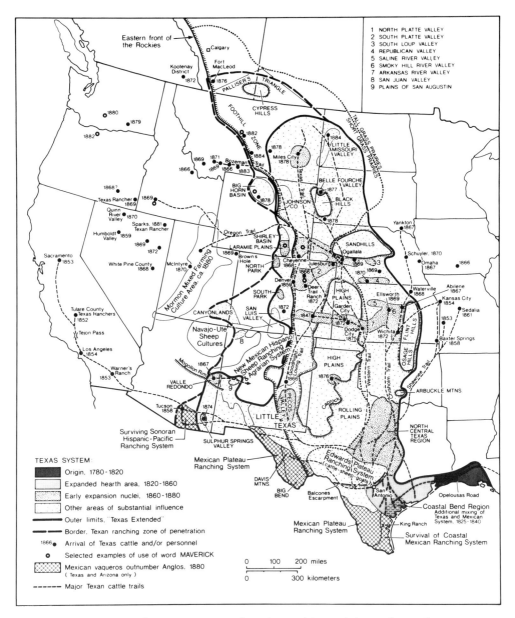

Figure 39. The Texas system of cattle ranching: origins and spread. Born on the Gulf coastal prairies of a mixing of South Carolinian and Mexican herding traditions, the Texas system achieved its greatest diffusion in the Great Plains. Sources: U.S. Census population schedules, 1880; Gordon, "Report on Cattle" (see note 22); and sources cited in note 1.

of origin, has been greatly overstated. Instead each of the three culturally distinct ranching outposts by 1850, in the Midwest, California, and Texas, played a major role during the final decades of the North American cattle frontier. Moreover the Anglo-Texan contribution to the cattle-ranching industry, spectacular and colorful though it was, remained largely confined to certain parts of the West and, in the long run, proved less substantial than that derived from the Ohio Valley and Midwest.[2]

The Texas System

The "Texas system" of cattle ranching, as suggested in the preceding two chapters, evolved on the coastal prairies of southwestern Louisiana and adjacent southeastern Texas as a result of contact between pine-barrens southerners, Louisiana French, and *tejanos*. The already creolized Carolina-Jamaican system absorbed Mexican influences derived mainly from the Veracruz-Tamaulipan Gulf Coast tradition, in the process of adapting to an open grassland environment. A second, lesser developmental phase occurred in the Texas Coastal Bend region between 1825 and 1840, in Martín de León's colony, where southern Anglos and their slaves once again encountered the Mexican coastal cattle-horse culture.

The resultant Texas system of ranching clearly displayed the cultural inputs of both Carolina and Tamaulipas. From both sources came the essential trait of the Texas system: the subtropical practice of allowing cattle to care for themselves year-round in stationary pastures on the free range, without supplementary feeding or protection. Through such self-maintenance, the herds should not merely survive, but reach a grass-fattened maturity, ready for market. The humid subtropical prairies, canebrakes, and salt marshes of coastal Texas and Louisiana were even better suited to this careless system than had been the Andalusian marshes, yielding a still more profound neglect of the livestock.

Mexican and Carolinian Contributions

From Mexicans the Texas system received Iberian longhorn cattle, reinforcing Spanish bloodlines that had earlier entered the pine-barrens complex by way of Florida and French Louisiana. More purely Mexican in origin were equestrian skills, including roping, the horned saddle, the whirligig, hemp or maguey ropes, and sundry

Figure 40. Perhaps the earliest illustration of Anglo-Texan cowboys at work, this drawing, entitled "branding cattle on the prairies of Texas," appeared in 1867. Note the lassos and horned saddle. The cowboys' clothing, including the hat, is not that of the Mexican vaquero. Source: sketch by James E. Taylor, published in Frank Leslie's Illustrated Newspaper, vol. 24, no. 613 (June 25, 1867), p. 232.

loanwords such as "lariat," "corral," "remuda," and "tank" (figure 40). Also from Mexico came the doctrine of "accustomed range," by which ownership of unbranded stock was awarded to the owner of the pasture. The more precise points along the coastal prairie where each of these adoptions occurred have earlier been enumerated, but Louisiana served as the first and most important arena of exchange (figure 29).[3]

At the same time, a virulent and enduring prejudice against Mexicans prevailed among Anglo-Texans beginning in the early 1830s, serving to retard additional and more pervasive borrowing of Hispanic traits and personnel. Anglo cowboys rejected many elements of vaquero equipment and garb, including the sharpened great-rowel spur (see figure 49). Hispanic sheep raising in conjunction with cattle was so resoundingly rejected by Anglo-Texan cattlemen that an anti-sheep bias became a hallmark of the Texas system. Moreover some

Mexican traits initially adopted by Anglos failed to accompany the great northward expansion of the Texas system, including the whirligig and certain Spanish loanwords such as *caporal* ("range boss") and *mott*. Other Mexican contributions underwent major changes under Anglo influence, including the broad-brimmed sombrero and horned saddle. The bulky, single-cinched Tamaulipan saddle, with its thick tree-branch horn and large "doghouse" closed stirrups, was modified to a lighter design under Anglo-Texan influence, featuring a much slimmer horn and double-cinching to hold down the back of the sleeker saddle during roping (see figure 49). Some Texans began tying the lariat to the saddle fork, due to the weakness of the redesigned horn.[4]

In general the tendency has been to overstate the magnitude of the Mexican contribution to the Texas system. Efforts to find Hispanic antecedents for a variety of other elements of Anglo-Texan ranching, from stockmen's associations, brand designs, and earmarks to folk songs are generally misguided. In the final analysis, the only consequential Mexican contribution lay in the more sophisticated use of equestrian skills, particularly roping.[5]

Prejudice also dictated that the Mexican vaquero would never play a substantial role in the diffusion of the Texas ranching system. The tejano cowhand failed to advance in any consequential numbers northward from the border districts where Hispanics formed, then and now, a majority of the total population (figure 39). While one should not underestimate the potential for diffusion represented, for example, in the lone vaquero, "Mexican John," who worked alongside 27 Anglos and one black on the "D. Waggoner & Sons Cattle Ranche" in Wichita County, north Texas, in 1880, the truth is that only 1.5 percent of all cowboys resident north of 32° latitude in West Texas by that year were of Hispanic origin. On the expanses of the Great Plains, beyond Texas, very few Hispanic cowboys could be found (table 1).[6]

The vaquero, then, remained largely confined to the south, below the Nueces River. As a result, that region (the old Nueces Strip) did not immediately participate in the evolving, expanding Texas system. Instead the older Mexican ranching cultures of both Gulf Coast and interior plateau continued to prevail in South Texas (figure 39). Many ranches there remained under Mexican ownership, and the Anglos who ventured south of the Nueces did not implant the Texas system, but instead adopted the "Mexican plan" of ranching. They employed a vaquero work force, took the honorific or affectionate titles of *don, doña,* or *tío,* and simply replaced the earlier Hispanic landed aristocracy. The famous King Ranch on the Wild Horse Prairie in the Nueces Strip was precisely such an enterprise in its formative years (figure 39). Only later, largely in the twentieth century, would modernized Anglo-Texan cattle-ranching influences penetrate the border region, spreading far beyond into northern Mexico.[7]

When the Carolina-derived coastal southern herders entered the Louisiana prairies and adopted certain Mexican pastoral traits, they retained much of their older cattle-raising culture. Carolinian and even Jamaican influences survived in the West in vocabulary such as *dogie, cowhunt, pen,* and *cowboy,* the latter becoming a word diagnostic of Texas influence in the West; in the use of black-slave and "cracker" poor-white herders; in the custom of only two roundups, held in the spring and fall; in periodic "riding line" around the perimeter of a range; in the continuing use of whips, especially during droving and "cutting"; in regular calf castration; in the practice of animal wrestling, or "bulldogging"; in the involvement of wealthy planters and other absentee entrepreneurs in the raising of open-range cattle; in the unimportance of sheep raising; in the focus upon beef production rather than merely hides and tallow; in annual range firing; in marketing by means of often lengthy overland droving of grass-fattened cull steers; in the reliance upon livestock statutes based in English common law, officially adopted by the Republic of Texas in 1840; and in the British-derived block-letter and numeral brands (figure 34). So complete was the Anglo-Texan rejection of Spanish brand designs that the term "with Mexican brand" appeared in some newspapers as an adequate description of strayed animals. Consistent with coastal, Carolina-derived custom, too, was the pervasive neglect of livestock in the Texas system, including the practice of stationary pasturing, without any attempt to reserve special winter ranges. Symptomatic of such carelessness and neglect, the Texas term "maverick," appropriately, would be derived from the surname of an Anglo-Texan of South Carolina ancestry.[8]

At the same time, certain other coastal southern practices would not long endure in the evolving Texas system. Hog raising in conjunction with cattle herding, calf-capture and the related British milking complex, and the use of Afro-American cowboys would generally not survive in the western plains. Neither would the use of herding dogs, although some local exceptions can be noted. *Cowpen* would surrender to *ranch* and *pond* to *tank,* although the term *cowpen herd* survived in the western meaning of cattle belonging to a

TABLE 1. CULTURAL-ETHNIC MAKEUP OF THE COWBOY WORK FORCE IN SELECTED RANCHING AREAS, 1880

Ranching Area	Number of Cowboys in Counties Surveyed	Texas-born Anglo, %	Coastal southern Anglo,[13] %
Hispanic South Texas[1]	673	4	2
Expanded hearth area of Texas ranching system[2]	377	42	16
North-central Texas[3]	500	50	16
Pecos Valley, New Mexico[4]	21	43	14
Arkansas River Valley[5]	367	22	5
Platte River Valley[6]	498	7	4
Eastern Montana[7]	19	0	0
Black Hills[8]	38	13	11
Southern Arizona–New Mexico[9]	205	4	3
Colorado Mountain Parks[10]	71	0	3
High Desert[11]	146	0	1
Columbia Basin[12]	56	0	0

Source: U.S. Census, MS population schedules

1. As represented by Duval, Encinal, Nueces, and Webb counties, Texas.
2. See figure 39. Represented by Atascosa, Frio, Live Oak, and McMullen counties, Texas.
3. Callahan, Coleman, and McCulloch counties, Texas.
4. Lincoln County, New Mexico.
5. Bent County, Colorado; Ford and Kearny counties, Kansas.
6. Weld County, Colorado; Keith County, Nebraska; Laramie County, Wyoming.
7. Custer and Dawson counties, Montana.
8. Custer and Lawrence counties, Dakota Territory.
9. Grant County, New Mexico, and Pima County, Arizona.
10. Grand and Park counties, Colorado.
11. Grant County, Oregon, and Owyhee County, Idaho.
12. Yakima County, Washington.
13. White natives of Alabama, Florida, Georgia, Louisiana, Mississippi, and South Carolina.
14. White natives of Illinois, Indiana, Iowa, Kentucky, Missouri, and Ohio.

small-scale rancher. Penned-over gardening and herd control through means of salting would, similarly, not enter the West. Still other southern traits and practices underwent substantial modification in the Anglo-Texas system, a transformation best exemplified by the boot. In the late 1870s, cobbler H. J. Justin of Spanish Fort, on the Red

Midwestern Anglo,[14] %	California-born Anglo, %	Oregon-born Anglo, %	Hispanic-American %	African-American %
1	0	0	91	<1
9	0	0	13	3
18	0	0	<1	<1
5	0	0	10	5
34	1	0	3	2
46	<1	<1	<1	<1
37	0	5	0	0
24	0	0	3	0
6	2	<1	64	<1
44	0	0	1	0
42	7	19	1	0
26	0	25	0	0

River in north Texas, responded to the complaints of trail-droving cowboys by modifying the British-derived riding shoe. He produced a stiff-countered, pointed-toe boot that could more easily be inserted into the stirrup, elevated the heel to hold better when the rider braced against a taut rope, and inserted a steel-shank arch for more adequate riding support. Such boots are still today manufactured in the nearby town of Nocona, Texas, and their popularity was such that the Anglo-Texan boot subsequently spread through most of the West.[9]

Early Spread of the Texas System

From its nucleus in the coastal prairies of Louisiana and southeastern Texas, the creolized Carolina-Mexican cattle-ranching system spread initially westward after 1820, clinging to the belt of humid, lush grasslands bordering the Gulf, to enter the Coastal Bend region (figure 39). By 1840 the largest cattle herds in Texas ranged between the lower Trinity and Colorado rivers, and by 1860 the focus had shifted to the Coastal Bend, where the census takers enumerated

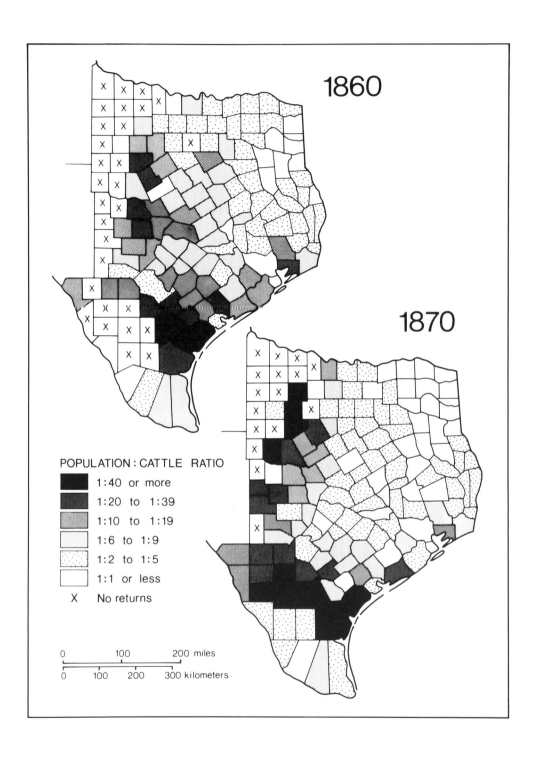

Figure 41. *Population:cattle ratios in Texas, 1860 and 1870. The two separate focal regions of the Texas system during this crucial decade are clearly shown: the southern coastal nucleus of the Texas system and the north-central Texas region. Sources: Jordan,* Trails *(see note 1), pp. 126, 128.*

forty times as many cattle as people (figure 41). A fourth of all the cattle in Texas, almost 900,000 head, grazed in the Coastal Bend region by the outbreak of the Civil War, ranging through lands along the lower Guadalupe, San Antonio, and Nueces rivers. Overrun in the process were the older Hispanic complexes in the Coastal Bend and San Antonio Valley. In the wake of this early expansion frontier of the Texas system lay badly overgrazed pastures in the eastern coastal prairies. Even very early in its development and expansion, in a humid environment, the system, with its huge free-ranging herds, lacked ecological sustainability.[10]

We can glimpse the workings of the Texas ranching system in its crucial early Coastal Bend period, thanks to a highly informative, first-hand account written in 1860.[11] The anonymous informant, resident in Refugio County, Texas, described how Anglos called "cow boys," beginning about 1838, rounded up largely feral cattle that had belonged to the earlier, departed Mexican residents and also stole some domestic herds. These longhorns increased at a net rate of 25 percent per annum, yielding more profitability "in proportion to the capital employed and the labor and attention required than any other business of which I have had a knowledge." As a result, many who "commenced on small beginnings . . . are now . . . comparatively rich." The "cattle are permitted to range indiscriminately over a large surface of country, thirty, forty, and even fifty miles in extent from north to south and east to west." It was "no easy task to hunt up and mark and brand the calves of a large stock; still it is done, and with tolerable accuracy." The "cattle hunting" took place "twice in the year—in the spring and fall," accomplished by cowboys organized into "crowds" of ten, twelve, or fifteen men. Each herder had a "lasso at saddle-bow," as well as a pistol and bowie knife, and they drove the cattle "into close herds" at "places most convenient to a pen." They then "cut out" from the herd "such cattle as belong to the men who compose the crowd," put them "into the pen, and mark, brand, and alter the calves."

"The young men that follow this 'Cow-Boy' life, notwithstanding its hardships and exposures, generally become attached to it. . . .

Many of them are not inferior to the best Mexican vaqueros in the management of their fiery steeds," and "they 'rope,' or throw the lasso also with great dexterity and precision." The author mentioned no other livestock being raised, aside from horses and mules, and crops were apparently altogether absent.

From the very first, the Texas system was linked by drove trails to distant beef markets. For a time New Orleans, reached by old Spanish trails along the coastal prairies, remained the primary destination, but by the 1850s Coastal Bend cattle were also being trailed to feeder areas in the Midwest, in Illinois, Iowa, and Missouri, finding an eventual market in New York and Philadelphia, as well as to California. Some livestock were even exported by sea. A few slaughter yards for extracting hides and tallow developed, as at Houston in the 1840s, a reminder of a more Hispanic enterprise, but most Coastal Bend cattle were raised for beef.[12]

Continued Expansion in Texas

The Texas system of cattle raising had little spatial opportunity for contiguous expansion after 1860. Wedged in between the vaquero-based Mexican herding systems to the south, the rocky, hilly, dissected Edwards Plateau region to the north beyond the Balcones Escarpment, and Old Mexico to the west, bearers of the Texas system could only seize a small additional cul-de-sac corridor west of San Antonio (figures 39, 41). Representative of the Texas system in this narrow expansion corridor was the large enterprise of one William Perryman in Frio County, southwest of San Antonio. Perryman, an Alabamian by birth with roots in the pine-barrens cattle complex of the coastal south, by 1870 referred to his place as a "ranche" and controlled some 10,000 hectares (25,000 acres) of land. His herds reputedly numbered some 25,000 open-range longhorns, from which he culled about 4,500 annually, and Perryman also owned 1,000 horses, employing as many as a hundred cowboys. In riverine woodlands, he kept 500 to 600 pigs, still maintaining the traditional Carolinian livestock combination. Perryman tilled "but few acres" for subsistence crops, still echoing the eastern cowpen garden model. In spite of his wealth, bachelor Perryman's ranchstead consisted merely of several pens, a garden, and a few modest houses (figure 42).[13]

The Texas system, in fact, never spread completely unmodified beyond the confined southern belt. Both cultural and environmental factors conspired to change the coastal-based system when its bear-

Figure 42. The William Perryman Ranch in Frio County, Texas, about 1870, typical of the southern hearth area of the Texas ranching system. The ranchstead complex is very modest, even though Perryman owned 25,000 head of cattle. Source: McCoy, Historic Sketches (see note 3), p. 12.

ers escaped from their southern confinement. Even the development of marketing trails to the Midwest initiated the process of change by ending the era of local grass fattening, demoting the South Texas steers to the status of mere lean stockers.

A key to the dilution and modification of the Texas system lay in repeated encounters with other, rival herding cultures. North of the Balcones Escarpment, in the Edwards Plateau region of Texas, the initial settlers, many of whom were of German or British birth, developed in the 1850s a labor-intensive ranching system based as much in sheep as in cattle (figure 39). Shepherds, often of Hispanic origin, labored on these Hill Country ranches, and the whole herding system, beautifully suited to the rugged terrain, owed much to the Mexican plateau model. The Texas system penetrated the Edwards Plateau along the paths of the market trails in the 1860s, but succeeded only in adding longhorn cattle and equestrian skills to a ranching culture that retained its diversity and underlying ethnic distinctiveness.[14]

To the north, beyond the Edwards Plateau in north-central Texas, lay a gentler hilly region of tallgrass bluestem prairies, oak savannas, and cedar brakes, different parts of which are variously called the Cross Timbers, Grand Prairie, and "Heart of Texas" (figure 39). This region would soon become the principal focus of the expanding Texas system, producing "more notable cattle ranchers than any other region of comparable size in America" (figure 41). Legendary names such as Chisum, Goodnight, Loving, Hittson, Slaughter, and Waggoner are linked to the north-central Texas region. Oddly very few ranchers from the coastal hearth of the Texas system settled in this north-central part of the state. Instead the local ranchers were generally of upland southern stock and had entered north-central Texas from the Blackland Prairie to the east (see figure 43). The influence of the Texas ranching system seems to have filtered north along the market trails in the absence of any substantial migration, borne more by contagious than relocationary diffusion.[15] Cowboys may have been more mobile than ranch owners and possibly moved in substantial numbers from the coastal plains northward into the hills, but their genealogies are generally lost to history. By whatever mechanism, the essential traits of the Texas system—equestrian skills, roping, and livestock neglect—spread into north-central Texas about 1860 along the axes of the drove trails. By the outbreak of the Civil War, the core counties of the region—Erath, Comanche, and Palo Pinto—each had over 25,000 cattle, dwarfing the human population. Palo Pinto had twenty-three times as many cattle as people. Herds peaked in size between 1865 and 1870, and by the latter year Erath County housed over 57,000 range cattle, while adjacent Eastland County had 159 head per inhabitant (figure 41). As the primary cattle-ranching section of the state, the north-central region served as the principal staging ground from which the Texas ranching system launched its spectacular diffusion through the Great Plains. As the cattle raisers left north-central Texas in the early 1870s, fleeing pasture depletion, the numbers of range livestock plummeted, and by 1874 no herds of consequence remained in the core counties. In less than two decades, the cattle-ranching frontier had come and gone.[16]

The Texas system had not achieved its passage through the north-central region of the state unaltered. Perhaps most notably, the Afro-American cowboy became far less common beyond the coastal belt, perhaps in large part because the abolition of slavery coincided with the diffusion of the Texas system to the interior of North America. Runnels and Callahan counties, representative of the north-central Texas region, by 1880 housed, respectively, 118 and 177

Anglo-American cowboys, but not a single black cowhand (table 1). Wichita County, farther north on the Red River, in that year boasted the greatest concentration of black cowboys on the Texas cattle frontier, a total of 15, but even there the African-Americans were greatly outnumbered by the 67 Anglo herdsmen. In all of West Texas in 1880, only 4 percent of all cowboys, excluding camp cooks, were black, while Anglo-Americans accounted for about 9 out of every 10 (figure 34).[17] Subsequently black ranch hands never became common anywhere in the West, nor did much African influence survive in Anglo-Texan cowboy techniques, in spite of exaggerated claims to the contrary (table 1).[18] Only a few local exceptions can be found, such as the famous "bulldogger" Texan Bill Pickett of Oklahoma's 101 Ranch; "Nigger Henry," one of several well-known cowhands in southern Idaho; and John Ware, a celebrated black cowboy of turn-of-the-century Alberta.[19]

Ascending the Western Plains, 1866–1885

The spread northward through the Great Plains by Texas cattle and many basic elements of the Texas ranching system represents one of the most rapid episodes of frontier advance in the Euroamerican occupation of the continent. The original motivation for the northward thrust lay in the search for new markets. New Orleans after the Civil War gave way with surprising suddenness to Chicago, which provided the meat demanded by the growing urban-industrial market of the American northeast. Trailing Texas cattle to such a distant market left them in substandard condition, and the ranchers became dependent on Corn Belt feeders to "finish" the animals before the final rail journey to the packing houses. Certainly without the Chicago-northeastern market, Texans could not have expanded northward into the Great Plains.

A prototype for the northern drives and expansion can be found in the small-scale movement of Texas cattle herds to midwestern feeder areas as early as the 1840s and 1850s, mainly along the Shawnee Trail (figure 39). Sedalia in Missouri served as a notable early northern "cowtown" for such drives.[20] Soon finding the Shawnee route plagued by thieves and hostile farmers, Texans after 1866 began following trails farther west, skirting the edge of the plow frontier, to reach railroad shipping pens in Kansas and Nebraska, as well as mining districts in the Rocky Mountains and beef-hungry Indian agencies in the northern plains and New Mexico. Many different braided routes found favor, but the most popular were the Chisholm, Western, and Goodnight-Loving trails (figure 39). The deepest roots of these famous cattle tracks lay in the southern coastal hearth of the original Texas ranching system, but in fact the trail blazers and most active livestock traders came from the previously mentioned north-central part of Texas. Perhaps their midwestern/Ohio Valley roots, coupled with their ancestral tradition of seeking feeder markets in that region, caused them to regard trails leading north as the logical ones along which to drive their herds, setting into motion the whole northward expansion. In any case, by 1859 cattlemen in the Cross Timbers brought Texas cattle to the Colorado gold towns, by 1866 to the western Montana mines and New Mexico Indian agencies, and by 1867 to the Abilene railhead in Kansas and the reservation tribes of Dakota Territory (figure 39). In all over five million Texas cattle were reportedly driven north between 1866 and 1884, involving the largest short-term geographical shift of domestic herd animals in the history of the world.[21]

As had happened a century earlier in the Carolina Piedmont and Appalachians, the trails northward to market soon also became routes of migration and range stocking. At least by 1869 a substantial demand for Texas breeding stock had arisen in parts of the central and northern Great Plains. Texas ranchers and cowboys also began migrating north, attracted by the splendid grasslands they had previously encountered en route to the markets.[22] To the naive Texan eye, these immense plains offered an environment "throughout which cattle could be raised and fattened on the open range, seeking their own food, water, and shelter without any aid from men."[23] Displacing both the bison and the plains tribes dependent upon them, the Texas cattle and their keepers spilled all the way into the prairie provinces of Canada within a decade or two.[24]

Although it looked appealing to the Anglo-Texan eye, the physical environment of the Great Plains was in fact very different from the southern lands they had previously known. The vast plains they now entered were far drier and colder than the humid subtropical hearth area of the Texas system. It was true that the western margins of the tallgrass prairies still lay open in 1865, offering rather familiar, nutritious bluestem pastures along the axis of the Chisholm Trail, but wheat farmers stood ready to compete for those lands.[25]

Instead the principal floral niche left open to the northward expanding Texans lay in the shortgrass prairies, west of about the

100th meridian (figure 39). Two distinct ecosystems, or biomes, occurred there. In the southern part of the shortgrass plains, sometimes called the Kansan biotic province and reaching about as far northward as the valley of the North Platte River, grama grasses and buffalo grass provided the grazing base. The main forage plants cured standing and remained available to cattle through the fall and winter. Texans quickly abandoned their traditional practice of range burning, in order not to destroy late-winter foraging.[26]

Beyond the North Platte, ranchers entered the so-called Saskatchewan biotic province. Needle- and wheatgrass prevailed there, although the southern blue grama and buffalo grass did not altogether disappear. More ominous for the Texas system was the fact that a grazier could not reasonably expect to pasture livestock year-round in the northern shortgrass districts, a fact of which the Texans remained ignorant. Besides being much colder, the northern plains presented a more broken and rolling surface, which at least had the advantage of providing some winter shelter for the stock.[27]

Throughout the shortgrass prairies, the flora clearly warned of semiaridity. Deficient precipitation presented another new challenge for the Texans, even though an unusually severe drought experienced in north-central Texas in the late 1850s gave the ranchers from the south some chance to adapt to the problem. Fortunately for the Texans, a quarter-century wet period commenced in the plains about 1860, accompanied by a warming, allowing bearers of the Texas system to venture far to the north without immediate catastrophe.[28]

The Texan spread northward through these inviting yet treacherous plains did not occur in an orderly, wavelike manner. Instead scattered early enclaves appeared, leaving large unoccupied areas in the rear (figure 39). Western Texas, for example, was settled rather late, and ranchers from the north-central region, confronted by the formidable Comanches, required fifteen years to spread through the Rolling Plains, ascend the Caprock escarpment, and finally begin, by 1880, the occupation of the table-flat High Plains, or Llano Estacado, where surface water deficiency further retarded their advance (figure 43). In fact the Pecos Valley of eastern New Mexico received central Texas ranchers such as John Chisum, following the Goodnight-Loving route, almost a decade earlier than did the High Plains.[29] Southeastern New Mexico, bordering the Pecos, still today bears the popular appellation "Little Texas." So rapid and spatially fragmented was the Anglo-Texan expansion that some ranches in intermediary

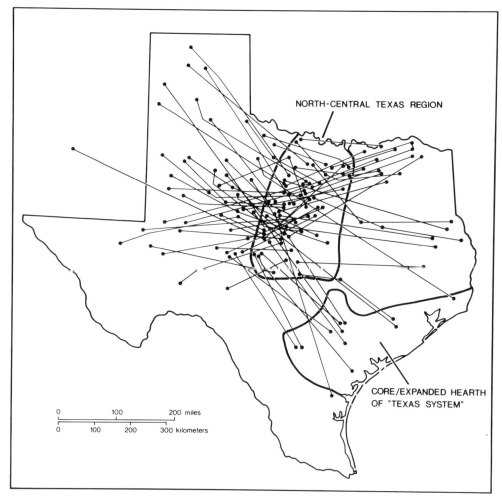

Figure 43. Documented migrations of cattle ranchers to the north-central Texas region and to the Great Plains of West Texas. Notice the relative unimportance of the old southern coastal nucleus of the Texas system as a source of migrants. The north-central region served as a far more important funnel and source of Great Plains ranchers, and it had drawn more heavily upon northeastern Texas for its original herding tradition. Source: Jordan, Trails (see note 1), pp. 131, 138, 140, 142.

locations, such as the Two Circles Bar in West Texas, originated as stock-resting places for herds ultimately headed much farther north.[30]

The western half of Oklahoma, due to its status as an Indian reserve, had an even later beginning in cattle raising than did the Texas High Plains, causing it to be dubbed "the ranchman's last frontier."[31] Texans often drove herds through Oklahoma, usually pausing to fatten their cattle before reaching the Kansas railroad towns, but they did not initially attempt to settle the area. Only Greer County in southwest Oklahoma, long administered as a part of Texas, began acquiring a resident population of ranchers relatively early. Gradually Texan and Kansan cattle raisers started squatting on Cherokee, Cheyenne, Arapaho, Wichita, and Caddo lands, usually making informal agreements with the chiefs for use of the pastures. Some such leasing arrangements, especially in the Cherokee Strip in the north and the Cheyenne-Arapaho lands in the west, became formalized in the late 1870s and early 1880s, but the federal government eventually intervened to harass and expel many of the Texas ranchers.[32]

Kansas, by contrast, acquired a very early implantment of the Texas ranching system, centered mainly in the central and southwestern parts of the state along the Arkansas River valley from near Garden City downstream to the Oklahoma border south of Wichita, as well as along the Smoky Hill, Cottonwood, and Saline rivers. Texas ranchers began operating in Kansas about 1870, successfully "wintering over" their longhorn stock.[33] Neighboring Colorado, to which Texas herds had been directed as early as 1859 or even 1847, began attracting resident Texan ranchers by the early 1870s, both along the Arkansas River and, more importantly, in the South Platte Valley (figure 39). John Hittson from the Palo Pinto country in north-central Texas was among the early ranchers in the South Platte drainage, establishing his Deer Trail Ranch in 1872. The state reputedly housed a million head of cattle as early as 1869, many of them "Colorado Texans," or longhorns.[34]

In Nebraska the first Texas breeding stock arrived at the forks of the Platte in 1869, where the cowtown of Ogallala soon arose. Early Texas-based ranching districts in western Nebraska developed along the North Platte, its tributary, the South Loup, and the Republican River by 1871 (figure 39). The fine tallgrass ranges of the Nebraska Sandhills, where cattle could also feed on soapweed and "sand cherries," were initially regarded as unsafe for livestock, but when herds fleeing a blizzard drifted into the hills in 1878 and survived in good condition, stock raisers quickly occupied the area.[35]

In parts of neighboring eastern Wyoming, the Texas system also attained some early footholds, most notably between Cheyenne and the North Platte. Texas cattle apparently first came to Cheyenne in 1868, the year Indian treaties opened the area to ranching, arriving by way of the Goodnight-Loving Trail. By 1871 some forty ranches, with eighty-six thousand cattle, had been established in this southeastern corner of the state.[36]

In the Dakotas, aside from early shipments of longhorns to Indian agencies, the Texan presence came relatively slowly. As late as 1875 no cattle at all grazed west of the Missouri River, which would soon mark the eastern limits of ranching in the Dakotas. The early absence of railroads retarded the development of the grazing industry, as did the sizable Plains Indian presence. A mining bonanza in the Black Hills attracted the first ranchers in the late 1870s, and within a half-decade the grassy margins of the Hills had some sixty ranches holding over a quarter-million cattle. To the north, Texans began settling the grasslands along the Little Missouri River in far western Dakota Territory between 1880 and 1884.[37]

Greater Texan developments occurred in the Montana Great Plains, adjacent to Dakota on the west, even though an early lack of railroads also hurt that territory, as did Indian wars. By 1878 a major focus of the Texas system had begun to develop in Custer County around Miles City on the Yellowstone River. Ten thousand Texas cattle arrived in Custer County in 1880 alone, and this region long retained a vivid Texan imprint, as did northeastern Montana, where Texas-inspired open range survived into the present century. To the west the foothill zone of Montana, below the Rocky Mountain front, remained largely outside the Texan zone of influence, preempted by a rival herding culture, but some cowboys and ranchers of Texas origin did settle there.[38]

A relative handful of Texas ranchers, apparently seeking an Ultima Thule, continued northwestward from Montana, crossing the international border to enter the shortgrass prairies of Alberta and Saskatchewan. In the words of geographer Simon Evans, these prairies "were occupied briefly by the last survivors of a colorful company of men who had ridden the trails and followed the grass up from Texas." By the late 1870s, some Texans were among the American ranchers around Fort Macleod in southwestern Alberta, and a few others reached the Bow River Valley near Calgary. However, Texan influence in that foothill and western plains region of Alberta remained as inconsequential as it had been in the Montana foothills. If places such as Calgary today exhibit some measure of Texanity in their cultural makeup, we should look to the latter-day oil boom rather than to a few shivering southern cowboys of frontier days for an explanation. Greater Texan influence was felt in "Palliser's Triangle," the driest and most broken terrain in the Canadian prairie provinces. There, particularly in the Cypress Hills, a ranching economy survives to the present, most towns hold rodeos, and Texan-derived words such as *lariat* are used. Even in the Triangle, however, the Texan imprint remained weak, and the region should not be considered as part of Texas Extended. Instead it was occupied mainly by foothill ranchers moving eastward. As Evans has documented, southern influences prevailed nowhere in western Canada. He exaggerated in describing the 49th parallel as "an institutional fault line," but it is true that the Texan advance largely spent itself short of the border, around Miles City in Montana.[39]

While Anglo-Texans may not have added Palliser's Triangle to their extended culture area, their reach northward remains remarkable. Consider that the pastoral occupation of the Texas High Plains began about the same year that other Texans reached Alberta. A period of less than fifteen years following the American Civil War had witnessed the establishment of every noteworthy foothold of the Texas system in the Great Plains, from New Mexico to Montana, across almost the entire breadth of the United States. To be sure, the growing urban-industrial markets in the East helped fuel the Texan reach north, and the railroads of the dawning modern age had allowed unprecedented ease of access to those consumers, but neither these fortunate conditions nor the fact that the Texans' geographical reach far exceeded their eventual grasp dim the magnitude of the achievement.

Texans in the Western Mountains

The spread of Anglo-Texan cattle ranchers westward from the Great Plains into the valleys and intermontane basins of the cordilleran section of North America proved far less successful than their diffusion north through the plains. They would win few new areas for Texas Extended in the montane West.

Within the mountain region of Texas itself, the Anglo cattle raisers succeeded in wresting the elevated grama-tobosa grassland of the Davis Mountains and Big Bend Country away from the handful of local Hispanic sheep raisers, beginning in the late 1870s. The most unusual aspect of this expansion was its achievement by ranchers from the original hearth of the Texas system in the southern coastal part of the state, representing the only noteworthy migration by that group (figure 44). Always a people of the southern periphery, they knew their proper latitude and rarely deviated from it.[40]

By contrast the high basin-and-range country beyond the Río Grande in southwestern New Mexico and southeastern Arizona received its Texans preponderantly from the north-central region, including the Cross Timbers and Grand Prairie. The choicest sites, lying between about 1,200 and 1,800 m. (4,000–6,000 ft.) elevation, offered year-round grazing on extensive grama grasslands very similar to the western plateau pastures of Mexico, which had much earlier supported the spectacular northward expansion of cattle raisers from El Bajío. Anglo-Texans had initially become familiar with the southern Arizona–New Mexico pastures in the 1850s, during remarkable cattle drives to the California goldfields and to the army posts of the Gadsden Purchase. Soon some sought out southern Arizona as a place to settle. The first substantial implantment of the Texas system occurred in 1872, when H. C. Hooker held some ten thousand Texas longhorns in the Sulphur Springs Valley, prior to delivering them to the army, and two years later he began breeding operations there, at his Sierra Bonita Ranch. A few Texans established ranches farther west, in the San Pedro and Salt River valleys, but they did not overwhelm the indigenous Sonoran ranching and succeed in placing a Texas imprint there.[41]

Following the same southern route to the goldfields, individual Texans established ranches in the Central Valley of California as early as 1852, especially in Tulare County. There reportedly "nearly every one of the cowboys and most of the cattlemen" came from Texas, only to disperse within a decade or so, leaving no trace.[42]

Farther north in Arizona and western New Mexico, above the Mogollon Rim, lies the Colorado Plateau country (figure 39). In spite of territorial conflict with highland Hispano sheep herders, Texas ranchers gained footholds there, especially in the Plains of San Agustín and the Valle Redondo, and by the early 1870s longhorns moved east to market along the Magdalena Livestock Driveway, remnants of which are still visible, to the railroad at Socorro. Probing farther northward, a few Texans found a chink between the Hispano and Navajo-Ute sheep cultures and moved into the San Juan River

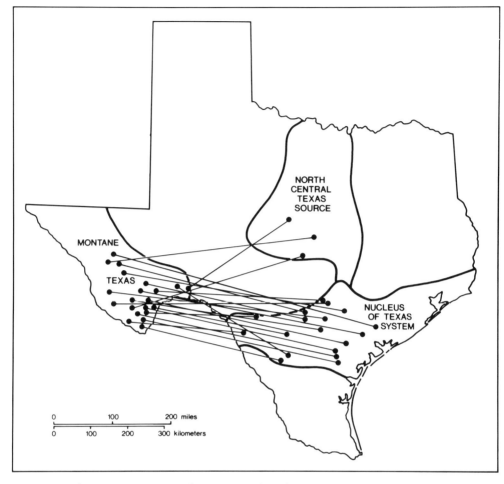

Figure 44. Documented migrations of cattle ranchers to montane West Texas, especially the Davis Mountains and Big Bend country. Compare this map to figure 43. Note the far greater contribution of the southern coastal nucleus of the Texas system. Sources: Jordan, Trails (see note 1), pp. 131, 140, 142; Utley, "Range Cattle," (see note 40), pp. 419–41; McAfee, Cattlemen (see note 40), pp. 21–22.

valley of northwestern New Mexico. From there they advanced into southeastern Utah, helping to implant a gentile ranching presence at the very gates of Mormon Deseret. Texas longhorns, if not ranching personnel, had even earlier played a role in the stocking of the Mormon communities. In fact a few Texans with previous ranching experience became Mormons and developed such longhorn-raising enterprises as McIntyre's Ranch in the Great Basin country of Juab County, Utah (figure 39). Subsequently Mormon cattle raisers, perhaps bearing some Texan influences, spread southward through Arizona's Colorado Plateau in the late 1870s, descended the Mogollon Rim, and occupied the upper Tonto Basin.[43]

Even the most remote corners of the Great Basin, in the high desert of northern Nevada and southwestern Idaho, felt some minimal Texan influences. Drawn by the Idaho mines, Texas drovers brought longhorns to the territory as early as 1866. Three years later the first southern cattle reached Owyhee County in southwestern Idaho, which also received in 1871 a large herd from the major source region in north-central Texas. About the same time, Texas cattle reached the nearby Quinn River valley of Nevada and other destinations in that state (figure 39). By the early 1880s, at least one cattleman with roots in north-central Texas had settled in northern Nevada, following a residential stopover near Cheyenne in the Wyoming Great Plains.[44]

In spite of these diverse and persistent penetrations of the Southwest and Great Basin, Texans achieved relatively little presence or influence there. Even Arizona, which reputedly received the deepest Texas cultural imprint of any area west of the continental divide, remained for the most part an outpost of the Sonoran Mexican ranching system. The Anglo-Texan presence in Arizona as late as 1880 never attained significant numbers.[45] The United States census population schedules for 1870 listed only six Anglos of Texas birth or removal living in Arizona who engaged in any aspect of the ranching industry, including "stock raisers," "herders," "cattle drivers," and "stock traders."[46] A decade later only fourteen Texas-born cattle ranchers resided in the territory.[47] Claims by the historian James A. Wilson concerning Texan influences in the Arizona cattle industry seem exaggerated, at least in the frontier era.[48]

The Anglo-Texan pastoral entry into the high valleys and "parks" of the Rocky Mountain chain proved equally unimpressive, with some local exceptions. Ranchers practicing the Texas system entered the Laramie Plains and Shirley Basin of Wyoming, west of the Medicine Bow Range, by about 1870, and a series of mild winters in the latter part of that decade lured others to ascend both forks of the Platte River to settle the North and South parks, high grassy basins behind the Front Range of Colorado (figure 39). By 1880 North Park alone held some thirty thousand cattle. Nearby lesser parks and the northern reaches of the expansive San Luis Valley also acquired

Texas cowmen.[49] Throughout these central Colorado valleys, "many southerners" established ranches, and the "early day roundups . . . were predominantly southern."[50] Representative of these immigrants was Dave Parker, a Texan who ranched in South Park after having worked earlier as a cowboy in the Pecos Valley cattle country of eastern New Mexico.[51]

Elsewhere in the Rockies, especially in the northern part of the range in Montana and Idaho, Texan ranching influence remained negligible. Texas longhorn cull steers reached some of the mountain mining-boom towns, such as Virginia City and even the Kootenay District in the Canadian Rockies, in the late 1860s and early 1870s, and Texas cowhands worked in Wyoming's Big Horn Basin, but no attempt was made to establish Texas-style ranches or bloodlines there (figure 39).[52] Instead the thriving cattle-ranching culture of the basins and high mountain valleys had very different, non-Texan origins, as will be discussed in chapters 8 and 9.

Ranching Practices in Texas Extended

In spite of formidable environmental challenges and the diluting or blocking effect of encounters with rival cultures, Texans succeeded in dispersing many key elements of their ranching system through Texas Extended, centered in the Great Plains (figure 39). The minimal amount of labor and capital required to raise cattle in the Texan manner, coupled with the rather considerable potential profits, made their system attractive to neophytes. At the same time, the very rapidity of the Texan expansion northward meant that the southerners had no real chance to change their ways. When they encountered rival systems, such as the sheep-raising culture of Hispanic New Mexico, the Mormon village-based mixed farming, or the Ohio Valley–Midwestern prairie-cattle complex, the Texans tended to avoid, assault, or displace them. As a result Texas influence dispersed widely in the American western heartland, bearing with it words such as *dogie, maverick,* and *lariat.*[53] Throughout Texas Extended, service centers and hamlets bearing names like Texas Creek (Colorado), Roundup (Montana), and even Ranchester (Wyoming) soon dotted the land, suggestive of the southern cultural intrusion.

One obvious measure of Texan influence in the expanded culture area was the breed of cattle. The Texas system, correctly, has been identified with the Gulf Coast type of Iberian longhorn, millions of which went north after 1865. As would be expected, longhorn blood decreased in importance with increasing distance from the coastal Texas source region of these animals. Their presence in the Great Plains was never as pervasive as the mythmakers would have us believe. Not only was some longhorn blood diluted in north-central Texas, but also perhaps four-fifths of the animals sent north were market steers, not range-stocking cows and bulls. From the very first, a strong northern prejudice against longhorns, based partly in the diseases they bore, was encountered, and through the 1870s and early 1880s the amount of longhorn blood on the ranges of Texas Extended was systematically reduced by crossbreeding, castration, and culling. Walter von Richthofen, for example, spoke of recent noticeable changes in cattle appearance in Kansas in about 1880.[54] Appropriately the coastal Texan-Carolinian, and originally Afro-Jamaican word *dogie* came to mean, in Texas Extended, a poor specimen or an animal with too much "southern" blood. Longhorns achieved, at best, an early numerical dominance in some northern ranges.[55]

Cowboy equipment and herding practices provide far better indices of Texan influence than do bovine bloodlines. Indeed the very word *cowboy*, before the ascendance of American popular culture, remained distinctive to the Texas system of ranching. This appellation, together with equestrian skills (in particular the casting of hemp lariats), went northward all the way into the Canadian prairies, and in the long run these techniques provided about the only durable Texan contribution to the plains cattle industry. Also diagnostic of the Gulf coastal Texas system were both the ability of the mounted cowboy and his agile quarter horse to detach individual animals from a herd and the use of the term "cut out" to describe this practice (figure 45). Even on the coastal prairie of Texas, cutting out from horseback had rendered the southern herder dog, a mainstay of the Jamaica-Carolina system, obsolete. Only occasionally in Texas Extended was the cow dog present, as in the mountain parks of central Colorado, where the majority of "early day cow men" reportedly kept one or more "catch dogs" to help round up cattle. While many observers acknowledged the superior riding skills of the Anglo-Texas cowboys on the Plains, some felt their roping skills were inferior to those of the Californians. Clearly the agile, intelligent quarter horses made a big contribution to the Texans' abilities, and one critic has even suggested that Texas horses knew more about working cattle than did their riders![56]

The mounted Texas cowboy of the Great Plains employed far more than his Mexican lasso in herd control, contrary to the popular

Figure 45. Texas cowboys "cutting out" cattle on the Great Plains, about 1870. Note the use of a whip. Cutting out from horseback made herder dogs unnecessary. Source: McCoy, Historic Sketches (see note 3), p. 80.

image. Contemporary sketches and descriptions often reveal the use of the Mexican "heavy cowhide quirt," or *cuarta*, a short-handled substitute for the old Carolina-British bullwhip that could more easily be wielded from horseback. These whips were employed both in overland driving and in "cutting out" cattle (figure 45). Some early illustrations also reveal a goad, perhaps a latter-day Mexican garrocha or vara, and the nicknames "cowpoke" and "cowpuncher" became associated with the Great Plains Texans.[57] Also ascending northward with Texas Extended was the Louisiana-derived, corrupted Spanish loanword *cavvyyard*, meaning a string of rotated saddle horses. In fact this is probably the single most diagnostic word in revealing the early Carolina-Hispanic contacts achieved in the Louisiana coastal prairies, and its usage can be compared to the synonym *remuda*, an uncorrupted loanword acquired much later, in South Texas. In Texas Extended, *cavvyyard* eventually lost out to *remuda*. Of coastal southern origin, also, was the practice in Texas Extended of weaning calves by slitting their tongues.[58]

The cowboys who plied these skills and spoke this jargon in Texas Extended consisted, for the most part, of an underclass of old-stock southern Anglo-Americans, including many Civil War veterans. That traditional stereotype, at least, seems valid. However, many or even most of them had no roots in the cattle country, having acquired the necessary skills as neophytes after coming out West, and the cowboy population included a rather cosmopolitan and even exotic mixture, from Texan to Yankee, Swiss, and Ceylonese. In Canada some Indian and half-breed, or Métis, cowhands worked on the early ranches.[59]

The annual round of labor required of these cowboys followed the traditional calendar of the southern system, keyed to the prevailing careless practices of keeping stationary stock on the native free grasses of unfenced pastures, without care, supplementary feed, or shelter—a method called "rawhiding" in some parts of Texas Extended.[60] The minimally tended cattle often drifted far from their home ranges, particularly in winter. Kansas herds, emulating the bison they had so recently supplanted, sometimes escaped as far as the Texas Panhandle, and South Park cattle in Colorado occasionally drifted during blizzards into the San Luis Valley.[61] With rare exceptions, early Texans erected "no fencing beside that of branding-corrals," and they engaged in "no hay cutting."[62]

In such a system, the old southern custom of two roundups each year also survived. The first major activity in the annual round, and "the great event of cattle management" on the open range, was the spring roundup, a cooperative and often highly organized undertaking by cowboys from all the ranches in a district. This greatest of cowhunts could last for as long as six weeks to two months and served the traditional Carolinian purposes of separating herds, branding the increase, and "cutting," or castrating, calves. Groups of cowboys typically encircled a creek watershed, taking advantage of the reluctance of cattle to graze along divides, then worked the "bunch" towards a centrally located branding pen or ground.[63] The corrals were usually built in the old southern style, with rails or poles stacked up between regularly spaced pairs of posts (figure 37), but occasionally the Mexican picket, or "stockade," type appeared. The "Colorado method," however, involved branding without penning, performed on the gathering grounds.[64] The brands burned onto the cattle at the spring roundup almost invariably displayed the British-derived designs of block letters and Arabic numerals, lacking any trace of Hispanic influence. By law, as in the South, they were registered in every county seat where the rancher's stock might normally be at pasture.[65]

Following the labors of the spring roundup, the cowboys "gave up close-herding" of the cattle and needed only to "ride line" periodically during the summer, patrolling the outer limits of the herd's normal range to keep them confined to their stationary pasture by turning back strays, just as coastal southerners had done for two centuries. To make this activity less difficult, the hands often resided in "line cabins"—log dugouts or shacks positioned at regular, wide intervals along the pasture perimeter.[66] Line riding increasingly took the place of calf-capture and salting in herd control. September usually brought the smaller fall, or "beef," roundup, lasting two or three weeks and intended mainly to cull fat steers for market. Animals that had earlier escaped branding, still called "mavericks," another word diagnostic of Texas influence, went under the iron.[67]

As winter approached, the cattle dispersed again, in hopes of finding sufficient forage, as well as shelter behind bluffs or in coulees. Line riding resumed, more sporadically, and some cattle escaped to drift southward with each icy wind. Cowboys had little work to do from January through April, aside from menial chores such as cutting poles for corrals and hunting for drifted stock during post-blizzard warm spells. Even as far north as Alberta, cowboys and ranchers pursuing the Texas system raised cattle "without any provision for their food and shelter beyond what nature afforded."[68]

The ranchers who directed the Texas system in its expansion phase included both large- and small-scale operators. Men who began as cowboys sometimes became ranchers, and probably more such social mobility existed in early Texas Extended than on any other cattle frontier in North America. In general, small ranchers were typical of the early expansion years, and the plains cattle kings became more important later. Ranchsteads reflected the differing scale of operation, ranging from a crude log cabin with adjacent pole corral for horses, at the lower end of the scale, to fine steads such as one in the Texas Panhandle about 1880, consisting of a five-room house, a freestanding kitchen-mess, a blacksmithy with attached sleeping quarters, a granary and tack room, a milk house, and a storage shed, in addition to a pen complex, fenced vegetable garden, and rye patch for horse feed.[69]

Large or small, the ranchers pursuing the Texas system normally used the "free grass" of the public domain, squatting on the land or at most homesteading a small property near a reliable water source. Nine of every ten early ranchers in the Texas Panhandle reportedly had no title to the pastures they used.[70] Only two sizable ranching regions of the Great Plains early developed a system of range leasing rather than free grass—the Indian reserves of western Oklahoma and the Canadian prairie provinces—and even there some squatting occurred in the first years.[71]

By the 1880s the big operators had mainly seized control of the ranching industry of Texas Extended. The eventual passing of the free-grass era and the trend toward leasing and landownership favored them. Some acquired huge land grants as their base of operation, most notably the famous XIT Ranch in the Texas Panhandle. Others, including numerous "land and cattle companies," depended upon capital flowing westward from the cities of the East or even Europe, particularly Britain. Speculating in land and livestock, these investors often realized huge profits in the "beef bonanza" of the early eighties. Some large enterprises formed "pools" to permit an even grander scale of operation. These big ranchers initiated, in effect, an agribusiness, one fully in harmony with the Industrial Revolution, with all the attendant benefits and curses. Large-scale ranching became little more than a risk-capital venture, more fully commercialized and specialized than before. Cattle raising became the sole activity on many of these big spreads.[72]

Fewer of the owners resided on their ranches as the role of risk capital grew. Though long typical of both Hispanic and southern cowpen herding cultures, absentee owners soon became the norm on the plains. Those relative few big operators who did choose to reside on the land increasingly took on the airs of a pastoral aristocracy. In Canada, where such gentry displayed more British tendencies, the mounted coyote hunt took the place of the Old World fox chase.[73]

Had all gone well on the ranges of Texas Extended, the dominance of the big, specialized ranchers might have become complete by the turn of the century, ushering in the age of agribusiness in the West well before its time. Instead the Texas system simply collapsed, bringing an end to one of the most colorful, remarkable North American cattle frontiers and wiping out many of the big operations.

The Collapse

The passing of the Texas system of cattle ranching had multiple causes, but at root it involved ecological maladaptation. Born in a mild, subtropical "lotus land" and based upon a breed of cattle that had never known the bite of a true continental midlatitude winter, the system was singularly unsuited to the far greater part of Texas

Extended in the north. Derived from a wet, even marshy coastal land of lush, tall grasses, Texas ranching proved ill-suited to cope with semiaridity. The system died of cold and drought.

The signs of maladaptation had been posted all along the trails north and west. Even the self-confident Texans must have noticed that calf yield in Kansas was only four-fifths of that typical on the central Texas ranges, in Nebraska two-thirds, and in Dakota but half. They must surely have entertained doubts about their methods when the severe winter of 1871–72 struck their earliest expansion nuclei in Kansas and Nebraska, killing half or more of the longhorns in the Cottonwood, Smoky Hill, and Arkansas river valleys of Kansas, permitting only a salvaging of their hides. Along the Republican River in southern Nebraska, only 40 percent of the longhorns in Franklin County survived that same winter, and in adjacent Harlan County, fewer than one in five. A clear and costly warning had been issued to the southerners, but they did not heed it. Had the Anglo-Texan system advanced northward gradually, a learning and readaptive process might have taken place, but their diffusion was explosive, coinciding with an interlude of relative warmth and wetness on the interior plains. They and their risk-capital sponsors learned nothing. The foolhardy subtropical methods persisted.[74]

The end came not suddenly, but gradually; not simultaneously throughout Texas Extended, but in different episodes in different regions. The winter die-offs of 1871–72 in the central Great Plains were but the first in a long succession of localized meteorological disasters, each of which weakened the foundations of the Texas system. A severe drought struck the plains of eastern Colorado in 1879 and 1880, and in Utah the winter spanning those two years proved particularly severe, with huge die-offs. The next winter brought large losses of cattle in Colorado and Nebraska.[75] Then, following a few good years, a harsh winter struck again, in 1884–85, bringing severe cold as far south as the Texas plains and herd mortality of up to 90 percent. The next winter, though equally severe in Kansas and the southern Great Plains at large, was mild in Wyoming. Drought recurred in 1886.[76]

The greatest disaster still lay ahead, but by the spring of 1886 one contemporary observer had learned enough to lay bare the folly of the Texas system. Where, he wondered rhetorically, were the million or more head of largely longhorn breeding stock that had been driven north from Texas during the previous ten years? "In a suitable climate these cattle would have been alive to-day. Where are they? The bones of thousands . . . lie bleaching on the wind-swept flanks of the foot-hills," he reckoned. "They pave the bottoms of miry pools; . . . they lie in disjointed, wolf-gnawed fragments on the arid, bunch-grass ranges; they are scattered over the short buffalo-grass . . ." Why had they perished? "They have died of hunger; they have perished of thirst, when the icy breath of winter closed the streams; they have died of starvation by the tens of thousands during the season when cold storms sweep out of the North and course over the plains, burying the grass under snow." They have "frozen into solid blocks during blood-chilling blizzards."[77]

The system, then, was in acknowledged collapse before the calamity of 1886–87, which is often blamed for the demise of open-range ranching. That bitter winter brought herd destruction all the way from Montana and Dakota in the north to West Texas. As many as 60, 80, and even 90 percent of the cattle perished. Few on the Great Plains would ever again trust the traditional Texan methods.[78] Farther west, in the mountains and basins, the final blow came a few years later. The winter of 1889–90 took up to 95 percent of the cattle in some montane valleys. Drought, not cold, largely ended the Texas system in Arizona in 1891 and 1892.[79]

While cattle raisers practicing more capital- and labor-intensive enterprises and raising British-derived stock also suffered herd losses in these climatic catastrophes, their attention to winter feeding mitigated the disaster. In the aftermath their methods provided an alternative to Texan ways. Even contemporary observers recognized that a rapid transition was occurring. They knew "the old system of the open range" had passed, that "the cow-boy of to-day, especially on the northern ranges, is of entirely different type from the original cowboy of Texas."[80] Gone was the longhorn, gone the free grass, open range, stationary pasturing, and neglect of winter feeding. Also gone, eventually at least, was the freedom and individuality of the plains ranchers and cowboys. Gone, in fact, was the cattle frontier of the great inland plains. In some remote pockets, such as northeastern Montana and southeastern Alberta, vestiges of the old system lingered a bit longer, and some unreconstructed Texans resisted change by cutting the new wire enclosing the pastures. A few belated attempts to revive something akin to the old system, offering a last hurrah for the Texans, occurred here and there on the Plains, as on the newly opened Sioux-reservation leases in Dakota after 1900. Other Texans wishing to ranch in the old way found refuges in southern Arizona and northern Chihuahua as late as 1910.[81]

In the long run, only the myth survived—the irrational belief or desperate hope that, surely, somewhere out West the colorful old-style Texas ranchers still plied their trade. So powerful is this myth that even in 1990, a full century after the Texas system perished, an inarticulate man posing as an old-fashioned cowboy could very nearly be elected governor of Texas, a state whose population is over four-fifths urban.

Climatic maladaptation alone would have destroyed the Texas ranching system, but the end perhaps came even faster because other causal factors also played a role. The free-grass policy permitted in the United States, except in Oklahoma, greatly encouraged overstocking, as did a serious misreading of the pastoral capacity of the fragile shortgrass plains and the speculation-fueled, hypercommercialized cattle boom of the early 1880s. The resultant cattle glut both severely damaged the ranges and, by 1886, led to a crash of beef prices. Livestock dumped on the market because the depleted pastures could no longer support them further depressed prices. Even so, thousands of additional cattle died due to the deteriorated condition of the ranges.[82] The Texas system was not merely maladapted to the Great Plains; it was not sustainable in any environment and would have collapsed even in the lushest and mildest of settings.

To make matters still worse, the Texans also had to cope with encroachment, particularly by farmers. The tallgrass prairie, destined to be the world's most prolific granary, was early ceded to wheat farmers from the East and Europe, including the Russian Germans. The ranchers could retain only sterile rocky or sandy refuges scattered through the tallgrass region, such as the Nebraska Sandhills, Arbuckle Mountains, and Flint-Osage Hills, which became pastoral islands in a sea of wheat. The advance of the farmers through the tallgrass prairies usually progressed gradually, but in Oklahoma the plow frontier went forward in a series of great bursts, when entire newly opened districts were seized in a matter of days. Unwisely, farmers continued westward into the semiarid shortgrass plains in the wake of the Texas collapse, laying the basis for the eventual Dust Bowl.[83]

The range deterioration associated with the Texas system opened the way for another encroachment, by sheep raisers. Following behind cattle overstocking, sheep could more effectively utilize what forage remained. The bovine dominance of the Great Plains never again attained the levels of the early years, and the sequence of the interior plateau of colonial Mexico was repeated.[84]

Besides the intrusions by the plow frontier and sheep raisers, the Texans faced, from the very first, competition in the West from rival systems of cattle herding, a competition in which the southerners fared poorly, both spatially and temporally. Practitioners of the midwestern/Ohio Valley herding system, as well as the more remote Californian system, achieved major expansions after about 1850, creating their own new cattle frontiers. The West never belonged, in anything approaching its entirety, to the Texans. In the following two chapters, the pastoral partitioning of the West by these rival systems will be described.

EIGHT

PASTORAL CALIFORNIA EXTENDED

THE PASTORAL TRANS-MISSISSIPPI WEST DISPLAYED CULtural diversity and internal sectionalism even in Spanish colonial times, and the intrusion of Anglo-Americans reinforced rather than obliterated these contrasts. During the development of the Anglo-Texan ranching system, a parallel but distinctive cultural exchange took place in California, yielding a second, fully independent nucleus of the western range-cattle industry.[1]

Origin of the System

The beginnings of the Anglo-Californian system came about because the Mexican government made extensive pastoral land grants to immigrants from the United States (figure 46). The first such grants date to the 1820s, but most were awarded in the 1840s, shortly before Mexico lost the territory. More than one-fifth of the 473 Alta California land grants in the period 1821–46 went to people with Anglo surnames, and at least a few score Anglo cattle raisers were active by 1848. The major concentration of these Anglo ranchers lay in the northern part of the Central Valley, also called the Sacramento Valley, and in lesser vales in the coastal hills north of San Francisco. This region on the settlement frontier of Mexican California became, by the late 1840s, the nucleus of the Anglo ranching system. At the census of 1850, the Anglo core area remained geographically distinct from the older Hispanic ranching focus in southern California (figures 32, 47).[2]

During this early formative period, Anglo-American cultural input remained minimal. In contrast to Texas, where a lengthy, bitter military conflict developed between ruling Mexicans and numerically superior Anglos, engendering enduring hatred and prejudice, California enjoyed a much more harmonious contact period, in which a benevolent Hispanic majority accommodated a small Anglo

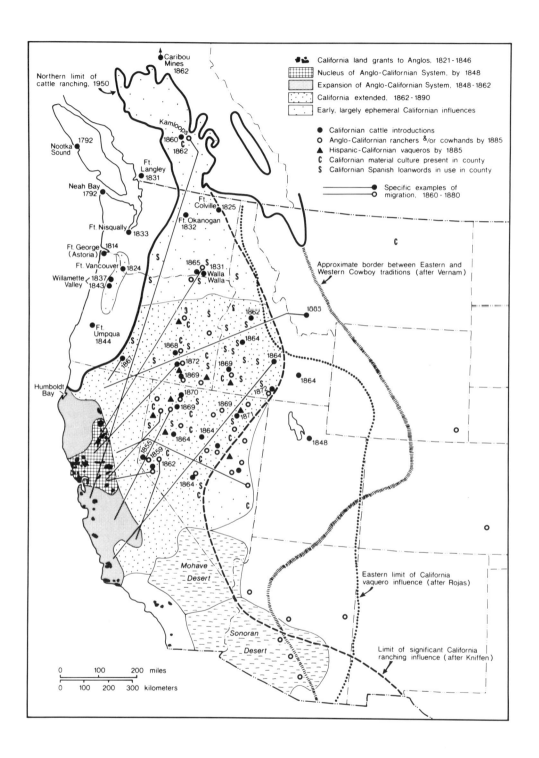

Figure 46. Origin and diffusion of the Anglo-Californian ranching system and the pastoral California Extended, 1825–90. From a hispanicized nucleus in the northern part of California's Central Valley, the system spread to much of the intermontane region of the West. The loanwords included are those listed in the text; the items of material culture include the beef wheel, Spanish windlass, snubbing post, single-rigged saddle, armitas, and Spanish brand designs. All symbols, except for cattle introductions and rancher migrations, represent one county unit. Sources: Kniffen, "Western Cattle Complex," p. 182 (see note 1); Vernam, Man on Horseback, p. 301 (see note 3); Rinschede, Wanderviehwirtschaft, pp. 91, 93 (see note 1); Oliphant, On Cattle Ranges, pp. 68, 87 (see note 4); Gordon, "Report on Cattle," p. 1098 (see note 12); Hornbeck, "Land Tenure," p. 389 (see note 2); Treadwell, Cattle King, endpaper (see note 11); Weir, Ranching in Southern, p. 18 (see note 42); Fletcher, Free Grass, p. 22 (see note 19); Rojas, Vaquero, pp. 8–9 (see note 26); and various other sources cited in this chapter.

minority. In this happier setting, most Anglo grant recipients simply adopted the previously described Hispanic-Californian rancho system. They took on the honorific titles "don" and "doña"; retained a work force consisting mainly, if not exclusively, of California mestizos and acculturated mission Indians; used Spanish equipment; generally employed Mexican ranch foremen; became bilingual; often married into Hispanic families; continued to produce mainly hides and tallow; and even dressed in the charro-inspired mode of the Mexican rancheros. Thus, for example, cattleman Patrick Murphy of San Luis Obispo became known as "don Patricio." The resultant ranching system, by the late 1840s, differed from that of Texas both because of the inconsequential Anglo cultural input and because the underlying Hispanic pastoral traditions in the two provinces had been in many ways distinctive.[3]

Early Diffusion

The influence of the California ranching system began to be felt beyond the state's borders during the early Anglo-Hispanic period. In fact small numbers of Iberian longhorn cattle were exported from California to the Pacific Northwest even earlier, in Spanish colonial times (figure 46). In the 1790s Spain's outpost settlements at Neah Bay on the Olympic Peninsula and Nootka Sound on Vancouver Island received "black cattle" from California, and in 1814 more went by sea to Fort George—the former Astoria, a British North West

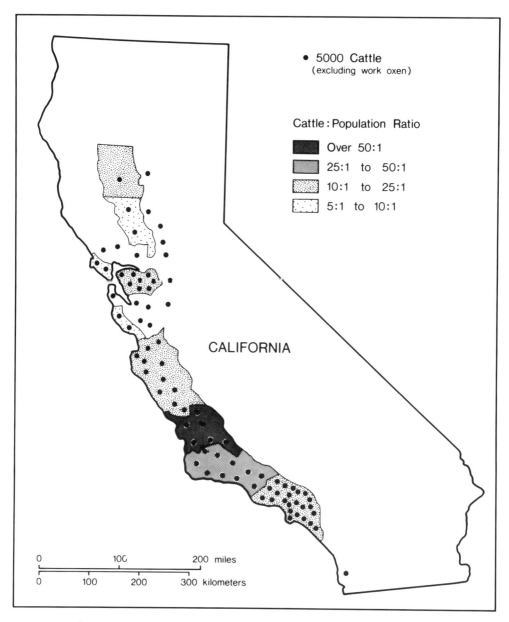

Figure 47. Cattle distribution and cattle:population ratio in California, *1852.* Work oxen are excluded. The evolving Anglo-Californian focus in the north is clearly discernible. *Source:* The Seventh Census of the United States, 1850 *(Washington, DC, 1853),* pp. *982–84.*

Company fur post at the mouth of the Columbia River. More important entries of California cattle into the Northwest came during the Mexican period, when the Hudson's Bay Company, attempting to achieve self-sufficiency in food supplies through its agricultural subsidiaries, stocked the post at Fort Vancouver in 1824. From that carefully managed Iberian herd, as well as from additional Californian imports, the Company stocked most of its trading posts in the Northwest, even distant Fort Langley on the lower Fraser River in British Columbia. The largest Company herd remained at Fort Vancouver, where the number of California-derived cattle grew from 17 in 1824 to 200 in 1829, 450 in 1836 when the first slaughter for beef occurred. Fort Vancouver can be regarded as the only consequential cattle-raising center operated by the Company, and as late as the early 1840s it was still the only major focus of such activity in the entire Pacific Northwest.[4]

By that time the Anglo-American agricultural colonization and Protestant missionization of the Pacific Northwest were underway. These settlers drew both upon the Hudson's Bay Company and Mexican California to found their herds of cattle and horses. In 1837 over 800 Iberian longhorns from California began an overland drive to the new settlements in Oregon's Willamette Valley, and 630 survived the trek. An even larger herd, numbering some 1,250, reached Oregon from California in 1843. In turn Iberian longhorns from the Willamette Valley went to stock Protestant missions at The Dalles in 1838 and Clatsop, near Astoria, in 1841. Into the middle 1840s, Spanish cattle of Californian origin remained the dominant breed in the Pacific Northwest.[5]

Similarly California Spanish cattle helped stock the newly founded Mormon settlements in the Great Basin. In 1848 several hundred longhorns were trailed east to the Salt Lake oasis, preceding the arrival of Ibero-Texan cattle there by almost two decades. Other drives of California livestock eastward before 1850 generally involved horses and mules.[6]

These early California-derived stockings, though widespread and repeated, generally bore relatively little cultural baggage. As D. W. Meinig observed, the cattle drives from California involved sporadic buying trips rather than regularly recurrent contact, with "no sustained intermingling" of pastoral cultures. Only in the Willamette Valley of early Oregon did some Hispanic-Californian herding methods gain a foothold. An account from the middle 1840s described Willamette cowhands who "went full speed amongst the herd and threw a rope with a almost dead certainty a round the horns or neck of the animal." This technique he referred to as "lassing," suggesting that California-derived Spanish loanwords were also coming into use along the Willamette.[7] Even so, most cattle exported from Mexican California did not become part of any ranching system, but instead fit into a diversified, mixed-farming economy designed to be self-sufficient.

The Anglo-Californian Cattle Boom

The gold discovery at Sutter's Mill on the eastern margin of the Central Valley of California in 1848 soon brought major changes to the Anglo-Californian ranching system. Most notably the mines provided for the first time a large local market for beef, precipitating a pastoral boom between 1849 and 1856. The boom, in turn, led to other major changes.[8]

In the process, two distinct pastoral regions developed in California, reinforcing a north-south distinction that first appeared, less overtly, in Spanish colonial times. Southern California remained largely Hispanic, an isolated, poor, rather lawless backwater that sent so many cattle to the goldfields as to cause range depopulation. The principal Anglo impacts in the south during the boom period of the 1850s remained confined to the confiscation or purchase of pastoral properties from Hispanic owners, a process facilitated by the United States takeover in 1848, and the rustling of cattle from Mexican ranches. Most new Anglo landowners in the south followed the model set in the period of Mexican rule and simply adopted the older Hispanic herding culture, becoming the new landed aristocracy but bringing no substantial changes to the system. Sometimes descendants of the original Hispanic grantees remained as ranch foremen under Anglo ownership, as occurred at Rancho Jesús María in Santa Bárbara County. Some Anglos, instead, converted the ranchos into farms or settlements. Even so, of the 130 large-scale cattle ranchers residing in southern California by 1860, almost two-thirds were still Hispanics residing upon their old land grants.[9]

By contrast, northern and interior California, including the great Central Valley, became considerably anglicized, serving both as the focus of the cattle boom and the center where the Anglo-Californian ranching system evolved in ways very different from the south. Anglos developed many new ranches in the remaining Mediterranean prairies, where excellent perennial bunchgrass pastures could still be found. By 1860 fully 40 percent of all California cattle grazed ranges in the Anglo-dominated Central Valley. Anglo stock raisers also spread northward in the coastal valleys, such as Clear Lake, reaching Humboldt Bay near the Oregon border (figure 46). Expansion in the Anglo interior and north was largely responsible for the three- or fourfold increase in California cattle population between 1850 and 1860.[10]

Many Hispanic herding practices survived in the north and interior during the boom, but equally important was the increasing Anglo cultural influence in cattle ranching there, producing a hybrid system. Not only did the industry in the north pass largely into Anglo hands during the 1850s, with the result that 81 percent of all large cattle ranches there were Anglo-owned by the end of the decade, but also many Anglos entered the cowboy work force. The abrupt shift from hide and tallow to beef production represented a break with Hispanic Californian tradition. Then too, the gold-rush market caused a radical change in the breed of cattle raised in the Anglo north. In fact the upgrading of herds there had begun even before 1848, but the massive culling of steers to feed the miners, coupled with large-scale importation of midwestern cattle in the 1850s, allowed an accelerated change in bloodlines. By the middle 1860s, Spanish cattle were fast disappearing in California. Even in Monterey County, on the borders of the more Hispanic south, only about one-third of all cattle remained "unimproved" by 1865.[11] The Anglo-Californian cattle-ranching system, then, evolved in the north and interior during the boom, becoming increasingly anglicized while retaining many Hispanic influences. The south, where Spanish pastoral California had begun in the previous century, became peripheral and of minimal importance to the evolving Pacific ranching culture.

California's cattle boom ended in the middle 1850s, when demand from the local goldfields diminished and the market became glutted. In response Anglo-Californian ranchers began driving herds to bonanza districts in other parts of the West, following the exploding mining frontier. They marketed steers in Oregon, Idaho, Nevada, and even British Columbia during the late 1850s and early 1860s, following often difficult paths to the new mining districts (figure 46).[12]

During that same period, a few cattle raisers also began stocking ranges outside California, sometimes even moving their entire ranching operations to other states and territories. The building of pastoral California Extended had begun. For example, Californian ranching operations became established in the Carson Valley of western Nevada at least by 1859. An Anglo-Californian ranching family, the Santa Clara County Harpers, relocated in 1862 to the Kamloops area of interior British Columbia. In that same year, other Californian herds went to help restock ranges in Oregon and Washington, where a severe winter had decimated the cattle.[13]

In spite of these exports and migrations, California pastures remained overstocked in the early 1860s, prompting *matanzas*, or slaughters, carried out voluntarily by cattle ranchers in the traditional Hispanic-Californian manner to help preserve the native grasslands. Such measures apparently had little effect, for some three million cattle grazed the open ranges of California by 1862. Extensive range damage occurred, especially in the Central Valley.[14]

Collapse in Mediterranean California

The newly formed Anglo-Californian cattle-ranching system collapsed in its area of origin in the 1860s, a quarter-century earlier than its counterpart in Texas. Overstocking certainly contributed to the demise, but the principal cause was meteorological. Floods in the winter of 1861–62 drowned many livestock, and this interlude of above-average precipitation was followed in the period 1862–65 by one of the most severe and prolonged droughts ever to strike the Mediterranean climate zone of California (figure 31). The winter rains failed for three years. Not even an increased, Hispanic-inspired shifting of cattle between seasonal pastures could avert disaster, and huge die-offs occurred. Hundreds of thousands of cattle perished. In Santa Bárbara County, on the borders of the south, 97,000 cattle grazed the dessicating range in the spring of 1863, but only 12,100 remained two years later. During the decade ending in 1870, the cattle population of California decreased by nearly half. Many ranges stood virtually denuded of palatable vegetation by the middle of the decade.[15]

The collapse, as later in Texas, brought fundamental changes and relocations. Anglicization in the postdrought ranching industry accelerated. Many or most of the remaining Hispanic ranch owners soon went out of business, the California longhorn all but vanished, and methods involving greater investment of labor and capital replaced the old ways. The vaquero class survived, as did the related equestrian skills, but increasingly these techniques passed to neophyte Anglo-American cowboys. The year 1865, more appropriately than 1848, marked the end of the Hispanic rancho era in California.[16]

Indeed Mediterranean California ceased to be a significant scene of cattle ranching. Herding gave way after 1865 to crops, including both small-scale farming and "bonanza" wheat-growing agribusinesses. The very word *ranch* came to mean a crop-centered enterprise in California. Pastoral California Extended became a hollow shell, its core largely evacuated by the Anglo stock raisers who had set its diffusion into motion. In the process, California proper forever shed its cowboy image, in a manner Texas never achieved. There would be no enduring range mythology to help shape the Californian popular culture. The building of pastoral California Extended continued, even hastened, after 1865, but the expansion was Turnerian, with cattle ranchers occupying an advancing frontier while simultaneously sacrificing their former hold over the Mediterranean core.[17] Not until the rise of Hollywood did the cowboy return.

Pastoral California Extended, 1864–1890

The calamitous drought and dispersing mining frontier set into motion the first major emigration of Anglo-Californian ranchers, a movement coinciding temporally almost precisely with the postbellum northward surge of Anglo-Texas cattle raisers. The two would share dominion over much of the interior West. Dispersal from California began in earnest when many ranchers, seeking to save their herds, fled to ranges outside the Mediterranean climate zone. Some went no farther than the previously unoccupied northern interior California counties of Modoc, Lassen, Siskiyou, and Shasta, or the Sierran eastern counties of Alpine, Inyo, and Moro, including the Owens Valley. Most of these new ranges lay east of the Sierra Nevada.[18]

Many other Anglo-Californian ranchers relocated their herds even farther afield, to western and northern Nevada, eastern Oregon, southern Idaho, and even Utah, Washington, southern British Columbia, and southwestern Montana (figure 46). In 1864 parts of northern Nevada were "invaded by stock driven in by old Californians," especially the Humboldt, Big Smoky, and Reese river valleys in present Nye, Churchill, Pershing, and Lander counties. In that same year, three hundred "Spanish" cattle from California arrived in western Idaho, followed the next year by a like number that reached the ranges in the Wood River drainage, part of the Snake River Plain.

Also in 1864, herds from San Luis Obispo County came to the Fort Hall area, farther upstream on the Snake. In 1865 some four hundred California cattle arrived in the Walla Walla Valley in the Columbia Basin of eastern Washington, and another four hundred were trailed from the Shasta Valley in northern California to the Beaverhead Valley in the Montana Rockies. Among the earliest cattle ranchers in British Columbia in the 1860s were Anglo-Californians such as Jerome Harper and Lewis Campbell.[19]

In this expansion, the Anglo ranchers left the formerly productive prairies of Mediterranean California, which had provided the major environmental base for Pacific coastal cattle ranching since the Spanish beginnings in the 1770s. Crossing the Sierras and Cascades, they developed pastoral California Extended in the much drier, elevated, and colder lands of the intermontane region of the West, most of which is called the Great Basin. Lying between the Rocky Mountain system on the east and the Sierra Nevada–Cascade complex on the west, this huge region of plains, basins, valleys, and parallel ranges was variously called the "High Desert," "Cold Desert," or "sage desert." In reality it constitutes a semiarid expanse of steppes lying north of the true deserts of southern Nevada, southwestern California, and western Arizona (figure 46). Nevertheless, the greater part of the intermontane region, forming the Great Basin, is characterized by interior drainage, its exotic streams ending in landlocked seas such as Carson Sink, Pelican Lake Marsh, Malheur Lake, and many other alkali depressions. Only the northern margins of the intermontane region, drained by the robust Snake-Columbia system, send waters to the ocean. Whether the streams ended in sinks or reached the Pacific made little difference to the invading Californians. In either case they situated in the valleys and plains along the rivers, often at the edge of the lowlands nestled up against the foothills.

Climatically the immigrating ranchers had to adjust to far colder winters and substantially less precipitation, although the familiar Mediterranean Californian regime of summer drought still prevailed east of the mountains. The ranges bordering the intermontane region and scattered regularly through it offered a complex vertical zonation of climates, another feature the ranchers had known in the California nucleus. Fortunately for the Californians, the Rocky Mountain complex bordering the intermontane region on the east serves as a great airmass dam, preventing most Canadian Arctic fronts from entering the area. Intermontane winters, as a result, are not as severe as those on the Great Plains, although they must have seemed very cold, indeed, to pastoral immigrants accustomed to the idyllic temperature regime of Mediterranean California.

In terms of vegetation, most of California Extended lay in the "sagebrush country." In the basins and valleys, a gray ocean of big sagebrush (*Artemisia tridentata*) and white sage (*Ceratoides lanata*) provided an open shrub overstory, beneath which grew palatable forbs and perennial, tall, coarse bunchgrasses, such as Great Basin wild rye and salt grass, which seasoned into good standing hay. The white sage, also called "winterfat," provided excellent cold-season browse, as did the willow, but the big sagebrush did not prove useful as cattle forage. Around the constantly fluctuating margins of the sinks and lakes, marshes similar to the tule lands of California's Central Valley provided both hay and winter grazing.

Seasonal vertical movement of the herds became essential in this setting, for the basins could not support many livestock through the parched summers. From winter ranges in the lowlands, cattle went in spring and early summer up into the piñon pine–juniper zone in the foothills, where both bunchgrasses and fescue abounded. In the heart of summer, the stock ranged higher in the mountains, into the oak-brush lands, where bunchgrasses were supplemented by browse; into the ponderosa pine–bunchgrass zone between about 2,100 and 2,750 m. (7,000–9,000 ft.) elevation; and into the alpine mountain pastures above, rich in tall Nevada bluegrass, meadow barley, forbs, and sedges.[20]

The Californian exodus prompted by the 1862–65 drought continued long after the winter rains resumed. For two decades following the Civil War, pastoral California Extended continued to expand, especially in the Great Basin. A market-driven Thünenian competition for land succeeded drought as the major force expelling the ranchers. Fencing laws in California proper, which had long favored stockraisers, were radically modified in the 1870s to the advantage of farmers, reflecting the land-use changes underway and adding to the flight of the graziers. Even though the main market for beef shifted from mining districts to the growing West Coast cities, the inland retreat of the ranchers continued.[21]

Following the eastward paths of earlier cattle raisers, the emigrant Anglo-Californians after 1865 relocated mainly in Nevada, Oregon, and Idaho.[22] Open ranges east of the Sierras in Nevada and the Cascades in Oregon, as far as the Mormon country and northward to the Blue Mountains of Oregon and the Salmon River in Idaho be-

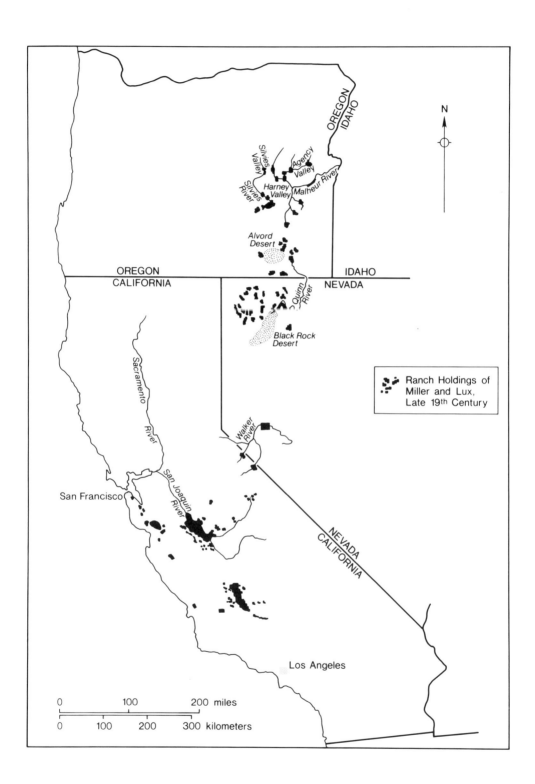

Figure 48. Ranches and other landholdings of the California-based enterprise of Henry Miller and Charles Lux. From their main base in the San Joaquin Valley, Miller and Lux expanded their cattle-ranching interests into western Nevada and southeastern Oregon in the 1870s and 1880s. While extraordinary in scale, the diffusion by Miller and Lux is representative of the Anglo-Californian cattle kings in the period 1870–90. Source: Treadwell, Cattle King, *endpaper map (see note 11).*

came, in large measure, "the pasture lands of Californians."[23] Intermontane cattle kings arose in the inland empire after about 1870, controlling huge expanses of rangeland and possessing enormous herds. Their roots lay in California and their investment capital flowed from that same source. Some started almost from scratch beyond the Sierra Nevada, but many, and perhaps most, simply transferred previously established ranching enterprises out of California (figures 46, 48).

Nowhere was the Californian pastoral presence felt more strongly than in southeastern Oregon. In trans-Cascadian counties such as Grant and Harney, a Californian cattle aristocracy developed, eventually coming into conflict with homesteaders of midwestern origin. As early as 1867, a Siskiyou County Californian brought two thousand cattle to ranges near Fort Klamath, below Crater Lake in southern Oregon. Within a few years, several other Californians had founded large ranches in Harney County, including John Devine from the Sacramento Valley, who, in partnership with W. B. Todhunter, established the well-known White Horse Ranch near Steens Mountain in Harney County.[24]

The greatest of the California-derived cattle empires in Oregon was that of Hugh Glenn and Peter French. Glenn had earlier developed a large cattle ranch, among other enterprises, in Colusa County northwest of Sacramento. Realizing that the pastoral period was ending in the Central Valley, he sent several thousand head from his sizable herds in 1872 to the Donner und Blitzen Valley of Harney County, Oregon, under the care and management of French and six California cowboys. French took this opportunity to establish his own ranch there, eventually controlling some 80,000 hectares (200,000 acres) of rangeland. He and others drove market steers overland to the railroad town of Winnemucca, Nevada.[25]

Glenn and Todhunter also acquired ranchland in Nevada, particularly in the Paiute Valley, representative of continuing Californian expansion in that direction. Other prominent Anglo-Californian cattle raisers moved into the valleys of the Humboldt, Quinn, and Walker rivers in Nevada, near the transcontinental railroad line. These included James Hardin, the "cattle king of Humboldt"; L. R. Bradley, Nevada's second governor, who had come over the Sierra Nevada with cattle in 1862; and the Henry Miller–Charles Lux enterprises, based originally in the San Joaquin Valley. Miller and Lux eventually controlled about 400,000 hectares (1,000,000 acres) of land and a million head of livestock in three states (figure 48), and their enterprise served the urban markets, especially San Francisco. Some Californians soon reached the borders of Mormon Utah, and cattle continued to cross the mountains into Nevada until about 1874, after which time California ceased to supply any substantial numbers. With Oregon and Nevada occupied, the California ranchers pushed over low divides into the southwestern section of Idaho, taking much of the Snake River Plain and the adjacent mountain valleys on the north. By the late 1880s, the pastoral geographical expansion of California Extended was essentially complete. The Californian imprint weakened greatly north of the Blue Mountains and east of the continental divide, prompting these areas to be excluded on the map of California Extended, but clearly some influences penetrated Washington, Montana, and British Columbia (figure 46).[26]

Less clear is the cultural relationship of southern Arizona to California in the latter half of the nineteenth century. Geographer Fred Kniffen, among others, regarded southern Arizona as an appendage of the Anglo-Californian ranching system (figure 46), but such claims remain unproven. In 1870 only two California-born cattle ranchers—one Anglo and one Hispanic—resided in the Arizona Territory, but by 1880 the total had risen to eighteen, including two Spanish-surnamed ranchers, as well as five Oregon natives. California ranked second as a state of birth for ranchers, Texas only sixth.[27]

Even so it would be as incorrect to regard southern Arizona in the 1870s and 1880s as part of California Extended as to link it to Texas. Instead Arizona remained basically Sonoran in its cattle-ranching tradition (figure 39). Contrary to the generally accepted notion, the older Sonoran system never completely died out during the Apache raids of the 1825–70 period, surviving in the Santa Cruz River valley south of Tucson (figure 27). While only some five thousand cattle remained in the entire territory by 1870, about a dozen small-scale Mexican ranchers, each owning fewer than seventy-five head, still resided in the valley. The Santa Cruz served as the nucleus of the revived Arizona cattle industry in the decade that followed, and by

1873 Hispanic ranchers began recolonizing the adjacent San Pedro Valley, to the east.[28]

The latter-day Sonoran influence derived not just from residual ranchers who had survived the period of Indian raids, but also from renewed Mexican immigration. In 1874, for example, the Redondo brothers from Sonora brought cattle herds to the Yuma area, dispersing them through the mesquite thickets along the lower Colorado and Gila rivers, where springtime overflows produced lush summer pastures. By the census of 1880, Spanish-surnamed people accounted for 28 percent of all Arizona cattle ranchers, up from 22 percent in 1870. In the two southern counties of Pima and Yuma, the proportion in 1880 stood at over half, and more than 70 percent of the Pima County cowboys in 1880 were Hispanic. Spanish-style brand designs were still abundant in the territory. In sum, southern Arizona remained the most Hispanicized of all the ranching frontiers in the United States in the late nineteenth century, its links to adjacent Sonora vital and intact. When anglicization did eventually come, neither Texas nor California provided the main impetus.[29]

Hispanic Influences in California Extended

California Extended, too, retained considerable Hispanic pastoral influence during the expansion into the intermontane region, exceeding the Texas system in this respect. D. W. Meinig claims that "few if any elements of Mexican ranching culture came north from California" and that any practices or regalia of Spanish origin found in the Pacific Northwest "were almost certainly derived from Texas rather than California" are easily disproved.[30] The key cause of the greater retention of Hispanic influence lay in the widespread use of Spanish Californian vaqueros, a marked contrast to the Texas system. The importance of the Hispanic cowhands was greatest in the early years of expansion into California Extended; in later decades, oddly, they seem to have virtually disappeared (table 1). Joining them in the early dispersal beyond the Sierra Nevada were even a few Spanish-surnamed ranch owners from California.[31]

Cattle king Peter French, who had learned to speak Spanish when working on a Sacramento Valley ranch in the 1860s, took a work force consisting exclusively of vaqueros when he settled in southeastern Oregon, and some of these *californios* later sent for relatives to join them. The far-flung ranches of Henry Miller and Charles Lux also employed Hispanic hands. Even as distant as eastern Nevada, cowboys with names like "Spanish Joe" worked the ranges. Throughout California Extended these men reputedly "enjoyed full camaraderie with the Anglos with whom they worked." Socialization facilitated the diffusion of equestrian techniques to the Anglos, who soon formed the large majority of the ranch work force. Similarly Chinese and Amerindian cowboys in the intermontane region readily adopted Hispanic herding techniques through close association with the Californian vaqueros. Actually the Hispanic work force in the interior region was more complex than simply Californians. While no Mexicans from Texas or Arizona were present, at least one Hispanic Costa Rican cowboy worked in Grant County, Oregon, by 1880, perhaps from the cattle-ranching region of Guanacaste province. The potential for diffusion represented by such individuals should never be dismissed.[32]

Spanish loanwords in the ranching vocabulary of greater pastoral California offer obvious evidence of Hispanic influence, as well as a clear distinction from the Texas tradition, since such borrowings differed substantially in the two ranching systems. Perhaps no more accurate geographical measure of the extent of Anglo western ranching exists than the isoglosses of uniquely Californian loanwords. In this category are *buckaroo* (from *vaquero*), originally "a cowboy of the Spanish-California type," and a word still widely used in Nevada, Idaho, and Oregon. Absent in the Texas ranch lingo, *buckaroo* remains a western, "especially Pacific" term, and earlier attempts to link it to the African West Indian and Carolina-derived slave-cowpen culture were misguided.[33]

Similarly *major domo* (from *mayordomo*) became the distinctive Anglo-Californian term for a ranch foreman, or boss. At least as early as 1850, *rodeo* was the preferred Pacific English word for a cowhunt or roundup, and it, too, spread through the far West as an indicator of non-Texas Hispanic influence in cattle ranching. Other Mexican loanwords belonging in the Anglo-Californian pastoral tradition include *tule* (bulrush marsh), *hackamore* (halter, from *jáquima*), *bosal* (a small braided rawhide hackamore), *romal* (leather knob at the end of the reins), *parada* (bunch of cattle to be cut out of a larger herd), *preatha* (group of animals bearing the same brand and already assembled during the rodeo), *concha* (literally "shell," a circular ornament for saddles and belts), *macardy* (a fiber rope, from *mecate*), *oreanna* (an unbranded animal), *taps* (leather hoods over stirrups, from *tapaderas*), *cavvy* (group of saddle horses, from *cavieda*), *riata* (rope, from *reata*), *lass* (from *lazo*), *theodore* (a device consisting of a hackamore and rope knotted to the romal, from *fiador*), *cosinera* (chuck wagon cook, from *cocinero*), *leppy* (motherless calf, from *la pepita?*), and *chinks* (short, fringed chaps reaching below the knee, from *chinquederos*). Even as far north as central British Columbia, Indian villages were known as "rancheries," from the Spanish *ranchería*, or shanty settlement.[34]

Material culture in the Anglo-Californian ranching system also reflected abundant Hispanic influence in distinctive ways not duplicated among the Texans. In his getup and gear, the buckaroo remained much more Spanish than his Texan counterparts. Riding equipment offers numerous examples, although Anglos modified many such items after about 1850. Inheriting the rather bulky, single-cinched Sonoran saddle with thick horn, Californians developed a smaller, less cumbersome, high-forked model with a tall, slender horn. They retained single cinching, providing a contrast to the double-rigged Anglo-Texan saddle (figure 49). Mexican immigrant Juan Martarell, a resident of the San Joaquin Valley, began marketing one version of this modified type as the "Visalia" saddle about 1869, and it soon became preferred throughout California Extended. Even as far afield as northeastern Montana, well within the purported zone of dominant Texas influence, some ranchers ordered Visalia saddles.[35] In Oregon "single-cinch saddles were mostly used," although "a Texas rider would sometimes drift in with a double rig, or two-cincher."[36]

Anglo-Californians, following the custom of Spanish times, continued to weave the bridle reins together, in contrast to the Texan custom of split reins held loosely together by a keeper. The braided reins terminated in the romal, made of several layers of light leather and used as a quirt. The distinctive Spanish-Californian leather armitas, also called chaparreras, or "chaps," which protected the rider's legs, passed into the Pacific Anglo tradition, proving very useful in the sagebrush country of the interior (figure 49). In California Extended, the Anglos began covering chaps with pelts of various kinds to provide warmth during the bitter intermontane winters. Fortified in this manner, the chaps acquired the nickname "woolies."[37]

Spur and boot development also took distinctive paths in Anglo California. While the stiff-countered Texan Nocona boot eventually spread throughout the West, it had not entered the Californian tradition as late as the 1880s. Accordingly the Spanish spur, designed to be worn over soft-leather, homemade shoes and ride low on the heel, survived in Anglo California. These spurs had large-diameter, spokelike,

Figure 49. *Selected items of equestrian material culture from the Californian and Texan traditions. Key: A = California great-rowel spur, 1830; B = Anglo-Californian great-rowel spur, 1870; C = spur, Texas Extended, 1880; D = single-cinched saddle with "taps," California Extended, 1910; E = early double-cinched Anglo-Texan saddle; F = Californian chaps, or armitas; G = Santa Barbara Spanish hinged*

258   CHAPTER EIGHT

snaffle, 1814; and H = Anglo-Californian hackamore. Source: selectively reproduced from Vernam, Man on Horseback *(see note 3), figures 132–36, 139, 152, by permission of HarperCollins, Publishers.*

sharpened rowels, straight heelbands, and long, drooping shanks. After 1850 Anglos shortened the length and reduced the diameter, but the resultant California spur remained larger than the Texan model, displaying more Mexican influence (figure 49).[38]

Substantial numbers of elaborate Hispanic brand designs survived in the Anglo-Californian tradition, in contrast to Texas, where Spanish brands were rarely if ever adopted. Examples survive even in the expansion states, especially Nevada, where some brands virtually identical to those of sixteenth-century Mexico and Las Marismas in Andalucía remain in use today (see figure 8).[39]

Several other form elements of ranching material culture common in California Extended but absent in the Texas tradition have Hispanic origin. An example is the "Spanish windlass," a device to extract mired cattle from marshy ground and used even as far north as interior British Columbia (figure 50). In the words of one Kamloops cattleman, "we cut a pole nine feet long and about six or eight inches thick," then, with an ax, "sharpened one end like a pencil and cut a groove right close to the other end all the way round the pole." The pointed end was jammed at a slant into the damp sod to "a depth of two or three feet, behind a willow root." Then, after the cowhand fit a rope into the groove at the top of the post, the other end "was taken back and tied as a guy rope to a willow or to another small stake driven into the ground." Using a second rope, the buckaroo cast a lasso over the horns of the mired animal, pulled it tight around the windlass post, and tied a bowline knot. "Through the loop formed by the bowline, we placed a good stout pole," resting behind the post. Then "we simply started walking round and round the upright post, causing it to turn and wind the rope onto it" and pulling the cow free. A horizontal log at the base of the windlass post kept the rope from sinking down into the marshy ground and drowning the stranded animal.[40]

Another wooden windlass of Spanish mission origin, the "beef wheel," stood in the killing pen and served to hoist the carcass of a slaughtered cow during butchering (figure 51). Found in various parts of California Extended, especially eastern Oregon, the beef wheel consisted of two vertical posts, upon which rested a windlass shaft, or

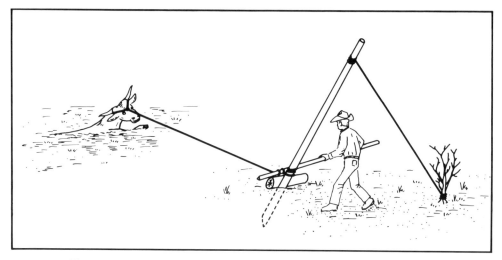

Figure 50 A "Spanish windlass," used to pull stranded cattle from marshy places, near Kamloops, British Columbia, about 1930. This device, whose operation is described in the text, was likely of Hispanic-Californian origin. Source: redrawn from Bulman, Kamloops Cattlemen, *following p. 96 (see note 40).*

drum. On one end of the shaft was mounted a wheel or square frame. Pulling a rope attached to the frame or pressing against the projecting wheel spokes turned the shaft, winding a rope and raising the gambrel bearing the carcass.[41]

Several Mexican types of livestock pens also became common in California Extended, along with the loanword *corral*. Picket, or palisado stock pens made of juniper and similar to those in central Texas were typical (see figure 30). Corrals in the Anglo-Californian tradition, following Mexican custom, were used principally for slaughtering cattle and for keeping horses, rather than for branding. Rodeos were held on open ground, without corrals. In the middle of the pen or out in the rodeo grounds stood a heavy "snubbing post," another Hispanic item of material culture that appears today as far north as southern British Columbia and southward through Sonora and at least to Guanacaste province, the cattle-ranching district of Costa Rica. Snubbing posts also survive on modern Española, providing a link to the earliest Spanish ranching in America. The animal destined for slaughter, breaking, or doctoring was roped, then restrained by snubbing the line around the post.[42]

Diverse herding techniques employed in the Anglo-Californian system also derived from Hispanic practices. Most of these borrowed

Figure 51. The Old Peter French beef wheel at Frenchglen, Oregon. This windlass device for hoisting slaughtered animals during butchering is a Hispanic item of material culture that passed to Anglos in the California ranching system. (Photo by the author, 1991.)

methods were in some way associated with management of cattle from horseback, the most enduring legacy of Spanish California. The old californios had been masters of *la reata larga,* a rawhide rope up to 30 m. (100 ft.) long, and this skill passed to the Anglos and other neophyte cowboys in the Pacific tradition. Under Anglo influence the typical length of the rope decreased to about 12 m. (40 ft.) and fiber, or macardy, ropes became more common, but the basic Hispanic technique survived. This was most obvious in "dally" roping, or "dally welta," a term derived from the Spanish *dar la vuelta,* or "to give a twist." The "dally man" wound half-hitched loops around the saddle horn, as slack to be released when needed as the rope became taut, a tricky technique which, if carelessly performed, could result in the loss of a thumb. Texans, by contrast, simply tied the rope to the horn or bow, letting the double-cinched saddle absorb the jolt of the suddenly tightened line.[43]

Hispanic, too, was the Anglo-Californian method of breaking wild horses. They used hackamores rather than bits for this purpose, unlike most Texans. The hackamore consists of the braided leather noseband, or bosal, held in place by a light leather headstall and a woven cotton cord. As a result, cow ponies, when tamed, reputedly remained "soft-mouthed," or more responsive to the distinctive California "high-port" bit and reins.[44]

One key difference between the Anglo herding methods of Texas and the Pacific was the previously mentioned Californian custom of shifting livestock seasonally between pastures, a custom present in the Pacific lowland coastal Hispanic tradition at least as far back as the Sonoran period and, of course, inherent in Andalusian marsh herding. The notion that transhumance began in California only after the great drought of the 1860s is false. The pronounced seasonality of rainfall in the Mediterranean climate zone of California had encouraged a vigorous revival of Iberian pasture shifting, since the prairie grasses of the northern Central Valley, some of which were annuals, could not provide any substantial grazing during the arid summer. Partly, too, the seasonal herd migration derived from the "instincts of the cattle" who, under open-range conditions, naturally drifted upslope in the summer, seeking forage. Representative of pasture shifting in the Anglo-Californian system was the Tejon Ranch, encompassing some 80,000 hectares (200,000 acres) north of Los Angeles, where "a change from summer to winter grazing is always made."[45] Subsequently, vertical transhumance in cattle ranching appeared through most parts of the mountain West, helping distinguish that region from the Texas system (figure 52). Stock drove trails have

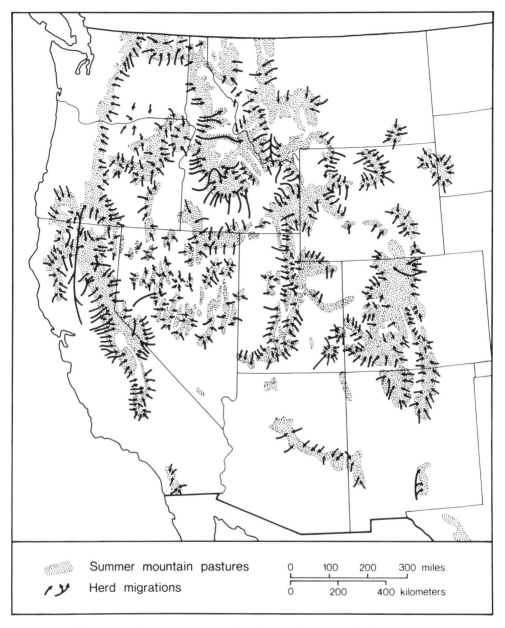

Figure 52. Routes of seasonal cattle transhumance in the western United States, 1977. Cattle are moved along roads and drove trails to summer high pastures on federal government lands. This vertical transhumance, a major feature of western-mountain cattle ranching, probably had its origins in the Hispanic Californian herding system. Source: Rinschede, Wanderviehwirtschaft, p. 158 (see note 1).

Figure 53. Transhumant cattle being driven to summer mountain pastures along a highway near Bondurant, Wyoming. (Photo by the author, 1987.)

been officially designated in many parts of the West to facilitate the seasonal movement of cattle between summer and winter pastures, and highways are also widely used for this purpose (figure 53). A plausible explanation for western cattle transhumance is that it derives at least in part from the Californian ranching tradition and has Pacific coastal Mexican roots. New Mexico highland herders may also have contributed to the practice, although their focus has always been on sheep.[46]

Nor did the Spanish cattle bloodlines become completely extinguished in pastoral California Extended. The "western" cattle that reached the Montana Rockies from Oregon in the 1870s were acknowledged to retain some Iberian traces, and the same was likely true throughout the areas west of the continental divide.[47] Diverse other Hispanic influences survived as well in the Anglo-Californian ranching system. The rise of "cattle kings" and the associated large scale of operation derived from the Mexican practice of awarding huge pastoral land grants, although most Anglo ranches were smaller than the Hispanic prototypes. Representative was eastern Oregon, where only five, largely California-derived cattle-raising enterprises, each owning twenty-five thousand head or more, dominated the local industry by the 1880s. Three of these five reportedly owned 60 percent of the range cattle in the area. In addition, the absence of an overt antisheep bias among Pacific ranchers may have its roots in Spanish practices.[48]

Anglo-American Influences

Anglicization of the western ranching system proceeded rather slowly, in stages, in part because the herding culture had been so thoroughly Hispanicized at the outset. As suggested earlier, the changes began in earnest with the American takeover of California and gold rush of the late 1840s, involving initially a shift to beef production and the associated decline of the Iberian breed of cattle. The 1862–65 drought episode in California brought more changes and caused the demise of the traditional Spanish ranching system in the Mediterranean climate zone. The following quarter-century, from 1865 to 1890, witnessed continued but gradual evolution away from the older system on the Intermontane pastoral frontier, where Anglo-Californian ranching had become centered. Another severe drought in 1887–89, partially overlapping unusually hard winters in 1888–90, destroyed the majority of intermontane cattle and with them most remaining vestiges of the traditional Californian system. These climatic disasters were coupled with extensive range damage in the intermontane region, further undermining the cattle-ranching industry and opening the way for sheep raising. Anglicization, then, progressed both gradually and in environmentally induced spurts. Temporally, both the diffusion and eventual demise of the California and Texas ranching systems coincided almost exactly.[49]

The growing Anglo influence took many forms. Hispanic cowboys, still a large majority in the 1840s, became a small minority by the 1880s, far outnumbered by Anglos. A pattern of two major rodeos, in the spring and fall, replaced the older Hispanic-Californian calendar, and branding pens replaced the open rodeo grounds. Pasture fencing brought an end to the open range, and natural hay from the streambank marshes and tule lands began to be collected as supplementary winter feed for the cattle. Irrigation enlarged many riverine meadows as early as the 1870s, greatly increasing hay production. Cowboys exercised greater vigilance in the care of herds as scrub cattle gave way to improved, British-derived breeds.[50]

These fundamental changes, and others as well, represented the rise to dominance in the West of the Ohio Valley/midwestern herding system. In fact, the decline and fall of both Californian and Texan ranching coincided with the ascendance of the more intensive herding system derived from the American heartland, involving much greater application of both labor and capital. Chapter 9 is devoted to this largely ignored, third Anglo-American ranching culture in the West.

NINE

THE MIDWEST TRIUMPHANT

JW THE ANGLO-CELTIC SYSTEM OF CATTLE HERDING, DERIVED from South Carolina by way of the Upland South and Ohio Valley, had spread into the midwestern prairie frontier by the 1830s. Having readapted their methods to the humid temperate grasslands, these graziers approached the Great Plains along a broad front from their major footholds in the Iowa wet prairie, Missouri River valley, Springfield Plain, and Blackland Prairie of northeastern Texas (figures 38, 54). Largely purged of Jamaican Hispanic influences, the midwestern herders would compete with the Texans and Californians for pastoral dominion over the West.[1] In the words of Peter Simpson, one of the few scholars along with Paul Henlein to recognize the existence of a distinctive American heartland herding system, "a different technique for cattle raising spread from the Midwest directly across the continent."[2] In some areas of the West, midwestern graziers led the way as pastoral pioneers, while in others they arrived as secondary settlers, and their herding system displaced earlier Texan or Californian ranching.

Fundamentally different from its hispanicized western competitors, the midwestern herding system, reflecting dominantly British influences, was distinguished mainly by greater attention to the welfare and quality of the livestock. Pursuing methods that were both more capital and labor intensive, the Midwesterners exercised greater diligence in the care of cattle than either the Texans or Californians, achieving in the process herd docility. They provided winter feed for the animals; strove to upgrade the bloodlines of their herds through selective breeding, even importing British stock; shifted livestock seasonally between different pastures, often so profoundly as to involve transhumance; early formed stockraisers' associations; possessed rather minimal equestrian skills; made extensive use of stock pens and erected, at their early convenience, pasture fences; derived some milk and butter from their herds; and produced lean cattle for

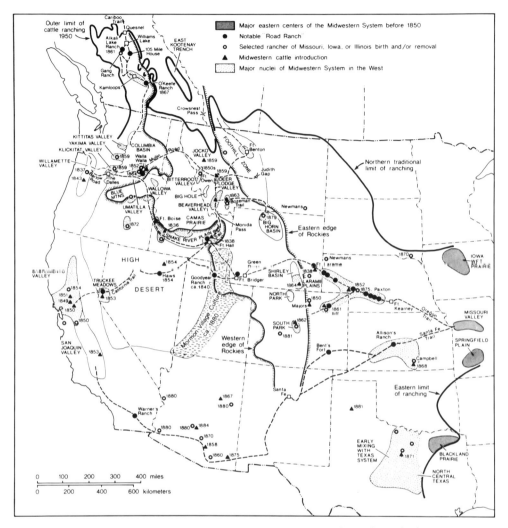

Figure 54. The diffusion of midwestern cattle ranching through the West. Data are selected and representative rather than comprehensive. Scarcely any part of the West escaped midwestern influence, although in some areas, as in much of Texas and California, the heartland system arrived in postfrontier times. The sources are too numerous to mention in full, but see in particular (in addition to relevant text passages): Yost, Call of the Range, pp. 27–36 (see note 9); Weir, Ranching in Southern, p. 18 (see note 78); Bailey, "Barlow Road" (see note 29); and Meinig, Great Columbia, map 1 (see note 31).

overland driving to areally segregated Corn Belt fattening districts, replicating the ancient British pattern of Celtic breeders and Saxon feeders. Few "cattle kings" arose among these midwestern producers of lean stockers, that honorific title belonging instead to certain large-scale feeders. Less specialized than either Texans or Californians, the midwesterners usually devoted considerable attention to crops and often raised sheep as well as cattle. Perhaps *ranching* is not the best word to describe their system, although they happily accepted that designation in the West. Contemporary observers often called their establishments "stock farms," "mixed farms," or "improved ranches."[3] *Frontier* may be an even more inappropriate word to use in conjunction with midwestern herders, since their methods generally rose to dominance in the capital-intensive, post-pioneer phase, but the record clearly reveals that in many areas the midwestern cattle-raising system prevailed from the very first.

Challenging Texas on the Great Plains

From the beginning, midwestern cattle raisers challenged the Texas system for possession of the Great Plains and for access to the Chicago meat packers. The outcome of that cultural contest is perhaps best revealed in the fact that virtually the entirety of the Plains today is regarded in the vernacular as part of the "Midwest," including even the northern reaches of the Texas Panhandle.[4]

In fact the midwestern challenge began in the very heart of the Lone Star State, in the region labeled in chapter 7 as north-central Texas, the staging ground from which the Texas ranching system spread through the Great Plains (figures 39, 54). Even before the Anglo-Texan ranching complex diffused northward from the coastal lands, the north-central region had been thinly occupied by cattle raisers largely of upland southern origin, moving westward from the Blackland Prairie, where, as described earlier, an enclave of the Ohio Valley cattle complex had been established before 1850 (figures 38, 43, 54). The herding techniques of these Texans, as late as 1860, still bore the stamp of the old midwestern system, based in British-derived scrub stock, herder dogs, whips, calf-capture, salting, hired Anglo cowhands, labor intensity, and herd docility.[5] Midwestern-born cowboys were well represented in north-central Texas, accounting for 18 percent of the ranch work force (table 39).

Dan Waggoner, one of the most famous and enduring cattlemen of north-central Texas, well exemplified the midwestern presence in the region. An upland southerner of South Carolinian ancestry, Waggoner came as a boy from Missouri to the Blackland Prairie of northeastern Texas. Following the example of his father, "Sol" Waggoner, Dan began raising range cattle there in the 1840s. A decade later he moved his growing business westward into north-central Texas, where he remained for the rest of his life, accumulating many thousands of cattle and huge acreages of pasture land. While he, in common with other local ranchers, was greatly influenced by the intrusive Texas system, Waggoner retained some midwestern traits. For example only about 1865 did he first notice a "slow change in appearance" in his British scrub cattle, a mystery solved when he discovered a lone, feral longhorn bull in his pasture.[6] Waggoner's herd was not unusual, for many other early north-central Texas ranchers also had shorthorn stock. Almost certainly they provided the source of the British-Texan cattle that went north up the trails with the more numerous longhorns. For example in the Little Laramie River valley of southeastern Wyoming, an 1870 account mentioned newly arrived "short horned Texan" herds, "a good lot of stock cattle."[7]

Several of the biggest cattle ranchers in north-central Texas, W. S. Ikard and "Lum" Slaughter, imported Kentucky shorthorns into the region in 1871 and Herefords as early as 1876, long before the collapse of the Texas system. Famous Panhandle rancher Charles Goodnight, whose pastoral roots went back into north-central Texas, acquired five hundred Kansas shorthorns in 1881 to help upgrade his herds.[8]

In many other parts of the Great Plains, as well, midwestern cattle raisers arrived before ranchers from the south implanted the Texas system. In so doing they placed their cultural stamp on the vast region from the very first and, in the long run, shrugged off Texan attempts to seize the Plains. In particular midwestern graziers entered the Great Plains along the axes of the Oregon and Santa Fe trails (figure 54), an east-to-west movement that countered the south-to-north migration of the Texans. The earliest pastoral pioneers founded numerous "road ranches" along the trails, especially in Nebraska, as early as the 1850s. Subsequently the road ranch became an occupance form closely linked to the spread of the midwestern cattle frontier in the West. Allison's Ranch, at the great bend of the Arkansas River in Kansas, was an example. The word "ranch" perhaps entered their vocabulary during the Mexican War of 1846–48 or, more likely, by way of the Santa Fe trade. The proprietors of the road ranches traded with migrants headed to the Pacific states, exchanging healthy cattle for the trail-weary, weakened stock belonging to the emigrants. The "pilgrim" cattle acquired in this manner were then restored to prime condition on the lush pastures along the Platte and Arkansas and traded to later contingents of Pacific-bound settlers. As early as 1852, a road ranch had been established near present Ogallala, long before that place acquired a railroad and became famous as a depot for Texas longhorns. Allison's ranch in Kansas dated to 1857 (figure 54).[9]

Similar road ranches appeared very early in eastern Wyoming and Colorado, developed by trail traders. Alexander Majors reportedly wintered cattle as early as 1850 at the eastern foot of the Medicine Bow Range, from the Cache la Poudre River near present Fort Collins northward to the Cheyenne area. By 1866, before the great Texas migrations, some twenty thousand midwestern-derived pilgrim cattle grazed the pastures along the South Platte in eastern Colorado.[10]

Midwestern cattle raisers continued to enter the central Great Plains after the Civil War, contemporaneous with the Texan arrival. Notable among them was Iowan W. E. "Shorthorn" Campbell, who in 1868 developed a large ranch south of Wichita, near the Oklahoma border, and introduced cattle from Missouri, Kentucky, and Iowa.[11] J. S. Smith, a feeder "cattle king" in the Sangamon Valley of Illinois, began wintering lean cattle in central and western Kansas in 1867–68, while Joseph McCoy, from the same Illinois valley, was instrumental in developing Wichita as a major cattle market.[12] Some of the Kansan graziers soon encroached upon Indian lands in neighboring Oklahoma, providing the foundation of midwestern cultural influence there.[13]

Farther north, midwestern herders continued to occupy lands along the Oregon Trail and its branches. John W. Iliff, born in 1831 on an Ohio stock farm, began operations on the Great Plains as an Oregon Trail road rancher and then built his famous cattle empire on the South Platte in Colorado in the 1860s. He crossbred Midwestern and Texan stock to supply beef for the military posts along the trail and its successor, the transcontinental railroad.[14] Bill Paxton, a Kentuckian by birth and Missourian by removal, in 1875 founded the well-known Keystone Ranch near Ogallala. Still farther north, Iowans began open-range ranching around Yankton in South Dakota about 1870.[15] Another Iowan, later to become a Wyoming cattle king, founded a ranch near the source of Chugwater Creek, north of Cheyenne, in 1873, and other midwestern cattlemen moved into these grasslands between the Nebraska Sandhills and the Rocky Mountains following the Sioux defeat in 1876. Included were the Missourian Newman brothers, who developed a huge ranching empire along the Niobrara Valley and later expanded into the Powder River country of southeastern Montana.[16]

If the ranchers of the central and northern Great Plains were predominantly of midwestern origin, so were the cowboys. One contemporary observer carefully distinguished two types of Plains cowhands, Texans and easterners "chiefly from Missouri," noting that the latter were a more orderly bunch. Even as far south as the Arkansas River valley of Kansas and Colorado, midwestern-born cowboys far outnumbered Texans by 1880, and in the Platte Valley nearly half were midwesterners, as contrasted to only 7 percent Texas-born (table 1). The 1880 census of Laramie County, Wyoming, revealed 386 cowboys, of whom 45 had been born in Missouri, 101 in the Ohio Valley states, and only 32 in Texas.[17] While many of these midwesterners no doubt learned some Texan techniques as neophytes on the Plains, others introduced the cattle-herding methods of their native American heartland.

Accordingly midwestern pastoral practices became established very early on the Great Plains, providing a marked contrast to the Texan methods. The most obvious such difference could be seen in the careful tending of stock inherent in the midwestern system, a practice which produced far tamer cattle. For example in early North Dakota many ranchers kept the livestock close to their headquarters, while in Kansas and Colorado, herders on some ranches rounded up the stock each evening during the summer to keep them tame and prevent straying. In the Arkansas River valley of eastern Colorado, an 1878 account described "constant handling" of the cattle, by which "they become thoroughly domesticated." The labor-intensive Plains practice of "blabbing" calves—attaching a nostril board to wean them—probably had midwestern origins and presented a marked contrast to the cruder Texan method of slitting the calf's tongue. Plains observers repeatedly noted that Texas longhorns were wilder or more "timid" than "northern" cattle, reflecting the essential difference in frequency of tending.[18]

Provision of winter feed, usually hay, also formed an essential aspect of the more labor- and capital-intensive midwestern system on the early Great Plains. The typical non-Texan rancher fenced in perhaps 20 to 40 hectares (50 to 100 acres) for hay meadow. Clarence Gordon observed in Wyoming in 1880 that "the man from Iowa or Missouri works well in the hay-field, but the Texan dislikes agriculture" and would "rather get his time than bother the ground."[19]

MAJ. J S. SMITHS' HERD—WINTERING ON HAY IN CENTRAL KANSAS.

Figure 55. *J. S. Smith, a midwestern-prairie feeder king based in central Illinois also bred lean cattle on the Kansas Great Plains. In good midwestern fashion, he furnished winter hay for these stockers, even in the early 1870s, providing an alternative to the careless Texan methods. Source: McCoy,* Historic Sketches, *p. 216 (see note 3).*

Natural meadows along the stream courses provided abundant wild hay. J. S. Smith had his Kansas hands mowing hay from the valley meadows in the central and western parts of the state as early as 1867, taking the precaution of "fire-guarding" the hay lands with a surrounding plowed or burned firestrip. In severe winter weather, Smith's cattle received hay from fenced-in stacks at various locations in the pastures (figure 55). By the middle 1880s, hay production on the Plains had been expanded by means of irrigation, as the midwestern graziers sought to recreate in a semiarid land the wet prairies they had known back East. Some early midwestern herders in southeastern South Dakota even fed cattle in the farmyard and provided sheds for shelter during blizzards. Winter feeding greatly reduced dieoffs of cattle and clearly revealed that the midwestern system was better adapted for the Great Plains than the subtropical Texan methods. In the long run, the midwesterners prevailed and Texans went to ruin largely because of the differences in winter care of livestock.[20]

Another midwestern technique that succeeded on the Great Plains, particularly north of the Platte, was seasonal pasture shifting.

In summer, "cow camps," distant from the ranch headquarters, often situated near springs, and accessible to high interfluvial areas, served as bases from which the herds ranged. On the western margins of the Great Plains and in the Black Hills, the initial penetration by ranchers into the mountains occurred as a result of the need for summer pastures. After the autumn roundup, the animals not culled returned to lower pastures along the streamcourses near the permanent ranchstead. This seasonal shift, a practice that had been part of the heartland herding system since the Carolina Piedmont episode, probably had highland British roots. Any hill Celt would have recognized the summer cow camps as shielings. While the cattle were at the camps, the ungrazed riverine lowlands yielded both mown meadow hay and a stand of pasture grass that cured into standing hay.[21]

Another very obvious index of midwestern influence in the cattle-ranching industry of the central and northern Great Plains can be found in bovine bloodlines. In addition to the pre-Texan "pilgrim," or British scrub stock, frequent and often rather early references to Durhams, Devons, "shorthorns," and even Herefords can be found. At least by the early 1870s, a substantial demand for improved breeds had arisen. The principal source of such stock remained the Ohio Valley states, although surprisingly often Great Plains ranchers, well before the Texan catastrophe of 1886–87, imported animals from Britain. One of the unexplained mysteries of the Midwestern-inspired era in western ranching was the eventual rise to dominance of the Hereford, a breed native to a temperate, well-watered English shire.[22] Contrary to popular image, the Iberian longhorn never dominated the central and northern Great Plains. Most Texas longhorn stock driven north did not become breeders, and even those that served that role were usually crossbred to reduce the "southern" blood. The whole controversy over Texas Fever suggests that the Plains cattle industry could do quite nicely without imported longhorns.[23]

Another sign of midwestern influence was the early and widespread appearance in the Plains states of ranchers' organizations, likely modeled after those of the Ohio Valley. Examples included the Cherokee Strip Live Stock Association and Wyoming Stock Growers' Association. These groups often helped promote midwestern-type interests, such as herd upgrading.[24]

Attention to fencing also came to the Great Plains with the midwesterners. To be sure the range remained largely open during the frontier years on the Plains, but even in that early period some progress toward enclosure was made, often with the support of the ranchers' associations. "Drift fences," built to discourage the native tendency of cattle to drift southward and stray during winter storms, represented the first major step in the direction of fully fenced range. A drift fence spanning the entire Texas Panhandle was constructed in the early 1880s, and others of lesser scale appeared elsewhere on the Plains. In the leased pasturelands of northern Oklahoma, range fencing began in 1882 under the direction of the Cherokee Strip Live Stock Association, and within a year about 1,600 km. (1,000 mi.) of wire fence had been completed there. Similarly midwesterners on the Plains devoted more attention to building cattle pens, for use during roundups, in contrast to the Texans, who followed Mexican custom and usually carried out the roundup duties without using pens.[25]

Midwestern influence, present from the earliest years on the Great Plains, grew steadily as the Texans experienced repeated setbacks due to climatic maladaptation. When the ill-advised Texas system finally collapsed in the 1880s, midwestern methods rose to unchallenged dominance throughout almost all of the Plains. Well before the end of the century, the midwestern preference for British-derived breeds and fenced pastures swept even through Texas itself, ending the day of the longhorn and open range in the very cradle of the Texas system.[26] The midwestern victory in Texas proper was well revealed in one particular Panhandle ranch of the early 1880s, which included a five-room house, freestanding kitchen–dining room, smithy, granary–saddle room, milking house, storage sheds, vegetable garden, barbed-wire enclosed horse pasture, fenced ryefield, and twenty-six hundred open-range cattle.[27]

The Willamette Valley

The greatest midwestern expansion occurred much farther west. The main goal of the Oregon Trail, in the early years, was the Willamette Valley, lying between the Coast Ranges and Cascades southward from Portland (figure 54). The modest, hispanicized beginnings of cattle raising there in the late 1830s soon yielded to heartland influences. Midwestern cattle first reached the Valley in the early 1840s, including about thirteen hundred head in 1843 alone, and the process of upgrading the Californian Spanish cattle began. Most of the early imports were "pilgrims" and other shorthorn scrubs, but some Durhams came at least by 1847 and Devons as well. Durham cattle

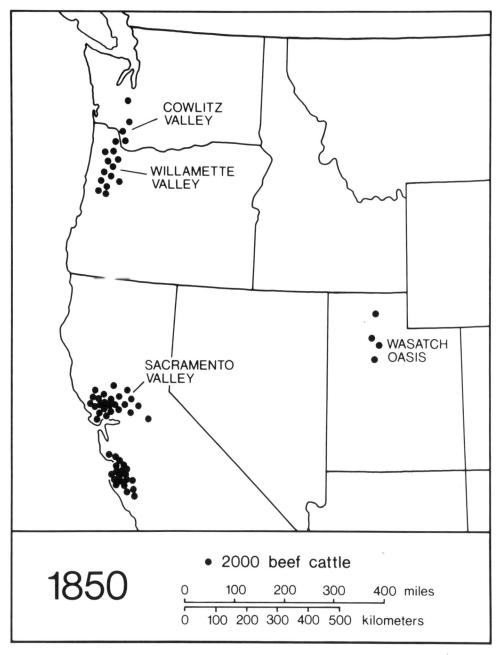

Figure 56. Beef cattle distribution in midwestern-influenced regions of the West, 1850. The Willamette Valley and northern California concentrations are apparent. Sources: The Seventh Census of the United States, 1850 (Washington, DC, 1853); and Gisbert Rinschede, Die Wan-

derviehwirtschaft im gebirgigen Westen des USA und ihre Auswirkungen im Naturraum *(Regensburg, Germany, 1984)*, p. 94.

reportedly withstood the rigors of the Oregon Trail better than other shorthorns, allowing them to become dominant in the Anglo-Pacific herds.[28]

In the early years, some cattle droves reaching the terminus of the Oregon Trail at The Dalles swam the Columbia to the north bank, passed through the gorge of the Cascades on the less-rugged Washington side, and reswam the river at Fort Vancouver to enter the Willamette Valley. Others were driven along difficult Indian trails through the Oregon Cascades south of Mount Hood, a route made much easier by the opening of the Barlow Road in the middle 1840s (figure 54).[29] During the decade of the 1840s, perhaps ten thousand cattle arrived in the Oregon settlements from the "Mississippi Valley" states, and stock raising spread through the Willamette and northward across the Columbia into Washington's Cowlitz Valley (figure 56). By the census of 1850, the Willamette country housed about twenty thousand beef cattle.[30]

Even so, the importance of the Willamette Valley in the evolution of Anglo-Western ranching has probably been overstated. Although the Willamette represented, in the structural geomorphological sense, a northern reemergence of the Central Valley of California, and in spite of its reputation as a "fine grazing country" where cattle thrived, the area never became a major center of range-livestock production.[31] Ranching scarcely existed in the Valley. At the 1850 census, only twelve settlers in the entire Willamette country possessed as many as a hundred beef cattle, and of these only two could be called "cattle kings," owning more than five hundred head (figure 57). Hamilton "Cow" Campbell, owner of twelve hundred largely Spanish cattle—the old Methodist mission herd—and resident on the lower Santiam River, a right-bank tributary of the Willamette, was the largest operator at the time of that census, and his very nickname revealed how unusual large-scale livestock raising was. Average herd size per farm in the Valley stood at only sixteen, reminiscent of the Appalachian farmers who sold stockers to the glade feeders a century earlier.[32]

Part of the reason for the relative unimportance of the Willamette Valley as a ranching frontier lay in its native vegetation. A fine forest of oak, fir, cedar, and hemlock covered most of the Valley,

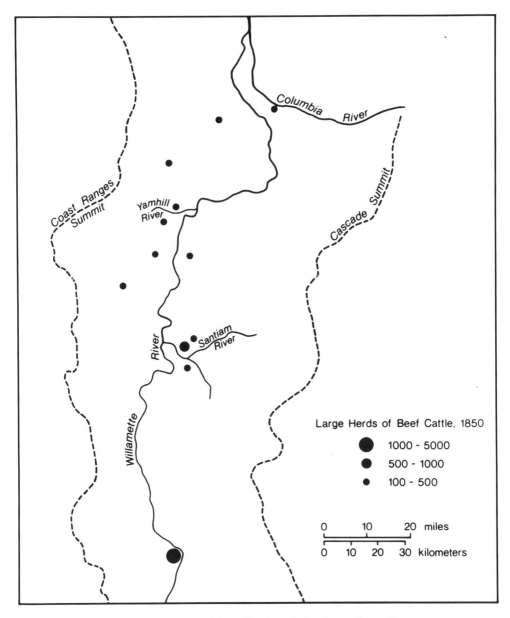

Figure 57. Distribution of large beef-cattle herds in the Willamette Valley, Oregon, 1850. Only fourteen settlers owned more than a hundred beef cattle, and ranching scarcely existed in this supposed "nursery" of the cattle industry in the Pacific Northwest. Source: adapted from Bowen, Willamette Valley, p. 81, who used manuscript U.S. census agricultural schedules.

giving way here and there to small prairies and wet meadows. Midwestern frontier graziers accustomed to the expansive tallgrass prairies of Missouri, Iowa, and Illinois surely found the wooded Valley rather confining.

In spite of the modest scale of operation typical of Willamette graziers, enough small cattle raisers lived there to allow a significant export of livestock by the late 1850s. During the 1860s most such livestock were steers destined for the mining districts of the Pacific Northwest. In the spring of 1861, more than a thousand Willamette Valley cattle were driven east through the Cascades, reversing the flow of the 1840s, and another four hundred went by boat up the Columbia as far as The Dalles. During the first eight months of 1862, another forty-six thousand head followed the same routes east. Some of these animals were breeders to help stock a budding trans-Cascadian ranching industry. In this sense the Willamette did, indeed, serve as one "nursery" of the interior ranching industry.[33] However, the Valley's contribution remained principally breeding stock, and it is misleading to depict a ranching frontier and culture surging eastward from the Willamette.

The Midwest Comes to California

Northern California, especially the Sacramento Valley, became a far more influential center of midwestern-inspired cattle ranching than the Willamette. In common with the Great Plains, California witnessed a cultural contest between British and Hispanic systems, with a similar outcome. The rise of an increasingly Americanized ranching system in California, built upon a Hispanic base, was almost entirely the result of midwestern immigration from the Ohio Valley and Missouri frontier.[34]

The midwestern presence in northern California became established during the period of Mexican rule, when local ranching retained a largely Hispanic character. Many early Anglo settlers who acquired ranchos in Mexican California had roots in the herding frontier of the Ohio and Missouri valleys. For example Kentucky-born William Wolfskill migrated to California in the 1830s from the cattle-rich Boonslick country of central Missouri.[35] As early as 1841, some of these midwesterners began bringing cattle with them to California.[36]

The Gold Rush served to increase the midwestern presence in California. Herds of "American" cattle, "the mongrel breeds of the western states," were driven overland from the Missouri frontier by way of the Oregon and California trails to the mining markets near Sacramento in the late 1840s and 1850s. In distance covered and difficulties encountered, these drives were far more impressive than the later, celebrated movements northward from Texas. They reflected the British-derived droving skills inherent in the heartland herding system since Carolina Piedmont times. In 1852 alone, some 90,000 cattle reportedly passed Fort Kearney, on the Oregon Trail in Nebraska, bound for the California markets, and in 1852–53 together, about 150,000 midwestern cattle supposedly reached California. Missourian Hugh Glenn, a legendary California rancher, drove his first herd to the Sacramento Valley in 1849 and subsequently made at least a dozen more trips back to his old home state to purchase additional livestock. Nor was the Oregon Trail the only route employed, for in 1850 a herd from Missouri's Springfield Plain reached California by way of Texas and the southern trail to the Pacific.[37]

The best-documented and most famous drive of cattle from Missouri to the Sacramento Valley was organized by Kentucky-born Walter Crow in his adopted home county of Pike, in the Salt River bottoms, a major herding region bordering the Mississippi River in northeastern Missouri (figure 38). Cyrus Loveland, who worked as a cowhand on this drive, left a diary describing the remarkable trek by the Crow party. Most of the drovers came from the Salt River country, but the Crows continued to acquire "gangs" of cattle and additional hands as they moved westward through the Missouri River valley.[38]

Most of the Missouri cattle were steers destined for butcher shops in the gold-mining camps, but others came as breeders to restock the depleted herds of California. For example Missouri breeders went to the tule bottoms of Yolo County as early as 1851, and two years later others arrived in Tulare County in the San Joaquin Valley.[39] As a result the bloodlines of California herds began a rapid change, paralleling the shift underway in the Willamette Valley. The resultant "western" cattle, basically Durham, or "American," with residual Iberian traces, later spread through most of the rangelands of the interior West. In California, Durham, Devon, and other shorthorn blood became so common that most stock had been upgraded by 1860. Twenty years later only about 3.5 percent of all California cattle remained "unimproved."[40] In part this shift occurred due to the midwesterners' desire to perpetuate the British milking complex, for the Spanish breed yielded very little milk.

Herding methods in California, too, soon reflected midwestern influences, transforming the industry. The cutting of hay to feed livestock during the summer dry season and the periodic droughts that struck California is perhaps the single best indicator of this influence. The Crow-Loveland party even cut hay for their cattle while still on the trail in Nevada. Provision of hay soon became an established practice among the more progressive Anglo-Californian ranchers, including big operators such as Miller-Lux, though many resisted this trend. During the catastrophic drought of 1863–65, hay sold for as much as $150 per ton in the Central Valley of California.[41]

Initially most hay came from naturally wet marshes and tule lands, but soon the midwestern stock raisers began enlarging these unimproved meadows by means of ditch irrigation, as on the Great Plains, a practice that became a hallmark of non-Texan ranching in the West at large.[42] The origin of this irrigation technology remains unclear. Perhaps it came from the long-established, if modest, California Hispanic garden irrigation, but some feel meadow watering derived from placer mining techniques in the early gold fields, which featured ditches and flumes constructed along the sides of gulches.[43] Another possible source lay among the nonagricultural Northern Paiute Indians, who, in locales such as the Owens Valley of California, since pre-Columbian times had constructed boulder and brushwood diversion dams, with associated ditches, to irrigate expanses of wild grasses and bulbs to enhance their gathering activities. Many Paiutes subsequently found work as buckaroos in the Anglo-Californian ranching system, placing them in a position to introduce meadow irrigation into the western ranching complex.[44]

In certain ways midwestern practices reinforced the older Hispanic pastoral ways. Most notably, both of these groups had traditionally shifted cattle between seasonal pastures, even to the extent of practicing vertical transhumance in some cases. To the present day, the summer highland settlements of herders in montane California bear the Ohio Valley generic name of "cow camp," suggesting the midwestern influence. Such modern California toponyms as Happy Camp (Modoc and Siskiyou counties), Saddle Camp (Tehama County), Meadows Camp (Glenn County), and Brown Cow Camp (Tulare County) suggest the midwestern imprint.[45]

The key to the midwesternization of northern California ranching lay in the fact that many immigrants from Missouri and the Ohio Valley states, including some of the drovers, developed ranches in their adopted home. Walter Crow, leader of the 1850 drive, settled in

Stanislaus County in the San Joaquin Valley, where he began ranching, while companion Cyrus Loveland did likewise near Santa Clara. Not only was Sacramento Valley cattle king Hugh Glenn a Missourian, but so was his protégé and eventual partner, Peter French, a native of the Missouri River valley county of Callaway. E. C. Singletary, a big rancher in the Colusa area, came from Sangamon County, in the heart of the Illinois cattle country, and one of his hands, who had come to California as a drover in 1854, was from Missouri's Springfield Plain, another major midwestern cattle district.[46] The fine open grasslands of the California Central Valley no doubt appealed greatly to those midwesterners who sought to pursue the traditional Ohio Valley prairie system of herding. Given the choice between Willamette's woodlands and the Sacramento Valley's prairies, midwesterners inclined to crop farming would have chosen the former, while pastoralists opted for California.

Seizing California Extended

In addition to northern California, midwestern herders and their distinctive system also spread through pastoral California Extended, in the intermountain region east of the Sierra Nevada and Oregon Cascades (figures 46, 58).[47] As in the Sacramento Valley, midwesternization came gradually to California Extended and was grafted onto the older, much-hispanicized base system. The heartland imprint beyond the mountains grew steadily more vivid in the 1870s and 1880s as Hispanic ways faded. The rise of midwestern herding practices in both California and the intermountain region should be regarded as an ongoing process that began in the Sacramento Valley in the 1840s and reached completion in the interior rangelands a half-century later.

The ranchers and buckaroos who, between 1860 and 1890, colonized the previously described "High Desert" of eastern Oregon and Nevada, as well as the Snake River Plain of southern Idaho, including the famous Camas Prairie, were largely of midwestern birth or parentage.[48] Fully 42 percent of all cowhands in the High Desert by 1880 declared midwestern birth (table 1). In spite of oft-stated claims that the Willamette Valley mothered the so-called Inland Empire, the more influential ranchers of the interior rangelands came from California (figures 48, 54).[49] To generalize, it is fair to say that the Willamette sent homesteaders while California contributed the ranchers to the "Inland Empire."[50]

In fact even the migration of herders directly from the Midwest to eastern Oregon, southern Idaho, and northern Nevada may have exceeded the eastward movement of graziers out of the Willamette. The 1880 census schedules reveal many such migrations. In Owyhee County, Idaho, for example, we find "stock raiser" Lafayette Moore, Sr., a native of Illinois whose father had been born in Virginia. In his employ was fifteen-year-old son Lafayette, Jr., of Idaho birth, engaged in "herding cattle."[51] In the same county, two other early ranchers had immigrated from Missouri's cattle-rich Springfield Plain.[52] For Grant County in eastern Oregon the 1880 census listed such typical stock raisers as W. Shuman of Iowa birth and Ohioan parentage; Missouri native J. W. Young, derived paternally from Kentucky; and F. Perry, a Kentuckian both by birth and parentage. Nor should we ignore the fact that a quite remarkable number of intermountain ranchers in California Extended claimed British highland birth, reflecting a migration of Irish, Welsh, and Scots that perhaps reinforced the dominantly British character of the midwestern herding system.[53]

Influence directly out of the Midwest in California Extended was also revealed in the very early introduction of "pilgrim" cattle along the Oregon and California trails and the associated development of road ranches at places such as Fort Hall, Haws Post, Fort Boise, and Truckee Meadows (figure 54). As early as 1836, migrants left pilgrims at Fort Boise, and several came to Fort Hall in 1838. These fur posts began a gradual diversification and evolution into road ranches. Along the California Trail, the Carson Valley and Truckee Meadows acquired midwestern cattle in the 1850s.[54]

In turn pilgrim stock in the intermountain region gave way by the 1870s to "westerns," Durhams, and other generic shorthorns, even including some Herefords. Selective breeding became an essential aspect of the local ranching system. In part this upgrading, as in California, reflected a shift in marketing patterns away from the undiscriminating miners, to whom beef of almost any quality could be sold, to the growing Pacific Coast urban market, where the consumers demanded a higher quality of meat. Similar pressures to upgrade herds were placed on those intermountain ranchers who chose, instead, to send their culls to the feeder districts of the ancestral Midwest.[55]

Ranching methods in intermontane California Extended also clearly reflected growing midwestern influence in the 1870–90 period. In his study of southeastern Oregon ranching, Peter Simp-

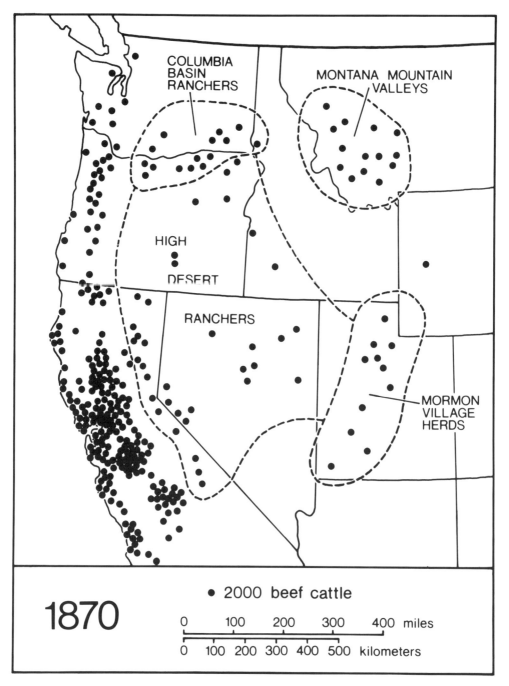

Figure 58. Beef-cattle distribution in midwestern-influenced districts of the West, 1870. The intermontane region is thinly occupied, and a major focus in the Rocky Mountains of southwestern Montana is

clearly evident. Sources: Ninth Census of the United States, 1870 *(Washington, DC, 1872), vol. 3;* Rinschede, Wanderviehwirtschaft, *p. 95 (see caption for figure 57).*

son concluded that midwestern techniques, or "stock farming," triumphed over Hispanic ways "in the end," by the late 1880s. He also observed that the great Anglo-Californian cattle "barons," such as Peter French, retained their dominance by changing and, in effect, becoming more midwestern. The term "improved ranch" was often employed by contemporary observers to describe the end product of the modernization process. One described three stages through which intermontane ranches passed in making the transition from largely Hispanic to more purely midwestern character.[56] Even so, many heartland herding traits were present from the very beginning in the High Desert, a result of the earlier mixing of the two traditions in northern California. For example Peter French of southeastern Oregon very early instructed his Spanish buckaroos to tend the herds carefully and to exercise "constant vigilance" in such matters as rescuing cattle that became mired in the marshes.[57]

Again winter feeding provides the best index to the rise of midwestern techniques. Most of the pioneer herders east of the Cascades and Sierra Nevada, in the "first stage" of ranching occupance, provided little or no winter feed, in spite of the much colder winters of the intermontane lands. Californian W. B. Todhunter wintered more than ten thousand cattle on Great Basin wildrye, which cures into standing hay, in Long Valley, northwestern Nevada, in the early days. If hay was mown at all in this first stage, it went to the ranch horses, and a barn for their protection was also erected, even while the rancher still lived in a log dugout.[58] The normally light snowcover of the High Desert, coupled with fine stands of the native wild-rye bunchgrass, allowed cattle to survive normal winters.

In the second stage of intermontane ranch development, the midwestern-inspired attention to hay cutting began in earnest. Peter French was the first rancher in Oregon's Harney Basin to place haystacks in the winter pastures, for use in case of heavy snow. Even earlier, by 1865, some Owyhee County, Idaho, ranchers had begun cutting hay by hand from natural riverine meadows, and the impetus may have come from cattle raisers fresh out of Iowa and Missouri rather than from the Californians. Often the same Great Basin wild rye that had earlier been sought by foraging cattle provided the principal source of mown hay.[59]

As early as the middle 1870s, the third stage of development, involving irrigation of meadows and the use of mechanical mowers, began on some intermontane ranches, greatly increasing hay production and the amount of winter feeding. The early diversion dams, made of rocks and wood, fed small, private ditches. One moderate-sized ranch in Klamath County, Oregon, produced 450 tons of hay annually by the turn of the century, employing extra hands from May to September to help in haying. Once again, midwesterners had artificially replicated the wet prairies of Iowa and Illinois in the semiarid West. The nickname "webfoot" for Oregonians may derive from their penchant for meadow irrigation work, a task generally despised by Hispanic Californian vaqueros and Texas cowboys. So profoundly did the development change cattle raising that Ernest Osgood, a respected student of the subject, distinguished two types of early western ranching—irrigated and nonirrigated. Wild hay continued to be the most common type, even well into the present century, but alfalfa and other imported hay crops made an appearance on some intermontane ranches by the middle 1880s, beginning the displacement of native bunchgrasses as the principal source of hay. After the severe winter of 1889–90, hay production soared. In Elko County, Nevada, for example, 6,500 hectares (16,000 acres) of hayland in 1890 had grown to almost 97,000 hectares (239,000 acres) by 1900.[60] Andalucía had at last yielded to highland Britain.

Gradually, too, the open range that prevailed initially in California Extended gave way to fencing. Haymaking, in fact, required enclosed meadows. By the late 1870s, range fencing had begun on some ranches, as was reflected in the 1880 census schedules of Grant County, Oregon, which listed among others a certain Anthony Charles, a "fence builder" of Iowa birth. A remarkable illustration of a cattle ranch on the Camas Prairie in southern Idaho about 1880 showed fully enclosed pastures, furnished with haystacks (figure 59). This place was owned by Missouri-born Benjamin Morris, whose parents were both of West Virginia birth, making the ranch both genealogically and morphologically the epitome of the midwestern–upland southern herding system in the West. Midwestern inspired, too, were the stock corrals made of willow branches stacked or woven between pairs of posts, a high-desert variation of the Anglo-American "straight-rail" fence of the East (figure 60).[61]

The intermontane environment encouraged vertical transhumance, and seasonal pasture shifts became part of the ranching system in California Extended from the very first. The distinctive mid-

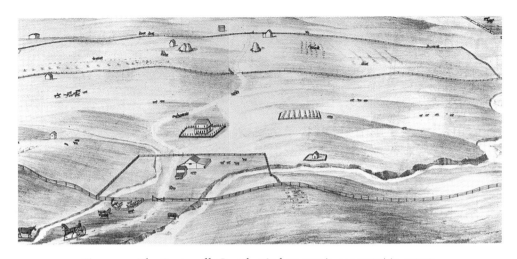

Figure 59. The Centerville Ranch, 580 hectares (1,440 acres) in extent, on the Camas Prairie in southern Idaho Territory, about 1880. Owned by Missourian Benjamin F. Morris, the ranch was a typical midwestern enterprise, with enclosed pastures, haymaking, and fenced-in haystacks. Source: lithograph by Wallace W. Elliott in History of Idaho Territory *(San Francisco, CA, 1884), following p. 278;* copy provided by the Idaho State Historical Society Library and Archives, Boise, and used with permission.

western contribution to this practice is suggested by the prevailing use of the term "cow camp," as in California proper. After the spring drive to the highlands, many buckaroos went immediately back to the home ranch for a long season of haying, leaving the herds to a skeleton crew.[62] In these and other cattle drives, intermontane cowhands used bullwhips, latter-day versions of the Ohio Centerville type.[63]

The Columbia Basin

North of the High Desert and Snake River Plain, beyond the Blue Mountains of Oregon, lay sections of the intermountain region that had received relatively little Californian pastoral influence. There, in the Columbia Basin and southern interior plateau of British Columbia, the cattle frontier from the first had a dominantly midwestern character. Attracted by mountain mining markets and pressed out by farmers in Iowa, Missouri, the Willamette Valley, and even northern California, midwestern cattle raisers began, shortly after midcentury, the occupation of these northern intermontane districts.[64]

Figure 60. A corral made of willow poles and branches stacked between pairs of posts, near Frenchglen, Oregon. This type of fence is a modification of the traditional eastern Anglo-American "straight rail" corral (see figure 37), to accommodate locally available building materials. (Photo by the author, 1991.)

Wreathed by mountains and plateaus, the Columbia Basin occupies nearly all of southeastern Washington, as well as smaller sections of Oregon and Idaho (figure 54). The Cascade Range to the west draws most of the precipitation from the moist Pacific marine airmasses coming onshore, causing a pronounced rain-shadow effect in the Columbia Basin. Indeed the Cascade crest presents one of the sharpest ecotones in North America, where forest gives way suddenly to grassland, humid climate to semiarid. As a result the Columbia Basin is covered by a huge expanse of bunchgrasses, mainly *Agropyrons*, "belly high on the horses," offering a fine cattle range that is palatable both green and as standing hay. The land surface of the Basin displays many irregularities, from the odd rounded loessial hills of the Palouse country in the east to the uninviting, channeled scablands of the center and fingered small valleys such as the Yakima, Kittitas, and Klickitat reaching into the Cascades along the western margin. The true "bunchgrass country" lies in the higher, somewhat moister, eastern parts of the Basin, while in the western half, lower and more severely affected by the Cascade rainshadow, big sagebrush of minimal value to cattle as browse competes with the grasses. Summer is the driest time of the year, echoing the climatic regime of Mediterranean California, but at that season the encircling highlands offer good pastures and the opportunity for vertical transhumance (figure 52).[65]

All things considered, the Columbia Basin looked familiar and appealing to cattle graziers arriving fresh from the valleys of northern California and western Oregon. Willamette cattle raisers found the Basin better suited for their purposes than the small, wooded valley they departed. The beginnings of cattle ranching in the Columbia Basin, aside from the small herds kept at Hudson's Bay Company posts such as Walla Walla and Fort Okanogan, occurred in the middle and late 1830s at Anglo-American Protestant missions, most notably The Dalles and Walla Walla. The latter place functioned as a road ranch, acquiring exhausted Oregon Trail pilgrim cattle as early as 1834. Through the mission system, as on the Spanish colonial frontier, cattle-herding skills passed to Indian neophytes, allowing native people to become the first notable ranchers in valleys such as the Yakima.[66] As late as 1880, some 14 percent of all cowhands in Yakima County were still *siwashes*, or Indians.[67]

Anglo-American secular ranchers began settling certain peripheral valleys in the Columbia Basin in the early 1850s, although effective occupance awaited pacification of the Indians later in that decade, ending a series of uprisings that had begun with the Whitman massacre at Walla Walla in 1847.[68] Three different valleys—the Klickitat, Yakima, and Walla Walla—experienced almost simultaneous settlement in the late 1850s (figure 54). Willamette families such as the Splawns and Thorpes entered the Klickitat Valley, across the Columbia River from The Dalles, in the southwestern corner of the Basin in 1859 and began ranching cattle. Following an upland southern custom with eighteenth-century roots, these herders at once sought other, nearby vales to use as seasonal pastures for their cattle, leading them north into the Yakima Valley. As in Appalachia, seasonal use of the Yakima soon led to migration and the establishment of permanent ranches there, such as the one developed by local cattle king Ben Snipes.[69] By about 1860 some thirty thousand cattle grazed the Klickitat and Yakima pastures. In turn Yakima cattle raisers soon sought summer ranges in the Kittitas Valley, still farther north, a prelude to the settlement, within a decade, of that area (figure 54).[70]

In the Walla Walla Valley, to the east, where the fur post and mission road ranch had been engaged in cattle raising since the 1830s, private ranchers began settling shortly after 1850, only to be harassed by the Indians. The influx resumed before 1860, a time when the Texans still sat on their coastal prairie, and by the outbreak of the Civil War the valley housed about ten thousand cattle.[71]

During the 1860s this northwestern cattle frontier spread through most of the remainder of the southern half of the Basin, especially the Umatilla Valley, spilling over into lowlands imbedded in the Blue Mountains, such as the Wallowa Valley (figures 54, 58). By the early 1870s, eastward-moving cattle ranchers crossed the territorial border to occupy the Paradise Valley around Moscow in Idaho. Others claimed the central and northern parts of the Basin, including the channeled scablands and Big Bend country. So rapid was this pastoral expansion and settlement progression that by about 1880 the range-cattle era was nearly over and ranchers began seeking refuges from an influx of wheat-growing homesteaders. The Turnerian principle of occupance stages reasonably well explains what happened in the Columbia Basin, and appropriately so, given the midwestern bias of his model.[72] Ranching today survives mainly in the drier western reaches of the Basin, where places such as Ellensburg in the Kittitas Valley still cultivate their ranching image, and along the southern margins, where Pendleton in the Umatilla Valley holds the nationally famous annual "Round Up" rodeo.

The cattle ranchers who settled the Columbia Basin were, overwhelmingly, of midwestern origin. Two of the most famous were Iowan Ben Snipes of the Yakima Valley and Dan Drumheller, a rancher from Missouri's Springfield Plain who settled in the Walla Walla Valley.[73] In many and perhaps most cases, these graziers had resided in the Willamette Valley before coming east to the Columbia Basin frontier. Snipes followed that migration pattern, as did his neighbor rancher, A. J. Splawn. The Basin, culturally and economically, has almost from the very first belonged as a hinterland to western Oregon and Washington, to Portland and Seattle.[74] However, this connection has often been overstated. Some migration came directly from the Midwest, and California also touched the Columbia Basin, as exemplified by both Drumheller, who worked on a Sacramento Valley cattle ranch before coming north, and Snipes, who lived for a time in California before coming to the Willamette.[75]

The cowhands who worked on the ranches in the Basin had similar origins. In Yakima County, Washington, in 1880, natives of the

Midwest and Oregon accounted for the majority of the work force, supplementing the sizable Indian contingent mentioned earlier (table 1). Not a single Hispanic, Anglo-Texan, or African-American cowhand resided in the county, although to the north, in the Kittitas Valley, the log cabin of Texas-born ranchhand Millard Saylor still stands on a back street in Ellensburg. In the local lore, Saylor's outspoken aversion to doing meadow irrigation work also survives. Another early cowboy, named "Texas Bill," worked on a ranch near Pasco in the Big Bend country along the Columbia River, but these southerners were exceptional.[76]

Midwestern cattle breeds quickly replaced the California Iberian foundation stock in the Columbia Basin. Methodist missionary Jason Lee, en route to the Willamette Valley, left a few exhausted "pilgrims" from Clay County in the Missouri Valley at Fort Walla Walla, the Hudson's Bay Company post, in 1834. In so doing Lee inadvertently began the process by which fur posts on or near the Oregon Trail became road ranches, a sequence very important in the spread of midwestern breeds in the interior West. Nine years later a herd of almost one thousand midwestern cattle reached the Whitman mission at Walla Walla, further diluting the Iberian bloodlines of Columbia Basin stock. The pioneer ranchers who spread through the Basin after 1858 introduced primarily "American" cattle, and even herds driven in from California in the 1860s were largely non-Iberian. Ben Snipes early upgraded his Yakima Valley cattle with Durhams.[77]

British Columbia

As in the Great Plains, the international boundary proved to be no barrier to the diffusion of the ranching frontier. The southern interior plateau of British Columbia acquired a range-cattle industry almost simultaneously with the occupance of the Columbia Basin, producing a remarkable northern appendage of ranching and its farthest contiguous extension poleward (figure 54). In physical environment, sources of pastoral immigrants, and markets served, this Canadian ranching region closely resembled the Columbia Basin. For all practical purposes of culture history, the two areas are one.[78]

The southern interior part of British Columbia, between the Rockies and the Coast Ranges, consists of a plateau between 900 and 1,500 m. (3,000–5,000 ft.) elevation above sea level, segmented into blocks by the deeply entrenched Fraser River and its tributaries. In the incised, low valleys, which lie as much as 800 or 900 m. (2,600–

3,000 ft.) below the plateau surface, a bunchgrass-sagebrush vegetation cover much like that of the Columbia Basin prevails, extending as far north as about 53° latitude and reflecting the rather meager precipitation received in this rain-shadow region, usually between 175 and 275 mm. (7 to 11 in.). In the cold steppe climate, bluebunch wheatgrass is particularly abundant and provides good grazing. The adjacent plateau surface, higher and wetter, is covered by coniferous and mixed forests that contain abundant "meadow openings," both dry and marshy, offering coarse grasses, rank sedges, forbs, vetches, and horsetail, all of value to cattle in the summer.[79]

Aside from early herds at a few Hudson's Bay Company posts, including five or six thousand head at Fort Kamloops by 1848, ranching in British Columbia began as a result of the mining frontier, as was true for most areas west of the continental divide. Gold discoveries in the Fraser Valley during the late 1850s, culminating in a bonanza in the Cariboo Mountains around Barkerville, almost immediately attracted cattle drives from Washington, Oregon, and even California. Bullwhip-wielding drovers followed the Okanogan Trail northward from the Columbia Basin, then ascended its extension, the famed Cariboo Trail, to reach the goldfields (figure 54). Between 1859 and 1870, some twenty-two thousand cattle legally crossed the international border on this route, a trade that peaked in the 1861–65 period. Such notable Columbia Basin cattlemen as Ben Snipes of Yakima, William Gates of The Dalles, "Spokane" Jackson, and Dan Drumheller from Walla Walla participated in these drives. The 1,300-km. (800-mi.) trek from the Yakima Valley to the Cariboo district made by Snipes, together with John Jeffries, represents one of the epic cattle drives of all time.[80]

These drovers not only delivered livestock to the mining-town butchers, but also prompted the establishment of a local, midwestern-inspired cattle industry. Road ranches provided the prototypes, repeating the developmental sequence that occurred on the Oregon Trail. The earliest example in British Columbia was apparently the Alkali Lake Ranch, founded in 1861 when the proprietor of a roadhouse on the trail to the Cariboo country purchased breeding stock from two Oregonian drovers (figure 54). More often the drovers themselves established road ranches after discovering that the herds being driven north could, with the help of local "chinook" winds, survive the boreal winter. During the winter of 1861–62, several Washington drovers, unable to sell their cattle, held them until spring on the "bunch grass benches" near Kamloops. That success, coming in

Figure 61. Log horse barn near 100 Mile House in the ranching country of the southern interior plateau of British Columbia. The structure contains stables for ranch horses, tie-stalls, mangers, box stalls, a tack room, and a huge hayloft. In carpentry and construction, it reveals links to the frontier of the Midwest. (Photo by the author, 1990.)

the very winter when huge die-offs occurred in the Columbia Basin, convinced some of the Americans to establish ranches in Canada.[81]

By the 1870s ranches were "strung like beads along a few roads and trails," including also the Okanogan Trough at the American end of the route north.[82] By the beginning of that decade, over ten thousand cattle grazed the British Columbian ranges. These road ranches, also called "mile houses," depended largely for their livelihood upon the trail traffic, including revenue from freighting and staging. Spaced at least 25 km. (15 mi.) apart, they offered inn services for travelers and draft animals for trail wagons, in addition to beef for the mines.[83] The present British Columbia highway 97 southwest from Williams Lake is lined with the often impressive remains of such road ranches, most notably the 105 Mile Post House national heritage site and the O'Keefe Ranch, as well as many thriving modern ranches (figures 54, 61).

Expansion of the British Columbia plateau cattle ranching continued until the middle 1880s, spreading away from the roads to areas such as the Nicola Basin south of Kamloops, the largest lowland in the region, until all of the desirable ranges were taken. By 1887 some 262 ranches existed in the southern interior plateau, although by that time five or six big operators, by consolidation and purchase, dominated the industry.[84] Ranching proved an enduring land-use system in British Columbia, giving way to fruit orchards only in the warmer southern valleys, and cattle raising remains viable through most of the region still today.

Americans of midwestern birth or ancestry dominated the early ranching era in British Columbia, although some personnel of the Hudson's Bay Company and other Canadians were active in the industry from the first. As one firsthand observer noted, "scratch a British Columbia rancher and you'll discover an American expatriate."[85] A link to the American heartland is well demonstrated by the Harper brothers, Jerome and Thaddeus, who developed ranches near Kamloops and Williams Lake, including the famous Gang Ranch (figure 54). Born upland southerners in the 1820s, sons of Adam Tucker, in the Glade country of Tucker County, West Virginia, the brothers had handled "thousands of head of beef cattle in Shenandoah valley" and along the Glady Fork of the Cheat River, a tributary of the Ohio, in the heart of the old Appalachian cattle country. In young manhood, the Harpers relocated to Santa Clara County, California, participating in the rise of the Anglo-Californian ranching system and absorbing some Hispanic influences. By 1859 they were in British Columbia, drawn by the mining market, and in following years their herds repeatedly ascended the Cariboo Trail to Barkerville. Soon they developed the Kamloops ranches and rose to become two of the most important cattlemen in the province. They linked Appalachia and the Ohio Valley to Anglo-California and British Columbia in one life span, demonstrating the remarkable speed and reach of diffusion achieved by the midwestern herding system.[86]

Many early cowhands north of the border also had midwestern genealogies. No references to Texans or Hispanic Californians working on ranches in early British Columbia have been found, but the widespread western practice of employing acculturated Indians as cowboys was common on the interior plateau.[87]

When the mining markets began to decline, British Columbian cattle ranchers sought other outlets for their beef. Vancouver, the great port city of Pacific Canada, became a reliable market, particularly after the completion of the Canadian Pacific railroad, but other opportunities also attracted the attention of plateau ranchers.

The Harpers, in the middle 1870s, successfully drove a herd of twenty-five hundred cattle to the railroad in northwestern Nevada, for shipment to San Francisco and the eastern United States. One particularly ill-advised effort by four British Columbian ranchers in 1898 to drive a herd 2,400 km. (1,500 mi.) to the Klondike goldfield in the Yukon Territory ended in complete failure, although several of their drovers, in an epic attempt, got as far as Teslin Lake, on the southern Yukon border near Whitehorse. Their failure revealed the remarkable pluck characteristic of drovers since colonial days in Mexico and Carolina, but it also announced a painful truth—the cattle frontier would not outlive the century. This truth must have dawned upon the unfortunate Yukon drovers when, after their horses also perished, they had to walk over the Coastal Ranges to Wrangel before catching a boat ride back home.[88]

Ranching Methods in the Northern Intermountain Region

The very early midwestern dominance of the Columbia Basin and the southern interior Canadian plateau led to an almost immediate prevalence of heartland herding methods, although, as suggested in chapter 8, Hispanic Californian influences were not altogether lacking. A disastrous winter in 1861–62 in the Columbia Basin further demonstrated, almost at the outset, the need for midwestern methods in the cattle industry. While some of the first ranchers, both in Washington and British Columbia, provided no winter feed, forcing the cattle to "rustle" for themselves, haying began at least by the middle 1860s in the Yakima Valley. Natural meadows became the most favored sites for ranch settlements in British Columbia. Around Kamloops ranchers cut wild hay of sedge and horsetail, after first breaking down beaver dams to facilitate the hand mowing. Haying was the first job found by Illinoisan Walter Pierce, later to become a prominent rancher and state governor, when he immigrated to northern interior Oregon.[89]

Fencing in eastern Washington and southern British Columbia progressed at about the same pace as in California Extended. Meadow enclosure began very early, followed in about fifteen years by pasture fencing. Selective breeding received early attention, both by castration and the spaying of inferior heifers, the latter a practice never found in the traditional Texan and Californian ranching systems.[90] Vertical transhumance, as in California Extended, occurred almost from the very first, and the midwestern term "cattle camp" appeared widely in these northern regions. Some contemporary observers felt that the upslope and downslope movement of the cattle with the changing seasons was natural and instinctive, requiring little human intervention.[91]

The weaker influence of Hispanic California in the north had other manifestations. In the Columbia Basin, for example, fewer "cattle kings" comparable to those of the High Desert arose, and the dominance by small ranchers might best be regarded as a midwestern characteristic.[92] Still, a transition toward more intensive methods, or midwesternization, did occur in the northern intermontane region. Some Canadian scholars view that process in British Columbia as a diminution of early "American" influence and a Canadianization of the industry, rather than as the ascendance of the Midwest. They point to a significant presence in Canada of ranchers and cowboys born in highland Britain, including Irish. Epitaphs in British Columbia churchyards, such as the one at Westwold in the bunchgrass Grand Prairie east of Kamloops, confirm this British presence. Some of them also came south into the Columbia Basin. Perhaps, then, the raising of range cattle in the northern intermontane districts could have received renewed British highland influences. The use of collie dogs to help herd cattle and horses on the British Columbian ranch of one immigrant from the north of England suggests as much.[93]

While it would be erroneous, then, to regard ranching in British Columbia as a mere extension of the Anglo-American heartland, cattle raising in the southern interior plateau was more nearly midwestern than anything else. The built environment of the surviving traditional ranchsteads reveals this influence. Midland American notched-log construction prevailed in British Columbia and is reflected in structures such as the massive horse barns still found in the region (figure 61). These buildings, both in architecture and details of carpentry, belong in the Anglo-American western-mountain tradition and speak persuasively of midwestern influence. So do the traditional log fences of the British Columbian ranching country, all of which derive ultimately from Pennsylvania and the midwestern frontier.[94]

The Midwest in Utah and Arizona

The settlement of the oases in Utah and southeastern Idaho by Mormons, especially given the midwestern roots of these intermontane agrarian colonists, has led some scholars to add Deseret to the western nuclei of midwestern-inspired cattle ranching. The 1850 census did, indeed, reveal over twelve thousand cattle in the Wasatch oasis, the Mormon hearth area (figure 56).[95] In fact, however, ranching played little role in the Mormon colonization system, particularly in frontier times. Instead their settlement was based in the agrarian irrigation village, producing a farming rather than a pastoral frontier, violating the Turnerian sequence. In the peripheries of Deseret and in certain distant outliers of Mormon culture, in later decades, some ranching did arise among the Latter Day Saints, usually to utilize extensive nonirrigable tracts, but in no respect could the early Mormon culture area be called a cattle-ranching frontier. Even today the rodeo is largely absent in their communities.[96]

In fact the Mormons deliberately chose, in devising their adaptive strategy for the intermontane region, not to emulate an early midwestern road ranch in their domain. In 1836 Miles Goodyear, who had come west with the Whitman party headed for Walla Walla, decided instead to establish a fort and trading post on the Weber River at the present site of Ogden (figure 54). Within a few years he had added road ranching to his activities, raising cattle, horses, and goats. Goodyear sold out to the Mormons in 1847 and moved to California. His road ranch soon became a farm village.[97] In the final analysis, then, midwestern pastoralists proved no more successful than Anglo-Texan ranchers in influencing or altering Mormon village society.

A far sounder claim for midwestern herding influence can be made for Arizona, the southernmost part of the intermountain region. Missouri-born cattle ranchers were active in Arizona at least by 1860, and a decade later natives of the Ohio Valley states plus Missouri and Iowa accounted for a quarter of the 45 cattle ranchers in the territory and were almost twice as numerous as Texans. The 1880 census enumerated 217 cattle ranchers in Arizona, over a third of whom were midwesterners, and they were also well represented in the cowboy work force by then (table 1).[98]

These ranchers and herdsmen had usually reached Arizona by way of the Santa Fe Trail and its extension westward toward southern California, although many had come through Texas instead. While the Santa Fe route and southern access to California never rivaled the Oregon Trail as a path of diffusion for midwestern cattle and herding practices, it still deserves attention (figure 54). As early as 1858, one hundred Illinois heifers and four bulls were purchased by a Tucson rancher from a drover headed for California, and in 1867 some fifty

"Missouri cows" were introduced to upgrade longhorn stock on the Little Colorado in eastern Arizona. Noted rancher H. C. Hooker used Illinois Devons and shorthorns to improve his Texas-derived herd in the Sulphur Springs Valley in the middle 1870s, and Iowan cattle reached the Salt River Valley in 1884 (figure 39). Even Oregonian cattle entered Arizona before 1880.[99] Midwestern influences in the cattle-ranching industry of Arizona may have been second only to those derived from Sonora.

Rocky Mountain Ranching

The remaining part of the West, between the Great Plains on the east, the intermountain region together with the Mormon villages to the west, and the Hispanic-Navajo sheep culture to the south, might best be called the Anglo-Gentile Rockies, stretching from Canada to Colorado. Its culture history is little known or understood, but this section of the Rocky Mountains also experienced the cattle frontier. Ranchers became established, often surprisingly early, in its myriad valleys, basins, "holes," and "parks." While some Texan and Californian herding influences reached the Rockies, the mountain ranching culture was dominated from the very first by the midwestern system, which underwent certain modifications to fit the highland environment. In the high valleys, above about 1,800 m. (6,000 ft.) elevation, ranching remains in place to the present day.[100]

The valleys strewn through the Rockies provided the ranchstead sites, winter pastures, and haylands. The basins, such as the Big Horn and Shirley, were large and offered expansive bunchgrass-sagebrush steppes much like those of the intermontane High Desert and Columbia Basin. "Parks"—higher, relatively small, saucer-shaped basins—presented much the same pattern in less grand dimensions and, interestingly, preserved the ancient French generic toponym *parc*, used to describe the high summer pastures of cattle raisers in Auvergne. Still smaller were the "holes" and elongated valleys, such as Jackson Hole in western Wyoming and the Bitterroot Valley of Montana, some of which were so elevated as to preclude crop cultivation (figure 54). Surrounding these basins and valleys were vertically zoned foothills and mountains, rich in forests, high meadows, and sources of meltwater for irrigation. The early Rocky Mountain ranchers settled near streams in the valleys, often homesteading a small plot of natural meadow or irrigable land near the margin of the bordering slopes.[101]

East of the main front of the Rockies, only in the north, a broad "foothill zone" provides a transition to the Great Plains (figure 54). The ranching history of the foothills is much more closely linked to the Rockies than to the Plains, and its cattle frontier will be considered as part of the mountain complex. Mainly a shortgrass prairie, the foothill zone also offered ready access to the wooded ridges and low ranges for which it is named, including the Little Belt, Big Snowy, and Crazy mountains, among others. As a result the valleys of the foothill zone, such as the Judith Basin, take on much the same appearance as the lowlands within the Rockies proper.[102]

The entry of the cattle frontier into the northern part of the Anglo-Gentile Rockies came very early, by way of certain Oregon Trail road ranches, most notably the one at Fort Hall in Idaho (figure 54). Because grass along the Trail grew scarce, due to the incessant demands of the stock accompanying emigrant wagon trains, the traders who owned the road ranches, relying upon their extensive knowledge of the adjacent mountains acquired in the fur business, began taking the pilgrim cattle north over Monida Pass into the lush valleys in western Montana for summer pasturage. Some believe this transhumance began as early as 1843 from Fort Hall. Soon successful attempts were made to winter the cattle in these same valleys. At least by the late 1850s, the Deer Lodge, Bitterroot, Jocko, Beaverhead, and several lesser valleys had acquired resident pilgrim herds. Cattle also came from the old fur post at the Green River crossing on the Oregon Trail in Wyoming. Fort Owen, south of present Missoula in the relatively low Bitterroot Valley, reputedly became the first cattle ranch in the northern Rockies.[103]

Initially the restored pilgrims were driven back to the Oregon Trail for resale to emigrants, but a mining boom in the Montana Rockies provided a new and much larger market in the 1860s. The valleys closest to the bonanzas quickly filled with ranches. The veteran road ranchers and their hardy pilgrims soon found competition for possession of the valleys, as cattle were driven in from Oregon, California, Kentucky, and Missouri. By the time the Texans stirred, after Appomattox, and joined the cattle rush to western Montana, a thriving midwestern-derived ranching industry had occupied the region. Deer Lodge County alone contained over seventy-five thousand cattle by the early 1870s, and the previous United States census had clearly revealed the major concentration of cattle in southwestern Montana (figure 58). As late as 1880, ranching remained concentrated in the southwestern part of the territory, and 45 percent of all Montana cattle grazed there. Even some of the high valleys were occupied, including the Big Hole, where the first cattle entered in 1874 for summer grazing, coming from the adjacent Deer Lodge and Beaverhead valleys (figure 54). Soon the Big Hole acquired its own resident ranchers, and their herds were being driven to market over the Beef Trail, by way of Deer Lodge Pass, to Butte slaughterhouses. As the gold and silver bonanza played out, some miners also became ranchers.[104]

To the north the pattern of the Montana Rockies was later repeated, on a far smaller scale. In the East Kootenay Trench in southeastern British Columbia, adjacent to another notable mining district, the northernmost ranching valley in the Rockies, forming an outlier, was occupied when the first six ranchers arrived about 1885 (figure 54).[105]

The Montana Rockies, rather than the Great Plains, provided most of the ranchers who colonized the foothill zone, both in central Montana and in Alberta, producing another eastward-moving cattle frontier like those of the High Desert and Columbia Basin. In the valleys of the Judith, Sun, and Musselshell rivers, the mountain-derived ranching frontier took root in the foothills during the 1870s and easily withstood a penetration by Texans in the following decade. Midwesternization of the foothill zone was assisted by the arrival of some cattle raisers direct out of Missouri and Iowa. An example was Kentucky-born Robert Ford, who came from Missouri to found one of the biggest ranches in the Sun River country, including the Sweet Grass Hills near the Canadian border.[106] Similarly ranchers and cattle herds arriving straight out of the heartland states helped reinforce the midwestern character of herding in the Montana Rockies proper. Even as early as 1863, the very year of gold discovery at Virginia City in Montana, the "Jessamine Stock Ranch" was in operation near that boom town, in the Madison Valley, raising Durham cattle imported from Kentucky, including some purebreds. Throughout the high valleys of the Rockies, Ohio was an important ancestral birth state of the mountain ranchers.[107]

The Midwest, then, came to the northern Rockies and foothill zone, both from the Oregon Trail road ranches and by way of migration directly from states such as Missouri and Iowa. A further, later reinforcement of midwestern influence reached these mountains from Oregon, Nevada, Washington, British Columbia, and other intermontane districts to the west. In the main this eastward movement involved only cattle herds. As early as 1867, Yakima Valley cattle were trailed to Montana's Bitterroot Valley, drawn by the mining market. For the following two decades, intermontane cattle continued to move east, supplying both steers for the miners and additional breeding stock for valley and foothill ranches. In fact such western cattle sometimes came as far as Nebraska and the Dakotas, well into the Great Plains. Perhaps as many as a quarter million Oregon cattle went to Montana alone in the 1869–75 period. Such stock were referred to as "westerns."[108] The stories of several such drives have been preserved in the literature.[109] A parallel movement, by way of Crowsnest Pass in the Canadian Rockies, brought cattle and horses from the interior plateau of British Columbia to the foothill zone of Alberta.[110]

Some intermontane ranchers and buckaroos migrated eastward from the intermontane region, taking up new residence in the northern Rockies and foothills. A few "Webfoot" cowhands from Oregon and other westerners even settled in the Great Plains, where they were said to be virtually indistinguishable from Missourians (table 1).[111] Representative of the ranchers moving east was John W. Chapman, a native of the Illinois Grand Prairie. Chapman had moved as a child with his family to Oregon in 1851. In the late 1870s, retracing his path, he brought a herd east to the Big Horn Basin of northern Wyoming and founded one of the earliest cattle ranches there.[112]

Ranching in the southern Rockies, south of the Oregon Trail, developed somewhat differently. The road ranches played a smaller role, an influx from the intermontane region never occurred in force, and Texans played a larger role. Occupance progressed more normally, from east to west. Midwesterners were dominant, but most simply moved in from the adjacent Great Plains. In the 1860s some ranchers below the Front Range of the Rockies began bringing cattle into mountain parks and basins for summer pasture. Sam Hartsel, who had previously worked for cattle raisers in Ohio and operated his own ranch in Iowa in the 1850s, brought the first cattle into Colorado's South Park in 1862. A Missourian of Kentucky birth wintered the first cattle, midwestern shorthorns, on the Laramie Plains in 1864–65. As earlier, in Appalachia and the Ohio Valley, summer cow camps often became permanent ranches when the herders migrated permanently to the former seasonal range. Some valleys were settled much later, as for example North Park, where even summer herds remained rare before 1878. Ranchers from the Laramie Plains wintered the first cattle in North Park in 1879, after which it soon filled with ranches. Valleys west of the continental divide in the southern

Rockies were settled mainly in the 1880s. Midwestern-born ranchers and cowboys dominated the southern Rockies from the very first (table 1).[113]

Mountain ranching methods, while readapted to the highland setting, were dominantly midwestern and echoed many Appalachian practices. The scattering of Texans in the Rocky Mountain basins and parks maintained their old ways at first, but soon they, too, bowed to necessity and accepted midwestern methods. The margin for error was small and the penalty for livestock neglect severe in the mountains, especially in the high valleys, giving the Ohio Valley system a preadaptive advantage.

Hay provided the key to successful mountain ranching, and the higher the valley, the more copious the amount required. Three to five months of winter feeding could be required in the mountains, and an individual animal could consume a ton or more during the winter. At first wild grass and sedge hay, cut from poorly drained natural meadows in the valleys, sufficed. The annual spring thaw caused meltwater-fed streams to spread out over the flats, providing good opportunities for haying. Abundant natural meadow was perhaps the single most important criterion in selecting ranchstead sites. Enclosure of the meadow with a zig-zag log fence normally occurred at the founding of the ranch (figure 62).[114] One early mountain settler summed it up best in the simple statement, "our principal crop was hay."[115] The wisdom and necessity of attention to winter feed, to "improved" ranching, was demonstrated in the Shirley Basin of Wyoming, where "the disastrous winter of 1886–1887 did not hurt Quealy as badly as it damaged some of his neighbors, because he was a *pasture and hay rancher.*"[116]

Wild hay very gradually gave way to improved varieties such as timothy, clover, and alfalfa, which spread through the mountains and foothills beginning in the 1870s and 1880s.[117] The areas of natural flooding were soon augmented by irrigation dams and ditches, which spread the meltwater over a larger expanse. Simple diversion dams and crude ditch systems could be constructed even by neophytes, and the legal principle of "prior appropriation," by which water rights remained vested in the first person to divert a stream, encouraged very early irrigation attempts. Even so, water disputes broke out by the middle 1870s in central Colorado, among other areas, precipitating the "Lake County War."[118] With irrigation the mountain cattle ranchers created an artificial wet-prairie ecosystem of meadows, and

Figure 62. A traditional "chock-and-log" fence encloses an irrigated haymeadow in the Big Hole, a Rocky Mountain high valley in southwestern Montana. This sturdy pioneer fence type, the origins of which go back through the Midwest to Pennsylvania, appears fairly widely in the mountain West and demonstrates the midwestern cultural roots of the region. (Photo by the author, 1987.)

pastures as well, in the valleys, not unlike the natural wet prairies they had known in Iowa, Illinois, and Indiana.[119]

Most likely, knowledge of irrigation methods came east into the Rockies from California Extended. Still the rapid spread of irrigation technology through both the agrarian and pastoral West has always been something of a mystery. How did a people from humid lands with no prior knowledge of irrigation so quickly acquire this skill? Independent invention may provide part of the answer. For example one pioneer rancher in Gunnison County, Colorado, described how he copied the dam-building technology of beavers, so skillfully that his diversion pond almost at once attracted a colony of the busy rodents, who unfortunately began damming his ditches. In southeastern Wyoming, an accidental logjam of railroad ties being floated

Figure 63. Traditional roofless and largely empty log haycrib, in the ranching country along the North Platte River near Saratoga, Carbon County, Wyoming, on the approach to North Park. Such structures are also known in Appalachia. (Photo by the author, 1987.)

down the Little Laramie River caused extensive meadow flooding and led the owner of the adjacent ranch to construct a similar dam to achieve the same effect.[120]

The heightened importance of hay in mountain cattle ranching was soon reflected in the cultural landscape—the built environment of the valleys. Some such features, such as a roofless log haycrib, had apparently been part of the midwestern herding system since Appalachian times, and some of these cribs can still be found in both the eastern and western mountains of North America (figure 63). Many other items of haying material culture represented Rocky Mountain innovations, reflecting the greatly increased role of winter feeding in the West. An example is the "Beaverslide" hay stacker, also known as the "Sunny Slope Stacker" (figure 64). Invented in the Big Hole in Beaverhead County, Montana, this stacker later spread through a wide area of the Rockies, foothill zone, and intermontane region. Many remain in use today, largely because the Beaverslide is a very efficient device for building the larger, taller haystacks needed in the areas with long winters. This improved stacker appeared in conjunc-

Figure 64. Three "Beaverslide" hay stackers in the Big Hole, the Montana valley where this device was invented. These stackers allowed the building of very large haystacks and spread to become part of the cultural landscape through much of the Anglo-Gentile Rocky Mountains. (Photo by the author, 1987.)

tion with mechanical mowers and a variety of horse-drawn rakes, all of which increased the amount of hay produced.[121]

Many mountain ranchers undertook very early pasture fencing, even before the advent of barbed wire. One South Park cattleman enclosed a large pasture in 1882, only a year after establishing his ranch. The Beaverhead Valley of Montana had fenced pastures at least as early as 1876. Most early pasture enclosures were of the "worm fence" type, well known on the wooded frontier of eastern Anglo-America. Made of lightly notched, round poles stacked in a zig-zag line, these fences, based in the tripod principle, could withstand a fair amount of shoving by cattle. Many survive to the present in the Rockies (figure 65).[122]

Vertical transhumance, traditional in midwestern and upland southern herding, also took on added importance in the Rockies (figure 52). Indeed, as has been described, the first intrusions of cattle raisers into the mountains occurred because of a search for summer high pastures. Stationary pasturing was simply not possible. In the

Figure 65. A sagebrush-bunchgrass cattle pasture enclosed by a pole "worm" fence, Beaverhead County, Montana. The earliest pasture fences in the Rockies were of this type, and it derives from the frontier Midwest and Pennsylvania. (Photo by the author, 1987.)

summer the valleys and basins became scenes of feverish haymaking, while cattle went with herders into high pastures, often some distance away (figure 53). The vertical zonation of vegetation allowed reasonably easy access to high meadows, prompting a herd movement that continues in many areas to the present day. Early ranchers quickly scouted the adjacent uplands, returning with knowledge of "an excellent place to summer cattle." Soon toponymic maps of the Rocky Mountains became dotted with "cow camps" and "meadows," as well as striped with "stock driveways."[123]

The Rocky Mountain cattle frontier and its preponderantly midwestern character were well exemplified by one ranch in the Beaverhead Valley of Montana, for which a rather detailed 1876 description survives.[124] At the ranch headquarters in the valley, about 200 hectares (500 acres) were enclosed, a fifth of which were cultivated and the rest in haymeadow. A good, solid dwelling and enormous horsebarn, the latter measuring 15 by 24 m. (50 by 78 ft.), dominated the ranchstead. About three thousand cattle, including some shorthorns, were raised on the place, in addition to a like number of sheep and some five hundred horses. Huge amounts of butter were produced each year, suggesting that the British-derived milking complex never completely disappeared in the Midwest Extended. Two wooden fences spanned the entire width of the valley, from slope to slope, enclosing an astounding 7,775 hectares (19,200 acres) of pasture, with a divider fence to separate winter and summer ranges. While this ranch may have been exceptional in some ways, it demonstrated the ascendance of midwestern ways in the western mountains.

Nor were other, less striking traces of midwestern influence absent. In the great cattle drives eastward to the Rockies, on both the Oregon Trail and Mullin Road, a belled cow led the way, as long before in the Carolina Piedmont (figure 54). The ancient British method of using salt as a herd-control device was employed, and range cattle in the mountain system remained rather tame, due to careful and frequent tending. All things considered, a herder of the eighteenth-century Appalachians or seventeenth-century highland Britain would have recognized Rocky Mountain cattle ranching as a familiar and kindred system.[125]

In the far greater part of the West, then, bearers of the Ohio Valley–upland-southern cattle complex prevailed in a frontier competition with the Texans and Californians. Was Missouri not, after all, called the "mother of the West?"[126] The midwestern system reached even into the very home states of the rival herding systems to challenge and ultimately replace them. The popular image of western cattle-ranching history must be revised to allow a far greater role for these highly successful, if less colorful herders from the American heartland. Their largely British-inspired system proved to be the most successful pastoral adaptive strategy in the West and, accordingly, the one that survived.

TEN

CONCLUSION

WHAT, IN THE FINAL ANALYSIS, ARE WE TO MAKE OF THIS temporal and geographical confusion of cattle-ranching frontiers, of multiple Old World sources, Antillean footholds, continental implantments, environmental settings, adaptive strategies, and repeated blendings? Can useful generalizations or principles be drawn from this seemingly chaotic array of cultures, diffusions, environments, and adaptations?

If this book has a single message, it is a warning against facile generalizations, against the assumption that a monolithic cattle-ranching frontier swept through the New World, that a single Old World prototype existed, that a particular physical environment or condition of market access housed and fostered ranching in North America. From diversity in the Atlantic fringe of the eastern hemisphere it sprang; in equal variety it found expressions in America. Simplistic acultural models such as that of Frederick Jackson Turner or Heinrich von Thünen simply do not work consistently when applied to the North American settlement frontiers, and Walter Prescott Webb's environmental determinism, similarly acultural, obscures more than it illuminates.

Rather the cattle frontier was decidedly pluralistic in character, drawing upon diverse antecedents, residing among contrasting New World cultures, entailing different local adaptations to an array of American physical settings. Each cattle frontier was unique and far more accidental than predictable, the result of chance juxtapositions of peoples and places. The multiplicity of ranching frontiers in North America is best understood in these idiographic terms. Only at hazard and amid attendant overgeneralization can the herding frontier be rendered into any sort of nomothetic explanatory model. If the true purpose of education and learning is, as I now in middle age darkly suspect, to generalize genuine chaos into artificial order, then we who study the ways of humankind should follow the likes of von

Thünen, Turner, and Webb: seek regularities or even laws of human behavior and function as scientists in a great real-world laboratory. But are we social scientists not at best foundlings before the doors of science? Does our true parentage not lie among the humanists, who tend to reject the general and seek to comprehend the unique? We would do better to approach the cattle frontiers as humble humanists than as confident scientists, if seeking the truth is our goal. No one gazing eastward from the American shores in 1492 could have predicted what happened. We should not, retroactively, pretend such powers of prognostication and explanation, for we deal with matters far more chaotic than steel balls rolling down inclined planes. It all unfolded as an unlikely series of unforeseeable accidents.

To a considerable degree, the North American cattle-ranching frontiers can be presented in terms of chance diffusion from the European and African sources. The respective importance of the Old World contributing cultures—Andalusian, Mesetan, hill British, and Sahelo-Sudanese—appears to have varied both spatially and temporally. In the Antillean episode and even as late as the 1700s in the various mainland implantments, the influence of the salt-marsh Andalusians remained disproportionately great (table 2). Compare the traits of cattle raising on the various continental ranching frontiers by 1750 with the practices known in the different sections of the Old World Atlantic fringe, from Galway to Gibraltar and Guinea, and the best match-up with widespread American practices, whether in lowland Carolina, Florida, Louisiana, or coastal Mexico, is found in Las Marismas of the Guadalquivir. Most of highland Mexico, it is true, had taken on a mesetan, or Extremeñan character, but even there the transitory cattle frontier, vanished by 1750, had been more nearly Andalusian.

To a considerable degree, cattle-ranching cultures had obeyed the flag, subject to the fickle fortunes of war. Every one of the frontiers in question, both insular and continental, save only Carolina, had at one time or another been under Spanish rule. The overwhelming importance of Sevilla and its environs as a source of early colonists, coupled with the accidentally preadapted traits of the commercialized, specialized Iberian marsh-cattle system to the tropics and subtropics of America, makes the initial Andalusian dominance understandable. Even Carolina lay close enough—genealogically, historically, and geographically—to Spanish Jamaica and Florida to receive Andalusian influences.

TABLE 2. HERDING PRACTICES IN THE OLD WORLD SOURCES AND IN THE AMERICAN MAINLAND IMPLANTMENTS, BY 1700

American trait, at least regionally	Salt Marsh Andalucía	Extremadura and Castilla	Highland Britain	African Cattle Fulani
Focus upon cattle	yes	rarely	yes	yes
Bos taurus, Iberian	yes	yes	no	no
Bos taurus, British	no	no	yes	no
Open range	yes	some	some	yes
Mounted herders	yes	some	rarely	rarely
Male herders only	yes	yes	no	yes
Herders not youths or aged	yes	no	no	no
Herders from social/professional underclass	yes	yes	yes	no
Lasso known	yes	?	no	?
Cattle staff or pike	yes	yes	no	yes
Cattle whip	no	no	yes	no
Herder dogs	no	no	yes	no
Germanic-derived livestock laws	no	no	yes	no
Branding cattle	yes	yes	some	no
Earmarking cattle	yes	yes	some	yes
Infrequent care of stock	yes	no	no	no
Transhumance/nomadism	some	yes	yes	yes
Overland drives to market	yes	some	yes	yes
Semiwild stock	yes	no	no	no
Athletic contests involving cattle	yes	yes	no	no
Range burning	yes	yes	yes	no
Haymaking	no	no	yes	no
Crops of secondary importance	yes	no	yes	yes
Market-oriented production	yes	no	yes	no
Large-scale operators	yes	few	yes	no
Beef important food	yes	no	yes	no
No consumption of cattle blood	yes	yes	no	yes
Bovine dairying unimportant	yes	yes	no	no

Change the flag to the Union Jack and the cattle-herding system reflected the shift. Jamaica became English, and its ranching took on a hybrid character, a process continued in the Carolina transplantment. Similarly, if we characterize the three Anglo-American cattle-ranching systems by 1850—Texan, Californian, and midwestern—it becomes clear that the most hispanicized was the Californian and the least the midwestern, a hierarchy that matches the chronology of extinction of Spanish and Mexican rule—1848 in California, 1800/1836 in Louisiana/Texas, and 1655 for Jamaica, a Carolinian-upland southern prototype.

Latitude and its implied ecological zonation also played a role, though far from determinant. The Andalusian marsh system was, in essence, best suited to the tropics and subtropics, well preadapted for areas lacking severe winters, where year-round grazing and foraging were available. At every point where the North American cattle-ranching frontier crossed into midlatitudinal zones afflicted with true winters—the Carolina Piedmont and Appalachians, the Great Plains, and the cold deserts of the intermontane West—the Andalusian-derived system faltered and failed. At each of these points, British-inspired pastoral influences replaced those of Iberian origin. The most crucial such metamorphosis occurred in the southern Appalachians, where the process was aided by immigrant British herders. The resultant upland-southern system, transplanted to the midwestern prairies, became the basis for modifying the Texas and California systems when their bearers blundered out of warm lotuslands into the wintry reaches of the continent. British ways allowed cattle ranching to survive and play an enduring role in the postfrontier West.

The role of Africans in all these events remains unclear and enigmatic. Did the African pastoralist-slave play a meaningful role in shaping one or more of the American cattle frontiers? Certainly their blood ran in the veins of vaqueros and even some ranchers in the Antilles, Mexico, Texas, Carolina, Florida, and Louisiana, just as it had in pre-Columbian times among the herders working in Las Marismas. A scattering of freedman African progeny labored on ranges as far afield as Idaho and Alberta. But were these black and mulatto cowboys of any greater cultural consequence than, say, the Chinese or Amerindians who served in the ranch work force in California Extended? No flag reinforced African influence, save that raised by the slave rebels in Haiti, and that hoisting came long after the ranching frontier had vanished from the Indies. I find no compelling evidence of meaningful African influence in the cultures and adaptive systems of the various American cattle frontiers (table 2). Perhaps a loanword or two survived the dreadful transition from Sahelo-Sudanic Africa to America, but little else. Seek the vivid New World impress of the African in plantation culture, not on the ranching frontier. If any substantial African influence is to be found in ranching, it entered Moorish Las Marismas and Andalucía with black slaves before the time of Columbus—a topic deserving additional research.

One cannot but be struck by how many North American ranching traits derived from the Old World Atlantic fringe, by the surpassing importance of diffusion. Events in America generally involved a mere exchange and reshuffling of preadapted British and Iberian practices. Frontier inventiveness remained rather minimal. Still, innovation occurred on all the cattle frontiers, causing American ranching to take on a character somewhat different from herding in the homelands across the Atlantic.

On the Spanish colonial frontiers, initially in the face of royal opposition, the private pastoral estate arose, beginning in the Antillean episode and culminating in the period of Mexican rule in the American Southwest. Similarly ecclesiastical ranching estates, part of the mission system, represented an innovation crucial to the cattle frontier in the Mesa Central, Pacific coast, and Florida. Herding techniques also underwent changes and refinement. Most notably American vaqueros developed and perfected the technique of casting the ancient Roman lasso from horseback, probably in seventeenth-century Mexico. This innovation, in turn, led to a lengthy and continuous evolution of the range saddle into forms unknown in the Old World. Indeed virtually every form element of equestrian material culture underwent modification in Spanish North America, a process still underway on the regional level in, for example, early nineteenth-century California. In the process of these Hispanic-American modifications, many old Iberian words such as *hato, estancia, reata, rodeo,* and *charro* took on rather different New World meanings, and a rich new ranching vocabulary was coined.

British-Americans, too, engaged in invention and modification on their pastoral frontiers. Anglo-Texans developed the double-cinched saddle and, more important, the stiff-countered, pointed-toe, high-heeled riding boot. Their horses, improbably, were taught to maneuver like herding dogs in managing cattle, and a race with mutton-eating ancestors learned to despise sheep. Midwestern Anglo cattle raisers, grown accustomed to the hay yields of the rank wet prairies of Indiana, Illinois, and Iowa, learned to create lesser artificial wet meadows all through the semiarid West by means of irrigation. Increasing their hay production in response to the more prolonged winters of the western mountains, these heartland herders invented an array of stackers and other haying equipment.

In that portion of North America traditionally (and still today) most closely associated with ranching—the West—the cattle frontier in its final and climactic episode witnessed a collision and competition for territory between three related, yet distinctive pastoral systems. Born of Old World diversity, the ranching frontier ended its days, fittingly, in a contest for survival of the fittest among three equally diverse herding cultures. Each of the three—Texan, Californian, and midwestern—possessed diagnostic traits that allow the latter-day observer to decipher who went where, when, and with what staying power. Recognize the Texans by their double-cinched saddle, hemp "lariat," "Nocona" boot, and canine-inspired horse; in their techniques of line riding, "cutting out," and tongue slitting; and by their use of words such as *maverick, cowboy,* and *cavvyyard.* Distinguish the Californians by their well-tooled, single-cinched Visalia saddles, extraordinary skills with the overly long rawhide "riata," abundant use of Hispanic "buckaroos," beef wheels, and other distinctive vocabulary such as *taps, major domo,* and *theodore.* Discern the midwesterners in their preference for British cattle breeds, obsession with haying and its paraphernalia, desire to build pasture fences and sizable barns, tendency to irrigate meadows and form stock raisers' associations, and practice of carefully tending cattle while instilling herd docility.

The outcome of this contest for the last cattle frontier was by no means simple. While the midwesterners should be judged the winners, conquering even the strongholds of their rivals, the mixing and borrowing and tinkering went on to the very end. The cattle rancher today in, say, Idaho, while speaking the dialect of Iowa and still fearful of what the next winter might bring, wears Texan boots and casts a Texan fiber lariat; he utters certain ranch words derived from Spanish California and mounts a Visalia inspired saddle.

Even so, nothing approaching cultural homogenization was ever achieved in the ranching West, even to the present day. Nothing like a single ranching way of life was imposed on that huge region. Even the most cursory modern field traverse reveals obvious regional differences. The high-valley rancher of Montana is not like the West Texan, nor do either duplicate the Cypress Hills cattle folk of Canada or the cattle raisers of Nevada's High Desert. Modern ranching in its postfrontier condition continues to reflect the multiplicity of ancestral American cattle frontiers and even the still more remote diversity of Old World roots. The spirit and uniqueness of region and place prevail now as then, confounding the model builders and lawgivers.

Notes

Chapter One

(See Bibliography for list of abbreviations.)

1. Simon M. Evans, "The Origins of Ranching in Western Canada: American Diffusion or Victorian Transplant?" *Great Plains Quarterly*, 3:2 (1983), p. 80 (quote); Marshall W. Fishwick, "The Cowboy: America's Contribution to the World's Mythology," *Western Folklore*, 11 (1952), pp. 77–92; Mody C. Boatright, "The American Myth Rides the Range: Owen Wister's Man on Horseback," *Southwest Review*, 36 (Summer 1951), pp. 157–63; Joe B. Frantz and Julian E. Choate, Jr., *The American Cowboy: The Myth and the Reality* (Norman, OK, 1955); Lewis Atherton, "Cattlemen and Cowboy: Fact and Fancy," *Montana*, 11 (Oct. 1961), pp. 2–17; Jack Weston, *The Real American Cowboy* (New York, NY, 1985), pp. 209–52; Don D. Walker, *Clio's Cowboys: Studies in the Historiography of the Cattle Trade* (Lincoln, NE, 1981).

2. Lewis Atherton, *The Cattle Kings* (Bloomington, IN, 1961); Arnold Strickon, "The Euro-American Ranching Complex," in Anthony Leeds and Andrew P. Vayda, eds., *Man, Culture, and Animals* (Washington, DC, 1965), pp. 229–32; Bernd Andreae, *Betriebsformen in der Landwirtschaft* (Stuttgart, Germany, 1964), pp. 181–84; John W. Bennett, *Northern Plainsmen: Adaptive Strategy and Agrarian Life* (Chicago, IL, 1969), p. 194.

3. Richard W. Slatta, *Cowboys of the Americas* (New Haven, CT, 1990); William W. Savage, Jr., ed., *Cowboy Life: Reconstructing an American Myth* (Norman, OK, 1975); Strickon, "Euro-American Ranching," pp. 229, 241; Fishwick, "Cowboy," pp. 77, 81; Boatright, "American Myth," p. 157.

4. Otto Jessen, "Cosacos, cowboys, gauchos, boers y otros pueblos a caballos propios de las estepas," *Runa*, 5:1/2 (1952), pp. 171–86.

5. Walter P. Webb, *The Great Plains* (Boston, MA, 1931); Théodore Monod, ed., *Pastoralism in Tropical Africa* (London, 1975), pp. 103–6; Strickon, "Euro-American Ranching"; Bennett, *Northern Plainsman*, p. 194.

6. William A. Galbraith and E. William Anderson, "Grazing History of the Northwest," *Journal of Range Management*, 24 (1971), p. 6 (quote); Lincoln Ellison, "Influence of Grazing on Plant Succession of Rangelands," *Botanical Review*, 26 (1960), pp. 4, 5, 61–65; William C. Robison, "Grazing

as a Natural Ecological Factor," *GR*, 51 (1961), pp. 308–10; J. L. Davidson, "Growth of Grazed Plants," *Australian Grasslands Conference 1968, Proceedings* (Melbourne, Australia, 1969), pp. 126, 134.

7. D. B. Edmond, "The Influence of Animal Treading on Pasture Growth," *Proceedings of the X International Grasslands Congress* (Helsinki, 1966), pp. 453–54; P. J. Vickery, "Grazing and Net Primary Production of a Temperate Grassland," *Journal of Applied Ecology*, 9 (1972), pp. 307–14; M. B. Alcock, "The Physiological Significance of Defoliation on the Subsequent Regrowth of Grass-Covered Mixtures and Cereals," in D. J. Crisp, ed., *Grazing in Terrestrial and Marine Environments* (Oxford, UK, 1964), pp. 25–41; Walter Deshler, "Cattle in Africa: Distribution, Types, and Problems," *GR*, 53 (1963), p. 54; L. T. Burcham, *California Range Land: An Historico-Ecological Study of the Range Resource of California* (Sacramento, CA, 1957), pp. 176–214; William B. Drew, "Floristic Composition of Grazed and Ungrazed Prairie Vegetation in North-Central Missouri," *Ecology*, 28 (1947), pp. 26–41; Jeffrey A. Gritzner, *The West African Sahel: Human Agency and Environmental Change* (Chicago, IL, 1988), p. 55; Monod, *Pastoralism*, pp. 109–10; Ellison, "Influence of Grazing," pp. 3, 48, 54, 55, 63–64; Robison, "Grazing as Natural," p. 309; Davidson, "Growth of Grazed," pp. 125, 128.

8. Johann Heinrich von Thünen, *Von Thünen's Isolated State*, trans. Carla M. Wartenberg (Elmsford, NY, 1966); Richard Peet, "The Spatial Expansion of Commercial Agriculture in the Nineteenth Century: A von Thünen Explanation," *EG*, 45 (1969), pp. 283–301; Ursula Ewald, "The von Thünen Principle and Agricultural Zonation in Colonial Mexico," *JHG*, 3 (1977), pp. 123–33; Ernst Griffin, "Testing the von Thünen Theory in Uruguay," *GR*, 63 (1973), pp. 500–16; Terry G. Jordan, "The Origin and Distribution of Open-Range Cattle Ranching," *Social Science Quarterly*, 53 (1972), pp. 105–21; Terry G. Jordan, *Trails to Texas: Southern Roots of Western Cattle Ranching* (Lincoln, NE, 1981), pp. 18–23.

9. Frederick J. Turner, "The Significance of the Frontier in American History," *Annual Report of the American Historical Association*, 1893, p. 208.

10. Terry G. Jordan and Matti Kaups, *The American Backwoods Frontier: An Ethnic and Ecological Interpretation* (Baltimore, MD, 1989).

11. Terry G. Jordan, "Preadaptation and European Colonization in Rural North America," *AAAG*, 79 (1989), pp. 389–500.

12. Jo T. Bloom, "Cumberland Gap versus South Pass: The East or West in Frontier History," *WHQ*, 3 (1972), pp. 153–67.

13. Exhibits at the Israel Museum, Jerusalem, on the Canaanite sites at Tel Naglia and Shiloh; exhibits at the Egyptian Museum, Cairo, particularly those associated with the Dendera temple site; bas relief at the monuments in the Valley of the Queens, Thebes West, near Luxor in Upper Egypt; murals in the Mastaba of Ti and tomb of Manufer, Saqqara, ancient Memphis, Lower Egypt; murals in tomb no. 3 at Beni Hassan Necropolis, near El-Minya,

Egypt; exhibits at the Archaeological Museum of Iraklion, Crete; Manfred R. Wolfenstine, *The Manual of Brands and Marks* (Norman, OK, 1970), pp. 3, 4, and plate 1.

14. E. Estyn Evans, "The Atlantic Ends of Europe," *Advancement of Science*, 15 (1958), p. 54.

15. Evans, "Atlantic Ends," p. 55.

Chapter Two

1. The section on Iberian origins appeared earlier, in preliminary form, as Terry G. Jordan, "An Iberian Lowland/Highland Model for Latin American Cattle Ranching," *JHG*, 15 (1989), pp. 111–25.

2. C. Julian Bishko, "The Peninsular Background of Latin American Cattle Ranching," *HAHR*, 32 (1952), pp. 491–515.

3. Bishko, "Peninsular Background," pp. 494–97.

4. William E. Doolittle, "Las Marismas to Pánuco to Texas: The Transfer of Open Range Cattle Ranching from Iberia through Northeastern Mexico," *Yearbook of the Conference of Latin Americanist Geographers*, 13 (1987), pp. 3–11; Karl W. Butzer, "Cattle from Old to New Spain," *AAAG*, 78 (1988), pp. 29–56; Charles W. Towne and Edward N. Wentworth, *Cattle and Men* (Norman, OK, 1955), p. 103.

5. Otto Jessen, "Südwest-Andalusien," *Petermanns Mitteilungen, Ergänzungsheft*, 186 (1924), p. 34; André Vialles, "Camargue, the Cowboy Country of Southern France," *National Geographic*, 42 (1922), pp. 1–34; E. H. G. Dobby, "The Ebro Delta," *GJ*, 87 (1936), p. 456; a good general source on such marshlands is David R. Stoddart, ed., *Salt Marshes and Coastal Wetlands* (New York, NY, 1990).

6. S. Rivas-Martínez et al., "La vegetación de Doñana," *Lazaroa*, 2 (1980), pp. 43–45; Juan F. Ojeda Rivera, *Organización del territorio en Doñana y su entorno próximo (Almonte), siglos XVIII–XX* (Sevilla, Spain, 1987), pp. 17–25; M. de Terán, ed., *Geografía de España y Portugal* (Barcelona, Spain, 1952–67), I, p. 462; II, p. 125; Jessen, "Südwest-Andalusien," pp. 25–27, 34–35.

7. M. Granados Corona, et al., "Etude diachronique d'un écosystème à longue échelle: la pinède de Marismillas," *Mélanges de la Casa de Velázquez*, 20 (1984), p. 408; J. Valverde Alvarez, "El paisaje y los modos de vida en Sanlúcar de Barrameda," *Geográphica*, (1959), pp. 72–75; Rivas-Martínez, "La vegetación," pp. 141–60; Jessen, "Südwest-Andalusien," pp. 12–20, 36–39; Ojeda, *Organización del territorio*, pp. 17–27.

8. Jessen, "Südwest-Andalusien," pp. 35–36, 40.

9. Paul Gwynne, *The Guadalquivir: Its Personality, Its People and Its Associations* (London, UK, 1912), following p. 240; Richard W. Slatta, *Cowboys of the Americas* (New Haven, CT, 1990), p. 1.

10. Mercedes Borrero Fernández, *El mundo rural sevillano en el siglo XV: Aljarafe y Ribera* (Sevilla, Spain, 1983), p. 447.

11. *Censo de la ganadería de España segun el recuento verificado en 24 de Setiembre 1865* (Madrid, Spain, 1868), p. 202.

12. Gregorio de Tapia y Salzedo, *Exercicios de la gineta al príncipe nuestro Señor d. Baltasar Carlos* (Madrid, Spain, 1643), pp. 72–74; Fred B. Kniffen, "The Western Cattle Complex: Notes on Differentiation and Diffusion," *Western Folklore*, 12 (1953), p. 179; Glenn R. Vernam, *Man on Horseback* (New York, NY, 1964), pp. 107, 159, 194, 197, 238–39, 245; Richard E. Ahlborn, ed., *Man Made Mobile: Early Saddles of Western North America* (Washington, DC, 1980), p. 11; Harold B. Barclay, *The Role of the Horse in Man's Culture* (London, UK, and New York, NY, 1980), p. 212; Vialles, "Camargue," pp. 15–19; *Diccionario de la lengua castellana por la Real Academia Española*, 5th ed. (Madrid, Spain, 1817), p. 520; J. Corominas, *Diccionario crítico etimológico de la lengua castellana* (Madrid, Spain, 1954), III, p. 59; Towne and Wentworth, *Cattle and Men*, follows p. 50.

13. Luis L. Cortés y Vázquez, "Ganadería y pastoreo en Berrocal de Huebra," *Revista de Dialectología y Tradiciones Populares*, 8 (1952), p. 429; David E. Vassberg, *Land and Society in Golden Age Castille* (New York, NY, 1984), pp. 39–40; C. Julian Bishko, "The Andalusian Municipal Mestas in the 14th–16th Centuries: Administrative and Social Aspects," in *Andalucía Medieval* (Córdoba, Spain, 1978), vol. I, p. 371; Terry G. Jordan, *Trails to Texas: Southern Roots of Western Cattle Ranching* (Lincoln, NE, 1981), pp. 93–94; Granados, "Etude diachronique," p. 412; Borrero, *El mundo rural*, pp. 96–98.

14. Manuel Alvar, A. Llorente, and G. Salvador, *Atlas lingüístico y etnográfico de Andalucía* (Granada, Spain, 1961–65), II, plates 451, 486; Julio González, *Repartimiento de Sevilla* (Madrid, Spain, 1951), II, pp. 506–8; Bishko, "Peninsular Background," pp. 504–5.

15. A. Matilla Tascón, *La única contribución y el catastro de la ensenada* (Madrid, Spain, 1947), pp. 531–32; Alfonso Franco Silva, *El concejo de Alcalá de Guadaira a finales de la edad media* (Sevilla, Spain, 1974), p. 71; Eduardo Camacho Rueda, *Propiedad y explotación agrarias en el Aljarafe sevillano: el caso de Pilas, 1760–1925* (Sevilla, Spain, 1984), pp. 147–56; *Censo de la ganadería*, p. 142; Borrero, *El mundo rural*, p. 447; Ojeda, *Organización del territorio*, p. 294; Bishko, "Andalusian Municipal," p. 372.

16. Borrero, *El mundo rural*, pp. 316–17, 447; Franco, *El concejo de Alcalá*, pp. 69, 102; *Censo de la ganadería*, p. 205; Camacho, *Propiedad*, pp. 84–90.

17. Ojeda, *Organización del territorio*, pp. 77–79; Borrero, *El mundo rural*, pp. 34–56.

18. Manuel González Jiménez, "Aspectos de la economía rural andaluza en el siglo XV," in *Huelva en la Andalucía del siglo XV* (Huelva and Sevilla, Spain, 1976), p. 25; J. M. Borbujo, "Los toros guadalquiveños de las Marismas

bajas," in J. Rubiales et al., eds., *El río: el bajo Guadalquivir* (Madrid, Spain, 1985), pp. 151–54; Antonio Sancho Corbacho, "Haciendas y cortijos sevillanos," *Archivo Hispalense*, (1952), pp. 10–13; Vassberg, *Land and Society*, pp. 65, 90, 100, 169.

19. Antonio Alcalá Venceslada, *Vocabulario andaluz* (Madrid, Spain, 1951), p. 266; *Spanien 1:50 000* (map), [German] General-Stab des Heeres, 2nd ed., July 1941, sheets 1033–35, 1047–49; Francisco J. Santamaría, *Diccionario general de americanismos* (Mexico City, Mexico, 1942), I, p. 627.

20. *Diccionario de la lengua castellana*, pp. 393, 459; Bishko, "Peninsular Background," p. 511; Franco, *El concejo de Alcalá*, p. 72.

21. Sancho, "Haciendas y cortijos," pp. 12–13; Alcalá, *Vocabulario andaluz*, p. 314; Bishko, "Andalusian Municipal," pp. 359, 360, 365.

22. Alcalá, *Vocabulario andaluz*, p. 523.

23. *Spanien 1:50 000* (map).

24. Ojeda, *Organización del territorio*, p. 48; Borrero, *El mundo rural*, pp. 95, 96, 316; Vassberg, *Land and Society*, 33–34, 57, 67, 92; Bishko, "Andalusian Municipal," pp. 350, 351, 366.

25. Ruth Pike, "Sevillian Society in the Sixteenth Century: Slaves and Freedmen," *HAHR*, 47 (1967), pp. 344–46; Magnus Mörner, *Race Mixture in the History of Latin America* (Boston, MA, 1967), p. 16; Bishko, "Andalusian Municipal," pp. 364–65; Borrero, *El mundo rural*, pp. 95–96, 318; Franco, *El concejo de Alcalá*, pp. 69, 72, 101.

26. Bishko, "Andalusian Municipal," pp. 360, 371–73.

27. Antonio Collantes de Terán Sánchez, "La tierra de Huelva en el siglo XV," in *Huelva en la Andalucía del siglo XV* (Huelva and Sevilla, Spain, 1976), p. 47; Miguel A. Ladero Quesada, "Los señoríos medievales onubenses," ibid., p. 86; González, "Aspectos de la economía," p. 18; Borrero, *El mundo rural*, pp. 187–99, 437; Ojeda, *Organización del territorio*, p. 48; Vassberg, *Land and Society*, pp. 155–57; Camacho, *Propiedad*, p. 131; Bishko, "Andalusian Municipal," p. 363.

28. González, "Aspectos de la economía," p. 23.

29. Borrero, *El mundo rural*, pp. 34–56, 347–49, 393–94, 449–50; Bishko, "Andalusian Municipal," pp. 367–70.

30. Peter Boyd-Bowman, *Indice geobiográfico de cuarenta mil pobladores españoles de América en el siglo XVI* (Bogotá, Colombia, and Madrid, Spain, 1964–68), I, II.

31. Peter Boyd-Bowman, "Patterns of Spanish Emigration to the Indies until 1600," *HAHR*, 58 (1976), p. 603.

32. C. Julian Bishko, "The Castilian as Plainsman," in A. R. Lewis and T. F. McGann, eds., *The New World Looks at Its History* (Austin, TX, 1963), pp. 48–49.

33. Terán, *Geografía de España*, II, pp. 6–7, 280; Oskar Schmieder, "Die Sierra de Gredos," *Mitteilungen der Geographischen Gesellschaft in München*, 10 (1915), pp. 49–51.

34. Miguel A. Ladero Quesada, "Algunos datos para la historia económica de las ordenes militares de Santiago y Calatrava en el siglo XV," *Hispania*, 30 (1970), pp. 637–62; Matilla, *La única*, pp. 531–32.

35. *Censo de la ganadería*, p. 196.

36. James J. Parsons, "The Acorn–Hog Economy of the Oak Woodlands of Southwestern Spain," *GR*, 52 (1962), p. 215; Miguel Muñoz de San Pedro, *Extremadura* (Madrid, Spain, 1961), p. 19; Julius Klein, *The Mesta: A Study in Spanish Economic History* (Cambridge, MA, 1920).

37. David E. Vassberg, "Concerning Pigs, the Pizarros, and the Agro-Pastoral Background of the Conquerors of Peru," *Latin American Research Review*, 13:3 (1978), pp. 47–62; Evelio Teijón Laso, "Los modos de vida en la dehesa salmantina," *Estudios Geográficos*, 9 (1948), pp. 436–37; Muñoz, *Extremadura*, p. 19.

38. *Censo de la ganadería*, p. 197; Matilla, *La única*, pp. 531–32.

39. F. Santos Coco, "Vocabulario extremeño," *Revista del Centro de Estudios Extremeños*, 14 (1940), pp. 150–62; Ladero, "Algunos datos," pp. 644–47; Parsons, "Acorn–Hog Economy," p. 215; Vassberg, "Concerning Pigs," p. 47.

40. A. Gil Crespo, "Hoyos del Espino," *Boletín de la Real Sociedad Geográfica*, 97 (1961), p. 187; Gisela Fiedler, *Kulturgeographische Untersuchungen in der Sierra de Gredos/Spanien* (Würzburg, Germany, 1970), pp. 59, 173; *Spanien 1:50 000* (map), sheet 754; Teijón, "Los modos de vida," pp. 433–34.

41. Bishko, "Castilian as Plainsman," pp. 48, 64.

42. *Censo de la ganadería*; Fiedler, *Kulturgeographische*, pp. 173, 178, 281–85; Schmieder, "Sierra de Gredos," p. 58.

43. Teijón, "Los modos de vida," pp. 423, 433–34; Vassberg, *Land and Society*, p. 13.

44. Thomas J. Abercrombie and Bruno Barbey, "Extremadura: Cradle of Conquerors," *National Geographic*, 179:4 (1991), pp. 118, 134; A. Gil Crespo, "La mesta de carreteros del reino," *Anales de la Asociación Española para el Progreso de las Ciencias*, 22 (1957), pp. 207–30; Barbara Dresser, "The Sierra de Gredos," *SGM*, 74 (1958), pp. 178, 180; Gil, "Hoyos del Espino," pp. 188–90; Fiedler, *Kulturgeographische*, pp. 82–84, 91, 174.

45. Klein, *The Mesta*.

46. Cortés, "Ganadería y pastoreo," p. 427; *Spanien 1:50 000* (map), sheets 778, 802; Teijón, "Los modos de vida," p. 434 and fig. 3, following p. 441; *Censo de la ganadería*, pp. 22, 36, 205; Gil, "Hoyos del Espino," p. 195; Abercrombie and Barbey, "Extremadura," pp. 120–21, 124.

47. Teodoro Marañón, "Agro-Sylvo-Pastoral Systems in the Iberian Peninsula: *Dehesas* and *Montados*," *Rangelands*, 10 (Dec. 1988), pp. 255–58; Muñoz, *Extremadura*, p. 19; Teijon, "Los modos de vida," pp. 424–33.

48. Boyd-Bowman, *Indice geobiográfico*; Vassberg, *Land and Society*, p. 53.

49. Ladero, "Algunos datos," pp. 644–45; *Censo de la ganadería*, pp. 34–35.

50. Cortés, "Ganadería y pastoreo," pp. 427–30, 434, 437–44, 455–56, 590; José de Lamano y Beneite, *El dialecto vulgar salamantino* (Salamanca, Spain, 1915), pp. 598, 620; Tapia y Salzedo, *Exercicios*, p. 27.

51. Michael R. Weisser, *The Peasants of the Montes* (Chicago, IL, 1976), pp. 38–42, 45, 131; Bishko, "Castilian as Plainsman," p. 54; Vassberg, "Concerning Pigs," p. 50.

52. Grady McWhiney, *Cracker Culture: Celtic Ways in the Old South* (Tuscaloosa, AL, 1988), pp. 51–79; Grady McWhiney and Forrest McDonald, "Celtic Origins of Southern Herding Practices," *JSH*, 51 (1985), pp. 165–82; Jordan, *Trails to Texas*, pp. 8–10, 27–28, 31, 35–37; John S. Otto, "The Origins of Cattle-Ranching in Colonial South Carolina, 1670–1715," *South Carolina Historical Magazine*, 87 (1986), pp. 117–24; Ray August, "Cowboys v. Rancheros: The Origins of Western American Livestock Law," *SWHQ*, 96 (1993) forthcoming.

53. E. Estyn Evans, *Irish Folk Ways* (London, UK, 1957), p. 33; E. Estyn Evans, "The Atlantic Ends of Europe," *Advancement of Science*, 15 (1958), p. 56; R. N. Millman, *The Making of the Scottish Landscape* (London, UK, 1975), p. 48; K. B. Cumberland, "Livestock Distribution in Craven," *SGM*, 54 (1938), pp. 79–85; H. C. Darby and R. Welldon Finn, *The Domesday Geography of South-West England* (Cambridge, UK, 1967), pp. 290–94, 342; E. H. Carrier, *The Pastoral Heritage of Britain* (London, UK, 1936), following p. 32.

54. Alexander Fenton, "The Traditional Pastoral Economy," in M. L. Parry and T. R. Slater, eds., *The Making of the Scottish Countryside* (London, UK, 1980); Robert Trow-Smith, *A History of British Livestock Husbandry, 1700–1900* (London, UK, 1959); R. H. Kinvig, *The Isle of Man: A Social, Cultural, and Political History* (Liverpool, UK, 1975); Pierre Flatrès, *Géographie rurale de quatre contrées celtiques: Irlande, Galles, Cornwall & Man* (Paris, France, 1957), pp. 103–21; A. K. Hamilton Jenkin, *Cornwall and Its People* (London, UK, 1945), pp. 379–80; E. Estyn Evans, *The Personality of Ireland: Habitat, Heritage and History* (Cambridge, UK, 1973), p. 48; John O'Donovan, *The Economic History of Live Stock in Ireland* (Dublin and Cork, Ireland, 1940), pp. 13–15; Fynes Moryson, *An Itinerary Containing His Ten Yeeres Travell* (Glasgow, UK, 1907–8), IV, p. 193; Evans, "Atlantic Ends," pp. 54–55; Darby and Finn, *Domesday Geography*; Carrier, *Pastoral Heritage*, pp. 47–58.

55. Edmund Spenser, "A View of the State of Ireland," in James Ware, ed., *Two Histories of Ireland* (Dublin, Ireland, 1633), p. 35 (first two quotes); Evans, *Personality of Ireland*, p. 38 (third quote).

56. A. E. Smailes, *North England* (London and Edinburgh, UK, 1960), p. 81 (first quote); Trow-Smith, *History British Livestock*, pp. 3–4 (second quote).

57. Trow-Smith, *History British Livestock*, pp. 7–8, 18; Jenkin, *Cornwall*, pp. 379–80; Fenton, "Traditional Pastoral," p. 94.

58. Millman, *Making Scottish Landscape*, p. 101.

59. G. Whittington and I. D. Whyte, eds., *An Historical Geography of Scotland* (London, UK, 1983), p. 63; Millman, *Making Scottish Landscape*, pp. 99–100; Evans, "Atlantic Ends," p. 55; Fenton, "Traditional Pastoral," pp. 96, 106.

60. Horace Fairhurst, "Scottish Clachans," *SGM*, 76 (1960), pp. 67–76; V. B. Proudfoot, "Clachans in Ireland," *Gwerin*, 2 (1959), pp. 110–22; Gerhard Bersu, "Celtic Homesteads in the Isle of Man," *Journal of the Manx Museum* (1945–46), pp. 177–85; B. M. Swainson, "Rural Settlement in Somerset," *G*, 20 (1935), p. 123; Smailes, *North England*, p. 81; Evans, *Personality of Ireland*, pp. 50–57; Flatrès, *Géographie rurale*, pp. 88–89. One of the best surviving examples of a ring fort, perhaps twenty-five hundred years old, is Staigue Fort in County Kerry, southwestern Ireland.

61. A. R. B. Haldane, *The Drove Roads of Scotland* (London, UK, 1952), p. 7; Fenton, "Traditional Pastoral," pp. 94–95, 103; Evans, *Personality of Ireland*, pp. 38, 48, 57; O'Donovan, *Economic History*, p. 18.

62. K. J. Bonser, *The Drovers, Who They Were, and How They Went: An Epic of the English Countryside* (London, UK, 1970), pp. 50–52; Fenton, "Traditional Pastoral," p. 103; Cumberland, "Livestock Distribution," p. 81; William Thompson, "Cattle Droving between Scotland and England," *Journal of the British Archaeological Association*, n.s. 37 (1932), p. 173; Carrier, *Pastoral Heritage*, pp. 164–78.

63. Philip S. Robinson, *The Plantation of Ulster: British Settlement in an Irish Landscape, 1600–1670* (Dublin, Ireland, 1984), p. 35; O'Donovan, *Economic History*, p. 26; Smailes, *North England*, p. 65; Millman, *Making Scottish Landscape*, p. 100; Haldane, *Drove Roads*, p. 141; G. Kenneth Whitehead, *The Ancient White Cattle of Britain and Their Descendants* (London, UK, 1953).

64. Trow-Smith, *History British Livestock*, pp. 5, 9; O'Donovan, *Economic History*, p. 20; Alexander Fenton, "Cattle," in his *The Northern Isles: Orkney and Shetland* (Edinburgh, UK, 1978), p. 429.

65. Trow-Smith, *History British Livestock*, p. 4.

66. Bonser, *The Drovers*, p. 52; Moryson, *An Itinerary*, IV, p. 193.

67. Robinson, *Plantation of Ulster*, p. 35; Trow-Smith, *History British Livestock*, pp. 5, 10; Whittington and Whyte, *Historical Geography*, p. 65.

68. Malcolm D. MacSween, "Transhumance in North Skye," *SGM*, 75 (1959), p. 84; A. J. Kayll, "Moor Burning in Scotland," *Proceedings, Tall Timbers Fire Ecology Conference*, 6 (1967), pp. 28–39; Smailes, *North England*, p. 64; Millman, *Making Scottish Landscape*, pp. 48, 101; Cumberland, "Livestock Distribution," pp. 79–85, 89.

69. Moryson, *An Itinerary*, IV, p. 193.

70. Seamus Ó Duilearga, "Mountain Shielings in Donegal," *Béaloideas*, 9

(1939), pp. 295–97; Peter S. Gelling, "Medieval Shielings in the Isle of Man," *Medieval Archaeology*, 6/7 (1962/63), pp. 156–72; Mrs. J. M. Graham, "Transhumance in Ireland," *Advancement of Science*, 10 (1953), pp. 74–79; E. Cecil Curwen, "The Hebrides: A Cultural Backwater," *Antiquity*, 12 (1938), pp. 273–80; William A. Hance, "Crofting in the Outer Hebrides," *EG*, 28 (1952), p. 45; Michael Williams, *The Draining of the Somerset Levels* (Cambridge, UK, 1970), pp. xiii, 14–15, 32–33; Swainson, "Rural Settlement," pp. 116–17; E. Estyn Evans, *Mourne Country: Landscape and Life in South Down* (Dundalk, Ireland, 2nd ed., 1967), pp. 133–41; Evans, *Irish Folk Ways*, pp. 34–37; Evans, *Personality of Ireland*, pp. 79–80; Evans, "Atlantic Ends," p. 55; Fenton, "Traditional Pastoral," pp. 95, 99–101; Robinson, *Plantation of Ulster*, pp. 27, 33–34, 187; Jenkin, *Cornwall*, pp. 380–81; MacSween, "Transhumance," pp. 75–88; Smailes, *North England*, p. 65; Millman, *Making Scottish Landscape*, p. 100; Trow-Smith, *History British Livestock*, pp. 5, 10; O'Donovan, *Economic History*, p. 47; Spenser, "View of State," (1633), p. 35; F. H. A. Aalen, "Transhumance in the Wicklow Mountains," *Ulster Folklife*, 10 (1964), pp. 65–72; Norman J. G. Pounds, "Note on Transhumance in Cornwall," *G*, 27 (1942), p. 34; T. H. Bainbridge, "A Note on Transhumance in Cumbria," *G*, 25 (1940), pp. 35–36; F. H. A. Aalen, "Clochans as Transhumance Dwellings in the Dingle Peninsula, County Kerry," *Journal of the Royal Society of Antiquaries of Ireland*, 94 (1964), pp. 39–45; Carrier, *Pastoral Heritage*, pp. 59–65.

71. McWhiney and McDonald, "Celtic Origins," pp. 167–70.

72. Isobel M. L. Robertson, "The Head-Dyke: A Fundamental Line in Scottish Geography," *SGM*, 65 (1949), pp. 6, 14; Parry and Slater, *Making Scottish Countryside*, pp. 93–94; Fenton, "Traditional Pastoral," p. 95; Cumberland, "Livestock Distribution," p. 79.

73. Alexander Carmichael, "Grazing and Agrestic Customs of the Outer Hebrides," *Celtic Review*, 10 (1914/16), p. 46; E. Estyn Evans, *Irish Heritage: The Landscape, The People and Their Work* (Dundalk, Ireland, 1944), p. 55; Fenton, "Traditional Pastoral," p. 96; Aalen, "Transhumance," p. 69; Whittington and Whyte, *Historical Geography*, p. 67; Moryson, *An Itinerary*, IV, p. 193; Williams, *Draining*, pp. 32–33; Haldane, *Drove Roads*, pp. 7–9, following p. 162; Thompson, "Cattle Droving," p. 174.

74. Fenton, "Traditional Pastoral," pp. 94, 100; Haldane, *Drove Roads*, following p. 166.

75. Xavier de Planhol, "Le chien de berger: développement et signification géographique d'une technique pastoral," *Bulletin de l'Association de Géographes Français*, 370 (1969), pp. 355–68; Caroline Skeel, "The Cattle Trade between England and Wales from the Fifteenth to the Nineteenth Centuries," *Transactions of the Royal Historical Society*, 4th series, 9 (1926), p. 155; Bonser, *The Drovers*, pp. 25, 35–37; Haldane, *Drove Roads*, pp. 26–27; Evans, *Personality of Ireland*, pp. 11–12.

76. Georges Lutz, "Catahoula hog dog ou Catahoula cur: une opinion

européenne ou le retour aux origines du chien," *Journal d'Agriculture Traditionelle et de Botanique Appliquée,* 31 (1984), p. 157.

77. Bailey C. Hanes, *Bill Pickett, Bulldogger: The Biography of a Black Cowboy* (Norman, OK, 1977), p. 4.

78. McWhiney and McDonald, "Celtic Origins," pp. 171-72, to the contrary notwithstanding; Trow-Smith, *History British Livestock,* p. 4.

79. Moryson, *An Itinerary,* IV, p. 200 (quotation); Kinvig, *Isle of Man,* p. 124; Fenton, "Traditional Pastoral," p. 96; Trow-Smith, *History British Livestock,* p. 5; Fenton, "Cattle," p. 428; Hance, "Crofting," p. 45; Curwen, "Hebrides," p. 266.

80. Edmund Spenser, *A View of the State of Ireland as It was in the Reign of Queen Elizabeth* (Dublin, Ireland, 1763), pp. 251-52; Evans, *Irish Heritage* p. 55; R[ichard] Bradley, *The Gentleman and Farmer's Guide for the Increase and Improvement of Cattle* (London, UK, 1732), p. 75; Ian D. Whyte, *Agriculture and Society in Seventeenth-Century Scotland* (Edinburgh, UK, 1979), p. 83; Manfred R. Wolfenstine, *The Manual of Brands and Marks* (Norman, OK, 1970), pp. 4, 5, plate 61; Haldane, *Drove Roads,* pp. 10, 74; Ian Whyte, *Agriculture and Society in Seventeenth-Century Scotland* (Edinburgh, UK, 1979), p. 83.

81. August, "Cowboys v. Rancheros"; Wolfenstine, *Manual,* p. 248.

82. Harold Orton and Nathalia Wright, *A Word Geography of England* (London, UK, 1974), p. 261; Evans, *Irish Folk Ways,* p. 36.

83. Fenton, "Cattle," p. 435; Evans, *Irish Folk Ways,* p. 35.

84. Bonser, *The Drovers,* pp. 28-29 (first quote); Moryson, *An Itinerary,* IV, p. 201 (second quote); Ó Duilearga, "Mountain Shielings," p. 294.

85. Moryson, *An Itinerary,* IV, pp. 199-200 (quote); Evans, *Irish Heritage,* p. 53.

86. Jenkin, *Cornwall,* pp. 380-81; MacSween, "Transhumance," p. 84; Curwen, "Hebrides," pp. 273-80; O'Donovan, *Economic History,* p. 31; Fenton, "Traditional Pastoral," pp. 99-101; Evans, *Irish Heritage,* p. 53.

87. Evans, *Personality of Ireland,* p. 81.

88. Moryson, *An Itinerary,* IV, pp. 198-201.

89. Skeel, "Cattle Trade," p. 137; O'Donovan, *Economic History,* p. 23; Moryson, *An Itinerary,* IV, p. 193; Fenton, "Cattle," p. 435; Bonser, *The Drovers,* pp. 17-20; Haldane, *Drove Roads,* pp. 7, 12; Thompson, "Cattle Droving," p. 172; Arthur J. Hubbard and George Hubbard, *Neolithic Dew-Ponds and Cattle-Ways* (London, UK, 1905), pp. 51-70; Richard J. Colyer, *The Welsh Cattle Drovers: Agriculture and the Welsh Cattle Trade Before and During the Nineteenth Century* (Cardiff, UK, 1976), p. 43.

90. Fenton, "Traditional Pastoral," pp. 103-6.

91. O'Donovan, *Economic History,* pp. 25-26, 35, 46-76; Trow-Smith, *History British Livestock,* pp. 3-4, 7; Fenton, "Traditional Pastoral," pp. 104-5; Skeel, "Cattle Trade," p. 135; Haldane, *Drove Roads,* p. 169; Thompson, "Cattle Droving," pp. 174-77.

92. Thompson, "Cattle Droving," p. 179 (quote); Skeel, "Cattle Trade," pp. 135–36; Bonser, *The Drovers*, p. 23; Haldane, *Drove Roads*, p. 40; Fenton, "Traditional Pastoral," pp. 103–4; Trow-Smith, *History British Livestock*, pp. 5, 6, 12; Colyer, *Welsh Cattle Drovers*.

93. Thompson, "Cattle Droving," pp. 181–82; Bonser, *The Drovers*, p. 40; Trow-Smith, *History British Livestock*, pp. 5, 6, 12; Fenton, "Traditional Pastoral," pp. 103–4; O'Donovan, *Economic History*, p. 9; Haldane, *Drove Roads*, p. 40.

94. Bonser, *The Drovers*, pp. 20, 25, 35–37, 40, 47, 58–64, following p. 192; Haldane, *Drove Roads*, pp. 23–27, 35–40, 72, following p. 166, 183–84; Skeel, "Cattle Trade," p. 143; Fenton, "Traditional Pastoral," p. 103; Trow-Smith, *History British Livestock*, p. 9; Thompson, "Cattle Droving," pp. 178–79; Carrier, *Pastoral Heritage*, p. 211; Hugh R. Rankin, "Cattle Droving from Wales to England," *Agriculture*, 62 (1955), p. 220; John C. Lloyd, "The Black Cattle Droves," *Historical Memoranda of Breconshire* (1903), p. 53.

95. Trow-Smith, *History British Livestock*, p. 8; Fenton, "Traditional Pastoral," pp. 103–6.

96. Thompson, "Cattle Droving," pp. 176–77.

97. Robinson, *Plantation of Ulster*, pp. 61, 100, 108, 187, 202–8; Evans, *Irish Heritage*, pp. 54–55; Evans, *Irish Folk Ways*, p. 35.

98. Maurice le Lannou, *Géographie de la Bretagne* (Paris, France, 1952), vol. I, pp. 14–22; André Meynier, "Le Cantal, premier massif montagneaux français pour l'estivage bovin," in *Études géographiques offertes à Louis Papy* (Bordeaux, France, 1978), pp. 309–19; Philippe Arbos, *L'Auvergne* (Paris, France, 1952), pp. 176–89; André Fel, *Les hautes terres du Massif Central: Tradition paysanne et économie agricole* (Paris, France, 1962), pp. 30, 46–47, 128, 228; R. Cole Harris, ed., *Historical Atlas of Canada* (Toronto, ON, 1987), I, plate 45.

99. Vialles, "Camargue"; Lauren C. Post, "The Old Cattle Industry of Southwestern Louisiana," *MR*, 9 (1957), p. 45.

100. Peter H. Wood, "It Was a Negro Taught Them: A New Look at African Labor in Early South Carolina," *Journal of Asian and African Studies*, 9 (1974), pp. 168–69; Peter H. Wood, *Black Majority: Negroes in Colonial South Carolina from 1670 through the Stono Rebellion* (New York, NY, 1975), pp. 30–31; Jordan, *Trails*, pp. 14–15, 156; the exaggerated claims can be found in "Africans Shaped American Cattle Industry," *UCLA African Studies Center Newsletter* (Fall 1986), p. 13; and Arnold R. Rojas, *The Vaquero* (Charlotte, NC, and Santa Barbara, CA, 1964), p. 11 (second quote).

101. R. J. Harrison Church, *West Africa: A Study of the Environment and Man's Use of It*, 2nd ed. (London, UK, 1960), p. 64.

102. Jeffrey A. Gritzner, *The West African Sahel: Human Agency and Environmental Change* (Chicago, IL, 1988), pp. 5–9; Church, *West Africa*, pp. 17, 18, 24–25, 37; Reuben K. Udo, *The Human Geography of Tropical*

Africa (Ibadan, Nigeria, and London, UK, 1982), pp. 24, 28–29; Edmond Bernus, "Les Touareg du Sahel nigérien," *Cahiers d'Outre Mer*, 19 (1966), p. 5; Walter Deschler, "Cattle in Africa: Distribution, Types, and Problems," *GR*, 53 (1963), Plate 1 at end of issue.

103. Werner Fricke, *Die Rinderhaltung in Nordnigeria und ihre natur- und sozialräumlichen Grundlagen* (Frankfurt, Germany, 1969), p. 31; Johannes Nicolaisen, "Some Aspects of the Problem of Nomadic Cattle Breeding among the Taureg of the Central Sahara," *Geografisk Tidsskrift*, 53 (1954), pp. 68–70; Douglas L. Johnson, *The Nature of Nomadism: A Comparative Study of Pastoral Migrations in Southern Asia and Northern Africa* (Chicago, IL, 1969), pp. 121, 136; Derrick J. Stenning, "Transhumance, Migratory Drift, Migration: Patterns of Pastoral Fulani Nomadism," *Journal of the Royal Anthropological Institute of Great Britain and Ireland*, 87 (1957), p. 59; Church, *West Africa*, pp. 59, 76–77; Udo, *Human Geography*, pp. 26–29; Gritzner, *West African Sahel*, pp. 31–33.

104. C. Edward Hopen, *The Pastoral Fulbe Family in Gwandu* (London, UK, 1958), pp. 22, 39; F. W. de St. Croix, "Some Aspects of the Cattle Husbandry of the Nomadic Fulani," *Farm and Forest*, 5 (1944), p. 31; Deschler, "Cattle in Africa," p. 54; Fricke, *Rinderhaltung*, pp. 36–37, 92; Church, *West Africa*, p. 77.

105. Jean Gallais, *Le delta intérieur du Niger: étude de géographie régionale* (Ifan-Dakar, Senegal, 1967), I, pp. 11, 55; Udo, *Human Geography*, pp. 8, 13; Church, *West Africa*, pp. 14–15; R. Bluzet, "Le région de Tombouctou," *Bulletin de la Société de Géographie*, 7th series, 16 (1895), p. 375; Johnson, *Nature of Nomadism*, pp. 120, 133; Hopen, *Pastoral Fulbe*, p. 17; Johannes Nicolaisen, *Ecology and Culture of the Pastoral Taureg* (København, Denmark, 1963), p. 26.

106. Derrick J. Stenning, *Savannah Nomads: A Study of the Wodaabe Pastoral Fulani of Western Bornu Province, Northern Region, Nigeria* (London, UK, 1959), pp. 2–4, 14–16, 20–25, 38–41, 52–53; John A. Grayzel, "Markets and Migration: A Fulbe Pastoral System in Mali," in John G. Galaty and Douglas L. Johnson, eds., *The World of Pastoralism: Herding Systems in Comparative Perspective* (New York, NY, 1990), pp. 35–67; John A. Grayzel, "The Ecology of Ethnic-Class Identity among an African Pastoral People: The Doukoloma Fulbe," Ph.D. diss., University of Oregon, (Eugene, OR, 1977); Stenning, "Transhumance," p. 57; C. Daryll Forde, "The Cultural Map of West Africa: Successive Adaptations to Tropical Forests and Grasslands," in Simon Ottenberg and Phoebe Ottenberg, eds., *Cultures and Societies of Africa* (New York, NY, 1960), pp. 126, 132; Marguerite Dupire, *Peuls nomades: étude descriptive des Wodaabe du Sahel nigérien* (Paris, France, 1962), pp. 2, 5; Udo, *Human Geography*, p. 132; Gallais, *Delta intérieur*, I, pp. 87–93, 119–61; Hopen, *Pastoral Fulbe*, pp. 1–7; Horst Mensching, "Der Sahel in Westafrika," *Hamburger Geographische Studien*, 24

(1971), p. 69; W. H. Barker, "Historical Geography of West Africa," *Geographical Teacher*, 10 (1919–20), p. 55.

107. P. Borricand, "La nomadisation en Mauritanie," *Travaux de l'Institut de Recherches Sahariennes*, 5 (1948), pp. 86–88, 93; G. Doutressoulle, *L'élevage en Afrique occidentale* (Paris, 1947), p. 22; Wolfgang Creyaufmüller, *Nomadenkultur in der Westsahara* (Stuttgart, Germany, 1983), pp. 43, 52; Paul Marty, *Études sur l'Islam et les tribus maures* (Paris, France, 1921), pp. 1, 3.

108. Théodore Monod, ed., *Pastoralism in Tropical Africa* (London, UK, 1975), p. 242; Edmond Bernus, *Les Illabakan (Niger): une tribu touarègue sahélienne et son aire de nomadisation* (Paris, France, and The Hague, Netherlands, 1974), p. 71; Bernus, "Les Touareg," p. 7; Doutressoulle, *L'élevage*, p. 23; Mensching, "Der Sahel," p. 69; Nicolaisen, *Ecology and Culture*.

109. Michael M. Horowitz, "Ethnic Boundary Maintenance among Pastoralists and Farmers in the Western Sudan (Niger)," *Journal of Asian and African Studies*, 7 (1972), p. 109; Reuben K. Udo, *Geographical Regions of Nigeria* (Berkeley and Los Angeles, CA, 1970), pp. 191–94; Gallais, *Delta intérieur*, I, p. 87; Monod, *Pastoralism*, p. 350; Bernus, *Illabakan*, p. 71; Stenning, "Transhumance," p. 58.

110. Nicolaisen, *Ecology and Culture*, pp. 47–49; Church, *West Africa*, p. 128; Gallais, *Delta intérieur*, I, pp. 131–32; Deschler, "Cattle in Africa," p. 52; Bernus, *Illabakan*, p. 71.

111. Horowitz, "Ethnic Boundary Maintenance," pp. 106, 109–10; Stenning, *Savannah Nomads*, pp. 4, 6; Gallais, *Delta intérieur*, I, p. 131; II, p. 405; Nicolaisen, "Some Aspects," p. 64; Bernus, *Illabakan*, pp. 66–68, 73; Johnson, *Nature of Nomadism*, pp. 126–27, 144; Doutressoulle, *L'élevage*, pp. 22–23.

112. E. A. Brackenbury, "Notes on the Bororo Fulbe or Nomad Cattle Fulani," *JAS*, 23 (1923), pp. 272–73; J. R. Wilson-Haffenden, "Ethnological Notes on the Shuwalbe Group of the Bororo Fulani," *Journal of the Royal Anthropological Institute of Great Britain and Ireland*, 57 (1927), pp. 276–77; Hopen, *Pastoral Fulbe*, pp. 23, 26, 27; Gallais, *Delta intérieur*, I, p. 131; Doutressoulle, *L'élevage*, pp. 21–22; Udo, *Human Geography*, pp. 131–32; Horowitz, "Ethnic Boundary," pp. 110–11; Stenning, *Savannah Nomads*, pp. 20, 46.

113. S. M. Rafiullah, *The Geography of Transhumance* (Aligarh, India, 1966), p. 63; [?] Mazoudier, "Le rythme de vie indigène et les migrations saisonnières dans la colonie du Tchad," *Annales de Géographie*, 53/54 (1945), pp. 296–99; Claude Grandet, "La vie rurale dans le cercle de Goundam (Niger soudanais)," *Cahiers d'Outre Mer*, 11 (1958), p. 29; Borricand, "Nomadisation," pp. 86–87; Johnson, *Nature of Nomadism*, pp. 126–28, 145–48; Hopen, *Pastoral Fulbe*, pp. 22, 28, 32, 34; Stenning, "Transhumance," pp. 58–61, 67–69; Doutressoulle, *L'élevage*, p. 17; Gallais, *Delta*

intérieur, II, p. 368; Monod, *Pastoralism*, p. 367; Udo, *Human Geography*, p. 132; Church, *West Africa*, p. 128; Stenning, *Savannah Nomads*, p. 104 (quote).

114. Creyaufmüller, *Nomadenkultur*, p. 61; Bernus, *Illabakan*, p. 35; Stenning, "Transhumance," pp. 57, 61; Stenning, *Savannah Nomads*, pp. 38–41, 52–53; Gallais, *Delta intérieur*, I, pp. 121–23.

115. Borricand, "Nomadisation," pp. 86–87; Creyaufmüller, *Nomadenkultur*, p. 11; Hopen, *Pastoral Fulbe*, p. 13; Monod, *Pastoralism*, p. 395; Rafiullah, *Geography of Transhumance*, p. 63; Udo, *Human Geography*, pp. 108, 132; Bernus, "Les Touareg," p. 7; Stenning, *Savannah Nomads*, p. 105; Brackenbury, "Notes," pp. 213–15; Wilson-Haffenden, "Ethnological Notes," p. 276.

116. Hans-Joachim Koloss, *Die Haustierhaltung in Westafrika* (München, Germany, 1968), pp. 35–36; Dupire, *Peuls nomades*, pp. 83, 98–99, 280; Stenning, "Transhumance," p. 57; Stenning, *Savannah Nomads*, pp. 103, 112; Fricke, *Rinderhaltung*, p. 92; Nicolaisen, *Ecology and Culture*, pp. 139, 142, 151–52; Nicolaisen, "Some Aspects," pp. 64–65; Bernus, *Illabakan*, pp. 29, 65; Borricand, "Nomadisation," p. 86; Hopen, *Pastoral Fulbe*, p. 25; Horowitz, "Ethnic Boundary," p. 112; Forde, "Cultural Map," p. 126.

117. Z. Ligers, "Comment les Peul de Koa castrent leurs taureaux," *Bulletin de l'Institut Français d'Afrique Noire*, series B, 20 (1958), pp. 191–204; St. Croix, "Some Aspects," pp. 30–32; Stenning, *Savannah Nomads*, pp. 101–11; Doutressoulle, *L'élevage*, p. 39, and following pp. 16 and 17; Dupire, *Peuls nomades*, pp. 85, 86, 89, 93–98, and following p. 64; Hopen, *Pastoral Fulbe*, p. 25; Nicolaisen, *Ecology and Culture*, pp. 50–53, 135; Fricke, *Rinderhaltung*, p. 92; Brackenbury, "Notes," pp. 271, 273; Udo, *Human Geography*, p. 134; Bernus, *Illabakan*, p. 78; Gallais, *Delta intérieur*, II, figs. 21, 22.

118. Augustin Bernard, *Afrique septentrionale et occidentale* (Paris, France, 1939), part II, p. 437; Georges Doutressoulle, "Le cheval au Soudan Français et ses origines," *Bulletin de l'Institut Français d'Afrique Noire*, 2 (1940), pp. 342–46; Stenning, *Savannah Nomads*, preceding p. 113; Doutressoulle, *L'elevage*, pp. 23, 237, 240; Gallais, *Delta intérieur*, I, p. 131; Nicolaisen, *Ecology and Culture*, pp. 28, 113; Bernus, *Illabakan*, pp. 67–68; Forde, "Cultural Map," p. 131; Michael Mason, "Population Density and Slave Raiding—The Case of the Middle Belt of Nigeria," *Journal of African History*, 10 (1969), p. 551.

119. Dupire, *Peuls nomades*, pp. 96 (second quote), 97 (first quote); Ligers, "Comment les Peul," p. 192.

120. Dupire, *Peuls nomades*, p. 85, and following p. 64; Nicolaisen, *Ecology and Culture*, p. 50; Kaloss, *Haustierhaltung*, p. 28.

121. Stenning, *Savannah Nomads*, pp. 4–5, 101–3; Stenning, "Transhumance," p. 58; Hopen, *Pastoral Fulbe*, p. 13; Dupire, *Peuls nomades*, pp. 85, 87; Mazoudier, "Rythme de vie," p. 298; Nicolaisen, *Ecology and Culture*,

pp. 51–52, 221, 224, 227; Udo, *Human Geography*, p. 132; Wilson-Haffenden, "Ethnological Notes," pp. 280–81; Brackenbury, "Notes," pp. 214–15.

122. Creyaufmüller, *Nomadenkultur*, p. 69; Church, *West Africa*, p. 132; Hopen, *Pastoral Fulbe*, p. 13.

Chapter Three

1. Carl O. Sauer, *The Early Spanish Main* (Berkeley and Los Angeles, CA, 1966); David Watts, *The West Indies: Patterns of Development, Culture and Environmental Change since 1492* (Cambridge, UK, 1987). On adaptation, see Carl O. Sauer, "Man in the Ecology of Tropical America," in *Proceedings of the Ninth Pacific Science Congress, 1957* (Bangkok, Thailand, 1958), vol. 20, pp. 104–10; Richard K. Ormrod, "The Evolution of Soil Management Practices in Early Jamaican Sugar Planting," *JHG*, 5 (1979), p. 169; and David Harris, ed., *Human Ecology in Savanna Environments* (London, UK, 1980).

2. Edmundo Wernicke, "Rutas y etapas de la introducción de los animales domésticos en las tierras americanas," *Anales de la Sociedad Argentina de Estudios Geográficos*, 4 (1938), p. 77; Peter Boyd-Bowman, *Indice geobiográfico de más de 56 mil pobladores de la América hispánica*, vol. 1 (Mexico City, Mexico, 1985), p. xxxix; and, by the same author, *Indice geobiográfico de cuarenta mil pobladores españoles de América en el siglo XVI*, vol. 1 (Bogotá, Colombia, 1964); "Patterns of Spanish Emigration to the Indies until 1600," *HAHR*, 58 (1976), pp. 580–604; and "Spanish Emigrants to the Indies, 1595–98: A Profile," in Fredi Chiappelli, ed., *First Images of America: The Impact of the New World on the Old* (Berkeley, CA, 1976), pp. 723–35. See also Levi Marrero, *Cuba: economía y sociedad* (Río Piedras, PR, 1972), vol. I, pp. 138, 146.

3. Emilio R. Demorizi, *Los Dominicos y las encomiendas de indios de la isla Española* (Santo Domingo, Dominican Republic, 1971), pp. 16–23; Francisco Morales Padrón, *Jamaica española* (Sevilla, Spain, 1952), p. 267; John S. Otto and Nain E. Anderson, "The Origins of Southern Cattle-Grazing: A Problem in West Indian History," *Journal of Caribbean History*, 21 (1988), p. 144; Marrero, *Cuba*, I, pp. 211, 213, 217; Watts, *West Indies*, p. 103.

4. M. F. Ashley-Montagu, "The African Origins of the American Negro and his Ethnic Composition," *Scientific Monthly*, 58 (1944), pp. 59–63; Melville J. Herskovits, "On the Provenience of New World Negroes," *Social Forces*, 12 (1933–34), p. 251; Claude Meillassoux, "The Role of Slavery in the Economic and Social History of Sahelo-Sudanic Africa," in J. E. Inikori, ed., *Forced Migration: The Impact of the Export Slave Trade on African Societies* (London, UK, 1982), pp. 75–77, 87; Gabriel Debien et al., "Les origines des esclaves des Antilles," *Bulletin de l'Institut Français d'Afrique Noire*, Series B, vols. 23–29; Philip D. Curtin, *The Atlantic Slave Trade: A Census* (Madison, WI, 1969), pp. 98, 100, 194; Ruth Pike, "Sevillian Society in the Sixteenth Century: Slaves and Freedmen," *HAHR*, 47 (1967), pp. 344–47; Mag-

nus Mörner, *Race Mixture in the History of Latin America* (Boston, MA, 1967), p. 16; Gonzalo Aguirre Beltrán, "Tribal Origins of Slaves in Mexico," *Journal of Negro History*, 31 (1946), pp. 273, 285–88, 293, 350; Marrero, *Cuba*, I, p. 211; Morales, *Jamaica*, pp. 267–68.

5. David Dary, *Cowboy Culture: A Saga of Five Centuries* (New York, NY, 1981), pp. 5, 8; J. E. Rouse, *The Criollo: Spanish Cattle in the Americas* (Norman, OK, 1977), fig. 1; Sauer, *Early Spanish Main*, pp. 71, 76; Richard J. Morrisey, "The Establishment and Northward Expansion of Cattle Ranching in New Spain," Ph.D. diss., University of California (Berkeley, CA, 1949), pp. 10–18, 146; Alfred W. Crosby, Jr., *The Columbian Exchange: Biological and Cultural Consequences of 1492* (Westport, CT, 1972), p. 75.

6. Leo Waibel, "Place Names as an Aid in the Reconstruction of the Original Vegetation of Cuba," *GR*, 33 (1943), pp. 376, 395–96; G. I. Asprey and R. G. Robbins, "The Vegetation of Jamaica," *Ecological Monographs*, 23 (1953), p. 403; Sauer, *Early Spanish Main*, p. 4.

7. Carl Johannessen, *Savannas of Interior Honduras* (Berkeley and Los Angeles, CA, 1963), pp. 5, 19, 25; Raúl F. Alonso Olivé, "Pastos y forrages: una vista panorámica de su historia en Cuba," *Revista de Agricultura* (Cuba), 36:1 (1952–53), pp. 89, 90; George W. Bridges, *The Annals of Jamaica* (London, UK, 1968), I, p. 178; Pedro Henríquez Ureña, *El español en Santo Domingo* (Buenos Aires, Argentina, 1940), pp. 45, 124; Marrero, *Cuba*, I, pp. 30–35 and II, p. 96; Morales, *Jamaica*, pp. 277–78.

8. J. S. Beard, "The Savanna Vegetation of Northern Tropical America," *Ecological Monographs*, 23 (1953), pp. 165–71, 208–9; Peter R. Bacon, *Flora and Fauna of the Caribbean: An Introduction to the Ecology of the West Indies* (Port of Spain, Trinidad, 1978), pp. 239–40; A. S. Hitchcock and Agnes Chase, "Grasses of the West Indies," *Contributions from the United States National Herbarium*, 18 (1917), pp. 261–494; Ovidio García-Molanari, *Grasslands and Grasses of Puerto Rico* (Río Piedras, PR, 1952), pp. 42, 148; Sauer, "Man in the Ecology," p. 109.

9. Detail on Antillean grasslands can be found in Garcia-Molanari, *Grasslands*, pp. 7, 17, 42–43, 46, 76–102, 119–48; Bacon, *Flora and Fauna*, p. 240; Hitchcock and Chase, "Grasses," pp. 277, 280, 309, 312; Morales, *Jamaica*, p. 282; Waibel, "Place Names," pp. 379–82; Sauer, "Man in the Ecology," p. 110; and Sauer, *Early Spanish Main*, pp. 156–57.

10. Hans Sloane, *A Voyage to the Islands Madera, Barbados, Nieves, S. Christophers and Jamaica* (London, UK, 1707), vol. I, map following p. cliv; Dolores M. Nadal and Hugo W. Alberts, "The Early History of Livestock and Pastures in Puerto Rico," *AH*, 21 (1947), pp. 62, 63.

11. Médéric L. E. Moreau de Saint-Méry, *Descripción de la parte española de Santo Domingo* (Santo Domingo, Dominican Republic, 1944), p. 108; John M. Street, "Feral Animals in Hispaniola," *GR*, 52 (1962), p. 400; John M. Street, *Historical and Economic Geography of the Southwest Peninsula of*

Haiti (Berkeley, CA, 1960), p. 81; Asprey and Robbins, "Vegetation of Jamaica," p. 388; Waibel, "Place Names," p. 395; Sauer, "Man in the Ecology," pp. 104–10; Sloane, *Voyage*, vol. I, following p. cliv.

12. L. T. Burcham, *California Range Land: An Historico-Ecological Study of the Range Resource of California* (Sacramento, CA, 1957), p. 12; James Burney, *History of the Buccaneers of America* (London, UK, 1891), p. 32; Samuel Purchas, *Hakluytus Posthumus or Purchas his Pilgrimes* (Glasgow, UK, 1905–7), vol. XVI, pp. 91–93; Edward Long, *The History of Jamaica* (London, UK, 1970), vol. 3, p. 959; Charles H. Smith, *Mammalia: Dogs* (Edinburgh and London, UK, 1854–56), vol. II, pp. 120–21; Marrero, *Cuba*, vol. II, pp. 89, 90, 96; Morales, *Jamaica*, p. 278; Dary, *Cowboy Culture*, p. 8; Street, "Feral Animals," pp. 400–401; Street, *Historical and Economic*, p. 86; Rouse, *Criollo*, pp. 21, 34–37; Sauer, *Early Spanish Main*, pp. 156–57; Watts, *West Indies*, pp. 105, 108, 117, 125; Johannessen, *Savannas*, p. 38; A. O. Exquemelin, *The Buccaneers of America*, trans. Alexis Brown (Baltimore, MD, 1969), p. 49; Crosby, *Columbian Exchange*, pp. 76, 92.

13. James J. Parsons, "Spread of African Pasture Grasses to the American Tropics," *Journal of Range Management*, 25 (1972), pp. 12–17; James J. Parsons, "The Africanization of the New World Tropical Grasslands," *Tübinger Geographische Studien*, 34 (1970), pp. 141–53; Bridges, *Annals of Jamaica*, vol. I, p. 177; Crosby, *Columbian Exchange*, p. 75; Alonso, "Pastos y forrajes," p. 89; Watts, *West Indies*, pp. 119, 126; Sauer, *Early Spanish Main*, p. 156; Johannessen, *Savannas*, pp. 19, 48, 106; Morrisey, "Establishment," pp. 151–52.

14. Watts, *West Indies*, pp. 91, 103, 104.

15. Frank Cundall and Joseph L. Pietersz, *Jamaica under the Spaniards* (Kingston, Jamaica, 1919), pp. 18, 34; Bridges, *Annals of Jamaica*, vol. I, pp. 179–80; Watts, *West Indies*, pp. 101, 111, 122; Marrero, *Cuba*, vol. I, p. 129.

16. Richard B. Sheridan, *Sugar and Slavery: An Economic History of the British West Indies, 1623–1775* (Baltimore, MD, 1973), p. 92; Watts, *West Indies*, pp. 104, 109, 112; Long, *History of Jamaica*, vol. III, pp. 960–61.

17. Moreau, *Descripción de la parte*, p. 219; Purchas, *Hakluytus*, vol. XVI, p. 91; Street, "Feral Animals," p. 401; Morrisey, "Establishment," p. 19.

18. John S. Otto and Nain E. Anderson, "Cattle Ranching in the Venezuelan Llanos and the Florida Flatwoods: A Problem in Comparative History," *Comparative Studies in Society and History*, 28 (1986), pp. 674–76; Dave D. Davis, "The Strategy of Early Spanish Ecosystem Management on Cuba," *Journal of Anthropological Research*, 30 (1974), p. 294; Donald E. Chipman, *Nuño de Guzmán and the Province of Pánuco in New Spain, 1518–1533* (Glendale, CA, 1967), pp. 46–47; Marrero, *Cuba*, vol. I, p. 121, vol. II, pp. 96–99; Cundall and Pietersz, *Jamaica*, pp. 7, 22, 26; Morrisey, "Establishment," pp. 22–23; Morales, *Jamaica*, pp. 277–78; Mörner, *Race Mixture*, pp. 77–78,

119; Sauer, *Early Spanish Main*, p. 208; Purchas, *Hakluytus*, vol. XVI, p. 90; Moreau, *Descripción de la parte*, pp. 100, 102, 405; Demorizi, *Dominicos*, pp. 25–58; Bridges, *Annals of Jamaica*, vol. I, p. 180; field observation, Dominican Republic, Sep. 1992.

19. Elizabeth J. Reitz, "Vertebrate Fauna from Locus 39, Puerto Real, Haiti," *Journal of Field Archaeology*, 13 (1986), pp. 317, 319, 324; Irene A. Wright, *The Early History of Cuba, 1492–1586* (New York, NY, 1916), pp. 82, 265–66, 305–7; Watts, *West Indies*, pp. 123, 125; Morrisey, "Establishment," p. 33; Morales, *Jamaica*, pp. 282, 285; Purchas, *Hakluytus*, vol. XV, p. 127, vol. XVI, p. 91.

20. Sauer, *Early Spanish Main*, p. 157; Wright, *Early History*, p. 307; Reitz, "Vertebrate Fauna," pp. 317, 324; Morales, *Jamaica*, pp. 282, 285; Richard J. Morrisey, "Colonial Agriculture in New Spain," *AH*, 31 (1957), p. 25; Rouse, *Criollo*, pp. 40, 41, 44.

21. Pierre Deffontaines, *Contribution à la géographie pastorale de l'Amérique latine* (Rio de Janeiro, Brazil, 1964), pp. 79–92, 97–98.

22. James Lockhart, "Encomienda and Hacienda: The Evolution of the Great Estate in the Spanish Indies," *HAHR*, 49 (1969), pp. 412, 419–21; Duvon C. Corbett, "Mercedes and Realengos: A Survey of the Public Land System in Cuba," *HAHR*, 19 (1939), p. 263; David E. Vassberg, *Land and Society in Golden Age Castille* (New York, NY, 1984), pp. 7, 10–13, 65, 169; Nadal and Alberts, "Early History," p. 63; Watts, *West Indies*, p. 125; Davis, "Strategy," p. 298; Marrero, *Cuba*, vol. II, p. 99; Morales, *Jamaica*, p. 277; Dary, *Cowboy Culture*, p. 10; Cundall and Pietersz, *Jamaica*, p. 7; Sauer, *Early Spanish Main*, pp. 156, 157; Wright, *Early History*, pp. 62, 94, 305; Morrisey, "Establishment," pp. 23–28.

23. Harris, *Human Ecology*, p. 271; Constantino Suárez, *Vocabulario cubano* (Habana, Cuba, and Madrid, Spain, 1921), p. 283; Antonio Alcalá Venceslada, *Vocabulario andaluz* (Madrid, 1951), p. 266; *Diccionario de la lengua castellana por la Real Academia Española*, 5th ed. (Madrid, Spain, 1817), p. 462; Alonso, "Pastos y forrajes," p. 89; Marrero, *Cuba*, vol. I, p. 121; Morales, *Jamaica*, p. 277; Moreau, *Descripción de la parte*, pp. 83, 376; Otto and Anderson, "Cattle Ranching in Venezuelan Llanos," p. 673; Wright, *Early History*, pp. 266, 305; Cundall and Pietersz, *Jamaica*, pp. 7, 20; Bridges, *Annals of Jamaica*, vol. I, pp. 177, 179; Lockhart, "Encomienda," p. 421; Henríquez, *El español*, pp. 53, 211; Street, *Historical and Economic*, p. 110; Sauer, *Early Spanish Main*, pp. 156–57, 287; Corbitt, "Mercedes," pp. 263, 266–68; Vassberg, *Land and Society*, p. 61.

24. Lockhart, "Encomienda," p. 424; Watts, *West Indies*, p. 125; Sauer, *Early Spanish Main*, p. 157, 208; Cundall and Pietersz, *Jamaica*, p. 15; Morrisey, "Colonial Agriculture," p. 25.

25. Henríquez, *El español*, p. 249; Moreau, *Descripción de la parte*, pp. 101–5; Purchas, *Hakluytus*, vol. XV, p. 127; Bridges, *Annals of Jamaica*, vol. I, p. 179; Cundall and Pietersz, *Jamaica*, p. 15.

26. John G. Varner and Jeanette J. Varner, *Dogs of the Conquest* (Norman, OK, 1983); William Dampier, *A New Voyage Round the World* (London, UK, 1699–1703), vol. II, part 2, pp. 96–98; Xavier de Planhol, "Le chien de berger: développement et signification géographique d'une technique pastorale," *Bulletin de l'Association de Géographes Français* 370 (1969), p. 363; Marrero, *Cuba*, vol. II, p. 98; Burney, *Buccaneers*, p. 39; Purchas, *Hakluytus*, vol. XV, p. 127; Wright, *Early History*, pp. 265, 307; Dary, *Cowboy Culture*, pp. 18–19; Moreau, *Descripción de la parte*, pp. 100–105; Morrisey, "Establishment," p. 29; Cundall and Pietersz, *Jamaica*, p. 35.

27. Clarence H. Haring, *The Buccaneers in the West Indies in the XVII Century* (London, UK, 1910), pp. 63–67, 78; Martha McCulloch-Williams, *Dishes and Beverages of the Old South* (Knoxville, TN, 1988), p. 273; Street, "Feral Animals," p. 403; Burney, *Buccaneers*, pp. 39, 48; Exquemelin, *Buccaneers*, pp. 45, 54, 68.

28. Irene A. Wright, "The Spanish Resistance to the English Occupation of Jamaica, 1655–1660," *Transactions of the Royal Historical Society*, 4th series, 13 (1930), p. 117; Sheridan, *Sugar and Slavery*, p. 94; Sloane, *Voyage*, vol. I, preface, n.p.; Bridges, *Annals of Jamaica*, vol. I, p. 185; J. Harry Bennett, "William Whaley, Planter of Seventeenth-Century Jamaica," *AH*, 40 (1966), p. 113.

29. Frederic G. Cassidy and R. B. LePage, *Dictionary of Jamaican English*, 2nd ed., (Cambridge, UK, 1980), pp. 18, 28, 129, 245, 336, 337; Sloane, *Voyage*, vol. I., following p. cliv; B. W. Higman, *Jamaica Surveyed: Plantation Maps and Plans of the Eighteenth and Nineteenth Centuries* (Kingston, Jamaica, 1988), p. 11.

30. R. H. Whitbeck, "The Agricultural Geography of Jamaica," *AAAG*, 22 (1932), p. 25; John S. Otto, *The Southern Frontiers, 1607–1860* (Westport, CT, 1989), p. 32; Otto and Anderson, "Origins of Southern," pp. 145–47; Watts, *West Indies*, p. 162; Bennett, "William Whaley," p. 113; Morales, *Jamaica*, pp. 429–30; Long, *History of Jamaica*, vol. I, p. 345; Dampier, *New Voyage*, vol. II, pt. 2, p. 98; Higman, *Jamaica Surveyed*, p. 198.

31. B. W. Higman, "The Internal Economy of Jamaican Pens, 1760–1890," *Social and Economic Studies* (Jamaica), 38 (1989), pp. 62, 66; William Beckford, *A Descriptive Account of the Island of Jamaica* (London, UK, 1790), vol. II, p. 167, 170; A. P. Thornton, "Some Statistics of West Indian Produce, Shipping and Revenue, 1660–1685," *Caribbean Historical Review*, 4 (1954), pp. 259–65; U. B. Phillips, "A Jamaican Slave Plantation," *AHR*, 19 (1914), p. 544; Watts, *West Indies*, p. 162, 341, 345; Sheridan, *Sugar and Slavery*, pp. 93, 120, 222, 226; Sloane, *Voyage*, vol. I, p. lxxiii; Bridges, *Annals of Jamaica*, vol. I, p. 185; Otto and Anderson, "Origins of Southern," p. 146; Higman, *Jamaica Surveyed*, p. 205.

32. Sloane, *Voyage*, vol. I, pp. xvi, lviii; Bridges, *Annals of Jamaica*, vol. I, p. 178; Dampier, *New Voyage*, vol. II, pt. 2, p. 98; Sheridan, *Sugar and Slavery*, pp. 95; Otto and Anderson, "Origins of Southern," pp. 144–45.

33. Long, *History of Jamaica*, vol. I, p. 9; Sheridan, *Sugar and Slavery*,

pp. 95, 211, 368; Wright, "Spanish Resistance," p. 122; Bennett, "William Whaley," p. 113.

34. John Poyer, *The History of Barbados* (London, UK, 1971), pp. 7–9, 16–17; Vincent T. Harlow, *A History of Barbados* (Oxford, UK, 1926), pp. iii, 20–21, 39; Cassidy and LePage, *Dictionary*, pp. 129, 221, 345; Dampier, *New Voyage*, vol. II, pt. 2, p. 98; Sloane, *Voyage*, vol. I, following p. cliv; Otto and Anderson, "Origins of Southern," pp. 114, 147; Whitbeck, "Agricultural Geography," pp. 24–25; Beckford, *Descriptive Account*, vol. II, pp. 65, 167, 169, 175; Phillips, "Jamaica Slave," p. 548; Higman, *Jamaica Surveyed*, pp. 24, 45, 63, 197–230, 264; Higman, "Internal Economy," pp. 63, 77.

35. Sloane, *Voyage*, vol. I, following p. cliv; Beckford, *Descriptive Account*, vol. II, p. 172; Bridges, *Annals of Jamaica*, vol. I, p. 268; Watts, *West Indies*, p. 342; Bennett, "William Whaley," pp. 114, 115; Sheridan, *Sugar and Slavery*, pp. 13, 95, 208, 212; Otto and Anderson, "Origins of Southern," p. 145; Higman, *Jamaica Surveyed*, pp. 72, 197; Higman, "Internal Economy," p. 65.

36. Street, "Feral Animals," p. 401; Sloane, *Voyage*, vol. I, pp. xvii, xxviii; Cassidy and LePage, *Dictionary*, p. 346; Bennett, "William Whaley," p. 120; Otto and Anderson, "Origins of Southern," pp. 145–46; Beckford, *Descriptive Account*, vol. II, p. 169.

37. Smith, *Mammalia*, vol. II, pp. 122, 123, 154–55; Georges Lutz, "Catahoula hog dog ou Catahoula cur: une opinion européenne ou le retour aux origines du chien," *Journal d'Agriculture Traditionnelle et de Botanique Apliquée*, 31 (1984), p. 155; Planhol, "Le chien," p. 364; Sloane, *Voyage*, vol. I, p. lviii.

38. W. A. Claypole, "The Settlement of the Liguanea Plain between 1665 and 1673," *Jamaican Historical Review*, 10 (1973), pp. 11–15; Otto and Anderson, "Origins of Southern," pp. 145, 146.

39. J. L. Dillard, "The Maritime (Perhaps Lingua Franca) Relations of a Special Variety of the Gulf Corridor," *Journal of Pidgin and Creole Languages*, 2 (1987), pp. 246, 247; J. L. Dillard, "A Note on *Dogie*," MS "circulated for comments and criticism," copy in possession of T. G. J.; Cassidy and LePage, *Dictionary*, pp. 18, 129, 154; Long, *History of Jamaica*, vol. I, p. 345; Phillips, "Jamaica Slave," pp. 547, 555; Otto and Anderson, "Origins of Southern," pp. 145, 146; Higman, "Internal Economy," p. 74.

40. Exquemelin, *Buccaneers*, pp. 32, 68.

41. Médéric L. E. Moreau de Saint-Méry, *Description topographique, physique, civile politique et historique de la partie française de l'isle Saint-Domingue* (Paris, France, 1958), vol. I, pp. 15–16, vol. II, pp. 787, 795, 905, 911, 949; Street, *Historical and Economic*, p. 110; Exquemelin, *Buccaneers*, pp. 45, 49, 50, 54.

42. Aguirre, "Tribal Origins," pp. 285–86; Debien et al., "Les origines," vol. 25, pp. 240–41; Lutz, "Catahoula," pp. 150–51, 154–55; Moreau, *Description topographique*, vol. I, p. 48, vol. II, pp. 120–21.

43. Karl W. Butzer, "Cattle and Sheep from Old to New Spain: Historical Antecedents," *AAAG*, 78 (1988), pp. 29–56; William E. Doolittle, "Las Marismas to Pánuco to Texas: The Transfer of Open Range Cattle Ranching from Iberia through Northeastern Mexico," *Yearbook of the Conference of Latin Americanist Geographers*, 13 (1987), pp. 3–11; Grady McWhiney and Forrest McDonald, "Celtic Origins of Southern Herding Practices," *JSH*, 51 (1985), pp. 165–82.

Chapter Four

1. Charles W. Hackett, ed., *Historical Documents Relating to New Mexico, Nueva Vizcaya, and Approaches Thereto, to 1773* (Washington, DC, 1923–37), vol. 1, p. 41; François Chevalier, *Land and Society in Colonial Mexico—the Great Hacienda* (Berkeley, CA, 1963), pp. 2, 75, 85.

2. Lesley B. Simpson, *Exploitation of Land in Central Mexico in the Sixteenth Century* (Berkeley and Los Angeles, CA, 1952), p. 73.

3. Sources on Gulf Coast climate and vegetation include Robert C. West, ed., *Handbook of Middle American Indians*, vol. 1 (Austin, TX, 1964), pp. 245, 258, 377; Robert C. West, Norbert P. Psuty, and B. G. Thom, *Las tierras bajas de Tabasco en el sureste de México* (Villahermosa, Mexico, 1987), pp. 136–42, 253–354; and William E. Doolittle, "Las Marismas to Pánuco to Texas: The Transfer of Open Range Cattle Ranching from Iberia through Northeastern Mexico," *Yearbook of the Conference of Latin Americanist Geographers*, 13 (1987), pp. 5, 8–9; Carl [Christian] Sartorius, *Mexico about 1850* (Stuttgart, Germany, 1961), pp. 8–11, 41, 182.

4. Peter Gerhard, *A Guide to the Historical Geography of New Spain* (Cambridge, UK, 1972), p. 365; Peter Gerhard, *The North Frontier of New Spain* (Princeton, NJ, 1982), p. 214; Carl O. Sauer, "Man in the Ecology of Tropical America," in *Proceedings of the Ninth Pacific Science Congress, 1957* (Bangkok, Thailand, 1958), vol. 20, p. 109; Donald E. Chipman, *Nuño de Guzmán and the Province of Pánuco in New Spain, 1518–1533* (Glendale, CA, 1967), p. 212.

5. David Dary, *Cowboy Culture: A Saga of Five Centuries* (New York, NY, 1981), p. 7; J. E. Rouse, *The Criollo: Spanish Cattle in the Americas* (Norman, OK, 1977), pp. 40, 44, 50; Juan A. Hasler, "Situación y tareas de la investigación lingüística en Veracruz," *La Palabra y el Hombre*, 5 (1958), pp. 44, 49; Gerhard, *Guide*, p. 366; Chevalier, *Land and Society*, p. 93; Doolittle, "Las Marismas," p. 4.

6. William H. Dusenberry, *The Mexican Mesta: The Administration of Ranching in Colonial Mexico* (Urbana, IL, 1963), p. 29; Simpson, *Exploitation*, pp. 3, 4; Rouse, *Criollo*, p. 44.

7. Simpson, *Exploitation*, p. 4; Gerhard, *Guide*, pp. 364–66.

8. José Matesanz, "Introducción de la ganadería en Nueva España 1521–1535," *Historia Mexicana*, 14 (1964), pp. 541–42; Chevalier, *Land and So-*

ciety, p. 93; Doolittle, "Las Marismas," pp. 3–11; Simpson, *Exploitation*, pp. 2, 4; Rouse, *Criollo*, p. 44; Chipman, *Nuño de Guzmán*, pp. 52–53, 212; Gerhard, *Guide*, p. 214.

9. Peter Boyd-Bowman, *Indice geobiográfico de cuarenta mil pobladores españoles de América en el siglo XVI* (Bogatá, Colombia, and Mexico City, Mexico, 1964–68), 2 vols., data on settlers in Huasteca, Oxitipa, Pánuco, Santiago de los Valles, Tamazunchale, Tancuyalve, Teniztiquipac, Tlapaguatla, and Veracruz; Karl W. Butzer, "Cattle and Sheep from Old to New Spain: Historical Antecedents," *AAAG*, 78 (1988), pp. 34, 35; Elisée Reclus, *The Earth and Its Inhabitants: North America* (New York, NY, 1891), vol. II, p. 91.

10. Francisco J. Santamaría, *Diccionario de méjicanismos* (Mexico City, Mexico, 1959), p. 703; Francisco J. Santamaría, *Diccionario general de americanismos* (Mexico City, Mexico, 1942), vol. II, p. 253; Antonio Alcalá Venceslada, *Vocabulario andaluz* (Madrid, Spain, 1951), p. 392; Reclus, *The Earth*, vol. II, p. 105.

11. Gonzalo Aguirre Beltrán, "Tribal Origins of Slaves in Mexico," *Journal of Negro History*, 31 (1946), pp. 280–88, 292–94; Magnus Mörner, *Race Mixture in the History of Latin America* (Boston, MA, 1967), pp. 77–78, 119; Philip D. Curtin, *The Atlantic Slave Trade: A Census*, (Madison, WI, 1969), p. 98; Richard J. Morrisey, "The Establishment and Northward Expansion of Cattle Ranching in New Spain," Ph.D. diss., University of California (Berkeley, CA, 1949), pp. 116–20, 134, 167; Mario Gongora, "Vagabondage et société pastorale en Amérique latine," *Annales: Economies, Sociétés, Civilisations*, 21 (1966), pp. 159–77; Silvio R. D. Baretta and John Markoff, "Civilization and Barbarism: Cattle Frontiers in Latin America," *Comparative Studies in Society and History*, 20 (1978), pp. 589, 600; Robert Himmerich y Valencia, *The Encomenderos of New Spain, 1521–1555* (Albuquerque, NM, 1991); Sartorius, *Mexico*, pp. 10, 51, 181, 184; Chevalier, *Land and Society*, pp. 112–13; Dusenberry, *Mexican Mesta*, p. 89; Santamaría, *Diccionario general*, vol. III, p. 251.

12. Simpson, *Exploitation*, pp. 18–19; Matesanz, "Introducción," pp. 540–41; Morrisey, "Establishment," pp. 134, 143, 155, 250; Chevalier, *Land and Society*, p. 110; Dary, *Cowboy Culture*, pp. 19, 21; Baretta and Markoff, "Civilization," p. 588; Sartorius, *Mexico*, pp. 181–86.

13. José Alvarez del Villar, *Historia de la charrería* (Mexico City, Mexico, 1941), pp. 164–73, 346–59; Fred B. Kniffen, "The Western Cattle Complex: Notes on Differentiation and Diffusion," *Western Folklore*, 12 (1953), p. 179; Richard E. Ahlborn, ed., *Man Made Mobile: Early Saddles of Western North America* (Washington, DC, 1980), pp. 5–11, 14, 31–34; Harold B. Barclay, *The Role of the Horse in Man's Culture* (London, UK, and New York, NY, 1980), pp. 210–12; Glenn R. Vernam, *Man on Horseback* (New York, NY, 1964), pp. 155, 159, 194–95, 238, 245; Basil Hall, *Extracts from a Journal Written on the Coasts of Chile, Peru, and Mexico in the Years 1820, 1821, 1822* (Edin-

burgh, Scotland, 2nd ed., 1824), vol. I, pp. 146–52, 157–60; Dary, *Cowboy Culture*, pp. 18–21; Morrisey, "Establishment," pp. 138, 155, 241–42, 251; Gregorio de Tapia y Salzedo, *Exercicios de la gineta al príncipe nuestro señor d. Baltasar Carlos* (Madrid, Spain, 1643), follows p. 72, pp. 73–74; Sartorius, *Mexico*, plates following pp. 4, 180.

14. Hortense W. Ward, "Indian Sign on the Spaniard's Cattle," *Texas Folk-Lore Society Publication*, 19 (1944), pp. 94–105; Chevalier, *Land and Society*, p. 111; Morrisey, "Establishment," pp. 108, 134–35, 151–54; Alvarez, *Historia*, pp. 26, 33; Sartorius, *Mexico*, pp. 182–84.

15. Matesanz, "Introducción," pp. 536–39; Chipman, *Nuño de Guzmán*, pp. 93–94; West et al., *Tierras bajas*, p. 253; Sartorius, *Mexico*, p. 182; Morrisey, "Establishment," pp. 52, 143, 151, 251.

16. Leovigildo Islas Escárcega, *Diccionario rural de México* (Mexico City, Mexico, 1961), p. 120; Basil M. Bensin, "Agroecological Exploration in the Soto la Marina Region, Mexico," *GR*, 25 (1935), p. 288; David E. Vassberg, *Land and Society in Golden Age Castille* (New York, NY, 1984), pp. 10–13; Lesley B. Simpson, *The Encomienda in New Spain: The Beginning of Spanish Mexico* (Berkeley and Los Angeles, CA, 1982), p. 163; Morrisey, "Establishment," pp. 123, 126; Simpson, *Exploitation*, pp. 9, 17, 21–23; Sartorius, *Mexico*, pp. 167, 184.

17. Ursula Ewald, "The von Thünen Principle and Agricultural Zonation in Colonial Mexico," *JHG* 3 (1977), p. 127; Wolfgang Trautmann, *Der kolonialzeitliche Wandel der Kulturlandschaft in Tlaxcala* (Paderborn, Germany, 1983), p. 289; Samuel Purchas, *Hakluytus Posthumus or Purchas his Pilgrimes* (Glasgow, 1905–7), vol. XV, p. 127.

18. Richard J. Morrisey, "Colonial Agriculture in New Spain," *AH*, 31 (1957), p. 26; Robert C. West and John P. Augelli, *Middle America: Its Lands and Peoples* (Englewood Cliffs, NJ, 2nd ed., 1976), p. 269; Donald D. Brand, "The Early History of the Range Cattle Industry in Northern Mexico," *AH*, 35 (1961), p. 133; Leovigildo Islas Escárcega and Rodolfo García-Brava y Olivera, *Iconografía charra* (Mexico City, Mexico, 1969), pp. 19, 34; Chevalier, *Land and Society*, pp. 85, 93; Dusenberry, *Mexican Mesta*, p. 31; Dary, *Cowboy Culture*, pp. 8–9; Morrisey, "Establishment," pp. 46–47, 52–55, 96; Trautmann, *Der kolonialzeitliche*, p. 291.

19. Morrisey, "Establishment," pp. 56–57, 119–20, 241–52; Trautmann, *Der kolonialzeitliche*, pp. 290–91.

20. Chevalier, *Land and Society*, pp. 42, 84–114.

21. Jack A. Licate, *Creation of a Mexican Landscape: Territorial Organization and Settlement in the Eastern Puebla Basin, 1520–1605* (Chicago, IL, 1981), pp. 43, 112; Thomas M. Whitmore, "A Simulation of the Sixteenth-Century Population Collapse in the Basin of Mexico," *Annals of the Association of American Geographers*, 81 (1991), pp. 464, 467; Matesanz, "Introducción," p. 539; Chevalier, *Land and Society*, pp. 94–98; Simpson, *Exploitation*, pp. 4–5, 49, 53, 55; Morrisey, "Establishment," pp. 55–57.

22. Ewald, "von Thünen Principle," pp. 130, 133; Chevalier, *Land and Society*, pp. 59–60, 65.

23. Chevalier, *Land and Society*, pp. 103–4.

24. Terry G. Jordan, "An Iberian Lowland/Highland Model for Latin American Cattle Ranching," *JHG*, 15 (1989), p. 123.

25. David A. Brading, *Haciendas and Ranchos in the Mexican Bajío: León 1700–1860* (Cambridge, UK, 1978), pp. 11–12; Licate, *Creation*, pp. 114–15; Chevalier, *Land and Society*, pp. 59–60, 65, 84, 110; Matesanz, "Introducción," pp. 536–39; Trautmann, *Der kolonialzeitliche*, pp. 291–94.

26. Morrisey, "Establishment," pp. 46–50.

27. Morrisey, "Establishment," pp. 56–57, 134–35.

28. Carlos Rincón Gallardo, *El charro méxicano* (Mexico City, Mexico, 1939), pp. 1, 7–48, 81–90; José de Lamano y Beneite, *El dialecto vulgar salamantino* (Salamanca, Spain, 1915), pp. 370, 620; Jorge L. Tamayo, *Geografía general de México* (Mexico City, Mexico, 1962), vol. III, pp. 533, 547; Leovigildo Islas Escárcega, *Vocabulario campesino nacional* (Mexico City, Mexico, 1945), pp. 170–71; Santamaría, *Diccionario general*, vol. I, p. 474; Islas, *Diccionario*, p. 87; Islas and García-Brava, *Iconografía*; Alvarez, *Historia*, pp. 171, 291–325.

29. Ramon F. Adams, *Western Words: A Dictionary of the American West* (Norman, OK, 1968), p. 173; Lamano, *Dialecto*, p. 598; Rincón, *El charro*, pp. 77–78; Santamaría, *Diccionario general*, vol. III, pp. 18–19.

30. Pierre Deffontaines, *Contribution à la géographie pastorale de l'Amérique latine* (Rio de Janeiro, Brazil, 1964), pp. 73–75; Dusenberry, *Mexican Mesta*, pp. 111–12; Trautmann, *Der kolonialzeitliche*, pp. 294–96.

31. David E. Vassberg, *Land and Society in Golden Age Castille* (New York, NY, 1984), p. 13; Chevalier, *Land and Society*, pp. 57, 98; Simpson, *Exploitation*, p. 4; Trautmann, *Der kolonialzeitliche*, p. 295.

32. William H. Dusenberry, "Ordinances of the Mesta in New Spain," *The Americas*, 4 (Jan. 1948), pp. 345–50; Julius Klein, *The Mesta: A Study in Spanish Economic History, 1273–1836* (Cambridge, MA, 1920); Dusenberry, *Mexican Mesta*, pp. 45–51; Morrisey, "Establishment," pp. 106–8, 131, 141, 153, 233–54; Dary, *Cowboy Culture*, pp. 9–12.

33. Charles J. Bishko, "The Andalusian Municipal Mestas in the 14th–16th Centuries: Administrative and Social Aspects," in *Andalucía medieval* (Córdoba, Spain, 1978), vol. I, pp. 350–51, 360, 371; Dusenberry, *Mexican Mesta*, pp. 44, 50, 76, 89.

34. Peter Boyd-Bowman, *Patterns of Spanish Emigration to the New World (1493–1580)* (Buffalo, NY, 1973), pp. 6, 29; Reclus, *The Earth*, vol. II, p. 91.

35. Boyd-Bowman, *Indice*, vols. 1, 2.

36. Sources on the physical environment of Florida include A. S. Hitchcock and Agnes Chase, "Grasses of the West Indies," *Contributions from the United States National Herbarium*, 18 (1917), pp. 282, 284, 315; Brian G.

Boniface, "A Historical Geography of Spanish Florida, Circa 1700," M.A. thesis, University of Georgia (Athens, GA, 1971), pp. 116–20, 126, 131, 147, 149; George H. Dacy, *Four Centuries of Florida Ranching* (St. Louis, MO, 1940), pp. 26–27, 149, 221–24; W. Theodore Mealor and Merle C. Prunty, "Open-Range Ranching in Southern Florida," *AAAG,* 66 (1976), p. 363; Merle C. Prunty, "Some Geographic Views of the Role of Fire in Settlement Processes in the South," *Tall Timbers Fire Ecology Conference, Proceedings,* 3 (1964), pp. 161–68.

37. Charles Arnade, "Cattle Raising in Spanish Florida, 1513–1763," *AH,* 35 (1961), pp. 117–19; Joe A. Akerman, Jr., *Florida Cowman: A History of Florida Cattle Raising* (Kissimmee, FL, 1976), pp. 2–3; Robert E. Williams, "Cattle in the Southeastern United States, or the Original Wild West," *Corral Dust,* 10 (Spring 1965), p. 4; Dacy, *Four Centuries,* p. 17; Rouse, *Criollo,* p. 41; Santamaría, *Diccionario general,* vol. III, p. 251.

38. Arnade, "Cattle Raising," pp. 118–22; Akerman, *Florida Cowman,* pp. 3, 7–10; Dacy, *Four Centuries,* pp. 22, 26; Boniface, "Historical Geography," pp. 40, 140–48.

39. Hale Smith, "The Spanish Gulf Coast Cultural Assemblage, 1500–1763," in Ernest F. Dibble and Earle W. Newton, eds., *Spain and Her Rivals on the Gulf Coast* (Pensacola, FL, 1971), pp. 63–64; Alan K. Craig and Christopher S. Peebles, "Ethnoecologic Change among the Seminoles, 1740–1840," *Geoscience and Man,* 5 (1974), p. 87; Boniface, "Historical Geography," pp. 34–38, 46, 117, 123, 125, 138, 147–48; Akerman, *Florida Cowman,* p. 3.

40. Elizabeth J. Reitz, "Vertebrate Fauna from Locus 39, Puerto Real, Haiti," *Journal of Field Archaeology,* 13 (1986), p. 319; Akerman, *Florida Cowman,* p. 202; Boniface, "Historical Geography," pp. 125, 141, 146, 150, 169.

41. Smith, "Spanish Gulf Coast," pp. 64, 67; Boniface, "Historical Geography," pp. 44–46; Akerman, *Florida Cowman,* p. 10.

42. David H. Corkran, *The Creek Frontier, 1540–1783* (Norman, OK, 1967), p. 187; Louis De Vorsey, Jr., *The Indian Boundary in the Southern Colonies, 1763–1775* (Chapel Hill, NC, 1961), p. 219; Arnade, "Cattle Raising," p. 123; Dacy, *Four Centuries,* p. 22.

43. Mealor and Prunty, "Open-Range Ranching," pp. 364–65; Akerman, *Florida Cowman,* pp. 9, 12, 13, 23; James C. Bonner, "The Open Range Livestock Industry in Colonial Georgia," *Georgia Review,* 17 (1963), p. 87.

44. Craig and Peebles, "Ethnoecologic," pp. 84–85; Smith, "Spanish Gulf Coast," p. 68; Boniface, "Historical Geography," pp. 145, 150, 152; Akerman, *Florida Cowman,* p. 13.

45. William Bartram, *The Travels of William Bartram* (New Haven, CT, 1958), pp. 118–21; K. W. Porter, "The Founder of the Seminole Nation, Secoffee or Cowkeeper," *FHQ,* 27 (1949), pp. 362–84; K. W. Porter, "The Cowkeeper Dynasty of the Seminole Nation," *FHQ,* 30 (1952), pp. 341–49;

Craig and Peebles, "Ethnoecologic," pp. 85, 87; Dacy, *Four Centuries*, pp. 22–23; Akerman, *Florida Cowman*, p. 7.

46. Benjamin Hawkins, "A Sketch of the Creek Country in the Years 1798 and 1799," *Collections and Transactions, Georgia Historical Society*, 3:1 (1938), pp. 45–46; Michael F. Doran, "Antebellum Cattle Herding in the Indian Territory," *GR*, 66 (1976), pp. 48–58.

47. J. L. Dillard, *A History of American English* (London, UK, and New York, NY, 1992), p. 107; Lewis C. Gray, *History of Agriculture in the Southern United States to 1860* (New York, NY, 1941), vol. I, p. 151; John S. Otto, *The Southern Frontiers, 1607–1860* (Westport, CT, 1989), pp. 30–32; Terry G. Jordan, *Trails to Texas: Southern Roots of Western Cattle Ranching* (Lincoln, NE, 1981), pp. 38–43; Wesley N. Laing, "Cattle in Seventeenth-Century Virginia," *Virginia Magazine of History and Biography*, 67 (1959), p. 147; Gary S. Dunbar, "Colonial Carolina Cowpens," *AH*, 35 (1961), pp. 125–30; John S. Otto, "The Origins of Cattle-Ranching in Colonial South Carolina, 1670–1715," *SCHM*, 87 (1986), p. 117.

48. Old maps on display at the Barbados Museum in Bridgetown, in particular R[ichard] Ford, "A New Map of the Island of Barbadoes" (London, UK, 1680), and William Mayo, "A New & Exact Map of the Island of Barbadoes in America According to an Actual & Accurate Survey Made in the Years 1717 to 1721" (London, UK, 1722).

49. James. S. Maag, "Cattle Raising in Colonial South Carolina," M.A. thesis, University of Kansas (Lawrence, KS, 1964), pp. 5–10; Richard D. Brooks, "Cattle Ranching in Colonial South Carolina: A Case Study in History and Archaeology of the Lazarus/Catherina Brown Cowpen," M.A. thesis, University of South Carolina (Columbia, SC, 1988), p. 53; H. Roy Merrens, *The Colonial South Carolina Scene: Contemporary Views, 1697–1774* (Columbia, SC, 1977), p. 108; Alexander S. Salley, Jr., *Narratives of Early Carolina, 1650–1708* (New York, NY, 1911), p. 182; Donald W. Meinig, *The Shaping of America: A Geographical Perspective on 500 Years of History* (New Haven, CT, 1986), vol. I, p. 184; Dacy, *Four Centuries*, p. 20; Akerman, *Florida Cowman*, p. 12; Laing, "Cattle in Seventeenth," pp. 148–49.

50. Maurice Mathews, "A Contemporary View of Carolina in 1680," *SCHM*, 55 (1954), p. 157; Maag, "Cattle Raising," pp. 10, 12; Jordan, *Trails*, p. 27; Merrens, *Colonial South Carolina*, pp. 43, 47, 48; Salley, *Narratives*, pp. 145, 149, 167, 171, 184.

51. Converse D. Clowse, *Economic Beginnings in Colonial South Carolina, 1670–1730* (Columbia, SC, 1971), pp. 82, 184; Alexander S. Salley, Jr., *Warrants for Lands in South Carolina, 1672–1711* (Columbia, SC, 1973), p. 611; John S. Otto, "Livestock-Raising in Early South Carolina, 1670–1700: Prelude to the Rice Plantation Economy," *AH*, 61 (1987), p. 15; Maag, "Cattle Raising," pp. 15, 17, 39–43, 46; Jordan, *Trails*, pp. 39–40, 46; Merrens, *Colonial South Carolina*, p. 69; Otto, *Southern Frontiers*, pp. 30–31, 37; Otto, "Origins of Cattle," pp. 117, 123; Williams, "Cattle in Southeastern," p. 5.

52. Elizabeth Donnan, "The Slave Trade into South Carolina Before the Revolution," *AHR*, 33 (1928), p. 804; Richard S. Dunn, "The English Sugar Islands and the Founding of South Carolina," *SCHM*, 72 (1971), p. 83; Agnes L. Baldwin, *First Settlers of South Carolina, 1670–1680* (Columbia, SC, 1969), n.p.; John S. Otto and Nain E. Anderson, "The Origins of Southern Cattle-Grazing: A Problem in West Indian History," *Journal of Caribbean History*, 21 (1988), p. 140; Salley, *Narratives*, pp. 158, 368; Mathews, "Contemporary View," p. 159; Clowse, *Economic Beginnings*, p. 53; Otto, *Southern Frontiers*, p. 29; Otto, "Origins of Cattle," pp. 119–20.

53. *Names in South Carolina*, 13 (1966), p. 39; 18 (1971), p. 42; Merrens, *Colonial South Carolina*, p. 119; Meinig, *Shaping of America*, vol. I, p. 184; Brooks, "Cattle Ranching," p. 67; Otto, *Southern Frontiers*, pp. 28, 31; Salley, *Warrants*, pp. 596, 600, 611, 628; Otto and Anderson, "Origins of Southern," pp. 139, 150.

54. Erhard Rostlund, "The Myth of a Natural Prairie Belt in Alabama: An Interpretation of Historical Records," *AAAG*, 47 (1957), p. 397; Salley, *Narratives*, pp. 130, 131; Merrens, *Colonial South Carolina*, p. 119.

55. *Names in South Carolina*, 10 (1963), p. 3; 12 (1965), p. 12; 13 (1966), p. 22.

56. Mederic L. E. Moreau de Saint-Mery, *Description topographique, physique, civile politique et historique de la partie française de l'isle Saint-Domingue* (Paris, France, 1958), vol. I, p. 15; Maag, "Cattle Raising," p. 53; Salley, *Narratives*, p. 134; Frederic G. Cassidy and R. B. LePage, eds., *Dictionary of Jamaican English* (Cambridge, UK, 2nd ed., 1980), p. 336; Dacy, *Four Centuries*, p. 299; Akerman, *Florida Cowman*, p. 13; Williams, "Cattle in Southeastern," p. 6; Dillard, *History American English*, pp. 120, 123.

57. Brooks, "Cattle Ranching," p. 64.

58. Maag, "Cattle Raising," pp. 23–25; Otto, *Southern Frontiers*, p. 34; Otto, "Livestock-Raising," pp. 19–23; Otto and Anderson, "Origins of Southern," p. 149.

59. Otto and Anderson, "Origins of Southern," p. 148.

60. Peter H. Wood, *Black Majority: Negroes in Colonial South Carolina from 1670 through the Stono Rebellion* (New York, NY, 1975), pp. 30–32; Peter H. Wood, "It Was a Negro Taught Them: A New Look at African Labor in Early South Carolina," *Journal of African and Asian Studies*, 9 (1974), pp. 168–70; W. Robert Higgins, "The Geographical Origins of Negro Slaves in Colonial South Carolina," *South Atlantic Quarterly*, 70 (1971), p. 39; Melville J. Herskovits, "On the Provenience of New World Negroes," *Social Forces*, 12 (1933–34), pp. 251–52; "Africans Shaped American Cattle Industry," *UCLA African Studies Newsletter* (Fall 1986), p. 13; Donnan, "Slave Trade," pp. 804, 816; Otto, "Origins of Cattle," pp. 121–22; Otto and Anderson, "Origins of Southern," pp. 147–48.

61. Grady McWhiney and Forrest McDonald, "Celtic Origins of Southern Herding Practices," *JSH*, 51 (1985), pp. 165–82; John S. Otto, "Traditional

Cattle-Herding Practices in Southern Florida," *Journal of American Folklore*, 97 (1984), pp. 300–301; Alexander S. Salley, Jr., *The Early English Settlers of South Carolina* (n.p., 1946), pp. 5–7, 10, 15, 18; Baldwin, *First Settlers*, n.p.; Otto, "Livestock-Raising," p. 22; Otto, "Origins of Cattle," p. 120.

62. Ray August, "Cowboys v. Rancheros: The Origins of Western American Livestock Law," *SWHQ*, 96 (1993) forthcoming.

63. Otto, "Livestock Raising," pp. 15–16; Clowse, *Economic Beginnings*, pp. 31–32; Brooks, "Cattle Ranching," p. 57; Maag, "Cattle Raising," p. 28.

64. Robert M. Weir, *Colonial South Carolina: A History* (Millwood, NY, 1983), p. 37; Mart A. Stewart, "Whether Wast, Deodand, or Stray: Cattle, Culture, and the Environment in Early Georgia," *AH*, 65 (1991), pp. 6–7; Maag, "Cattle Raising," p. 26; Otto, "Livestock-Raising," pp. 15–16; Otto, *Southern Frontiers*, p. 42; Williams, "Cattle in Southeastern," p. 5.

65. Merrens, *Colonial South Carolina*, p. 48.

66. Otto, *Southern Frontiers*, pp. 37, 39.

67. Clowse, *Economic Beginnings*, p. 30; Weir, *Colonial*, p. 38, Merrens, *Colonial South Carolina*, pp. 43, 48; Otto, "Livestock-Raising," p. 15.

68. Salley, *Narratives*, p. 168; Brooks, "Cattle Ranching," p. 97; Otto, "Origins of Cattle," p. 118.

69. Merrens, *Colonial South Carolina*, p. 119; Otto, "Livestock-Raising," p. 15; Otto and Anderson, "Origins of Southern," p. 139.

70. Salley, *Narratives*, p. 131.

71. H. L. Leithead, L. L. Yarlett, and T. N. Shiflet, *100 Native Forage Grasses in 11 Southern States* (Washington, DC, 1971); Clowse, *Economic Beginnings*, p. 30; Brooks, "Cattle Ranching," p. 97; Hitchcock and Chase, "Grasses of West Indies," p. 284; Otto, "Traditional Cattle-Herding," p. 301.

72. Merrens, *Colonial South Carolina*, p. 43.

73. S. W. Greene, "Relation Between Winter Grass Fires and Cattle Grazing in the Longleaf Pine Belt," *Journal of Forestry*, 33 (1935), pp. 338–39; Maag, "Cattle Raising," p. 27; Otto, "Origins of Cattle," p. 118.

74. Mathews, "Contemporary View," pp. 157, 159; Salley, *Narratives*, p. 149; Brooks, "Cattle Ranching," p. 57; Otto and Anderson, "Origins of Southern," p. 138.

75. Brooks, "Cattle Ranching," pp. 67, 70–79; Jordan, *Trails*, pp. 28–29; Dunbar, "Colonial," pp. 125–30; Otto, "Livestock-Raising," p. 19.

76. Brooks, "Cattle Ranching," pp. 78–80; Jordan, *Trails*, p. 35.

77. Maag, "Cattle Raising," pp. 22, 36; Brooks, "Cattle Ranching," pp. 67, 71–73; Jordan, *Trails*, p. 26.

78. Brooks, "Cattle Ranching," pp. 55, 58; Maag, "Cattle Raising," pp. 18–19, 36; Otto, "Livestock-Raising," pp. 16, 19–20.

79. John S. Otto, "Cracker: The History of a Southeastern Ethnic, Economic, and Racial Epithet," *Names*, 35 (1987), pp. 28–39; Calvin L. Beale,

"American Triracial Isolates," *Eugenics Quarterly,* 4 (1957), pp. 187–96; Edward T. Price, "A Geographic Analysis of White-Negro-Indian Racial Mixtures in Eastern United States," *AAAG,* 43 (1953), pp. 138–55; Jordan, *Trails,* pp. 29–31, 49; Maag, "Cattle Raising," pp. 21–24; Stewart, "Whether Wast," p. 11.

80. Salley, *Narratives,* pp. 149, 172; Mathews, "Contemporary View," p. 157; Otto, "Livestock-Raising," pp. 18–19.

81. Brooks, "Cattle Ranching," pp. 59–62; Weir, *Colonial,* p. 142; Jordan, *Trails,* p. 26; Maag, "Cattle Raising," p. 20; Otto, "Origins of Cattle," pp. 118, 122; Otto, "Livestock Raising," p. 19; Otto, *Southern Frontiers,* p. 43; Laing, "Cattle in Seventeenth," p. 154.

82. Alexander S. Salley, Jr., "Stock Marks Recorded in South Carolina, 1695–1721," *SCHM,* 13 (1912), pp. 126–31, 224–28; Maag, "Cattle Raising," pp. 17, 30–34; Brooks, "Cattle Ranching," pp. 62, 75; Weir, *Colonial,* p. 142; Merrens, *Colonial South Carolina,* p. 43; Jordan, *Trails,* pp. 27, 28, 33, 35; Otto, "Origins of Cattle," p. 118; Otto, "Livestock-Raising," pp. 16, 18, 20.

83. Bailey C. Hanes, *Bill Pickett, Bulldogger: The Biography of a Black Cowboy* (Norman, OK, 1977); Joseph W. LeBon, Jr., "The Catahoula Hog Dog: A Folk Breed," *Pioneer America,* 3:2 (1971), pp. 35–45; Jordan, *Trails,* pp. 31–34; Brooks, "Cattle Ranching," p. 71; Maag, "Cattle Raising," pp. 21, 23, 29; Merrens, *Colonial South Carolina,* pp. 43, 49, 119.

84. Jordan, *Trails,* p. 34; Maag, "Cattle Raising," pp. 21, 29; Laing, "Cattle in Seventeenth," p. 152.

85. Mathews, "Contemporary View," p. 157; Weir, *Colonial,* p. 142; Salley, *Narratives,* pp. 172, 184, 368; Otto, *Southern Frontiers,* p. 39; Maag, "Cattle Raising," pp. 14–15, 66–68; Brooks, "Cattle Ranching," pp. 65, 67, 71, 72; Merrens, *Colonial South Carolina,* pp. 33, 43, 68, 69; Otto, "Origins of Cattle," p. 121; Otto, "Livestock-Raising," p. 20; Otto and Anderson, "Origins of Southern," p. 139.

86. Salley, *Narratives,* p. 182.

87. Merrens, *Colonial South Carolina,* p. 33; Salley, *Narratives,* p. 368; Otto and Anderson, "Origins of Southern," p. 139.

88. Otto, "Origins of Cattle," p. 123; Otto, "Livestock-Raising," p. 21; Maag, "Cattle Raising," p. 81.

89. John D. W. Guice, "Cattle Raisers of the Old Southwest: A Reinterpretation," *WHQ,* 8 (1977), p. 173; Lauren C. Post, "The Domestic Animals and Plants of French Louisiana as Mentioned in the Literature with Reference to Sources, Varieties and Uses," *Louisiana Historical Quarterly,* 16 (1933), pp. 560–65; Jack D. L. Holmes, "Livestock in Spanish Natchez," *Journal of Mississippi History,* 23 (1961), p. 15; Reuben G. Thwaites, ed., *The Jesuit Relations and Allied Documents* (Cleveland, OH, 1896–1901), vol. 69, p. 211.

90. "Resumen General que comprehende todos los Habitantes y Establecimientos de la Colonia de la Luisiana, Hecho el año de 1766," manuscript

in the Archivo General de Indias, Sevilla, Spain (copy courtesy Gregory Knapp); Jacqueline K. Voorhies, *Some Late Eighteenth Century Louisianans: Census Records of the Colony, 1758–1796* (Lafayette, LA, 1973), p. 163; Philippe Arbos, *L'Auvergne* (Paris, France, 1952), p. 178.

91. Jack D. L. Holmes, "Joseph Piernas and the Nascent Cattle Industry in Southwest Louisiana," *MR*, 17 (1966), p. 13; Lauren C. Post, "The Old Cattle Industry of Southwestern Louisiana," *MR*, 9 (1957), pp. 45–46, 49, 50; "Resumen General."

Chapter Five

1. Jerzy Rzedowski and Rogers McVaugh, *La vegetación de Nueva Galicia*, University of Michigan Herbarium, Contributions, 9:1 (Ann Arbor, MI, 1966), pp. 45–52, 80–81, map at end; David A. Brading, *Haciendas and Ranchos in the Mexican Bajío: León 1700–1860* (Cambridge, UK, 1978), pp. 14–15, François Chevalier, *Land and Society in Colonial Mexico—The Great Hacienda*, trans. Alvin Eustis (Berkeley, CA, 1963), p. 94.

2. José Matesanz, "Introducción de la ganadería en Nueva España 1521 1535," *Historia Mexicana*, 14 (1964), p. 539; Richard J. Morrisey, "Colonial Agriculture in New Spain," *AH*, 31 (1957), p. 27; Richard J. Morrisey, "The Establishment and Northward Expansion of Cattle Ranching in New Spain," Ph.D. diss., University of California (Berkeley, CA, 1949), p. 57; Brading, *Haciendas and Ranchos*, pp. 14, 16.

3. Lesley B. Simpson, *Exploitation of Land in Central Mexico in the Sixteenth Century* (Berkeley and Los Angeles, CA, 1952), p. 22; Donald D. Brand, "The Early History of the Range Cattle Industry in Northern Mexico," *AH*, 35 (1961), p. 133; Richard J. Morrisey, "The Northward Expansion of Cattle Ranching in New Spain, 1550–1600," *AH*, 25 (1951), pp. 116–20; David Dary, *Cowboy Culture: A Saga of Five Centuries* (New York, NY, 1981), pp. 10, 16; Morrisey, "Establishment," pp. 57, 58, 63–67, 76, 95; Matesanz, "Introducción," p. 540.

4. Chevalier, *Land and Society*, p. 121; Morrisey, "Establishment," pp. 52, 63, 71; Morrisey, "Northward Expansion," p. 121.

5. Simpson, *Exploitation*, p. 85; Brading, *Haciendas and Ranchos*, p. 16; Morrisey, "Establishment," p. 50; Morriscy, "Colonial Agriculture," p. 27; Morrisey, "Northward Expansion," p. 116.

6. Morrisey, "Establishment," p. 85; Morrisey, "Northward Expansion," pp. 116, 119, 120.

7. Silvo Zavala, "The Frontiers of Hispanic America," in Walker D. Wyman and Clifton B. Kroeber, eds., *The Frontier in Perspective* (Madison, WI, 1957), p. 47; Dary, *Cowboy Culture*, p. 35; Morrisey, "Establishment," pp. 114–15, 120–21.

8. Ramon M. Serrera Contreras, *Guadalajara ganadera: estudio regional novohispano, 1760–1805* (Sevilla, Spain, 1977), pp. 75–168; Brading, *Ha-*

ciendas and Ranchos, pp. 8, 11, 12, 17, 20, 22; Chevalier, *Land and Society,* pp. 95, 104; Morrisey, "Northward Expansion," p. 118; Charles W. Hackett, ed., *Historical Documents Relating to New Mexico, Nueva Vizcaya, and Approaches Thereto, to 1773* (Washington, DC, 1923–37), vol. 2, pp. 29–30, 85.

9. Robert C. West, *The Mining Community in Northern New Spain: The Parral Mining District* (Berkeley and Los Angeles, CA, 1949), pp. 57–59; Rzedowski and McVaugh, *Vegetación de Nueva Galicia,* pp. 45–53, map at end.

10. Robert C. West, "The Flat-Roofed Folk Dwelling in Rural Mexico," *Geoscience and Man,* 5 (1974), pp. 111–32; Morrisey, "Northward Expansion," p. 121.

11. Wolfgang Trautmann, "Geographical Aspects of Hispanic Colonization on the Northern Frontier of New Spain," *Erdkunde,* 40 (1986), p. 248; Peter Gerhard, *The North Frontier of New Spain* (Princeton, NJ, 1982), pp. 62–66; Enrique Florescano, ed., *Atlas histórico de México* (Mexico City, 2d ed., 1984), pp. 59, 77; Alfred W. Crosby, Jr., *The Columbian Exchange: Biological and Cultural Consequences of 1492* (Westport, CT, 1972), p. 88; Brand, "Early History," p. 134; Morrisey, "Northward Expansion," p. 121; Dary, *Cowboy Culture,* pp. 15, 23–25; Chevalier, *Land and Society,* p. 105; Morrisey, "Establishment," pp. 90–96; West, *Parral,* pp. 57, 61.

12. Howard S. Gentry, *Los pastizales de Durango: estudio ecológico, fisiográfico y florístico* (Mexico City, Mexico, 1957), pp. 25–94, map on p. 27; West, *Parral,* p. 58; Carl O. Sauer, *The Road to Cíbola* (Berkeley and Los Angeles, CA, 1932), p. 47; Lee C. Buffington and Carlton H. Herbel, "Vegetational Changes on a Semidesert Grassland Range from 1858 to 1963," *Ecological Monographs,* 35 (1965), p. 162; V. V. Parr, G. W. Collier, and G. S. Klemmedson, *Ranch Organization and Methods of Livestock Production in the Southwest* (Washington, DC, 1928), p. 31.

13. Donald D. Brand, "The Historical Geography of Northwestern Chihuahua," Ph.D. diss., University of California (Berkeley, CA, 1933), pp. 116, 120–21, 142; Brand, "Early History," pp. 134–35; West, *Parral,* pp. 58–62.

14. Ben F. Lemert, "Parras Basin, Southern Coahuila, Mexico," *EG,* 25 (1949), 94–96; Eugenio del Hoyo, *Señores de ganado, Nuevo Reino de León, siglo XVII* (Monterrey, Mexico, 1987), pp. 16, 24–28; Philip Wagner, "Parras: A Case History in the Depletion of Natural Resources," *Landscape,* 5 (1955), 19–28; Morrisey, "Northward Expansion," p. 121; Morrisey, "Establishment," pp. 96–97.

15. Manuel A. Machado, Jr., *The North Mexican Cattle Industry, 1910–1975* (College Station, TX, 1981), especially pp. xii, 108–9; del Hoyo, *Señores de ganado,* pp. 28–30.

16. West, *Parral,* pp. 60–62; Gerhard, *North Frontier,* pp. 63–65; Brand, "Historical Geography," p. 116; Brand, "Early History," pp. 134–35; Morrisey, "Establishment," pp. 197–99; Morrisey, "Northward Expansion," pp. 120–21; Dary, *Cowboy Culture,* p. 16.

17. Francis L. Fugate, "Origins of the Range Cattle Era in South Texas," *AH*, 35 (1961), p. 157; Gerhard, *North Frontier*, pp. 27, 65.

18. Jorge A. Vivó, *Geografía de México* (Mexico City, Mexico, and Buenos Aires, Argentina, 4th ed., 1958), pp. 82, 86; Basil M. Bensin, "Agroecological Exploration in the Soto la Marina Region, Mexico," *GR*, 25 (1935), pp. 287–91; Marshall C. Johnston, "Past and Present Grasslands of South Texas and Northeastern Mexico," *Ecology*, 44 (1963), pp. 456, 458–62; Robert C. West, ed., *Natural Environments and Early Cultures* (vol. 1 of *Handbook of Middle American Indians*) (Austin, TX, 1964), pp. 223, 377; David M. Vigness, trans., "Nuevo Santander in 1795: A Provincial Inspection by Félix Calleja," *SWHQ*, 75 (1972), pp. 470, 493; Morrisey, "Establishment," p. 97; Morrisey, "Northward Expansion," p. 121.

19. William E. Doolittle, "Las Marismas to Pánuco to Texas: The Transfer of Open Range Cattle Ranching from Iberia through Northeastern México," *Yearbook of the Conference of Latin Americanist Geographers*, 13 (1987), pp. 7–9; Oakah L. Jones, *Los Paisanos: Spanish Settlers on the Northern Frontier of New Spain* (Norman, OK, 1979), p. 71; Gerhard, *North Frontier*, pp. 27, 366; Morrisey, "Establishment," pp. 99–100.

20. Herbert E. Bolton, *Texas in the Middle Eighteenth Century* (Berkeley, CA, 1915), pp. 293–300; Alejandro Prieto, *Historia, geografía y estadística del estado de Tamaulipas* (Mexico City, Mexico, 1873), pp. 140–91; Gerhard, *North Frontier*, pp. 358, 366–67; Jones, *Paisanos*, p. 70; Vigness, "Nuevo Santander," p. 475.

21. Gerhard, *North Frontier*, p. 366; Herbert E. Bolton, "Tienda de Cuervo's Inspección of Laredo, 1757," *QTSH*, 6 (1903), pp. 191, 200–201; Valgene W. Lehmann, *Forgotten Legions: Sheep in the Rio Grande Plain of Texas* (El Paso, TX, 1969), pp. 13–16; Karl W. Butzer and Elisabeth Butzer, "The Huastec Frontier as a Prelude to Nuevo León," unpublished paper read at the symposium, Cultural Adaptation at the Edge of the Spanish Empire, University of Texas, Austin, February 23, 1991; Gaspar José de Solís, "Diary of a Visit of Inspection of the Texas Missions Made by Fray Gaspar José de Solís in the Years 1767–68," trans. Margaret K. Kress, *SWHQ*, 35 (1931), p. 35; del Hoyo, *Señores del ganado*, pp. 16, 25, 28; Prieto, *Historia*, pp. 157, 170, 174–75, 195; *Carta topográfica* of Mexico, 1:50,000, published by the Coordinación General del Sistema Nacional de Información, Mexico City—see for example sheets no. G14D51 (Burgos) and G14D71 (La Libertad).

22. J. Frank Dobie, *The Longhorns* (Boston, MA, 1941), pp. 9, 211; Warner P. Sutton, "Cattle-Breeding in Northern Mexico," in U.S. Consular Reports, *Cattle and Dairy Farming* (Washington, DC, 1888), vol. 2, p. 578 (quotation); Juan Fidel Zorrilla, "New Santander and the Integration of the Mexican Northeast," unpublished paper read at the symposium "Cultural Adaptation at the Edge of the Spanish Empire," University of Texas, Austin, February 23, 1991 (copy in possession of T. G. J.); Bensin, "Agroecological," p. 290; Vigness, "Nuevo Santander," p. 475; Prieto, *Historia*, p. 195.

23. Vigness, "Nuevo Santander," pp. 474–75, 477 (quotation); Jones, *Paisanos*, p. 71; Gerhard, *North Frontier*, p. 27; Bensin, "Agroecological," pp. 288, 290; *Carta topográfica*, sheets F14B74 and G14D16.

24. Bensin, "Agroecological," pp. 289, 296; Vigness, "Nuevo Santander," pp. 473, 475; Bolton, *Texas*, p. 300; Prieto, *Historia*, p. 195.

25. Vigness, "Nuevo Santander," p. 481.

26. Robert C. West and James J. Parsons, "The Topia Road: A Trans-Sierran Trail of Colonial Mexico," *GR*, 31 (1941), pp. 406–7; Sauer, *Road to Cíbola*, p. 29.

27. Forrest Shreve, "Lowland Vegetation in Sinaloa," *Bulletin of the Torrey Botanical Club*, 64 (1937), pp. 605–13; Howard S. Gentry, *Río Mayo Plants: A Study of the Flora and Vegetation of the Río Mayo, Sonora* (Washington, DC, 1942), pp. 24, 27–30; Roger Dunbier, *The Sonoran Desert: Its Geography, Economy, and People* (Tucson, AZ, 1968); James R. Hastings and Raymond M. Turner, *The Changing Mile: An Ecological Study of Vegetation Change with Time in the Lower Mile of an Arid and Semiarid Region* (Tucson, AZ, 1980); Frederick M. Wiseman, "The Edge of the Tropics: The Transition from Tropical to Subtropical Ecosystems in Sonora, Mexico," *Geoscience and Man*, 21 (1980), pp. 141–56; Glenn R. Vernam, *Man on Horseback* (New York, NY, 1964), p. 303; Jo Mora, *Californios: The Saga of the Hard-Riding Vaqueros, America's First Cowboys* (Garden City, NY, 1949), p. 99; Morrisey, "Establishment," pp. 91–94, 152; Rzedowski and McVaugh, *Vegetación de Nueva Galicia*, pp. 12–30, 80–81, map at end; Hackett, *Historical Documents*, vol. 3, p. 98 (quotation).

28. Carl O. Sauer, "Man in the Ecology of Tropical America," *Proceedings of the Ninth Pacific Science Congress of the Pacific Science Association, 1957* (Bangkok, Thailand, 1958), vol. 20, p. 109; Sauer, *Road to Cíbola*, pp. 6–9; J. J. Wagoner, *History of the Cattle Industry in Southern Arizona, 1540–1940* (Tucson, AZ, 1952), p. 8; Trautmann, "Geographical Aspects," p. 242; Morrisey, "Establishment," p. 94; Morrisey, "Northward Expansion," p. 121.

29. Serrera, *Guadalajara ganadera*, p. 107.

30. Hackett, *Historical Documents*, vol. 3, pp. 94–98, 122; Hastings and Turner, *Changing Mile*, pp. 28, 31; Dunbier, *Sonoran Desert*, pp. 118–31; Morrisey, "Establishment," p. 197; John F. Bannion, *The Mission Frontier of Sonora, 1620–1637* (New York, NY, 1955), pp. 1–5; Thomas E. Sheridan, *Where the Dove Calls: The Political Ecology of a Peasant Corporate Community in Northwestern Mexico* (Tucson, AZ, 1988), p. 7.

31. Dunbier, *Sonoran Desert*, pp. 118–31; Hastings and Turner, *Changing Mile*, p. 31; Morrisey, "Establishment," pp. 197–99; Philipp Segesser, "Document: The Relation of Philipp Segesser," trans. and ed. Theodore E. Treutlein, *Mid-America: An Historical Review*, ns. 27:3 (July 1945), pp. 184–86; Campbell W. Pennington, *The Pima Bajo of Central Sonora, Mexico: Volume I, the Material Culture* (Salt Lake City, UT, 1980), pp. 58–63; Bannion, *Mission Frontier*, pp. 22–38; Sheridan, *Where the Dove*, pp. 7–14.

32. Eusebio F. Kino, *Kino's Historical Memoir of Pimería Alta*, trans. Herbert E. Bolton (Berkeley and Los Angeles, CA, 1948), p. 65; Herbert E. Bolton, *Rim of Christendom: A Biography of Eusebio Francisco Kino, Pacific Coast Pioneer* (New York, NY, 1936); Morrisey, "Establishment," pp. 196–208; Herbert E. Bolton, *The Padre on Horseback* (San Francisco, CA, 1932), pp. 64–66; Bert Haskett, "Early History of the Cattle Industry in Arizona," *AZHR*, 6:4 (Oct. 1935), p. 4; David E. Doyel, "The Transition to History in Northern Pimería Alta," in David H. Thomas, ed., *Columbian Consequences* (Washington, DC, 1989), v. 1, p. 145.

33. Ignaz Pfefferkorn, *Sonora: A Description of the Province*, trans. Theodore E. Treutlein (Tucson, AZ, 1989), p. 98; Joseph Och, *Missionary in Sonora: The Travel Reports of Joseph Och, S.J., 1755–1767*, trans. Theodore E. Treutlein (San Francisco, CA, 1965), pp. 139–40; Robert C. West of Louisiana State University, Baton Rouge, letter to T. G. J. dated August 30, 1990; Sheridan, *Where the Dove*, p. 9; Wagoner, *History of Cattle*, pp. 24–26; Haskett, "Early History," pp. 4–5; Morrisey, "Establishment," p. 208; Pennington, *Pima Bajo*, p. 74.

34. Juan Nentvig, *Rudo Ensayo: A Description of Sonora and Arizona in 1764* (Tucson, AZ, 1980), p. 28; Pfefferkorn, *Sonora*, p. 102; Bert Haskett, "History of the Sheep Industry in Arizona," *AZHR*, 7:3 (July 1936), pp. 3–49; Segesser, "Document," p. 185; Morrisey, "Establishment," pp. 208–9; Haskett, "Early History," p. 7.

35. Pfefferkorn, *Sonora*, pp. 94–103; Nentvig, *Rudo Ensayo*, pp. 27–28, 43, 89; Segesser, "Document," pp. 184–86; Och, *Travel Reports*, pp. 139–43.

36. Pfefferkorn, *Sonora*, p. 95; Nentvig, *Rudo Ensayo*, pp. 27, 89; Segesser, "Document," pp. 165–70; Morrisey, "Establishment," pp. 211–12; Sheridan, *Where the Dove*, pp. 14–15.

37. Ray H. Mattison, "Early Spanish and Mexican Settlements in Arizona," *NMHR*, 21 (1946), pp. 285–327; Richard J. Morrisey, "The Early Range Cattle Industry in Arizona," *AH*, 24 (1950), p. 151; Haskett, "Early History," pp. 3–6; Wagoner, *History of Cattle*, pp. 20–21, 26–27; Sheridan, *Where the Dove*, p. 16.

38. Aurelio Martínez Balboa, *La ganadería en Baja California Sur* (La Paz, Mexico, 1981), v. 1, pp. 22–25, 105, 113; Peveril Meigs III, *The Dominican Mission Frontier of Lower California* (Berkeley, CA, 1935); Carl O. Sauer and Peveril Meigs III, "Lower California Studies, I: Site and Culture at San Fernando de Velicatá," *University of California Publications in Geography* 2 (1927), pp. 271–302; L. T. Burcham, *California Range Land: An Historico-Ecological Study of the Range Resources of California* (Sacramento, CA, 1957), pp. 48–49; Arnold R. Rojas, *The Vaquero* (Charlotte, NC, and Santa Barbara, CA, 1964), pp. 93–94; Robert G. Cleland, *The Cattle on a Thousand Hills: Southern California, 1850–1880* (San Marino, CA, 1951), pp. 235–37; Morrisey, "Establishment," pp. 204–5, 214–17; George F. Deasy and Peter

Gerhard, "Settlements in Baja California, 1768–1930," *GR*, 34 (1944), pp. 574–86; Kino, *Historical Memoir*, p. 223.

39. Oakah L. Jones, Jr., "Spanish Settlers of the Northern Borderlands: Origins and Occupations," *Proceedings of the Pacific Coast Council on Latin American Studies*, 7 (1980–81), pp. 12–13; Brand, "Early History," p. 134; Trautmann, "Geographical Aspects," p. 242; Morrisey, "Establishment," pp. 190–91.

40. John O. Baxter, *Las Carneradas: Sheep Trade in New Mexico, 1700–1860* (Albuquerque, NM, 1987), pp. x, 88; Haskett, "History of Sheep," pp. 4–5; Morrisey, "Establishment," pp. 192–95; Hackett, *Historical Documents*, vol. 3, pp. 71, 131.

41. J. W. Fewkes, "The Pueblo Settlements near El Paso," *American Anthropologist*, 4 (1902), pp. 57–72; Jones, "Spanish Settlers," p. 13; Brand, "Early History," p. 134; Trautmann, "Geographical Aspects," p. 242; Morrisey, "Establishment," p. 193; Brand, "Historical Geography," p. 121.

42. Baxter, *Carneradas*, pp. 21, 42, 90.

43. Richard Nostrand, "The Century of Hispano Expansion," *NMHR*, 62 (1987), pp. 361–86; Jones, "Spanish Settlers," p. 12; Harold Hoffmeister, "The Consolidated Ute Indian Reservation," *GR*, 35 (1945), p. 613.

44. Kathleen Gilmore, "The Indians of Mission Rosario: From the Books and from the Ground," in David H. Thomas, ed., *Columbian Consequences* (Washington, DC, 1989), v. 1, pp. 231–44; Jack Jackson, *Los Mesteños: Spanish Ranching in Texas, 1721–1821* (College Station, TX, 1986), pp. 12, 35–49, 485; Odie B. Faulk, "Ranching in Spanish Texas," *HAHR*, 45 (1965), pp. 257–58, 262–63; Sandra L. Myres, *The Ranch in Spanish Texas, 1691–1800* (El Paso, TX, 1969), p. 12; Charles Ramsdell, "Espíritu Santo: An Early Texas Cattle Ranch," *Texas Geographic Magazine*, 13:2 (Fall 1949), pp. 21–25; Carlos E. Castañeda, *Our Catholic Heritage in Texas, 1519–1936* (Austin, TX, 1939), vol. 4, pp. 6–15, 23–26, 30–32; Herbert E. Bolton, "The Founding of Mission Rosario: A Chapter in the History of the Gulf Coast," *QTSH*, 10 (1906), pp. 113–39; J. Frank Dobie, "The First Cattle in Texas and the Southwest, Progenitors of the Longhorns," *SWHQ*, 42 (1939), p. 177; David J. Weber, *The Mexican Frontier, 1821–1846: The American Southwest Under Mexico* (Albuquerque, NM, 1982), p. 53; James E. Ivey, "The Origins of the Mission Ranches of San Antonio," unpublished paper read at the meetings of the Texas State Historical Association in Austin, TX, 1992; Morrisey, "Establishment," pp. 175–87; Lehmann, *Forgotten Legions*, pp. 11–13; Solís, "Diary," pp. 39, 46.

45. Sandra L. Myres, "The Spanish Cattle Kingdom in the Province of Texas," *Texana*, 4 (1966), pp. 236, 240; Solís, "Diary," p. 47; Jones, "Spanish Settlers," p. 17; Robert H. Thonhoff, *The Texas Connection with the American Revolution* (Burnet, TX, 1981), frontispiece, pp. 13–17; Jackson, *Mesteños*, pp. 51–72, 91–96, 518–20; Myres, *Ranch in Spanish*, pp. 15–16, 22;

Faulk, "Ranching," pp. 260–62; Lehmann, *Forgotten Legions,* p. 20; Ivey, "Origins of Mission."

46. Anne A. Fox, "The Indians at Rancho de las Cabras," in David H. Thomas, ed., *Columbian Consequences* (Washington, DC, 1989), v. 1, pp. 259–67; James Ivey and Anne A. Fox, *Archaeological Survey and Testing at Rancho de las Cabras, Wilson County, Texas* (San Antonio, TX, 1981); Charles Ramsdell, "Spanish Goliad," unpub. manuscript, University of Texas Archives, p. 22; Jackson, *Mesteños,* pp. 41–45; Myres, *Ranch in Spanish,* p. 29; Dobie, *Longhorns,* p. 8; Myres, "Spanish Cattle Kingdom," p. 241; Ivey, "Origins of Mission."

47. C. Wayne Hanselka and Dan E. Kilgore, "The Nueces Valley: The Cradle of the Western Livestock Industry," *Rangelands,* 9:5 (Oct. 1987), pp. 196–97; Donald E. Worcester, *The Spanish Mustang: From the Plains of Andalusia to the Prairies of Texas* (El Paso, TX, 1986); LeRoy Graf, "The Economic History of the Lower Rio Grande Valley, 1820–1875); Ph.D. diss., Harvard University (Cambridge, MA, 1942); Tom Lea, *The King Ranch* (Boston, MA, 1957), vol. 1, pp. 378–79; Richard J. Russell, "Climates of Texas," *AAAG,* 35 (1945), pp. 37–52; Emilia S. Ramírez, *Ranch Life in Hidalgo County after 1850* (Edinburg, TX, 1971), n.p.; Myres, "Spanish Cattle Kingdom," p. 237; Jackson, *Mesteños,* pp. 34, 98–99, 445–49, 638–43; Lehmann, *Forgotten Legions,* pp. 44–48.

48. Victor M. Rose, *Some Historical Facts in Regard to the Settlement of Victoria, Texas* (Laredo, TX, 1883), pp. 10–15.

49. Jones, "Spanish Settlers," pp. 17–18; Faulk, "Ranching," p. 257; Lehmann, *Forgotten Legions,* pp. 1, 13–16, 44–48, 122; Dary, *Cowboy Culture,* pp. 37, 39.

50. Willard B. Robinson, "Colonial Ranch Architecture in the Spanish-Mexican Tradition," *SWHQ,* 83 (1979), pp. 131, 136–39, 144–46, 149; Lonn Taylor, "Rails, Rocks, and Pickets: Traditional Farmstead Fencing in Texas," in Francis E. Abernethy, ed., *Built in Texas* (Waco, TX, 1979), pp. 184–87; Dobie, *Longhorns,* p. 239.

51. Dan E. Kilgore, "The Spanish Missions and the Origins of the Cattle Industry in Texas," in Gilberto R. Cruz, ed., *Proceedings of the Second Annual Mission Research Conference* (San Antonio, TX, 1984), pp. 63, 67; Bolton, "Founding of Mission," p. 120; Doolittle, "Las Marismas," pp. 3–11.

52. Elmer B. Atwood, *The Regional Vocabulary of Texas* (Austin, TX, 1962), p. 109; J. Frank Dobie, "The Mexican Vaquero of the Texas Border," *Southwestern Political and Social Science Quarterly,* 8 (1927), p. 18; William Morris, ed., *The American Heritage Dictionary of the English Language* (Boston, MA, 1969), p. 857; Terry G. Jordan, "The Origin of *Mott* in Anglo-Texan Vegetational Terminology," in J. L. Dillard, *Perspectives on American English* (The Hague, Netherlands, 1980), pp. 163, 164, 170; Terry G. Jordan, "The Origin of Motte and Island in Texan Vegetational Terminology," *Southern Folklore Quarterly,* 34 (1972), p. 134. I was incorrect in earlier ascribing

mott to the Irish; it is now clear that they merely modified the Spanish word, using the spelling of a familiar Irish word describing a similar feature.

53. William A. McClintock, "Journal of a Trip through Texas and Northern Mexico in 1846–1847," *SWHQ*, 34 (1930–31), p. 156; Cornelius C. Cox, "Reminiscences of C. C. Cox," *QTSH*, 6 (1902–3), p. 206; Terry G. Jordan, *Trails to Texas: Southern Roots of Western Cattle Ranching* (Lincoln, NE, 1981), p. 151; Milford B. Mathews, *A Dictionary of Americanisms on Historical Principles* (Chicago, IL, 1951), vol. 2, p. 1355; Sutton, "Cattle-Breeding," p. 577; *Carta topográfica*, sheets no. G14D14, G14D51, G14D72.

54. J. H. Kuykendall, "Reminiscences of Early Texans," *QTSH*, 6 (1902–3), p. 253; Daniel Shipman, *Frontier Life: 58 Years in Texas* (n.p., 1879), pp. 40–42; Mrs. Thomas O'Connor, "Martín de León," in Walter P. Webb and H. Bailey Carroll, eds., *The Handbook of Texas* (Austin, TX, 1952), vol. 1, p. 484; Roy Grimes, ed., *300 Years in Victoria County* (Victoria, TX, 1968), pp. 60, 100–5; A. B. J. Hammett, *The Empresario Don Martín de León* (Waco, TX, 1973), p. 32; Mary V. Henderson, "Minor Empresario Contracts for the Colonization of Texas," *SWHQ*, 32 (1928), pp. 5–9; Jackson, *Mesteños*, pp. 505–6, 519, 549–50, 598; Rose, *Some Historical Facts*, pp. 11, 151–54; Prieto, *Historia*, pp. 156, 164–67, 206; Gerhard, *North Frontier*, p. 367.

55. Cox, "Reminiscences," pp. 208–9; Shipman, *Frontier Life*, pp. 40, 42; Rose, *Some Historical Facts*, p. 12; Grimes, *300 Years*, pp. 69–70; Hammett, *Empresario*, p. 28; Ramsdell, "Spanish Goliad," p. 22; Dobie, "Mexican Vaquero," p. 15.

56. Jackson, *Mesteños*, pp. 96–98, 114–17, 182–83; Myres, *Ranch in Spanish*, pp. 11–12, 16; Jordan, *Trails*, p. 108; Faulk, "Ranching," pp. 262–63.

57. Thonhoff, *Texas Connection*, pp. 46–72; Jordan, *Trails*, pp. 71, 133; Jackson, *Mesteños*, pp. 236–50; Rose, *Some Historical Facts*, p. 12; Ramsdell, "Spanish Goliad," p. 23; Myres, *Ranch in Spanish*, pp. 44–48; Dobie, "First Cattle," p. 178; Faulk, "Ranching," p. 265; Weber, *Mexican Frontier*, pp. 139, 332; Myres, "Spanish Cattle Kingdom," p. 244.

58. Fred B. Kniffen, "A Spanish Spinner in Louisiana," *Southern Folklore Quarterly*, 13 (1949), pp. 192–99; Fred B. Kniffen, "The Western Cattle Complex, Notes on Differentiation and Diffusion," *Western Folklore*, 12 (1953), pp. 179–81; Francis Baily, *Journal of a Tour in Unsettled Parts of North America in 1796 and 1797* (Carbondale, IL, 1969), p. 189; Antonio Alcalá Venceslada, *Vocabulario andaluz* (Madrid, Spain, 1951), p. 596; Atwood, *Regional Vocabulary*, pp. 55, 161.

59. Jackson, *Mesteños*, pp. 585–617; Hanselka and Kilgore, "Nueces Valley," pp. 195–202; Fugate, "Origins of Range," pp. 155–58.

60. Lea, *King Ranch*, vol. 2, pp. 499–500; Lehmann, *Forgotten Legions*, pp. 17–22, 121–23; Jordan, *Trails*, pp. 108–10, 133, 151–52.

61. Louise S. O'Connor, *Cryin' for Daylight: A Ranching Culture in the Texas Coastal Bend* (Austin, TX, 1989); William H. Oberste, *Texas Irish Empresarios and their Colonies* (Austin, TX, 1953); Jordan, *Trails*, pp. 133,

151; Grimes, *300 Years*, pp. 93–94, 213–15; Rose, *Some Historical Facts*, pp. 11, 15–16, 106–7, 115.

62. Julia G. Costello and David Hornbeck, "Alta California: An Overview," in David H. Thomas, ed., *Columbian Consequences* (Washington, DC, 1989), v. 1, pp. 303–31; Richard J. Russell, "Climates of California," *University of California Publications in Geography*, 2:4 (1926), pp. 73–84; John E. Kesseli, "The Climates of California According to the Köppen Classification," *GR*, 32 (1942), pp. 476–80; Hubert H. Bancroft, *California Pastoral, 1769–1848* (San Francisco, CA, 1888), p. 341; L. T. Burcham, "Cattle and Range Forage in California, 1770–1880," *AH*, 35 (1961), p. 147; Burcham, *California Range*, pp. 69–79, 140.

63. Burcham, *California Range*, p. 98.

64. Hazel A. Pulling, "Range Forage and California's Range-Cattle Industry," *Historian*, 7 (1944–45), pp. 114–18; Hazel A. Pulling, "A History of California's Range-Cattle Industry, 1770–1912," Ph.D. diss., University of Southern California (Los Angeles, CA, 1944), pp. 1–28; Joseph B. Davy, *Stock Ranges of Northwestern California: Notes on the Grasses and Forage Plants and Range Conditions* (Washington, DC, 1902), pp. 27–32; Roberta S. Greenwood, "The California Ranchero: Fact and Fancy," in David H. Thomas, ed., *Columbian Consequences* (Washington, DC, 1989), v. 1, p. 455; Burcham, *California Range*, pp. 7, 80, 86–92, 96, 122; Burcham, "Cattle and Range," p. 142.

65. R. Louis Gentilcore, "Missions and Mission Lands of Alta California," *AAAG*, 51 (1961), p. 65.

66. H. F. Raup, "Transformation of Southern California to a Cultivated Land," *AAAG*, 49:3, part 2 (1959), p. 58; Burcham, *California Range*, pp. 132–33; Gentilcore, "Missions," p. 65; Morrisey, "Establishment," pp. 216–20; Dary, *Cowboy Culture*, pp. 45–51; Weber, *Mexican Frontier*, pp. 48–49, 60; Bancroft, *California Pastoral*, pp. 183–247.

67. David Hornbeck, "Economic Growth and Change at the Missions of Alta California, 1769–1846," in David H. Thomas, ed., *Columbian Consequences* (Washington, DC, 1989), v. 1, pp. 423–33; James W. Thompson, *A History of Livestock Raising in the United States, 1607–1860* (Washington, DC, 1942), p. 113; Morrisey, "Establishment," pp. 219–20, 223; Burcham, *California Range*, pp. 132–33, 138; Burcham, "Cattle and Range," p. 141; Bancroft, *California Pastoral*, p. 339; Gentilcore, "Missions," p. 66; Pulling, "History of California's," pp. 32–59, Appendix A.

68. W. W. Robinson, "The Story of Rancho San Pasqual," *HSSCQ*, 38 (1955), p. 348; Gentilcore, "Missions," pp. 67–68; Burcham, "Cattle and Range," p. 144.

69. Jones, "Spanish Settlers," pp. 20–21; Burcham, *California Range*, p. 135; Gentilcore, "Missions," pp. 58, 69.

70. Arthur B. Perkins, "Rancho San Francisco: A Story of a California Land Grant," *HSSCQ*, 39 (1957), p. 106; Julia G. Costello, "Variability among

the Alta California Missions: The Economics of Agricultural Production," in David H. Thomas, ed., *Columbian Consequences* (Washington, DC, 1989), v. 1, pp. 435–49; Bancroft, *California Pastoral*, p. 339; Gentilcore, "Missions," pp. 59, 66–69.

71. Weber, *Mexican Frontier*, pp. 64–67; Bancroft, *California Pastoral*, p. 339.

72. Iris H. W. Engstrand, "California Ranchos: Their Hispanic Heritage," *SCQ*, 67 (1985), pp. 281–82; Dary, *Cowboy Culture*, p. 52; Bancroft, *California Pastoral*, pp. 248–59, 348; David Hornbeck, "Land Tenure and Rancho Expansion in Alta California, 1784–1846," *JHG*, 4 (1978), pp. 374–77; Gentilcore, "Missions," p. 68; Morrisey, "Establishment," pp. 220–22; Burcham, "Cattle and Range," p. 144; Greenwood, "California Ranchero," pp. 452–53; Pulling, "History of California's," pp. 39, 50.

73. Louis E. Guzman, "San Fernando Valley: Two Hundred Years in Transition," *California Geographer*, 3 (1962), p. 55; Cleland, *Cattle on Thousand*, pp. 12–17, 73; Morrisey, "Establishment," pp. 223–24.

74. Robert G. Cowan, *Ranchos of California: A List of Spanish Concessions, 1775–1822, and Mexican Grants, 1822–1846* (Fresno, CA, 1956), endpaper maps; Esther B. Black, *Rancho Cucamonga and Doña Merced* (Redlands, CA, 1975), pp. 187–207; Greenwood, "California Ranchero," pp. 454–61; Pulling, "History of California's," pp. 60–77; Hornbeck, "Land Tenure," pp. 384–85; Weber, *Mexican Frontier*, p. 60; Cleland, *Cattle on Thousand*, pp. 19–32.

75. Jan O. M. Broek, *The Santa Clara Valley, California: A Study in Landscape Changes* (Utrecht, Netherlands, 1932); Cleland, *Cattle on Thousand*, pp. 3–32; Greenwood, "California Ranchero," pp. 456–58; Hornbeck, "Land Tenure," pp. 371, 379, 384–88; Bancroft, *California Pastoral*, p. 342. For examples of individual ranches, see: Robert G. Cleland, *The Irvine Ranch of Orange County, 1810–1950* (San Marino, CA, 1952); William R. Cameron, "Rancho Santa Margarita of San Luis Obispo," *CHSQ*, 36 (1957), pp. 1–20; Robert G. Cleland, *The Place Called Sespe: The History of a California Ranch* (San Marino, CA, 1957); Robert Gillingham, *The Rancho San Pedro* (Los Angeles, CA, 1961); Perkins, "Rancho San Francisco," pp. 99–126; H. F. Raup, "The Rancho Palos Verdes," *HSSCQ*, 19 (1937), pp. 7–21; Lois J. Roberts, "Rancho Jesús María, Santa Barbara County," *SCQ*, 68 (1986), pp. 1–35; Robinson, "Story of Rancho," pp. 347–53; Black, *Rancho Cucamonga*.

76. Bancroft, *California Pastoral*, p. 180.

77. Richard H. Dana, *Two Years Before the Mast* (Boston, MA, 1884); Craig M. Carver, *American Regional Dialects: A Word Geography* (Ann Arbor, MI, 1987), pp. 224–25; Gentilcore, "Missions," pp. 61, 69–70; Burcham, *California Range*, pp. 135, 139; Bancroft, *California Pastoral*, pp. 335, 347; Raup, "Transformation," p. 61.

78. Rojas, *Vaquero*, p. 21; Mora, *Californios*, pp. 17, 99, 102; Vernam, *Man*

on Horseback, pp. 303, 313, 342, 344; Pulling, "History of California's," p. 102a; Kniffen, "Western Cattle," p. 185; Burcham, "Cattle and Range," p. 145; Bancroft, California Pastoral, pp. 231, 251, 289, 373–74, 379–80, 385–87, 531.

79. Manfred R. Wolfenstine, The Manual of Brands and Marks (Norman, OK, 1970), plate 59, following p. 353; Bancroft, California Pastoral, pp. 291, 340–45, 445, 531; Burcham, California Range, p. 135; Burcham, "Cattle and Range," p. 145; Thompson, History of Livestock, p. 114; Cleland, Cattle on Thousand, pp. 54–56, 72, 73; Rojas, Vaquero, p. 111; Cleland, Place Called Sespe, pp. 51–52; Pulling, "History of California's," pp. 75a, 102a.

80. Richard L. Nostrand, "The Santa Ynez Valley: Hinterland of Coastal California," HSSCQ, 48 (1966), pp. 41–42; Paul F. Starrs, "The Cultural Landscape of California Pastoralism: 200 Years of Changes," in W. James Clawson, ed., Landscape Ecology: Study of Mediterranean Grazed Ecosystems (Nice, France, 1989), pp. 49–61; Perkins, "Rancho San Francisco," p. 103; Gentry, Río Mayo Plants, p. 24; Burcham, California Range, pp. 185–88, 199–200; Burcham, "Cattle and Range," p. 145; Bancroft, California Pastoral, p. 341.

Chapter Six

1. John D. W. Guice, "Cattle Raisers of the Old Southwest: A Reinterpretation," WHQ, 8 (1977), pp. 167–87; Grady McWhiney, Cracker Culture: Celtic Ways in the Old South (Tuscaloosa, AL, 1988), pp. 51–79; Forrest McDonald and Grady McWhiney, "The Antebellum Southern Herdsman: A Reinterpretation," JSH, 41 (1975), pp. 147–66.

2. Louis DeVorsey, Jr., The Indian Boundary in the Southern Colonies, 1763–1775 (Chapel Hill, NC, 1961), pp. 154, 158; William P. Cumming, North Carolina in Maps (Raleigh, NC, 1966), map by Henry Mouzon; H. H. Biswell et al., "Native Forage Plants of Cutover Lands in the Coastal Plain of Georgia," Bulletin, Georgia Coastal Plain Experiment Station, 37 (1943), pp. 5–20.

3. Gary S. Dunbar, "Colonial Carolina Cowpens," AH, 35 (1961), pp. 125–30; H. Roy Merrens, The Colonial South Carolina Scene: Contemporary Views, 1697–1774 (Columbia, SC, 1977), p. 69; David Ramsay, The History of South-Carolina from Its First Settlement in 1670 to the Year 1808 (Charleston, SC, 1809), vol. 1, p. 207, vol. 2, p. 584; Converse D. Clowse, Economic Beginnings in Colonial South Carolina, 1670–1730 (Columbia, SC, 1971), pp. 184–86; Richard D. Brooks, "Cattle Ranching in Colonial South Carolina: A Case Study in History and Archaeology of the Lazarus/Catherina Brown Cowpen," M.A. thesis, University of South Carolina (Columbia, SC, 1988), pp. 8, 27; James S. Maag, "Cattle Raising in Colonial South Carolina," M.A. thesis, University of Kansas (Lawrence, KS, 1964), pp. 43, 45, 49, 51, 91; Richard L. Haan, "'The Trade Do's not Flourish as

Formerly': The Ecological Origins of the Yamassee War of 1715," *Ethnohistory*, 28 (1981), pp. 341, 343, 350; John S. Otto, *The Southern Frontiers, 1607–1860* (Westport, CT, 1989), p. 37; John S. Otto, "Livestock-Raising in Early South Carolina, 1670–1700: Prelude to the Rice Plantation Economy," *AH*, 61 (1987), p. 23; John S. Otto, "The Origins of Cattle-Ranching in Colonial South Carolina, 1670–1715," *SCHM*, 87 (1986), p. 123.

4. Maag, "Cattle Raising," p. 91; Otto, "Origins," p. 124; Merrens, *Colonial South Carolina*, pp. 180–81.

5. Walter Clark, ed., *The State Records of North Carolina*, vol. 23 (Goldsboro, NC, 1904), p. 59.

6. William L. Saunders, ed., *The Colonial Records of North Carolina* (Raleigh, NC, 1886–90), vol. 3, pp. 431–32, vol. 4, p. 53; Clark, *State Records*, vol. 23, pp. 167, 676; H. Roy Merrens, *Colonial North Carolina in the Eighteenth Century* (Chapel Hill, NC, 1964), pp. 138, 140; Gary S. Dunbar, *Historical Geography of North Carolina Outer Banks* (Baton Rouge, LA, 1958), pp. 18, 31; Harry J. Carman, ed., *American Husbandry* (New York, NY, 1939), pp. 240–41.

7. James C. Bonner, "The Open Range Livestock Industry in Colonial Georgia," *Georgia Review*, 17 (1963), pp. 85–92, and John H. Goff, "Cow Punching in Old Georgia," *Georgia Review*, 3 (1949), pp. 341–48.

8. Allen D. Chandler, ed., *The Colonial Records of Georgia* (Atlanta, GA, 1904–6), vol. 2, pp. 502, 520, vol. 4, pp. 160–61, vol. 6, p. 173; Goff, "Cow Punching," p. 341; Merrens, *Colonial South Carolina*, pp. 119–20.

9. William G. DeBrahm, *DeBrahm's Report of the General Survey in the Southern District of North America* (Columbia, SC, 1971), pp. 95, 96, 142; John S. Otto, "Traditional Cattle-Herding Practices in Southern Florida," *Journal of American Folklore*, 97 (1984), p. 302; Mart A. Stewart, "Whether Wast, Deodand, or Stray: Cattle, Culture, and the Environment in Early Georgia," *AH*, 65 (1991), pp. 2, 4, 11, 14, 22–23, 27–28.

10. Carman, *American Husbandry*, p. 343.

11. William Bartram, *The Travels of William Bartram* (New Haven, CT, 1958), pp. 196–98; Ann P. Malone, "Piney Woods Farmers of South Georgia, 1850–1900," *AH*, 60 (1986), pp. 59, 64.

12. Maag, "Cattle Raising," p. 48; Stewart, "Whether Wast," p. 15.

13. Timothy Flint, *A Condensed Geography and History of the Western States, or the Mississippi Valley* (Cincinnati, OH, 1828), vol. 1, p. 471; DeBrahm, *Report*, pp. 180–86, 212.

14. W. Theodore Mealor, Jr., "The Open Range Ranch in South Florida and its Contemporary Successors," Ph.D. diss., University of Georgia (Athens, GA, 1972); W. Theodore Mealor, Jr., and Merle C. Prunty, "Open-Range Ranching in Southern Florida," *AAAG*, 66 (1976), pp. 360–76; Frederic Remington, "Cracker Cowboys of Florida," *Harper's New Monthly Magazine*, 91 (1895), pp. 339–45; George H. Dacy, *Four Centuries of Florida Ranching* (St. Louis, MO, 1940), pp. 25–27; Jim B. Tinsley, *Florida Cow Hunter: The Life*

and Times of Bone Mizell (Orlando, FL, 1990); and the following articles by John S. Otto: "Florida's Cattle Ranching Frontier," *FHQ,* 63 (1984–85), pp. 71–83 and 64 (1985–86), pp. 48–61; "Open-Range Cattle Herding in Antebellum South Florida," *Southeastern Geographer,* 26 (1986), pp. 55–67; "Reconsidering the Florida 'Cracker'," *Journal of Regional Cultures,* 4 (1984), pp. 7–14.

15. Thomas D. Clark and John D. W. Guice, *Frontiers in Conflict: The Old Southwest, 1795–1830* (Albuquerque, NM, 1989), pp. 99–116, 279; Terry G. Jordan, *Trails to Texas: Southern Roots of Western Cattle Ranching* (Lincoln, NE, 1981), pp. 48, 51; Jack D. L. Holmes, "Livestock in Spanish Natchez," *Journal of Mississippi History,* 23 (1961), pp. 15, 31, 35–36; Robert J. Baxter, "Cattle Raising in Early Mississippi," *Mississippi Folklore Register,* 10 (1976), pp. 1–23; Kenneth D. Israel, "The Cattle Industry of Mississippi, Its Origin and Its Changes through Time up to 1850," Ph.D. diss., University of Southern Mississippi (Hattiesburg, MS, 1970); Sam B. Hilliard, *Hog Meat and Hoe Cake: Food Supply in the Old South, 1840–1860* (Carbondale and Edwardsville, IL, 1972), pp. 116–22; Cecil Johnson, *British West Florida, 1763–1783* (New Haven, CT, 1943), p. 170; Guice, "Cattle Raisers," pp. 175, 177; Flint, *Condensed Geography,* vol. 1, p. 490.

16. Fred B. Kniffen, "The Western Cattle Complex: Notes on Differentiation and Diffusion," *Western Folklore,* 12 (1953), p. 183; Jordan, *Trails,* pp. 112–14.

17. Hilliard, *Hog Meat,* p. 117; Jordan, *Trails,* p. 115.

18. Solomon A. Wright, *My Rambles as East Texas Cowboy, Hunter, Fisherman, Tie-Cutter* (Austin, TX, 1942), pp. 1–2; Otto, "Traditional Cattle-Herding," pp. 293–94; Otto, "Florida's Cattle," pp. 79–80; Baxter, "Cattle Raising," pp. 1–2; Jordan, *Trails,* pp. 115–16.

19. J. Crawford King, "The Closing of the Southern Range," *JSH,* 48 (1982), pp. 53–70; Alan Gallay, *The Formation of a Planter Elite* (Athens, GA, 1989), pp. 84–90; Joe A. Akerman, Jr., *Florida Cowman: A History of Florida Cattle Raising* (Kissimmee, FL, 1976), pp. 99, 168–70, 193–201; Robert B. Douglas, "Antebellum Paired Plantations along Coastal Georgia," *The American South: Proceedings, Sixth International Conference of Historical Geographers* (Baton Rouge and New Orleans, LA, 1986), n.p.; S. W. Greene, "Relation Between Winter Grass Fires and Cattle Grazing in the Longleaf Pine Belt," *Journal of Forestry,* 33 (1935), p. 338; Joseph W. LeBon, Jr., "The Catahoula Hog Dog: A Cultural Trait of the Upland South," M.A. thesis, Louisiana State University (Baton Rouge, LA, 1970); Joseph W. LeBon, Jr., "The Catahoula Hog Dog: A Folk Breed," *Pioneer America,* 3:2 (1971), pp. 35–45; Georges Lutz, "Catahoula hog dog ou Catahoula cur: une opinion européene ou le retour aux origines du chien," *Journal d'Agriculture Traditionnelle et de Botanique Appliquée,* 31 (1984), pp. 147–69; Terry G. Jordan, "Cowboys," in Randall M. Miller and John D. Smith, eds., *Dictionary of Afro-American Slavery* (New York, NY, 1988), pp. 152–53; Bartram, *Travels,*

pp. 13, 196–97; Jordan, *Trails*, pp. 118–23; Dacy, *Four Centuries*, pp. 65–68, 149, 180; Baxter, "Cattle Raising," pp. 4, 9, 11, 13–15; Bonner, "Open Range," pp. 87–89; Goff, "Cow Punching," pp. 343–46; Stewart, "Whether Wast," pp. 17, 26, 27.

20. E. Bagby Atwood, *The Regional Vocabulary of Texas* (Austin, TX, 1962), pp. 156, 162; Ramon F. Adams, *Western Words: A Dictionary of the American West* (Norman, OK, 1968), p. 80; Bonner, "Open Range," p. 91; Dacy, *Four Centuries*, pp. 180, 299; Goff, "Cow Punching," p. 348; Akerman, *Florida Cowman*, pp. ix, 13, 163; De Vorsey, *Indian Boundary*, p. 158; Stewart, "Whether Wast," pp. 3, 14.

21. Terry G. Jordan and Matti Kaups, *The American Backwoods Frontier: An Ethnic and Ecological Interpretation* (Baltimore, MD, 1989), pp. 105–15, 138–39, 186–87; Jordan, *Trails*, pp. 35, 118; Baxter, "Cattle Raising," p. 12; Goff, "Cow Punching," p. 343.

22. Akerman, *Florida Cowman*, pp. 22–23, 29–31.

23. Norman A. Graebner, "History of Cattle Ranching in Eastern Oklahoma," *CO*, 21 (1943), pp. 300–311; Michael F. Doran, "Antebellum Cattle Herding in the Indian Territory," *GR*, 66 (1976), pp. 48–58; Brad A. Bays, "The Historical Geography of Cattle Herding among the Cherokee Indians, 1761–1861," M.S. thesis, University of Tennessee (Knoxville, TN, 1991); Jordan, *Trails*, pp. 110–11.

24. Milton B. Newton, Jr., "Folk Material Culture of the Lower South, 1870 to 1940," in Lucius F. Ellsworth and Linda V. Ellsworth, eds., *The Cultural Legacy of the Gulf Coast, 1870–1940* (Pensacola, FL, 1976), pp. 76–77; Jordan, *Trails*, pp. 46–49, 108–12; Guice, "Cattle Raisers," p. 171; Holmes, "Livestock," p. 18; Flint, *Condensed Geography*, v. 1, p. 471.

25. William Foster-Harris, *The Look of the Old West* (New York, NY, 1960), p. 237; Dacy, *Four Centuries*, p. 32; Holmes, "Livestock," pp. 18–20; Flint, *Condensed Geography*, v. 1, p. 490; Baxter, "Cattle Raising," pp. 4–5; Otto, "Traditional," p. 303; Stewart, "Whether Wast," pp. 5, 16.

26. Frederic G. Cassidy, ed., *Dictionary of American Regional English* (Cambridge, MA, 1985), v. 1, p. 497; J. L. Dillard, "The Maritime (Perhaps Lingua Franca) Relations of a Special Variety of the Gulf Corridor," *Journal of Pidgin and Creole Languages*, 2 (1987), pp. 244–49; J. L. Dillard, *A History of American English* (London, UK, and New York, NY, 1992), p. 129. Elizabeth A. H. John, "Portrait of a Wichita Village, 1808," *CO*, 60 (1982–83), pp. 419, 435; Francis Baily, *Journal of a Tour in Unsettled Parts of North America in 1796 and 1797* (Carbondale and Edwardsville, IL, 1969), p. 155.

27. R. J. Russell and H. V. Howe, "Cheniers of Southwestern Louisiana," *GR*, 25 (1935), pp. 449, 455; Milton B. Newton, Jr., *Atlas of Louisiana* (Baton Rouge, LA, 1972), pp. 19, 27, 32, 35, 38–39; Terry G. Jordan, "The Origin of Anglo-American Cattle Ranching in Texas: A Documentation of Diffusion From the Lower South," *EG*, 45 (1969), pp. 67–70; Jordan, *Trails*, pp. 59–62.

28. Jordan, *Trails*, pp. 62, 70.

29. Sandra L. Myres, *The Ranch in Spanish Texas, 1691–1800* (El Paso, TX, 1969), p. 25; Francis B. Lubbock, *Six Decades in Texas, or Memoirs* (Austin, TX, 1900), p. 137; Kniffen, "Western Cattle," p. 183; Atwood, *Regional Vocabulary*, pp. 55, 161; Jordan, *Trails*, p. 73.

30. Frederick L. Olmsted, *A Journey through Texas, or, a Saddle-Trip on the Southwestern Frontier* (Austin, TX, 1978), pp. 356, 365–71, 391, 393.

31. Walter P. Webb, *The Great Plains* (Boston, MA, 1931), pp. 205–69; Francis L. Fugate, "Origins of the Range Cattle Era in South Texas," *AH*, 35 (1961), p. 155.

32. Lauren C. Post, "The Old Cattle Industry of Southwestern Louisiana," *MR*, 9 (1957), pp. 43–55; Lauren C. Post, "Cattle Branding in Southwest Louisiana," *MR*, 10 (1958), pp. 101–17; Jack D. L. Holmes, "Joseph Piernas and the Nascent Cattle Industry of Southwest Louisiana," *MR*, 17 (1966), pp. 13–26; Ameda R. King, "Social and Economic Life in Spanish Louisiana from 1763 to 1783," Ph.D. diss., University of Illinois (Urbana, IL, 1931), pp. 74, 164, 167; Robert C. West, *An Atlas of Louisiana Surnames of French and Spanish Origin* (Baton Rouge, LA, 1986), p. 5.

33. Lauren C. Post, *Cajun Sketches from the Prairies of Southwest Louisiana* (Baton Rouge, LA, 1962), pp. 50, 59; Flint, *Condensed Geography*, v. 1, pp. 544–47; Guice, "Cattle Raisers," p. 176; West, *Atlas*, p. 7.

34. William F. Gray, *From Virginia to Texas, 1835* (Houston, TX, 1909), p. 171; Jordan, *Trails*, pp. 77–79.

35. Jordan, *Trails*, pp. 66, 71, 74, 79–80.

36. Charles W. Towne and Edward N. Wentworth, *Cattle and Men* (Norman, OK, 1955), p. 141.

37. Michel-Guillaume St. Jean de Crèvecoeur, *Journey into Northern Pennsylvania and the State of New York*, trans. C. L. Bostelmann (Ann Arbor, MI, 1964), pp. 323–26, 336; Gary C. Goodwin, *Cherokees in Transition: A Study of Changing Culture and Environment Prior to 1775* (Chicago, IL, 1977), p. 135; Maag, "Cattle Raising," pp. 38–39; Bays, "Historical Geography."

38. Clark, *State Records*, v. 23, p. 676; Saunders, *Colonial Records*, vol. 5, p. 322.

39. Charles Woodmason, *The Carolina Backcountry on the Eve of the Revolution*, ed. Richard J. Hooker (Chapel Hill, NC, 1953), p. 214; Maag, "Cattle Raising," pp. 53, 55–62, 87.

40. Eugene J. Wilhelm, Jr., "Animal Drives in the Southern Highlands," *Mountain Life and Work*, 45:2 (1966), pp. 8, 11; James W. Thompson, *A History of Livestock Raising in the United States, 1607–1860* (Washington, DC, 1942), p. 63; Peter Kalm, *Travels into North America*, trans. John R. Forster (Barre, MA, 1972), p. 110; Brooks, "Cattle Ranching," p. 64; Maag, "Cattle Raising," pp. 52, 77.

41. Jordan, *Trails*, pp. 42, 51.

42. Goff, "Cow Punching," pp. 347–48; Thompson, *History of Livestock*, p. 62; Towne and Wentworth, *Cattle and Men*, pp. 141–43.

43. Edmund C. Burnett, "Hog Raising and Hog Driving in the Region of the French Broad River," *AH*, 20 (1946), pp. 86–103; Jordan and Kaups, *American Backwoods*, pp. 119–23; Crèvecoeur, *Journey*, pp. 325–26, 328, 331–37.

44. Philip J. Gersmehl, "A Geographic Approach to a Vegetation Problem: The Case of the Southern Appalachian Grassy Balds," Ph.D. diss., University of Georgia (Athens, GA, 1970), pp. 2–4, 14, 25, 30, 36, 43, 131, 354, 404; Philip J. Gersmehl, "Factors Leading to Mountaintop Grazing in the Southern Appalachians," *Southeastern Geographer*, 10 (1970), pp. 67–72; Crèvecoeur, *Journey*, pp. 337–38; Frederick L. Olmsted, *A Journey in the Back Country in the Winter of 1853–4* (New York, NY, 1860), pp. 3–7.

45. Horace Kephart, *Our Southern Highlanders* (New York, NY, 1916), p. 76; Crèvecoeur, *Journey*, pp. 325–31, 333, 337–38, 342.

46. Crèvecoeur, *Journey*, pp. 337–38.

47. Bartram, *Travels*, pp. 120–21.

48. Wilhelm, "Animal Drives," p. 7.

49. Archer B. Hulbert, ed., *Braddock's Road and Three Relative Papers* (Cleveland, OH, 1903), pp. 143–45.

50. Paul C. Henlein, *Cattle Kingdom in the Ohio Valley, 1783–1860* (Lexington, KY, 1959), p. 2; Jordan, *Trails*, pp. 53–54.

51. Gilbert Imlay, *A Topographical Description of the Western Territory of North America* (London, UK, third edition, 1797), pp. 175, 518.

52. Hilliard, *Hog Meat*, p. 194; Ramsay, *History of South-Carolina*, vol. 2, p. 577; Wilhelm, "Animal Drives," pp. 8, 11.

53. Hans Kurath, *Word Geography of the Eastern United States* (Ann Arbor, MI, 1949), figures 26, 61; Goff, "Cow Punching," p. 343; Henlein, *Cattle Kingdom*, p. 2.

54. Henlein, *Cattle Kingdom*, p. 2.

55. R. L. Jones, "The Beef Cattle Industry in Ohio Prior to the Civil War," *Ohio Historical Quarterly*, 64 (1955), p. 170.

56. David L. Wheeler, "The Beef Cattle Industry in the Old Northwest, 1803–1860," *PPHR*, 47 (1974), pp. 31–33; Jones, "Beef Cattle," p. 174; Henlein, *Cattle Kingdom*, p. 5.

57. Henlein, *Cattle Kingdom*, p. 4.

58. Samuel N. Dicken, "The Kentucky Barrens," *Bulletin of the Geographical Society of Philadelphia*, 33 (1935), pp. 42–51; Darrell H. Davis, *The Geography of the Jackson Purchase* (Frankfort, KY, 1923), pp. 54–56; Wheeler, "Beef Cattle," pp. 31–33; Henlein, *Cattle Kingdom*, pp. 5–7, 14, 56; Jordan, *Trails*, pp. 54–57.

59. Paul C. Henlein, "Shifting Range-Feeder Patterns in the Ohio Valley Before 1860," *AH*, 31 (1957), pp. 1–12; Paul C. Henlein, "Early Cattle Ranges of the Ohio Valley," *AH*, 35 (1961), pp. 150–54; Wheeler, "Beef Cattle," pp. 33–34; Jones, "Beef Cattle," pp. 174–76, 181; Henlein, *Cattle Kingdom*, pp. 5, 8–9, 12–17, 52.

60. Paul W. Gates, "Cattle Kings in the Prairies," *MVHR*, 35 (1948), pp. 379–412; Paul W. Gates, "Hoosier Cattle Kings," *Indiana Magazine of History*, 44 (1948), pp. 1–24; Richard L. Power, *Planting Corn Belt Culture: The Impress of the Upland Southerner and Yankee in the Old Northwest* (Indianapolis, IN, 1953), p. 95; Jones, "Beef Cattle," pp. 179, 180; Henlein, *Cattle Kingdom*, pp. 12, 55, 65.

61. Charles T. Leavitt, "Attempts to Improve Cattle Breeds in the United States, 1790–1860," *AH*, 7 (1933), pp. 53–55, 58, 61, 63; George F. Lemmer, "The Spread of Improved Cattle through the Eastern United States," *AH*, 21 (1947), pp. 79, 86–87; William W. Savage, Jr., "Stockmen's Associations and the Western Range Cattle Industry," *JOW*, 14:3 (1975), p. 53; Henlein, *Cattle Kingdom*, pp. 21, 25; Wheeler, "Beef Cattle," pp. 35, 43.

62. Harry S. Drago, *Great American Cattle Trails* (New York, NY, 1965), pp. 2–9; Garnet M. Brayer and Herbert O. Brayer, *American Cattle Trails, 1540–1900* (Bayside, NY, 1952), p. 21; Thompson, *History of Livestock*, pp. 93–95; Paul C. Henlein, "Cattle Driving from the Ohio Country, 1800–1850," *AH*, 28 (1954), pp. 83–95.

63. Edgar N. Transeau, "The Prairie Peninsula," *Ecology*, 16 (1935), pp. 423–37; Steve Packard, "Just a Few Oddball Species: Restoration and Rediscovery of the Tallgrass Savanna," *Restoration and Management Notes*, 6:1 (1988), pp. 13–22; Scott L. Collins and Linda L. Wallace, eds., *Fire in North American Tallgrass Prairies* (Norman, OK, 1990); Allan G. Bogue, *From Prairie to Corn Belt* (Chicago, IL, 1963), p. 1, chapter 5.

64. Henlein, *Cattle Kingdom*, p. 19.

65. Thompson, *History of Livestock*, p. 91; Leslie Hewes, "The Northern Wet Prairie of the United States: Nature, Sources of Information, and Extent," *AAAG*, 41 (1951), pp. 307–10, 315, 322; Wheeler, "Beef Cattle," p. 39; Gates, "Hoosier," p. 7; Henlein, *Cattle Kingdom*, pp. 11–13, 18, 53, 62.

66. Harlan H. Barrows, *Geography of the Middle Illinois Valley* (Urbana, IL, 1910), pp. 64–82; Wheeler, "Beef Cattle," p. 44; Henlein, *Cattle Kingdom*, pp. 12, 19–20, 53, 62; Bogue, *Prairie to Corn Belt*, pp. 1, 3, 58, 101.

67. Walter A. Schroeder, "Environment and Settlement in the Historic Ste. Genevieve District (Missouri)," *A[ssociation of] A[merican] G[eographers], Annual Meeting, Program and Abstracts, April 19–22, 1990, Toronto* (Washington, DC, 1990), p. 220; "Resumen General, que comprehende todos los Habitantes y Establecimientos de la Colonia de la Luisiana, Hecho el año de 1766," MS, Archivo General de Indias, Sevilla, Spain (copy courtesy Gregory Knapp); Lauren C. Post, "The Domestic Animals and Plants of French Louisiana as Mentioned in the Literature, with Reference to Sources,

Varieties and Uses," *Louisiana Historical Quarterly*, 16 (1933), p. 564; Jordan, *Trails*, pp. 57–58; Wheeler, "Beef Cattle," p. 44.

68. Thomas Nuttall, *A Journal of Travels into the Arkansas Territory, During the Year 1819* (Philadelphia, 1821), p. 78; Imlay, *Topographical*, p. 422; Crèvecoeur, *Journey*, p. 458; Jordan, *Trails*, pp. 47–48.

69. "Petitesas Plains—Saline County," *Missouri Historical Quarterly*, 26 (1931–32), pp. 329–30; Henlein, *Cattle Kingdom*, pp. 169–71; Jordan, *Trails*, p. 55.

70. Clifford D. Carpenter, "The Early Cattle Industry in Missouri," *Missouri Historical Review*, 47 (1953), pp. 201–15; William B. Drew, "Floristic Composition of Grazed and Ungrazed Prairie Vegetation in North-Central Missouri," *Ecology*, 28 (1947), pp. 26–41; Carl O. Sauer, *The Geography of the Ozark Highland of Missouri* (Chicago, 1920), p. 161; Henlein, *Cattle Kingdom*, pp. 9, 169–71; Jordan, *Trails*, pp. 55, 57.

71. Frank S. Popplewell, "St. Joseph, Missouri, as a Center of the Cattle Trade," *Missouri Historical Review*, 32 (1938), p. 445; Cardinal Goodwin, "The American Occupation of Iowa, 1833 to 1860," *Iowa Journal of History and Politics*, 17 (1919), p. 83; James C. Malin, *The Grassland of North America: Prolegomena to its History* (Ann Arbor, MI, 1947), pp. 274–76; George F. Lemmer, "Early Leaders in Livestock Improvement in Missouri," *Missouri Historical Review*, 37 (1942–43), pp. 29–39; Joseph G. McCoy, *Historic Sketches of the Cattle Trade of the West and Southwest* (Kansas City, MO, 1874), p. 171; David Dary, *Cowboy Culture: A Saga of Five Centuries* (New York, NY, 1981), p. 92; J. H. Atkinson, "Cattle Drives from Arkansas to California Prior to the Civil War," *Arkansas Historical Quarterly*, 28 (1969), pp. 275–81; Carpenter, "Early Cattle," p. 202; Henlein, *Cattle Kingdom*, pp. 9, 169.

72. W. W. Baldwin, "Driving Cattle from Texas to Iowa, 1866," *Annals of Iowa*, 14 (1924), p. 244 (quote); Leslie Hewes, "Some Features of Early Woodland and Prairie Settlement in a Central Iowa County," *AAAG*, 40 (1950), p. 51; Leslie Hewes and Phillip E. Frandson, "Occupying the Wet Prairie: The Role of Artificial Drainage in Story County, Iowa," *AAAG*, 42 (1952), p. 34; James W. Whitaker, *Feedlot Empire: Beef Cattle Feeding in Illinois and Iowa, 1840–1900* (Ames, IA, 1975); Bogue, *Prairie to Corn Belt*, p. 101; Goodwin, "American Occupation," pp. 83–102.

73. Terry G. Jordan, "Early Northeast Texas and the Evolution of Western Ranching," *AAAG*, 67 (1977), pp. 70–72.

74. Henlein, *Cattle Kingdom*, p. 181; Jordan, "Early Northeast," pp. 73–75.

75. Wayne Gard, "The Shawnee Trail," *SWHQ*, 56 (1953), pp. 359–77; Jordan, "Early Northeast," pp. 75–77, 82.

76. Terry G. Jordan, "The Imprint of the Upper and Lower South on Mid-Nineteenth-Century Texas," *AAAG*, 57 (1967), pp. 667–90; Jordan, "Early Northeast," pp. 81–82.

Chapter Seven

1. David Dary, *Cowboy Culture: A Saga of Five Centuries* (New York, NY, 1981), chapter 4, "The Texian Culture"; Terry G. Jordan, *Trails to Texas: Southern Roots of Western Cattle Ranching* (Lincoln, NE, 1981), pp. 59–82, 125–34; Walter P. Webb, *The Great Plains* (Boston, MA, 1931), chapter 6; T. R. Havins, "Livestock and Texas Law," *WTYB*, 36 (1960), pp. 18–32; Daniel E. McArthur, "The Cattle Industry of Texas, 1685–1918," M.A. thesis, University of Texas (Austin, TX, 1918); Barbara E. H. Sparks, "The History of Grazing in Texas; An Analytical Inventory of the Findings of the Historical Records Survey," M.A. thesis, Southwest Texas State University (San Marcos, TX, 1973); Clifford P. Westermeier, "The Cowboy in his Home State," *SWHQ*, 58 (1954–55), pp. 218–34; J. Frank Dobie, *The Longhorns* (Boston, MA, 1941).

2. Fred B. Kniffen, "The Western Cattle Complex: Notes on Differentiation and Diffusion," *Western Folklore*, 12 (1953), pp. 179–85.

3. Fred Arrington, *A History of Dickens County* (Quanah, TX, 1971), p. 98; Joseph G. McCoy, *Historic Sketches of the Cattle Trade of the West and Southwest* (Kansas City, MO, 1874), pp. 11, 79; Dane Coolidge, *Texas Cowboys* (New York, NY, 1937), p. 14; Frederick L. Olmsted, *A Journey through Texas, or, a Saddle-Trip on the Southwestern Frontier* (Austin, TX, 1978), pp. 281, 288; Ray August, "Cowboys v. Rancheros: The Origins of Western American Livestock Law," *SWHQ*, 96 (1993), forthcoming; Jordan, *Trails*, pp. 150–55; Kniffen, "Western Cattle," p. 185.

4. Harold B. Barclay, *The Role of the Horse in Man's Culture* (London, UK, and New York, NY, 1980), pp. 212–13; Glenn R. Vernam, *Man on Horseback* (New York, NY, 1964), pp. 317–32, 337; Terry G. Jordan and Lester Rowntree, *The Human Mosaic: A Thematic Introduction to Cultural Geography* (New York, NY, 5th ed. 1990), p. 158; Louise S. O'Connor, *Cryin' for Daylight: A Ranching Culture in the Texas Coastal Bend* (Austin, TX, 1989), pp. 57, 295; John K. Rollinson, *Wyoming Cattle Trails* (Caldwell, ID, 1948), following p. 110, p. 215, following p. 290; William Foster-Harris, *The Look of the Old West* (New York, NY, 1960), pp. 205, 242, 244; Don Worcester, *The Texas Cowboy* (Ft. Worth, TX, 1986), pp. 98–99; Andy Adams, *The Log of a Cowboy* (Boston, MA, 1931), p. 15; Coolidge, *Texas Cowboys*, pp. 13, 19; Jordan, *Trails*, pp. 145, 153–54; Kniffen, "Western Cattle," p. 185.

5. Américo Paredes, "The Bury-Me-Not Theme in the Southwest," *Texas Folk Lore Society Publication*, 29 (1959), pp. 88–92; Sandra L. Myres, "The Ranching Frontier: Spanish Institutional Backgrounds of the Plains Cattle Industry," in Harold M. Hollingsworth and Sandra L. Myres, eds., *Essays on the American West* (Austin, TX, 1969), pp. 28–30; Hortense W. Ward, "Ear Marks," *Texas Folk Lore Society Publication*, 19 (1944), pp. 106–16; William Dusenberry, "Constitutions of Early and Modern American Stock Growers' Associations," *SWHQ*, 53 (1949–50), pp. 255–75; Jordan, *Trails*, p. 153.

6. United States Census, manuscript population schedules, Texas, 1880, and Lincoln County, New Mexico, 1880; Lillie M. Hunter, *The Book of Years: A History of Dallam and Hartley Counties* (Hereford, TX, 1969), p. 119; Nellie S. Yost, *The Call of the Range: The Story of the Nebraska Stock Growers Association* (Denver, CO, 1966), p. 72; Ed Gould, *Ranching in Western Canada* (Saanichton, BC, and Seattle, WA, 1978), p. 39; O'Connor, *Cryin' for Daylight*, pp. 74–79, 99; Jordan, *Trails*, p. 146.

7. Tom Lea, *The King Ranch*, 2 vols. (Boston, MA, 1957); Toni Frissell and Holland McCombs, *The King Ranch, 1939–1944; A Photographic Essay* (Dobbs Ferry, NY, and Ft. Worth, TX, 1975); Emilia S. Ramirez, *Ranch Life in Hidalgo County after 1850* (Edinburg, TX, 1971); Cornelius C. Cox, "Reminiscences of C. C. Cox," *QTSH*, 6 (1902–3), pp. 206–9; McCoy, *Historic Sketches*, p. 13; Frank Goodwyn, *Life on the King Ranch* (New York, NY, 1951).

8. Donald E. Worcester, *The Texas Longhorn* (College Station, TX, 1987), p. 30; Paula M. Marks, *Turn Your Eyes toward Texas: Pioneers Sam and Mary Maverick* (College Station, TX, 1989); Arnold R. Rojas, *The Vaquero* (Charlotte, NC, and Santa Barbara, CA, 1964), p. 4; Elizabeth S. Bright, *A Word Geography of California and Nevada* (Berkeley and Los Angeles, CA, 1971), pp. 121, 168; *Neu-Braunfelser Zeitung* (New Braunfels, TX), vol. 1, no. 44 (Sept. 23, 1853), p. 3; Olmsted, *Journey*, pp. 261, 370; O'Connor, *Cryin' for Daylight*, pp. 86–91; August, "Cowboys v. Rancheros."

9. Ramon F. Adams, *Western Words: A Dictionary of the American West* (Norman, OK, 1968), p. 80; Bailey C. Hanes, *Bill Pickett, Bulldogger: The Biography of a Black Cowboy* (Norman, OK, 1977); Roy Grimes, ed., *300 years in Victoria County* (Victoria, TX, 1968), p. 93; Gus L. Ford, ed., *Texas Cattle Brands* (Dallas, TX, 1936); W. H. Jackson and S. A. Long, *The Texas Stock Directory, or Book of Marks and Brands* (San Antonio, TX, 1865); Kniffen, "Western Cattle," p. 185; Jordan, *Trails*, pp. 141–50; McCoy, *Historic Sketches*, p. 9; Vernam, *Man on Horseback*, pp. 316–17, 346.

10. Ralph S. Jackson, *Home on the Double Bayou: Memories of an East Texas Ranch* (Austin, TX, 1961); Margaret R. Warburton, "A History of the O'Connor Ranch, 1834–1939," M.A. thesis, Catholic University of America, Washington, DC, 1939; A. Ray Stephens, *The Taft Ranch: A Texas Principality* (Austin, TX, 1964); Victor M. Rose, *Some Historical Facts in Regard to the Settlement of Victoria, Texas* (Laredo, TX, 1883), pp. 15–16, 35, 106–7, 115; Terry G. Jordan, "The Origin of Anglo-American Cattle Ranching in Texas: A Documentation of Diffusion from the Lower South," *EG*, 45 (1969), pp. 63–87; Grimes, *300 Years*, pp. 93–94, 213–15; Jordan, *Trails*, pp. 62–70, 126–31; Olmsted, *Journey*, pp. 100–101.

11. "Stock-Raising," *Texas Almanac for 1861* (Galveston, TX, 1860), pp. 148–52, written by an anonymous informant in Lamar, Refugio Co., Texas, and dated June 17, 1860.

12. Francis R. Lubbock, *Six Decades in Texas, or Memoirs* (Austin, TX,

1900), p. 138; Joe B. Frantz and Julian E. Choate, Jr., *The American Cowboy: The Myth and the Reality* (Norman, OK, 1955), pp. 27–28; Charles W. Towne and Edward N. Wentworth, *Cattle & Men* (Norman, OK, 1955), p. 157; Virginia S. Hutcheson, "Cattle Drives in Missouri," *Missouri Historical Review*, 37 (1943), pp. 286–88; "Stock-Raising," p. 149; Cyrus C. Loveland, *California Trail Herd* (Los Gatos, CA, 1961), pp. 22–23; James G. Bell, "A Log of the Texas-California Cattle Trail, 1854," *SWHQ*, 35 (1931–32), pp. 208–37, 290–316; 36 (1932–33), pp. 47–67; M. H. Erskine, "A Cattle Drive from Texas to California: the Diary of M. H. Erskine, 1854," *SWHQ*, 67 (1963–64), pp. 397–412; Olmsted, *Journey*, pp. 258, 273–75.

13. Jimmy M. Skaggs, "John Thomas Lytle: Cattle Baron," *SWHQ*, 71 (1967–68), pp. 46–60; James H. Cook, *Longhorn Cowboy* (Norman, OK, 1984); McCoy, *Historic Sketches*, pp. 11–14.

14. William T. Chambers, "Edwards Plateau: A Combination Ranching Region," *EG*, 8 (1932), pp. 67–80; George Syring, "A Geographic Analysis of Diversified Ranching Operations on the Edwards Plateau: A Case Study of Edwards County, Texas," Ph.D. diss., University of Kansas (Lawrence, KS, 1975); Jordan, *Trails*, p. 129.

15. Rupert N. Richardson, *The Frontier of Northwest Texas, 1846 to 1876* (Glendale, CA, 1963), p. 153; Donald W. Meinig, *Imperial Texas: An Interpretive Essay in Cultural Geography* (Austin, TX, 1969), pp. 66–68; Jordan, *Trails*, pp. 127, 135, 138–42, 147–49; James T. Padgitt, "Colonel William H. Day, Texas Ranchman," *SWHQ*, 53 (1949–50), pp. 347–66; Jordan and Rowntree, *Human Mosaic*, pp. 14–16.

16. Jordan, *Trails*, pp. 128–29; manuscript tax lists for Comanche, Eastland, Erath, Palo Pinto, Shackelford, and Stephens counties, TX, for the period 1857–80, in Texas State Archives, Austin.

17. U.S. Census manuscript population schedules, Callahan, Runnels, and Wichita counties, Texas, 1880; Jordan, *Trails*, p. 146.

18. Philip C. Durham, "The Negro Cowboy," *American Quarterly*, 7 (1955), pp. 291–301; Philip C. Durham and Everett L. Jones, *The Negro Cowboys* (New York, NY, 1965); "The Texas Cow-Boy," *Frank Leslie's Illustrated Newspaper*, 57:1 (Dec. 1, 1883), p. 229; Rojas, *Vaquero*, p. 11.

19. Adelaide Hawes, *The Valley of Tall Grass* (Bruneau, ID, 1950), pp. 141–42; Mildretta Adams, *Owyhee Cattlemen, 1878–1978, 100 Years in the Saddle*, 2nd ed., (Homedale, ID, 1979), p. 23; Lewis G. Thomas, *Ranchers' Legacy: Alberta Essays* (Edmonton, AB, 1986), p. 66; Gould, *Ranching in Western*, p. 99; U.S. Census 1880, manuscript population schedules for Lincoln Co., NM, on Chisum Ranch; Hanes, *Bill Pickett*.

20. Wayne Gard, "The Shawnee Trail," *SWHQ*, 56 (1952–53), pp. 359–77; Ernest S. Osgood, *The Day of the Cattleman* (Chicago, IL, 1929), pp. 21, 27, 30–31, 48, 54; Louis Pelzer, *The Cattlemen's Frontier: A Record of the Trans-Mississippi Cattle Industry* (Glendale, CA, 1936), p. 37.

21. Robert R. Dykstra, *The Cattle Towns* (New York, NY, 1968); George L.

Cushman, "Abilene, First of the Kansas Cow Towns," *KHQ*, 9 (1940), pp. 240–58; Wayne Gard, *The Chisholm Trail* (Norman, OK, 1954); Harry S. Drago, *Great American Cattle Trails* (Norman, OK, 1920); Ralph H. Brown, "Texas Cattle Trails: Notes on Three Important Maps," *Texas Geographic Magazine*, 10:1 (1946), pp. 1–6; J. Marvin Hunter, *The Trail Drivers of Texas* (Nashville, TN, 1925); Joseph Nimmo, Jr., *Range and Ranch Cattle Traffic* (Washington, DC, 1885); Ronald B. Jager, "The Chisholm Trail's Mountain of Words," *SWHQ*, 71 (1967–68), pp. 61–68; T. U. Taylor, *The Chisholm Trail and Other Routes* (San Antonio, TX, 1936); T. C. Richardson, "Cattle Trails of Texas," *Texas Geographic Magazine*, 1:2 (1937), pp. 16–29; McCoy, *Historic Sketches*; Pelzer, *Cattlemen's Frontier*, pp. 48–50.

22. W. Baillie Grohman, "Cattle Ranches in the Far West," *Fortnightly Review*, n.s. 28 (1880), pp. 438–39; Edward E. Dale, *The Range Cattle Industry: Ranching on the Great Plains from 1865 to 1925* (Norman, OK, 1930); Arrell M. Gibson, "Ranching on the Southern Great Plains," *JOW*, 6 (1967), pp. 135–53; Harold Briggs, "The Development and Decline of Open Range Ranching in the Northwest," *MVHR*, 20 (1934), pp. 521–36; Walter von Richthofen, *Cattle-Raising on the Plains of North America* (Norman, OK, 1964); Maurice Frink et al., *When Grass Was King: Contributions to the Western Range Cattle Industry Study* (Boulder, CO, 1956), pp. 3–132; Webb, *Great Plains*, pp. 205–69; Osgood, *Day of Cattleman*, p. 79; McCoy, *Historic Sketches*, p. 189.

23. Joseph Nimmo, Jr., "The American Cowboy," *Harper's New Monthly Magazine*, 73 (Nov. 1886), pp. 881, 883.

24. Osgood, *Day of Cattleman*, p. 79.

25. Pelzer, *Cattlemen's Frontier*, p. 47.

26. Clarence Gordon, "Report on Cattle, Sheep, and Swine, Supplementary to Enumeration of Live Stock on Farms in 1880," *Report on the Productions of Agriculture as Returned at the Tenth Census* (Washington, DC, 1883), p. 998; Reginald Aldridge, *Ranch Notes in Kansas, Colorado, the Indian Territory and Northern Texas* (London, UK, 1884), pp. 27, 80; J. A. Rickard, "Hazards of Ranching on the South Plains," *SWHQ*, 37 (1933–34), pp. 313–19; Mont H. Saunderson, *Western Stock Raising* (Minneapolis, MN, 1950), pp. 7–11; James I. Culbert, "Cattle Industry of New Mexico," *EG*, 17 (1941), pp. 157, 160; Richthofen, *Cattle-Raising*, p. 14; Dary, *Cowboy Culture*, p. 227.

27. Hazel A. Pulling, "History of the Range Cattle Industry of Dakota," *South Dakota Historical Collections*, 20 (1940), p. 513; Saunderson, *Western Stock Raising*, pp. 3–7; Dary, *Cowboy Culture*, p. 227.

28. David W. Stahle and Malcolm K. Cleaveland, "Texas Drought History Reconstructed from 1698 to 1980," *Journal of Climate*, 1 (1988), pp. 59–74; Webb, *Great Plains*, pp. 17–26.

29. Seymour V. Connor, "Early Ranching Operations in the Panhandle: A Report on the Agricultural Schedules of the 1880 Census," *PPHR*, 27 (1954),

pp. 47–69; J. Evetts Haley, *Charles Goodnight: Cowman & Plainsman* (Boston, MA, and New York, NY, 1936); David J. Murrah, *C. C. Slaughter: Rancher, Banker, Baptist* (Austin, TX, 1981); Dulcie Sullivan, *The L S Brand: The Story of a Texas Panhandle Ranch* (Austin, TX, 1968); Harley T. Burton, *A History of the J A. Ranch* (Austin, TX, 1928); Floyd B. Streeter, "The Millet Cattle Ranch in Baylor County, Texas," *PPHR,* 22 (1949), pp. 65–83; David B. Gracy II, "George Washington Littlefield: Portrait of a Cattleman," *SWHQ,* 68 (1964–65), pp. 237–58; C. Boone McClure, "A History of the Shoe Nail Ranch," *PPHR,* 9 (1938), pp. 69–83; Harwood P. Hinton, Jr., "John Simpson Chisum, 1877–84," *NMHR,* 31 (1956), pp. 177–205, 310–37; 32 (1957), pp. 53–65; Mrs. J. Lee Jones and Rupert N. Richardson, "Colorado City, the Cattlemen's Capital," *WTYB,* 19 (1943), pp. 36–63; J. W. Williams, *The Big Ranch Country* (Wichita Falls, TX, 1954); Charles L. Kenner, "The Great New Mexico Cattle Raid, 1872," *NMHR,* 37 (1962), pp. 243–59; Culbert, "Cattle Industry," pp. 155–68; Gibson, "Ranching," pp. 135–53.

30. Josie Baird, "Ranching on the Two Circles Bar," *PPHR,* 17 (1944), pp. 8–67.

31. Edward E. Dale, "The Ranchman's Last Frontier," *MVHR,* 10 (1923), pp. 34–46.

32. Edward E. Dale, "History of the Ranch Cattle Industry in Oklahoma," *Annual Report of the American Historical Association,* 1920, pp. 309–22; Edward E. Dale, "The Cherokee Strip Live Stock Association," *CO,* 5 (1927), pp. 58–78; Edward E. Dale, "Ranching on the Cheyenne-Arapaho Reservation, 1880–1885," *CO,* 6 (1928), pp. 35–59; Melvin Harrel, "Oklahoma's Million Acre Ranch," *CO,* 29 (1951), pp. 70–78; Ellsworth Collings and Alma M. England, *The 101 Ranch* (Norman, OK, 1938).

33. Aldridge, *Ranch Notes;* C. W. McCampbell, "W. E. Campbell, Pioneer Kansas Livestockman," *KHQ,* 16 (1948), pp. 245–73; Cushman, "Abilene," pp. 240–58; Floyd B. Streeter, "Ellsworth as a Texas Cattle Market," *KHQ,* 4 (1935), pp. 388–98; Robert Dykstra, "Ellsworth, 1869–1875: The Rise and Fall of a Kansas Cowtown," *KHQ,* 27 (1961), pp. 161–92; Dary, *Cowboy Culture,* p. 235; Gordon, "Report on Cattle," p. 998; McCoy, *Historic Sketches.*

34. Clara M. Love, "History of the Cattle Industry in the Southwest," *SWHQ,* 19 (1915–16), p. 387; Charles Kenner, "A Texas Rancher in Colorado: The Last Years of John Hittson," *WTYB,* 42 (1966), pp. 28–40; Ora B. Peake, *The Colorado Range Cattle Industry* (Glendale, CA, 1937), pp. 22–26, 35; George C. Everett, *Cattle Cavalcade in Central Colorado* (Denver, CO, 1966), pp. 7–17; Edgar C. McMechen, "John Hittson, Cattle King," *Colorado Magazine,* 11 (1934), pp. 164–70; Dary, *Cowboy Culture,* p. 122; Gordon, "Report on Cattle," p. 1005; Osgood, *Day of Cattleman,* p. 39.

35. William D. Aeschbacher, "Development of Cattle Raising in the Sandhills," *NBH,* 28 (1947), pp. 41–64; Norbert R. Mahnken, "Early Nebraska Markets for Texas Cattle," *NBH,* 26 (1945), pp. 3–25, 91–103; Norbert R. Mahnken, "Ogallala," *NBH,* 28 (1947), pp. 85–109; James C. Dahl-

man, "Recollections of Cowboy Life in Western Nebraska," *NBH*, 10 (1927), pp. 335–39; James H. Cook, "Trailing Texas Long-Horn Cattle through Nebraska," *NBH*, 10 (1927), pp. 339–43; Hiram Latham, *Trans-Missouri Stock Raising: The Pasture Lands of North America: Winter Grazing* (Omaha, NE, 1871), pp. 23–26; Gordon, "Report on Cattle," p. 1011; Pelzer, *Cattlemen's Frontier*, p. 47; Yost, *Call of Range*, p. 72.

36. John M. Kuykendall, "The First Cattle North of the Union Pacific Railroad," *Colorado Magazine*, 7:2 (1930), p. 71; John Clay, *My Life on the Range* (Norman, OK, 1962); Eugene C. Mather, "The Production and Marketing of Wyoming Beef Cattle," *EG*, 26 (1950), pp. 81–82; James A. Young and B. Abbott Sparks, *Cattle in the Cold Desert* (Logan, UT, 1985), pp. 77–80; Briggs, "Development and Decline," pp. 521–22; Gordon, "Report on Cattle," p. 1015; Louis Pelzer, "A Cattleman's Commonwealth on the Western Range," *MVHR*, 13 (1926), pp. 30–49; Osgood, *Day of Cattleman*, pp. 42, 47–48, 87; Rollinson, *Wyoming Cattle*, pp. 60, 71, 221.

37. August H. Schatz, *Longhorns Bring Culture* (Boston, MA, 1961), p. 33; Harold E. Briggs, "Ranching and Stock-Raising in the Territory of Dakota," *South Dakota Historical Collections*, 14 (1928), pp. 417–66; Ray H. Mattison, "Ranching in the Dakota Badlands," *North Dakota History*, 19 (1952), pp. 93–128, 167–206; Bertha M. Kuhn, "The W-Bar Ranch on the Missouri Slope," *Collections, State Historical Society of North Dakota*, 5 (1923), pp. 155–66; Dick Williams and Bob Lee, *Last Grass Frontier: The South Dakota Stock Grower Heritage* (Sturgis, SD, 1964); George F. Shafer, "Cattle Ranching in McKenzie County, North Dakota," *North Dakota Historical Quarterly*, 1 (1926), pp. 55–61; Pulling, "History of Range," pp. 467–521; Gordon, "Report on Cattle," p. 1019.

38. John Leakey, *The West that Was: From Texas to Montana* (Dallas, TX, 1958), pp. 91–92; Michael Kennedy, "Judith Basin Top Hand: Reminiscences of William O. Burnett, an Early Montana Cattleman," *Montana Magazine of History*, 3:2 (1953), pp. 18–23; Robert Fletcher, *History of the Range Cattle Business in Eastern Montana* (Washington, DC, 1928); Michael S. Kennedy, ed., *Cowboys and Cattlemen: A Roundup from Montana* (New York, NY, 1964), pp. 103–44; Mont H. Saunderson and D. W. Chittenden, *Cattle Ranching in Montana* (Bozeman, MT, 1937); Donald H. Welsh, "Pierre Wibaux, Cattle King," *North Dakota History*, 20 (1953), pp. 5–23; Walt Coburn, *Pioneer Cattleman in Montana: The Story of the Circle C Ranch* (Norman, OK, 1968); Briggs, "Development and Decline," pp. 523–24, 536; Osgood, *Day of Cattleman*, pp. 53–57, 255; Gordon, "Report on Cattle," p. 1027; Rollinson *Wyoming Cattle*, p. 225.

39. John R. Craig, *Ranching with Lords and Commons, or, Twenty Years on the Range* (Toronto, ON, 1903), p. 273; Simon M. Evans, "American Cattlemen on the Canadian Range, 1874–1914," *Prairie Forum*, 4 (1979), pp. 121–35; Simon M. Evans, "The Origins of Ranching in Western Canada: American Diffusion or Victorian Transplant?," *Great Plains Quarterly*, 3:2

(1983), pp. 79–91; A. A. Lupton, "Cattle Ranching in Alberta, 1874–1910: Its Evolution and Migration," *Albertan Geographer*, 3 (1966–67), pp. 48–58; Lewis G. Thomas, "The Ranching Period in Southern Alberta," M.A. thesis, University of Alberta (Calgary, AB, 1935); David H. Breen, *The Canadian Prairie West and the Ranching Frontier, 1874–1924* (Toronto, ON, 1983), pp. 6–9, 25; Edward Brado, *Cattle Kingdom: Early Ranching in Alberta* (Vancouver, BC, and Toronto, ON, 1984), pp. 21, 48; L. V. Kelly, *The Range Men: The Story of the Ranchers and Indians of Alberta* (New York, NY, 1965), p. 150; Henry C. Klassen, "The Conrads in the Alberta Cattle Business, 1875–1911," *AH*, 64 (1990), pp. 31–59; John W. Bennett, *Northern Plainsmen: Adaptive Strategy and Agrarian Life* (Chicago, IL, 1969), chapter 6; Boyd M. Anderson, *Beyond the Range: A History of the Saskatchewan Stock Growers Association* (Saskatoon, SK, 1988), pp. 2, 3, 5, 9, 14–15, 36–37; Gould, *Ranching in Western*, pp. 1–107, 147; Richard W. Slatta, *Cowboys of the Americas* (New Haven, CT, 1990), pp. 25–27; Ed Gould, *Ranching in Western Canada* (Saanichton, BC, and Seattle, WA, 1978), p. 52.

40. W. R. McAfee, *The Cattlemen* (Alvin, TX, 1989); Robert M. Utley, "The Range Cattle Industry in the Big Bend of Texas," *SWHQ*, 69, (1965–66), pp. 419–41.

41. J. J. Wagoner, *History of the Cattle Industry in Southern Arizona, 1540–1940*, (Tucson, AZ, 1952), pp. 32, 36, 42; Gerald Baydo, "Cattlemen's Associations in New Mexico Territory," *JOW* 14:3 (1975), pp. 60–71; Bert Haskett, "Early History of the Cattle Industry in Arizona," *AZHR*, 6:4 (1935), pp. 15, 23, 26, 41; James R. Hastings and Raymond M. Turner, *The Changing Mile: An Ecological Study of Vegetational Change with Time in the Lower Mile of an Arid and Semiarid Region* (Tucson, AZ, 1980), p. 40; James A. Wilson, "West Texas Influence on the Early Cattle Industry of Arizona," *SWHQ*, 71 (1966–67), pp. 26–36; Jack Parsons and Michael Earney, *Land and Cattle: Conversations with Joe Pankey, a New Mexico Rancher* (Albuquerque, NM, 1978); Dane Coolidge, *Old California Cowboys* (New York, NY, 1939), pp. 67–72, following p. 84; Rojas, *Vaquero*, p. 12; Saunderson, *Western Stock*, pp. 21–23; Gordon, "Report on Cattle," pp. 1047, 1049.

42. G. L. Chester, "What Became of the Texans?," *Frontier Times*, 6:1 (1928), p. 45; Bell, "A Log of the Texas-California"; Erskine, "A Cattle Drive."

43. Don D. Walker, "Longhorns Come to Utah," *UTHQ*, 30 (1962), pp. 135–47; Don D. Walker, "The Cattle Industry of Utah, 1850–1900: An Historical Profile," *UTHQ*, 32 (1964), pp. 185–86; Don D. Walker, "The Carlisles: Cattle Barons of the Upper Basin," *UTHQ*, 32 (1964), p. 269; Neal Lambert, "Al Scorup: Cattleman of the Canyons," *UTHQ*, 32 (1964), pp. 301–20; Virginia N. Price and John T. Darby, "Preston T. Nutter: Utah Cattleman, 1886–1936," *UTHQ*, 32 (1964), pp. 232–51; James H. Beckstead, *Cowboying: A Tough Job in a Hard Land* (Salt Lake City, UT, 1991); Richard J. Morrisey, "The Early Range Cattle Industry in Arizona," *AH*, 24 (1950),

p. 153; Bert Haskett, "History of the Sheep Industry in Arizona," *AZHR*, 7:3 (1936), p. 35; Gordon, "Report on Cattle," p. 1051: Brado, *Cattle Kingdom*, pp. 136–37.

44. Martin F. Schmitt, ed., *The Cattle Drives of David Shirk: From Texas to the Idaho Mines, 1871 and 1873* (Portland, OR, 1956); J. R. Keith, "When the Long-Horned Cattle of Texas Came to Idaho Territory," *Idaho State Historical Society, Biennial Report*, 16 (1937–38), pp. 41–49; Byron D. Lusk, "Golden Cattle Kingdoms of Idaho," M.S. thesis, Utah State University (Logan, UT, 1978), pp. 29–30, 33, 38; Edna B. Patterson et al., *Nevada's Northeast Frontier* (Sparks, NV, 1969), pp. 208–12, 413; Mike Hanley and Ellis Luca, *Owyhee Trails: The West's Forgotten Corner* (Caldwell, ID, 1973), p. 81; Young and Sparks, *Cattle in Cold*, pp. 77–82; Gordon, "Report on Cattle," pp. 1059, 1098; Loveland, *California Trail*, p. 38; Adams, *Owyhee*, pp. 27–28.

45. Terry G. Jordan, "Texas Influence in Nineteenth-Century Arizona Cattle Ranching," *JOW*, 14:3 (1975), pp. 15–17.

46. United States Senate, [*1870*] *Federal Census, Territory of New Mexico and the Territory of Arizona* (Washington, DC, 1965), pp. 125–253.

47. United States Census, manuscript population schedules, Arizona Territory, 1880.

48. Wilson, "West Texas Influence," pp. 26–36.

49. John M. Crowley, "Ranches in the Sky: A Geography of Livestock Ranching in the Mountain Parks of Colorado," Ph.D. diss., University of Minnesota (Minneapolis, MN, 1964); Latham, *Trans-Missouri*, pp. 32–35; Everett, *Cattle Cavalcade*, pp. 6, 116, 149, 151, 163, 180, 199–201, 319, 329; Glen Barrett, "Stock Raising in the Shirley Basin, Wyoming," *JOW*, 14:3 (1975), pp. 18–24; Gordon, "Report on Cattle," p. 1003.

50. Everett, *Cattle Cavalcade*, p. 180.

51. Everett, *Cattle Cavalcade*, p. 200.

52. Robert Fletcher, *Free Grass to Fences: The Montana Range Cattle Story* (New York, NY, 1960), p. 24; Rollinson, *Wyoming Cattle*, pp. 116, 191, 194; Schmitt, *Cattle Drives*; Gordon, "Report on Cattle," p. 1097; Briggs, "Development and Decline," p. 522; Osgood, *Day of Cattleman*, pp. 21, 54; Dary, *Cowboy Culture*, pp. 230–32.

53. Kenner, "Great New Mexico," pp. 243–59.

54. W. J. Redmond, "The Texas Longhorn on Canadian Range," *Canadian Cattlemen*, 1 (Dec. 1938), pp. 112, 140; Schatz, *Longhorns Bring*; Fletcher, *Free Grass*, p. 24; Aldridge, *Ranch Notes*, pp. 83, 203; Richthofen, *Cattle-Raising*, pp. 23, 66; Everett, *Cattle Cavalcade*, pp. 116, 196, 348; Gordon, "Report on Cattle," p. 998.

55. Walker, "Cattle Industry," pp. 185–86; Welsh, "Pierre Wibaux," p. 10; Kelly, *Range Men*, p. 54; Everett, *Cattle Cavalcade*, pp. 163–64, 196, 348.

56. Adams, *Log of Cowboy*; Mark H. Brown and W. R. Felton, *Before Barbed Wire: L. A. Huffman, Photographer on Horseback* (New York, NY,

1956), pp. 35, 36, 198; James E. McCauley, *A Stove-Up Cowboy's Story* (Dallas, TX, 1943); Charles A. Siringo, *A Texas Cowboy* (Chicago, IL, 1885); C. C. Post, *Ten Years a Cowboy* (Chicago, IL, 1901); Philip A. Rollins, *The Cowboy: His Characteristics, His Equipment, and His Part in the Development of the West* (New York, NY, 1922); Foster-Harris, *Look of Old*, p. 250; Craig, *Ranching with Lords*, p. 102; Coolidge, *Texas Cowboys*, p. 14; Coolidge, *Old California*, p. 77; Rojas, *Vaquero*, p. 4; Grohman, "Cattle Ranches," p. 447; Everett, *Cattle Cavalcade*, pp. 61, 70–71; Aldridge, *Ranch Notes*, pp. 78, 89–91, 129; McCoy, *Historic Sketches*, p. 81; Nimmo, "American Cowboy," pp. 880–84.

57. Frederic G. Cassidy, ed., *Dictionary of American Regional English* (Cambridge, MA, 1985), vol. 1, p. 817; Rufus Phillips, "Early Cowboy Life in the Arkansas Valley," *Colorado Magazine*, 7 (1930), p. 169; Dary, *Cowboy Culture*, pp. 81, 108; Nimmo, "American Cowboy," p. 884; Kelly, *Range Men*, p. 15; Mahnken, "Early Nebraska," following p. 16; Dahlman, "Recollections," p. 336; Worcester, *Texas Cowboy*, p. 6; Grohman, "Cattle Ranches," p. 455; Kennedy, "Judith Basin," p. 23.

58. Phillips, "Early Cowboy," p. 171; O'Connor, *Cryin' for Daylight*, p. 76; Rollinson, *Wyoming Cattle*, p. 107; Cassidy, *Dictionary*, vol. 1, pp. 497, 575; Hawes, *Valley of Tall*, p. 101; Everett, *Cattle Cavalcade*, pp. 151, 201, 344; McCoy, *Historic Sketches*, pp. 11, 79.

59. Slatta, *Cowboys*, p. 27; Pelzer, *Cattlemen's Frontier*, p. 87; Aldridge, *Ranch Notes*; Cook, *Longhorn Cowboy*.

60. Frank Wilkeson, "Cattle-Raising on the Plains," *Harper's New Monthly Magazine* 72 (April 1886), p. 789; Everett, *Cattle Cavalcade*, p. 175.

61. Dahlman, "Recollections," p. 337; Aldridge, *Ranch Notes*, pp. 64–66; Everett, *Cattle Cavalcade*, p. 200.

62. Gordon, "Report on Cattle," p. 998 (quote); Gould, *Ranching in Western*, pp. 59–60.

63. Harley T. Burton, "A History of the J A Ranch," *SWHQ*, 31 (1927–28), p. 363; Ernest S. Osgood, "The Cattleman in the Agricultural History of the Northwest," *AH*, 3 (1929), p. 119; Gordon, "Report on Cattle," pp. 998, 1015; Aldridge, *Ranch Notes*, p. 65; Richthofen, *Cattle-Raising*, pp. 18, 28–29; Pelzer, *Cattlemen's Frontier*, pp. 74–76, 97–98; Everett, *Cattle Cavalcade*, pp. 315, 322, 343–45; Brown and Felton, *Before Barbed Wire*, pp. 173–79; Shafer, "Cattle Ranching," pp. 56–57.

64. Gordon, "Report on Cattle," pp. 972, 973; Aldridge, *Ranch Notes*, pp. 146–47; Brown and Felton, *Before Barbed Wire*, pp. 99, 139, 152–59; Everett, *Cattle Cavalcade*, p. 215.

65. Richard H. Cracroft, "The Heraldry of the Range: Utah Cattle Brands," *UTHQ*, 32 (1964), pp. 217–31; Aeschbacher, "Development," pp. 41–45; Richthofen, *Cattle-Raising*, p. 16; Everett, *Cattle Cavalcade*, pp. 166–67; Pelzer, *Cattlemen's Frontier*, appendix (n.p.).

66. Harrel, "Oklahoma's Million," p. 73; Dahlman, "Recollections," p. 336; Gordon, "Report on Cattle," pp. 973, 998; Aldridge, *Ranch Notes*, pp. 129–30, 144.

67. Phillips, "Early Cowboy," p. 167; Kennedy, "Judith Basin," p. 21; Burton, "A History," p. 363; Gordon, "Report on Cattle," p. 1016; Aldridge, *Ranch Notes*, p. 265; Richthofen, *Cattle-Raising*, pp. 28–29; McCoy, *Historic Sketches*, p. 9; Hawes, *Valley of Tall*, p. 120; Brown and Felton, *Before Barbed Wire*, p. 197; Everett, *Cattle Cavalcade*, p. 200; Shafer, "Cattle Ranching," p. 57; Mattison, "Ranching in Dakota," p. 103.

68. Craig, *Ranching with Lords*, p. 9 (quote); Dahlman, "Recollections," p. 336; Aldridge, *Ranch Notes*, pp. 64–66.

69. Coburn, *Pioneer Cattleman*, p. 8; Shafer, "Cattle Ranching," p. 58; Everett, *Cattle Cavalcade*, p. 157; Gordon, "Report on Cattle," p. 972; Aldridge, *Ranch Notes*, p. 171; Saunderson, *Western Stock*, p. 26; Breen, *Canadian Prairie*, pp. 11, 14; Kuhn, "W-Bar," p. 156; Osgood, *Day of Cattleman*, pp. 49, 238.

70. Gordon, "Report on Cattle," p. 968; Osgood, *Day of Cattleman*, p. 49; Pelzer, *Cattlemen's Frontier*, p. 94.

71. William W. Savage, Jr., "Leasing the Cherokee Outlet: An Analysis of Indian Reaction, 1884–1885," *CO*, 46 (1968), pp. 285–92; Breen, *Canadian Prairie*, pp. 21, 31, 46.

72. Richard Graham, "The Investment Boom in British-Texan Cattle Companies, 1880–1885," *Business History Review*, 34 (1960), pp. 421–45; Frink et al., *When Grass Was King*; Lester F. Sheffy, *The Francklyn Land and Cattle Company: A Panhandle Enterprise* (Austin, TX, 1963); Estelle D. Tinkler, "Nobility's Ranche: A History of the Rocking Chair Ranche," *PPHR*, 15 (1942), pp. 1–88; J. Evetts Haley, *The XIT Ranch of Texas and the Early Days of the Llano Estacado* (Norman, OK, 1953); William C. Holden, *The Espuela Land and Cattle Company: A Study of a Foreign-Owned Ranch in Texas* (Austin, TX, 1970); Gene M. Gressley, "Teschemacher and de Billier Cattle Company: A Study of Eastern Capital on the Frontier," *Business History Review*, 33 (1959), pp. 121–37; Mary Einsel, "Some Notes on the Comanche Cattle Pool," *KHQ*, 26 (1960), pp. 59–66; Lewis Nordyke, *Cattle Empire* (New York, NY, 1949); Jerome O. Steffen, *Comparative Frontiers: A Proposal for Studying the American West* (Norman, OK, 1980), pp. 57–66; Pelzer, *Cattlemen's Frontier*, pp. 73–79, 119–33, 138; Breen, *Canadian Prairie*, pp. 11–12, 15–25, 46; Klassen, "Conrads"; Kuhn, "W-Bar," pp. 159–60.

73. Breen, *Canadian Prairie*, pp. 13–14; McCoy, *Historic Sketches*, p. 7.

74. Rickard, "Hazards of Ranching," pp. 313–19; Averlyne M. Hatcher, "The Water Problem of the Matador Ranch," *WTYB*, 20 (1944), pp. 51–76; Young and Sparks, *Cattle in Cold* pp. 121–22; Pulling, "History of Range," p. 475; Mahnken, "Early Nebraska," pp. 94–95; Wilkeson, "Cattle-Raising," p. 789.

75. Gordon, "Report on Cattle," pp. 1004, 1005; Pelzer, *Cattlemen's Frontier*, pp. 148, 197; Aeschbacher, "Development," pp. 57–58; Walker, "Cattle Industry," pp. 186–89.

76. David L. Wheeler, "The Blizzard of 1886 and Its Effect on the Range Cattle Industry in the Southern Plains," *SWHQ*, 94 (1990–91), pp. 415–32; Thadis W. Box, "Range Deterioration in West Texas," *SWHQ*, 71 (1967–68), pp. 37–45; Pelzer, *Cattlemen's Frontier*, p. 141; Steffen, *Comparative Frontiers*, p. 66; Hatcher, "Water Problem," pp. 51–76; Young and Sparks, *Cattle in Cold*, p. 122; Gould, *Ranching in Western*, pp. 63–72.

77. Wilkeson, "Cattle-Raising," p. 793.

78. Robert S. Fletcher, "That Hard Winter in Montana, 1886–1887," *AH*, 4 (1930), pp. 123–30; Alfred T. Larson, "The Winter of 1886–1887 in Wyoming," *Annals of Wyoming*, 14 (1942), pp. 5–17; Box, "Range Deterioration," p. 43; Pelzer, *Cattlemen's Frontier*, p. 113; Osgood, *Day of Cattleman*, pp. 125–26, 215; Pulling, "History of Range," pp. 499–503; Briggs, "Ranching and Stock," pp. 456–60; Shafer, "Cattle Ranching," p. 58; Kuhn, "W-Bar," p. 158.

79. Young and Sparks, *Cattle in Cold*, pp. 121, 134; Morrisey, "Early Range," pp. 155–56.

80. Nimmo, "American Cowboy," p. 883; Osgood, *Day of Cattleman*, pp. 125–26.

81. T. R. Havins, "The Passing of the Longhorn," *SWHQ*, 56 (1952–53), pp. 51–58; Robert S. Fletcher, "The End of the Open Range in Eastern Montana," *MVHR*, 16 (1929), pp. 188–211; Simon M. Evans, "The Passing of a Frontier: Ranching in the Canadian West, 1882–1912," Ph.D. diss., University of Calgary (Calgary, AB, 1976); Simon M. Evans, "The End of the Open Range Era in Western Canada," *Prairie Forum*, 8:1 (1983), pp. 71–87; Wayne Gard, "The Fence-Cutters," *SWHQ*, 51 (1947–48), pp. 1–15; Ike Blasingame, *Dakota Cowboy: My Life in the Old Days* (New York, NY, 1958); James A. Wilson, "Cattlemen, Packers, and the Government: Retreating Individualism on the Texas Range," *SWHQ*, 74 (1970–71), pp. 525–34; Briggs, "Development and Decline," p. 536; Coolidge, *Texas Cowboys*, pp. 19, 33; Coolidge, *Old California*, pp. 67–69, 150.

82. Pelzer, *Cattlemen's Frontier*, pp. 119–33, 139; Briggs, "Development and Decline," pp. 530–32; Box, "Range Deterioration," pp. 37–45; Osgood, *Day of Cattleman*, p. 95; Herbert C. Hanson, L. Dudley Love, and M. S. Morris, *Effects of Different Systems of Grazing by Cattle upon a Western Wheat-Grass Type of Range near Fort Collins, Colorado* (Ft. Collins, CO, 1931).

83. Walter M. Kollmorgen, "The Woodsman's Assaults on the Domain of the Cattleman," *AAAG*, 59 (1969), pp. 215–39; Walter M. Kollmorgen and David S. Simonett, "Grazing Operations in the Flint Hills–Bluestem Pastures of Chase County, Kansas," *Annals of the Association of American Geographers*, 55 (1965), pp. 260–90; John W. Morris, "Arbuckle Moun-

tain Ranching Area," *EG,* 23 (1947), pp. 190–98; Dale, "History of Ranch," pp. 321–22; Aeschbacher, "Development," pp. 41–64; Osgood, *Day of Cattleman,* p. 243.

84. Young and Sparks, *Cattle in Cold,* p. 135.

Chapter Eight

1. Fred B. Kniffen, "The Western Cattle Complex: Notes on Differentiation and Diffusion," *Western Folklore,* 12 (1953), pp. 181–82; Gisbert Rinschede, *Die Wanderviehwirtschaft im gebirgigen Westen der USA und ihre Auswirkungen im Naturraum* (Regensburg, Germany, 1984), p. 91; Donald W. Meinig, "American Wests: Preface to a Geographical Interpretation," *AAAG,* 62 (1972), pp. 159–84.

2. David Hornbeck, "Land Tenure and Rancho Expansion in Alta California, 1784–1846," *JHG,* 4 (1978), pp. 388–89; Iris H. W. Engstrand, "California Ranchos: Their Hispanic Heritage," *SCQ,* 67 (1985), p. 288; H. F. Raup, "Transformation of Southern California to a Cultivated Land," *AAAG,* 49:3, supplement (1959), p. 62; L. T. Burcham, *California Range Land: An Historico-Ecological Study of the Range Resources of California* (Sacramento, CA, 1957), pp. 141–42; Hazel A. Pulling, "A History of California's Range Cattle Industry, 1770–1912," Ph.D. diss., University of Southern California (Los Angeles, CA, 1944), p. 78.

3. Hubert H. Bancroft, *California Pastoral, 1769–1848* (San Francisco, CA, 1888), pp. 347, 373–74, 379–80; Glenn R. Vernam, *Man on Horseback* (New York, NY, 1964), p. 302; Iris H. Wilson, *William Wolfskill, 1798–1866: Frontier Trapper to California Ranchero* (Glendale, CA, 1965); Jan O. M. Broek, *The Santa Clara Valley, California: A Study in Landscape Changes* (Utrecht, Netherlands, 1932); Richard W. Slatta, *Cowboys of the Americas* (New Haven, CT, 1990), p. 7; Burcham, *California Range,* pp. 141–42; Rinschede, *Wanderviehwirtschaft,* p. 94; Pulling, "History of California's," pp. 103–4.

4. J. Orin Oliphant, *On the Cattle Ranges of the Oregon Country* (Seattle, WA, 1968), pp. 3–19; Michael L. Olsen, "Transplantation of Domestic Plants and Animals in the Pacific Northwest," *JOW,* 14:3 (1975), p. 42; C. S. Kingston, "Introduction of Cattle into the Pacific Northwest," *Washington Historical Quarterly,* 14 (1923), pp. 163–71.

5. S. A. Clarke, *Pioneer Days of Oregon History* (Cleveland, OH, 1905), vol. 1, pp. 304–9; William A. Bowen, *The Willamette Valley: Migration and Settlement on the Oregon Frontier* (Seattle, WA, 1978), p. 80; James W. Thompson, *A History of Livestock Raising in the United States, 1607–1860* (Washington, DC, 1942), p. 122; Oliphant, *On Cattle Ranges,* pp. 16, 20–22; Kingston, "Introduction," pp. 179, 181; Olsen, "Transplantation," p. 44; Burcham, *California Range,* p. 140.

6. Don D. Walker, "The Cattle Industry of Utah, 1850–1900: An Histor-

ical Profile," *UTHQ*, 32 (1964), p. 183; Robert M. Denhardt, "Driving Livestock East from California prior to 1850," *CHSQ*, 20 (1941), pp. 341–47.

7. Donald W. Meinig, *The Great Columbia Plain: A Historical Geography, 1805–1910* (Seattle, WA, 1968), pp. 497–98; William A. Bowen, *The Willamette Valley: Migration and Settlement on the Oregon Frontier* (Seattle, WA, 1978), p. 80.

8. Robert G. Cleland, *The Cattle on a Thousand Hills: Southern California, 1850–1880* (San Marino, CA, 1951), pp. 51, 64, 102; L. T. Burcham, "Cattle and Range Forage in California, 1770–1880," *AH*, 35 (1961), p. 143; Burcham, *California Range*, p. 142; Broek, *Santa Clara*, pp. 60–61, 79–80.

9. William R. Cameron, "Rancho Santa Margarita of San Luis Obispo," *CHSQ*, 36 (1957), p. 14; James M. Jensen, "Cattle Drives from the Ranchos to the Gold Fields of California," *Arizona and the West*, 2 (1960), pp. 341–52; Lois J. Roberts, "Rancho Jesús María, Santa Barbara County," *SCQ*, 68 (1986), pp. 7, 15; Cleland, *Cattle on Thousand*, pp. 3, 51, 64, 75–76, 97, 102–8; Pulling, "History of California's," pp. 93–94.

10. Joseph B. Davy, *Stock Ranges of Northwestern California: Notes on the Grasses and Forage Plants and Range Conditions* (Washington, DC, 1902), pp. 35, 77; Hazel A. Pulling, "Range Forage and California's Range-Cattle Industry," *Historian*, 7 (1944–45), pp. 114–116, 123–24; Clara M. Love, "History of the Cattle Industry in the Southwest," *SWHQ*, 19 (1915–16), pp. 374–81; Cleland, *Cattle on Thousand*, p. 3; Burcham, *California Range*, pp. 143, 150, 152.

11. Edward F. Treadwell, *The Cattle King: A Dramatized Biography* (New York, NY, 1931), pp. 332–34; Peter K. Simpson, "Studying the Cattleman: Cultural History and the Livestock Industry in Southeastern Oregon," *Idaho Yesterdays*, 28:2 (1984), p. 5; Daniel M. Drumheller, *Uncle Dan Drumheller Tells Thrills of Western Trails in 1854* (Spokane, WA, 1925), pp. 1–5, 65; Pulling, "History of California's," pp. 79, 80, 85, 93–94; Jensen, "Cattle Drives," p. 351; Burcham, *California Range*, pp. 142–43, 151, 154, 157.

12. Clarence Gordon, "Report on Cattle, Sheep, and Swine, Supplementary to Enumeration of Live Stock on Farms in 1880," *Report on the Productions of Agriculture as Returned at the Tenth Census* (Washington, DC, 1883), p. 1058; Oliphant, *On Cattle Ranges*, p. 68; Cleland, *Cattle on Thousand*, p. 110.

13. F. W. Laing, "Some Pioneers of the Cattle Industry," *British Columbia Historical Quarterly*, 6 (1942), pp. 266–75; Love, "History of Cattle," pp. 381–82; Oliphant, *On Cattle Ranges*, pp. 68–69.

14. Hazel A. Pulling, "California's Fence Laws and the Range-Cattle Industry," *Historian*, 8 (1945–46), p. 144; Burcham, *California Range*, p. 152; Burcham, "Cattle and Range," p. 148; Cleland, *Cattle on Thousand*, pp. 107–8.

15. Pulling, "California's Fence," p. 144; Burcham, *California Range*, pp. 152–53; Cleland, *Cattle on Thousand*, pp. 110, 130–33; Burcham, "Cat-

tle and Range," pp. 143, 146; Pulling, "History of California's," pp. 111–17, 138.

16. J. M. Guinn, "The Passing of the Cattle Barons of California," *Historical Society of Southern California, Publications*, 8 (1909–10), p. 59; Treadwell, *Cattle King*, pp. 332–34; Broek, *Santa Clara*, pp. 60–61; Cleland, *Cattle on Thousand*, p. 165; Burcham, *California Range*, pp. 153–57; Burcham, "Cattle and Range," p. 143.

17. Howard F. Gregor, "A Sample Study of the California Ranch," *AAAG*, 41 (1951), pp. 285–306; Peter K. Simpson, *The Community of Cattlemen: A Social History of the Cattle Industry in Southeastern Oregon, 1869–1912* (Moscow, ID, 1987), p. 20; Cleland, *Cattle on Thousand*, pp. 3, 165; Pulling, "California's Fence," pp. 140, 154; Pulling, "History of California's," pp. 127–28, 136–38.

18. Rinschede, *Wanderviehwirtschaft*, p. 95; Burcham, *California Range*, p. 156; Pulling, "History of California's," pp. 113, 118, 127.

19. James A. Young and B. Abbott Sparks, *Cattle in the Cold Desert* (Logan, UT, 1985), p. 48; C. W. Vrooman, "A History of Ranching in British Columbia," *Economic Annalist*, 11 (1941), p. 20; Robert Fletcher, *Free Grass to Fences: The Montana Range Cattle Story* (New York, NY, 1960), p. 22; Gordon, "Report on Cattle," pp. 1059 (quote), 1071, 1097, 1098; Oliphant, *On Cattle Ranges*, p. 68.

20. Dana Yensen, *A Grazing History of Southwestern Idaho with Emphasis on the Birds of Prey Study Area* (Moscow and Boise, ID, revised ed., 1982), pp. 1–10; A. L. Lesperance et al., "Great Basin Wildrye," *Rangeman's Journal* 5:4 (1978), pp. 125–27; Thomas R. Vale, "Presettlement Vegetation in the Sagebrush-Grass Area of the Intermountain West," *Journal of Range Management*, 28 (1975), pp. 32–36; W. A. Galbraith and W. E. Anderson, "Grazing History of the Northwest," *Journal of Range Management*, 24 (1971), pp. 6–12; Mont H. Saunderson, *Western Stock Raising* (Minneapolis, MN, 1950), pp. 4, 16–21; Giles French, *Cattle Country of Peter French* (Portland, OR, 1964), pp. 9, 44–45; Sheldon D. Ericksen, *Occupance in the Upper Deschutes Basin, Oregon* (Chicago, IL, 1953), pp. 1, 15; Cyrus C. Loveland, *California Trail Herd: The 1850 Missouri-to-California Journal of Cyrus C. Loveland* (Los Gatos, CA, 1961), pp. 97, 109; Otis W. Freeman et al., "Physiographic Divisions of the Columbia Intermontane Province," *AAAG*, 35 (1945), p. 69; Oliphant, *On Cattle Ranges*, p. 80; Young and Sparks, *Cattle in Cold*.

21. Simpson, *Community*, p. 20; Pulling, "California's Fence," 140–41, 146–50, 154.

22. Howard W. Marshall and Richard E. Ahlborn, *Buckaroos in Paradise: Cowboy Life in Northern Nevada* (Washington, DC, 1980); Byron D. Lusk, "Golden Cattle Kingdoms of Idaho," M.S. thesis, Utah State University, Logan, UT, 1978, pp. 39–40; Velma S. Truett, *On the Hoof in Nevada* (Los Angeles, CA, 1950), pp. 46–47; Oliphant, *On Cattle Ranges*, p. 69.

23. Oliphant, *On Cattle Ranges,* pp. 134–35.

24. Margaret J. LoPiccolo, "Some Aspects of the Range Cattle Industry of Harney County, Oregon, 1870–1890," M.A. thesis, University of Oregon (Eugene, OR, 1962); French, *Cattle Country,* pp. 12, 43, 55, 94, 118–20; Simpson, *Community,* pp. 4–5, 9–11, 19; Oliphant, *On Cattle Ranges,* pp. 69, 77, 87–88, 93, 137–47, 188–91.

25. French, *Cattle Country,* pp. 12–31, 41–42, 57–59; Oliphant, *On Cattle Ranges,* pp. 96–97, 189–90.

26. Dale L. Morgan, *The Humboldt: Highroad of the West* (New York, NY, and Toronto, ON, 1943), pp. 311–16; Effie O. Read, *White Pine Lang Syne: A True History of White Pine County, Nevada* (Denver, CO, 1965), pp. 51, 180, 214–15; Edna B. Patterson et al., *Nevada's Northeast Frontier* (Sparks, NV, 1969), pp. 213, 326, 327, 345, 354, 356, 361, 368, 387, 419; Arnold R. Rojas, *The Vaquero* (Charlotte, NC, and Santa Barbara, CA, 1964), pp. 8–9, 51, 113–16; Sessions S. Wheeler, *The Black Rock Desert* (Caldwell, ID, 1978), pp. 153, 157–59; T. Alex Bulman, *Kamloops Cattlemen: One Hundred Years of Trail Dust* (Sidney, BC, 1972), following p. 96; Dane Coolidge, *Old California Cowboys* (New York, NY, 1939), pp. 37–46; Drumheller, *Uncle Dan,* p. 67; Gordon, "Report on Cattle," p. 1059; Truett, *On the Hoof,* p. 47; Treadwell, *Cattle King,* endpaper map, p. 179; French, *Cattle Country,* p. 21; Lusk, "Golden Cattle," pp. 39–40; Oliphant, *On Cattle Ranges,* pp. 134–35, 190–92.

27. Sharlet M. Hall, "Old Range Days and New in Arizona," *Out West,* 28 (Marsh 1908), pp. 182, 186; Dane Coolidge, *Arizona Cowboys* (New York, NY, 1938), p. 13; United States Census, 1880, manuscript population schedules, Arizona Territory; United States Senate, [1870] *Federal Census, Territory of New Mexico and the Territory of Arizona* (Washington, DC, 1965), pp. 125–253; Kniffen, "Western Cattle," p. 182; Coolidge, *Old California,* preceding p. 81.

28. J. J. Wagoner, *History of the Cattle Industry in Southern Arizona, 1540–1940* (Tucson, AZ, 1952), p. 36; James R. Hastings and Raymond M. Turner, *The Changing Mile: An Ecological Study of Vegetation Change with Time in the Lower Mile of an Arid and Semiarid Region* (Tucson, AZ, 1980), pp. 32–33, 40; Gordon, "Report on Cattle," p. 1047; Coolidge, *Arizona,* p. 55.

29. Bert Haskett, "Early History of the Cattle Industry in Arizona," *AZHR,* 6:4 (1935), pp. 11, 24, 31–35; Richard J. Morrisey, "The Early Range Cattle Industry in Arizona," *AH,* 24 (1950), p. 152; U.S. Senate, *Federal Census;* U.S. Census 1880, manuscript population schedules, Arizona Territory; Wagoner, *History of Cattle,* pp. 84–85, 101–2.

30. Meinig, *Great Columbia,* pp. 497, 498.

31. Truett, *On the Hoof,* p. 541; Rojas, *Vaquero,* p. 9; Marshall and Ahlborn, *Buckaroos,* p. 15.

32. Mildretta Adams, *Owyhee Cattlemen, 1878–1978: 100 Years in the Saddle,* 2nd ed. (Homedale, ID, 1979), pp. 9, 81, 82; Louie W. Attebery, "Celts

and Other Folk in the Regional Livestock Industry," *Idaho Yesterdays*, 28:2 (1984), pp. 23–24 (quote); Adelaide Hawes, *The Valley of Tall Grass* (Bruneau, ID, 1950), pp. 100–103, 112; Herman Oliver and E. R. Jackman, *Gold and Cattle Country* (Portland, OR, 1962), p. 159; J. Orin Oliphant, "The Cattle Herds and Ranches of the Oregon Country, 1860–1890," *AH*, 21 (1947), p. 230; Oliphant, *On Cattle Ranges*, pp. 211–12; French, *Cattle Country*, pp. 21, 23, 49–50; Treadwell, *Cattle King*, p. 250; Read, *White Pine*, p. 60; U.S. Census 1880, manuscript population schedules, Grant Co., Oregon; Marshall and Ahlborn, *Buckaroos*, pp. xii, 7, 12–15, 28; Patterson et al., *Nevada's Northeast*, pp. 216, 388–89; Rojas, *Vaquero*, p. 25; Drumheller, *Uncle Dan*, p. 123.

33. Elizabeth S. Bright, *A Word Geography of California and Nevada* (Berkeley and Los Angeles, CA, 1971), pp. 93, 193; Julian Mason, "The Etymology of *Buckaroo*," *American Speech*, 25 (1960), pp. 51–55; Frederic G. Cassidy, "Another Look at Buckaroo," *American Speech*, 53 (1978), pp. 49–51; Frederic G. Cassidy, ed., *Dictionary of American Regional English* (Cambridge, MA, 1985), vol. 1, p. 411; Terry G. Jordan, *Trails to Texas: Southern Roots of Western Cattle Ranching* (Lincoln, NE, 1981), pp. 14–15; Vernam, *Man on Horseback*, p. 316; Oliver and Jackman, *Gold and Cattle*, p. 132; Simpson, *Community*, p. 24; Truett, *On the Hoof*, p. 41; Marshall and Ahlborn, *Buckaroos*.

34. Annie R. Mitchell, *The Way it Was: The Colorful History of Tulare County* (Fresno, CA, 1976), p. 37; William A. Craigie and James R. Hulbert, *A Dictionary of American English on Historic Principles* (Chicago, IL, 1944), vol. 4, p. 1965; Jo Mora, *Trail Dust and Saddle Leather* (New York, NY, 1946), pp. 67–68; Craig M. Carver, *American Regional Dialects: A Word Geography* (Ann Arbor, MI, 1987), p. 225; George C. Everett, *Cattle Cavalcade in Central Colorado* (Denver, CO, 1966), pp. 151, 344; John K. Rollinson, *Wyoming Cattle Trails: History of the Migration of Oregon-Raised Herds to Mid-Western Markets* (Caldwell, ID, 1948), p. 115; Harry Marriott, *Cariboo Cowboy* (Sidney, BC, 1966), pp. 75; Louie W. Attebery, interview with T. G. J. Caldwell, ID, August 1, 1991; Louie W. Attebery, letters to T. G. J. dated January 22 and February 1, 1991; Marshall and Ahlborn, *Buckaroos*, pp. 15, 38, 61, 63, 80; Adams, *Owyhee*, pp. 7, 17; Cassidy, *Dictionary*, v. 1, pp. 497, 575; Patterson et al., *Nevada's Northeast*, p. 225; Rojas, *Vaquero*, pp. 4, 8, 26; Wheeler, *Black Rock*, p. 156; Vernam, *Man on Horseback*, pp. 311–13; Slatta, *Cowboys*, p. 7; Young and Sparks, *Cattle in Cold*, p. 109; Bright, *Word Geography*, pp. 108, 168, 174; Oliphant, *On Cattle Ranges*, p. 210; Oliphant, "Cattle Herds," pp. 227–30; Gordon, "Report on Cattle," pp. 1032–33, 1083; Hawes, *Valley of Tall*, pp. 115–20; Oliver and Jackman, *Gold and Cattle*, pp. 137–38, 152; Attebery, "Celts and Other," p. 24; Simpson, *Community*, p. 27.

35. William Foster-Harris, *The Look of the Old West* (New York, NY, 1960), pp. 207, 242, 244; David Dary, *Cowboy Culture: A Saga of Five Centuries* (New York, NY, 1981), p. 51; Walt Coburn, *Pioneer Cattleman in*

Montana: The Story of the Circle C Ranch (Norman, OK, 1968), p. 132; Mora, *Trail Dust*, p. 93; Vernam, *Man on Horseback*, pp. 302–6, 310; Mitchell, *Way it Was*, pp. 45–48; Hawes, *Valley of Tall*, p. 101; Attebery, "Celts and Other," p. 24; Rojas, *Vaquero*, pp. 29–33, 54; Marshall and Ahlborn, *Buckaroos*, pp. 1, 56.

36. Oliver and Jackman, *Gold and Cattle*, p. 135.

37. Vernam, *Man on Horseback*, pp. 303–4, 313; Attebery, "Celts and Other," p. 24; Oliver and Jackman, *Gold and Cattle*, pp. 134–37; Adams, *Owyhee*, p. 186; Foster-Harris, *Look of Old*, pp. 208–9, 214; Marshall and Ahlborn, *Buckaroos*, pp. 37–38, 54.

38. Mora, *Trail Dust*, p. 110; Vernam, *Man on Horseback*, pp. 301, 314–16; Foster-Harris, *Look of Old*, pp. 205, 207.

39. Truett, *On the Hoof*, pp. 496, 521, 527, 535, 539, 542, 543; Rojas, *Vaquero*, p. 111.

40. Bulman, *Kamloops*, following p. 96 and pp. 111–12.

41. French, *Cattle Country*, p. 38; Oliver and Jackman, *Gold and Cattle*, p. 161; Louie W. Attebery, letters to T. G. J. dated January 22 and February 16, 1991. A Spanish beef wheel is also found in the open-air museum at El Rancho de las Golondrinas, near Santa Fe, NM.

42. Thomas R. Weir, *Ranching in the Southern Interior Plateau of British Columbia*, revised ed. (Ottawa, ON, 1964), p. 129; "Breaking a Mustang," *Frank Leslie's Illustrated Newspaper*, 38:964 (March 21, 1874), drawing by Arthur Lemon on p. 29; Lawrence Kinnaird, ed., *The Frontiers of New Spain: Nicolas de Lafora's Description, 1766–1768* (Berkeley, CA, 1958), plate 9; historic Hacienda Santa Rosa ranchstead, now part of Santa Rosa National Park, near Liberia, Guanacaste province, Costa Rica; J. Sanford Rikoon, "Traditional Fence Patterns in Owyhee County, Idaho," *Pioneer America Society, Transactions*, 7 (1984), pp. 61–64; Attebery letter, February 16; Marshall and Ahlborn, *Buckaroos*, p. 46; Coolidge, *Old California*, p. 27; Lesperance et al., "Great Basin," p. 126; Oliphant, *On Cattle Ranges*, pp. 200, 214–15; French, *Cattle Country*, pp. 34–35, 59–60; Oliver and Jackman, *Gold and Cattle*, p. 161; Hawes, *Valley of Tall*, pp. 25, 101; field observations, Dominican Republic, September 1992.

43. Ramon F. Adams, *Western Words: A Dictionary of the American West* (Norman, OK, 1968), p. 173; Mora, *Trail Dust*, pp. 60, 63; Hawes, *Valley of Tall*, pp. 142–43; Oliver and Jackman, *Gold and Cattle*, p. 136; Attebery, "Celts and Other," p. 24; Dary, *Cowboy Culture*, p. 157; Vernam, *Man on Horseback*, pp. 306, 311; Carver, *American Regional*, pp. 224–25; Oliphant, "Cattle Herds," p. 230; Adams, *Owyhee*, p. 88; Coolidge, *Arizona*, p. 13; Coolidge, *Old California*, pp. 28, 85, and preceding p. 81; Foster-Harris, *Look of Old*, pp. 217, 250; Rollinson, *Wyoming Cattle*, p. 216.

44. Vernam, *Man on Horseback*, pp. 311–14; Dary, *Cowboy Culture*, pp. 48, 157; Mora, *Trail Dust*, pp. 67–68; Attebery, "Celts and Other," p. 24;

Oliver and Jackman, *Gold and Cattle*, p. 137; Coolidge, *Arizona*, p. 13; Rojas, *Vaquero*, pp. 34–38, 55; Foster-Harris, *Look of Old*, p. 253.

45. Richard L. Nostrand, "The Santa Ynez Valley: Hinterland of Coastal California," *HSSCQ*, 48 (1966), pp. 41–42; Love, "History of Cattle," p. 377; Simpson, *Community*, p. 27; Gordon, "Report on Cattle," p. 1032 (quote); Saunderson, *Western Stock*, pp. 23–26; Burcham, *California Range*, p. 153; Burcham, "Cattle and Range," pp. 142, 145, 147; Rojas, *Vaquero*, p. 111.

46. Rinschede, *Wanderviehwirtschaft*; Simpson, *Community*, p. 27; Ericksen, *Occupance*, p. 39; Young and Sparks, *Cattle in Cold*, pp. 53, 60, 147; French, *Cattle Country*, p. 44.

47. Fletcher, *Free Grass*, p. 39.

48. George E. Carter, "The Cattle Industry of Eastern Oregon, 1880–90," *ORHQ*, 67 (1966), pp. 143–44; Mark H. Brown and W. R. Felton, *Before Barbed Wire: L. A. Huffman, Photographer on Horseback* (New York, NY, 1956), p. 69; French, *Cattle Country*, p. 21; Gordon, "Report on Cattle," p. 1080; Simpson, *Community*, p. 40; Burcham, *California Range*, pp. 159–62; Oliphant, "Cattle Herds," p. 223; Pulling, "History of California's," p. 98.

49. Truett, *On the Hoof*, p. 40; Oliphant, "Cattle Herds," p. 236; Treadwell, *Cattle King*, p. 100; Lusk, "Golden Cattle," p. 161; Hawes, *Valley of Tall*, pp. 105–6; Morgan, *Humboldt*, pp. 315–16; French, *Cattle Country*, p. 96.

50. Young and Sparks, *Cattle in Cold*, pp. 111–12, following p. 119; Oliphant, *On Cattle Ranges*, pp. 212–15; Lusk, "Golden Cattle," pp. 104, 106, 163; Simpson, *Community*, pp. 24–25; French, *Cattle Country*, pp. 21, 60, 138; Oliver and Jackman, *Gold and Cattle*, p. 132; Gordon, "Report on Cattle," pp. 1032, 1080–81; Treadwell, *Cattle King*, p. 100; Burcham, *California Range*, p. 153.

Chapter Nine

1. Jo T. Bloom, "Cumberland Gap versus South Pass: The East or West in Frontier History," *WHQ*, 3 (1972), p. 159; Terry G. Jordan, "Early Northeast Texas and the Evolution of Western Ranching," *AAAG*, 67 (1977), pp. 66–87.

2. Peter K. Simpson, "Studying the Cattleman: Cultural History and the Livestock Industry in Southeastern Oregon," *Idaho Yesterdays*, 28:2 (1984), p. 7; Paul C. Henlein, *Cattle Kingdom in the Ohio Valley, 1783–1860* (Lexington, KY, 1959).

3. Ernest S. Osgood, *The Day of the Cattleman* (Chicago, IL, 1929), pp. 126–27; Peter K. Simpson, *The Community of Cattlemen: A Social History of the Cattle Industry in Southeastern Oregon, 1869–1912* (Moscow, ID, 1987), pp. 25, 117; Henlein, *Cattle Kingdom*, p. 181.

4. James R. Shortridge, *The Middle West: Its Meaning in American Culture* (Lawrence, KS, 1989); James R. Shortridge, "The Vernacular Middle

West," *AAAG*, 75 (1985), p. 50; Joseph W. Brownell, "The Cultural Midwest," *Journal of Geography*, 59 (1960), p. 83; Wilbur Zelinsky, "North America's Vernacular Regions," *AAAG*, 70 (1980), p. 14.

5. Terry G. Jordan, *Trails to Texas: Southern Roots of Western Cattle Ranching* (Lincoln, NE, 1981), pp. 127–29, 134–39; William C. Holden, *A Ranching Saga: The Lives of William Electious Halsell and Ewing Halsell*, 2 vols. (San Antonio, TX, 1976); Floyd F. Ewing, "James H. Baker: Cattleman and Trail-Driver," *WTYB*, 43 (1967), pp. 3–17; Henlein, *Cattle Kingdom*, p. 181.

6. Jordan, *Trails*, pp. 89, 134–36, 147–49.

7. Hiram Latham, *Trans-Missouri Stock Raising: The Pasture Lands of North America, Winter Grazing* (Omaha, NE, 1871), p. 40.

8. T. R. Havins, "The Passing of the Longhorn," *SWHQ*, 56 (1952–53), pp. 51–58; Lauren C. Post, "The Upgrading of Beef Cattle on the Great Plains," *California Geographer*, 2 (1961), p. 28; Jordan, *Trails*, p. 149.

9. Nellie S. Yost, *The Call of the Range: The Story of the Nebraska Stock Growers Association* (Denver, CO, 1966), pp. 27–36; Maurice Frink et al., eds., *When Grass Was King: Contributions to the Western Range Cattle Industry Study* (Boulder, CO, 1956), p. 337; Latham, *Trans-Missouri*, p. 43.

10. Osgood, *Day of Cattleman*, pp. 11–18, 39; Latham, *Trans-Missouri*, p. 31.

11. C. W. McCampbell, "W. E. Campbell, Pioneer Kansas Livestockman," *KHQ*, 16 (1948), pp. 246, 251–52.

12. Joseph G. McCoy, *Historic Sketches of the Cattle Trade of the West and Southwest* (Kansas City, MO, 1874), pp. 214–24; Henlein, *Cattle Kingdom*, p. 180.

13. Edward E. Dale, "The Cherokee Strip Live Stock Association," *CO*, 5 (1927), p. 62; Edward E. Dale, "History of the Ranch Cattle Industry in Oklahoma," *American Historical Association Annual Report* (1920), p. 316.

14. Agnes W. Spring, "A Genius for Handling Cattle: John W. Iliff," in Frink et al., *When Grass Was King*, pp. 333–450; W. Baillie Grohman, "Cattle Ranches in the Far West," *Fortnightly Review*, n.s. 28 (1880), pp. 450–53; George C. Everett, *Cattle Cavalcade in Central Colorado* (Denver, CO, 1966), pp. 12–13; John M. Kuykendall, "The First Cattle North of the Union Pacific Railroad," *Colorado Magazine*, 7:2 (1930), p. 70; Latham, *Trans-Missouri*, p. 41.

15. Harold Briggs, Ranching and Stock-Raising in the Territory of Dakota," *South Dakota Historical Collections*, 14 (1928), p. 418; Yost, *Call of Range*, pp. 73–74.

16. Herbert O. Brayer, "The L7 Ranches: An Incident in the Economic Development of the Western Cattle Industry," *Annals of Wyoming*, 15 (1943), pp. 6–7; Robert H. Burns, "The Newman Ranches: Pioneer Cattle Ranches of the West," *NBH*, 34 (1953), pp. 21–32; Robert H. Burns, "The Newman Brothers: Forgotten Cattle Kings of the Northern Plains," *Montana*, 11 (Oct.

1961), pp. 28–36; Briggs, "Ranching and Stock-Raising," p. 441; Osgood, *Day of Cattleman*, p. 47; Henlein, *Cattle Kingdom*, p. 182.

17. Walker D. Wyman, *Nothing But Prairie and Sky: Life on the Dakota Range in the Early Days* (Norman, OK, 1988); United States Census, 1880, manuscript population schedules; Grohman, "Cattle Ranches," p. 447.

18. Reginald Aldridge, *Ranch Notes in Kansas, Colorado, the Indian Territory and Northern Texas* (London, UK, 1884), pp. 64–66, 183, 204; George F. Shafer, "Cattle Ranching in McKenzie County, North Dakota," *North Dakota Historical Quarterly*, 1 (1926), p. 58; McCoy, *Historic Sketches*, p. 82; Everett, *Cattle Cavalcade*, pp. 10, 17.

19. Clarence Gordon, "Report on Cattle, Sheep, and Swine, Supplementary to Enumeration of Live Stock on Farms in 1880," *Report on the Production of Agriculture as Returned at the Tenth Census* (Washington, DC, 1883), p. 1016; Grohman, "Cattle Ranches," p. 441; Wyman, *Nothing But Prairie*, p. 114.

20. Frank Wilkeson, "Cattle-Raising on the Plains," *Harper's New Monthly Magazine*, 72 (April 1886), p. 794; Shafer, "Cattle Ranching," p. 58; Briggs, "Ranching and Stock," p. 419; Aldridge, *Ranch Notes*, p. 204; McCoy, *Historic Sketches*, pp. 216–24; Osgood, *Day of Cattleman*, p. 204.

21. Shafer, "Cattle Ranching," pp. 58–59.

22. James C. Malin, *The Grasslands of North America: Prolegomena to its History* (Lawrence, KS, 1947), p. 275; Walter von Richthofen, *Cattle-Raising on the Plains of North America*, ed. E. E. Dale (Norman, OK, 1964), p. 67; Gordon, "Report on Cattle," p. 998; Osgood, *Day of Cattleman*, pp. 92–93; Latham, *Trans-Missouri*, p. 42; Aldridge, *Ranch Notes*, p. 203; McCoy, *Historic Sketches*, p. 243; McCampbell, "W. E. Campbell," pp. 251–52; Briggs, "Ranching and Stock," p. 419; Post, "Upgrading," pp. 23–30.

23. Latham, *Trans-Missouri*, pp. 38–41; Richthofen, *Cattle-Raising*, pp. 23, 66.

24. William W. Savage, Jr., *The Cherokee Strip Live Stock Association* (Columbia, MO, 1973); Boyd M. Anderson, *Beyond the Range: A History of the Saskatchewan Stock Growers Association* (Regina, SK, 1988); William D. Aeschbacher, "Development of Cattle Raising in the Sandhills," *NBH*, 28 (1947), pp. 45–48; Hazel A. Pulling, "History of the Range Cattle Industry of Dakota," *South Dakota Historical Collections*, 20 (1940), p. 489; Gerald Baydo, "Cattlemen's Associations in New Mexico Territory," *JOW*, 14:3 (1975), pp. 60–71; C. Boone McClure, "The Laws and Customs of the Open Range," *PPHR*, 10 (1937), pp. 64–79; W. Turrentine Jackson, "The Wyoming Stock Growers' Association: Political Power in Wyoming Territory," *MVHR*, 33 (1947), pp. 571–94; Dale, "Cherokee Strip," pp. 58–78.

25. Louis Pelzer, *The Cattlemen's Frontier: A Record of the Trans-Mississippi Cattle Industry from Oxen Trains to Pooling Companies, 1850–1890* (Glendale, CA, 1936), pp. 74–79; Dale, "Cherokee Strip," p. 65; McCoy, *Historic Sketches*, p. 81; Aeschbacher, "Development of Cattle," p. 60.

26. R. D. Holt, "The Saga of Barbed Wire in the Tom Green Country," *WTYB*, 4 (1928), pp. 32–49; Post, "Upgrading," p. 28; Havins, "Passing of Longhorn"; Aeschbacher, "Development of Cattle," pp. 57–58.

27. Aldridge, *Ranch Notes*, p. 171.

28. Robert Fletcher, *Free Grass to Fences: The Montana Range Cattle Story* (New York, NY, 1960), p. 39; Simpson, "Studying Cattleman," p. 5.

29. Walter Bailey, "The Barlow Road," *ORHQ*, 13 (1912), pp. 287–88.

30. J. Orin Oliphant, *On the Cattle Ranges of the Oregon Country* (Seattle, WA, 1968), p. 30; Clifford D. Carpenter, "The Early Cattle Industry in Missouri," *Missouri Historical Review*, 47 (1953), p. 202; C. S. Kingston, "Introduction of Cattle into the Pacific Northwest," *Washington Historical Quarterly*, 14 (1923), pp. 181–85; Henlein, *Cattle Kingdom*, p. 182.

31. Donald W. Meinig, *The Great Columbia Plain: A Historical Geography, 1805–1910* (Seattle, WA, 1968), p. 498; William A. Bowen, *The Willamette Valley: Migration and Settlement on the Oregon Frontier* (Seattle, WA, 1978), p. 80; David Dary, *Cowboy Culture: A Saga of Five Centuries* (New York, NY, 1981), p. 232; Oliphant, *On Cattle Ranges*, p. 78; Kingston, "Introduction," p. 179; Simpson, "Studying Cattleman," p. 7.

32. Bowen, *Willamette*, p. 80.

33. Earle K. Stewart, "Transporting Livestock by Boat up the Columbia, 1861–1868," *ORHQ*, 50 (1949), pp. 253–55; Oliphant, *On Cattle Ranges*, p. 78.

34. Simpson, "Studying Cattleman," p. 7.

35. Robert G. Cleland, *The Cattle on a Thousand Hills: Southern California, 1850–1880* (San Marino, CA, 1951), p. 113; Iris H. Wilson, *William Wolfskill, 1798–1866: Frontier Trapper to California Ranchero* (Glendale, CA, 1965), p. 19.

36. Edna B. Patterson, et al., *Nevada's Northeast Frontier* (Sparks, NV, 1969), p. 207.

37. Hazel A. Pulling, "A History of California's Range-Cattle Industry, 1770–1912," Ph.D. diss., University of Southern California (Los Angeles, CA, 1944), pp. 99–103, 342 (quote); Frank S. Popplewell, "St. Joseph, Missouri, as a Center of the Cattle Trade," *Missouri Historical Review*, 32 (1968), p. 445; L. T. Burcham, *California Range Land: An Historico-Ecological Study of the Range Resources of California* (Sacramento, CA, 1957), p. 143; Giles French, *Cattle Country of Peter French* (Portland, OR, 1964), pp. 17–18; James M. Jensen, "Cattle Drives from the Ranchos to the Gold Fields of California," *Arizona and the West*, 2 (1960), p. 351; Annie R. Mitchell, *The Way it Was: The Colorful History of Tulare County* (Fresno, CA, 1976), p. 36; Cleland, *Cattle on Thousand*, p. 108; Dary, *Cowboy Culture*, pp. 92–97; Carpenter, "Early Cattle," p. 202; Patterson et al., *Nevada's Northeast*, p. 207.

38. Cyrus C. Loveland, *California Trail Herd: The 1850 Missouri-to-California Journal of Cyrus C. Loveland*, ed. Richard H. Dillon (Los Gatos, CA, 1961), pp. 43–45, 49, 51, 81, 125–26.

39. Loveland, *California Trail*, p. 46; Mitchell, *Way it Was*, p. 36.

40. Burcham, *California Range*, pp. 151, 205; Loveland, *California Trail*, p. 44; Pulling, "History of California's," pp. 102, 348; Simpson, "Studying Cattleman," p. 5.

41. Joseph B. Davy, *Stock Ranges of Northwestern California: Notes on the Grasses and Forage Plants and Range Conditions* (Washington, DC, 1902), p. 33; Edward F. Treadwell, *The Cattle King: A Dramatized Biography* (New York, NY, 1931), pp. 100, 332–34; Pulling, "History of California's," p. 133; Loveland, *California Trail*, p. 112.

42. James A. Young and B. Abbott Sparks, *Cattle in the Cold Desert* (Logan, UT, 1985), pp. 145–48; Sheldon D. Ericksen, *Occupance in the Upper Deschutes Basin, Oregon* (Chicago, IL, 1953), p. 40; Simpson, *Community*, pp. 27, 92; French, *Cattle Country*, p. 138.

43. Osgood, *Day of Cattleman*, p. 231; Young and Sparks, *Cattle in Cold*, p. 142.

44. Julian Steward, "Irrigation without Agriculture," *Papers of the Michigan Academy of Science, Arts and Letters*, 12 (1930), p. 150; Arnold R. Rojas, *The Vaquero* (Charlotte, NC, and Santa Barbara, CA, 1964), p. 25; Howard W. Marshall and Richard E. Ahlborn, *Buckaroos in Paradise: Cowboy Life in Northern Nevada* (Washington, DC, 1980), pp. 13, 28.

45. Ronald B. Taylor (Los Angeles *Times* Service), "Sierra Cattle Drive Races Winter across Mountains," Austin (TX) *American-Statesman*, Sunday, January 4, 1987, p. G36; Davy, *Stock Ranges*, pp. 24–25, 32–33.

46. Daniel M. Drumheller, *Uncle Dan Drumheller Tells Thrills of Western Trails in 1854* (Spokane, WA, 1925), pp. 1–5, 51; McCoy, *Historic Sketches*, p. 365; Loveland, *California Trail*, pp. 43–44; French, *Cattle Country*, pp. 15–18.

47. Byron D. Lusk, "Golden Cattle Kingdoms of Idaho," M.S. thesis, Utah State University (Logan, UT, 1978), pp. 12–13; C. W. Hodgson, *Idaho Range Cattle Industry* (Moscow, ID, 1948); Marshall and Ahlborn, *Buckaroos*, p. 18; Mont H. Saunderson, *Western Stock Raising* (Minneapolis, MN, 1950), pp. 4, 16–21; Oliphant, *On Cattle Ranges*, pp. 75–114.

48. Mildretta Adams, *Owyhee Cattlemen, 1878–1978: 100 Years in the Saddle*, 2nd ed. (Homedale, ID, 1979), p. 16; Adelaide Hawes, *The Valley of Tall Grass* (Bruneau, ID, 1950), pp. 194–97, 219–21; M. Alfreda Elsensohn, *Pioneer Days in Idaho County* (Caldwell, ID, 1947), vol. I, p. 5; Simpson, *Community*, p. 20; Meinig, *Great Columbia*, p. 221; Lusk, "Golden Cattle," pp. 32, 41, 45.

49. Edward Gray, *An Illustrated History of Early Northern Klamath County, Oregon* (Bend, OR, 1989), p. 76; Oliphant, *On Cattle Ranges*, pp. 78, 90–97; Simpson, "Studying Cattleman," p. 9.

50. Simpson, *Community*, pp. 4–5, 19.

51. United States Census, 1880, manuscript population schedules, Owyhee County, Idaho.

52. Hawes, *Valley of Tall*, pp. 219–21.

53. Louie W. Attebery, "Celts and Other Folk in the Regional Livestock Industry," *Idaho Yesterdays*, 28:2 (1984), p. 21; Patterson, et al., *Nevada's Northeast*, pp. 335, 339, 353–54, 374, 408; U.S. census 1880, manuscript population schedules, Grant County Oregon.

54. Michael S. Kennedy, ed., *Cowboys and Cattlemen: A Roundup from Montana* (New York, NY, 1964), p. 28; Gordon, "Report on Cattle," p. 1058; Oliphant, *On Cattle Ranges*, pp. 13, 17–18; Lusk, "Golden Cattle," pp. 22, 28, 37; Patterson, et al., *Nevada's Northeast*, pp. 207–8; Dary, *Cowboy Culture*, p. 228.

55. Ericksen, *Occupance*, p. 38; Simpson, *Community*, pp. 4, 39, 52; French, *Cattle Country*, pp. 56–60; Patterson, et al., *Nevada's Northeast*, p. 229.

56. George E. Carter, "The Cattle Industry of Eastern Oregon, 1880–1890," *ORHQ*, 67 (1966), p. 157; Jill A. Chappel, "Homestead Ranches of the Fort Rock Valley: Vernacular Building in the Oregon High Desert," M.A. thesis, University of Oregon (Eugene, OR, 1989); Oliphant, *On Cattle Ranges*, p. 195; Gordon, "Report on Cattle," p. 1080; Simpson, *Community*, pp. 51, 117; Simpson, "Studying Cattleman," pp. 7–8

57. French, *Cattle Country*, pp. 47, 60; Simpson, *Community*, p. 21.

58. A. L. Lesperance, et al., "Great Basin Wildrye," *Rangeman's Journal*, 5:4 (1978), p. 126; Sessions S. Wheeler, *The Black Rock Desert* (Caldwell, ID, 1978), pp. 155–57; Ericksen, *Occupance*, p. 40; Gordon, "Report on Cattle," p. 1080.

59. James A. Young, "Hay Making: The Mechanical Revolution on the Western Range," *WHQ*, 14 (1983), p. 312; Gordon, "Report on Cattle," p. 1080; Lusk, "Golden Cattle," p. 163; Hawes, *Valley of Tall*, pp. 219–21; Gray, *Illustrated History*, pp. 81–83, 136; French, *Cattle Country*, p. 60; Simpson, *Community*, p. 4.

60. Herman Oliver and E. R. Jackman, *Gold and Cattle Country* (Portland, OR, 1962), follows p. 161; Young, "Hay Making," pp. 312–14; Gray, *Illustrated History*, p. 81; French, *Cattle Country*, p. 138; Ericksen, *Occupance*, p. 40; Young and Sparks, *Cattle in Cold*, pp. 145–48; Patterson, et al., *Nevada's Northeast*, pp. 332, 364, 408; Simpson, *Community*, pp. 27, 52, 92; Gordon, "Report on Cattle," p. 1080; Adams, *Owyhee Cattlemen*, p. 99; Osgood, *Day of Cattleman*, p. 231.

61. J. Sanford Rikoon, "Traditional Fence Patterns in Owyhee County, Idaho," *Pioneer America Society, Transactions*, 7 (1984), pp. 61–65; U.S. census, 1880, manuscript population schedules, Grant Co., Oregon and Idaho Co., Idaho Terr.; Simpson, *Community*, pp. 4, 24–25; Ericksen, *Occupance*, p. 39; Gordon, "Report on Cattle," pp. 1080–81; Oliphant, *On Cattle Ranges*, p. 197.

62. John K. Rollinson, *Wyoming Cattle Trails: History of the Migration of Oregon-Raised Herds to Mid-Western Markets* (Caldwell, ID, 1948), p. 89;

Gray, *Illustrated History*, p. 268 and the chapter entitled "Mayfield Cattle Drives, 1906–1935," pp. 81–84; Adams, *Owyhee Cattlemen*, p. 17; Simpson, *Community*, p. 27; Ericksen, *Occupance*, p. 39; French, *Cattle Country*, p. 44; Young and Sparks, *Cattle in Cold*, pp. 53, 60.

63. Lusk, "Golden Cattle," p. 111.

64. Gregory E. G. Thomas, "The British Columbia Ranching Frontier, 1858–1896," M.A. thesis, University of British Columbia (Vancouver, BC, 1976), p. x; Meinig, *Great Columbia*, pp. 222, 236; Oliphant, *On Cattle Ranges*, pp. 39–74.

65. Rexford E. Daubenmire, "An Ecological Study of the Vegetation of Southeastern Washington and Adjacent Idaho," *Ecological Monographs*, 12 (1942), pp. 55–65; William W. Galbraith and E. William Anderson, "Grazing History of the Northwest," *Journal of Range Management*, 24 (1971), p. 9; Otis W. Freeman, et al., "Physiographic Divisions of the Columbia Intermontane Province," *AAAG*, 35 (1945), pp. 59–65; Daniel M. Drumheller, *Uncle Dan Drumheller Tells Thrills of Western Trails in 1854* (Spokane, WA, 1925), p. ix; Oliphant, *On Cattle Ranges*, pp. 76, 79–80; Meinig, *Great Columbia*, pp. 3–6, 120.

66. A. J. Splawn, *Ka-Mi-Akin: The Last Hero of the Yakimas* (Portland, OR, 1917), p. 17; Galbraith and Anderson, "Grazing History," p. 7; Oliphant, *On Cattle Ranges*, pp. 9, 12–17, 33; Kingston, "Introduction of Cattle," pp. 170, 171, 175.

67. U.S. census, 1880, manuscript population schedules, Yakima Co., Washington.

68. Meinig, *Great Columbia*, p. 201; Oliphant, *On Cattle Ranges*, pp. 32–33.

69. Roscoe Sheller, *Ben Snipes, Northwest Cattle King* (Portland, OR, 1957); Gretta Gossett, "Stock Grazing in Washington's Nile Valley: Receding Ranges in the Cascades," *Pacific Northwest Quarterly*, 55 (1964), pp. 119–20; Splawn, *Ka-Mi-Akin*, pp. 129, 141, 146; Meinig, *Great Columbia*, pp. 202, 204, 220–21; Oliphant, *On Cattle Ranges*, pp. 82–83, 98–100.

70. Gossett, "Stock Grazing," p. 120; Splawn, *Ka-Mi-Akin*, p. 299; Meinig, *Great Columbia*, p. 236.

71. Galbraith and Anderson, "Grazing History," p. 7; J. Orin Oliphant, "The Cattle Herds and Ranches of the Oregon Country, 1860–1890," *AH*, 21 (1947), p. 220; Meinig, *Great Columbia*, pp. 202, 220; Oliphant, *On Cattle Ranges*, pp. 12, 17.

72. Meinig, *Great Columbia*, pp. 220–21, 236, 284–93; Oliphant, *On Cattle Ranges*, pp. 90, 93, 97, 100, 102.

73. Drumheller, *Uncle Dan*, pp. 1, 91–101, 120; Sheller, *Ben Snipes*, p. 11.

74. Sheller, *Ben Snipes*, pp. 11–13; Splawn, *Ka-Mi-Akin*, p. 129; Meinig, *Great Columbia*, pp. 202, 220; Oliphant, *On Cattle Ranges*, pp. 78, 90, 93, 97, 134–35.

75. Drumheller, *Uncle Dan*, pp. 1, 91–101; Sheller, *Ben Snipes*, p. 13.

76. U.S. Census, 1880, manuscript population schedules, Yakima County, Washington; Splawn, *Ka-Mi-Akin*, p. 290.

77. Michael Kennedy, ed., "Cowboy and Cattleman's Issue," *Montana: The Magazine of Western History*, 11:4 (1961), p. 22; Kingston, "Introduction of Cattle," p. 175; Sheller, *Ben Snipes*, p. 103; Oliphant, *On Cattle Ranges*, p. 20.

78. Thomas R. Weir, *Ranching in the Southern Interior Plateau of British Columbia* (Ottawa, ON, 1964); F. W. Laing, "Some Pioneers of the Cattle Industry," *British Columbia Historical Quarterly*, 6 (1942), pp. 257–75; John S. Lutz, "Interlude or Industry? Ranching in British Columbia, 1859–1885," *British Columbia Historical News*, 13 (1980), pp. 2–11; H. Lavington, *The Nine Lives of a Cowboy* (Victoria, BC, 1982); Harry Marriott, *Cariboo Cowboy* (Sidney, BC, 1966); G. D. Brown, "Historical Notes on Early Cattle and Horse Raising in the Kamloops District," *Canadian Cattlemen*, 4 (Dec. 1940), p. 492; Ed Gould, *Ranching in Western Canada* (Saanichton, BC, and Seattle, WA, 1978); Thomas, "British Columbia Ranching."

79. T. Alex Bulman, *Kamloops Cattlemen: One Hundred Years of Trail Dust* (Sidney, BC, 1972), p. 11; E. W. Tisdale, "The Grasslands of the Southern Interior of British Columbia," *Ecology*, 28 (1947), pp. 346–82; Griffith Taylor, "British Columbia: A Study in Topographic Control," *GR*, 32 (1942), pp. 372–402; Weir, *Ranching*, pp. 17–52.

80. William C. Brown, "Old Fort Okanogan and the Okanogan Trail," *Quarterly of the Oregon Historical Society*, 15 (1914), pp. 1–38; Weir, *Ranching*, pp. 80, 84, 88, 90; Thomas, "British Columbia Ranching," p. 15; Drumheller, *Uncle Dan*, pp. 66–70, 99–100, 123–31; Lutz, "Interlude," pp. 3–4; Splawn, *Ka-Mi-Akin*, pp. 160–80; Sheller, *Ben Snipes*, pp. 15–16, 58–80; Laing, "Some Pioneers," pp. 258–62; Oliphant, *On Cattle Ranges*, p. 68.

81. Bulman, *Kamloops*, p. 18; Weir, *Ranching*, pp. 86–87, 90–91; Drumheller, *Uncle Dan*, pp. 67–70, 99–100; Thomas, "British Columbia Ranching," pp. 26, 42; Laing, "Some Pioneers," pp. 263, 265.

82. Weir, *Ranching*, pp. 84, 91 (quote); Thomas, "British Columbia Ranching," p. 77; Oliphant, *On Cattle Ranges*, pp. 104–5.

83. Weir, *Ranching*, pp. 86–87.

84. Weir, *Ranching*, pp. 21, 91–94; Thomas, "British Columbia Ranching," p. 70.

85. Weir, *Ranching*, pp. 90–91; Thomas, "British Columbia Ranching," p. 43; Gould, *Ranching in Western*, p. 139.

86. C. W. Vrooman, "A History of Ranching in British Columbia," *The Economic Analyst*, 11 (April 1941), pp. 20, 22; Edward Brado, *Cattle Kingdom: Early Ranching in Alberta* (Vancouver, BC, and Toronto, ON, 1984), p. 31; Weir, *Ranching*, p. 91; Laing, "Some Pioneers," pp. 266–75; Lutz, "Interlude," pp. 4–5; Drumheller, *Uncle Dan*, pp. 126–29.

87. Bulman, *Kamloops*, p. 16; Laing, "Some Pioneers," p. 270.

88. J. W. G. MacEwan, "The Great Cattle Drive Northward from Chil-

cotin, 1898," *Canadian Cattleman*, 3 (December 1939); Drumheller, *Uncle Dan*, p. 128; Laing, "Some Pioneers," p. 271; Lutz, "Interlude," p. 8; Vrooman, "History of Ranching," p. 22; Gould, *Ranching in Western*, pp. 121–22.

89. Thomas R. Weir, "The Winter Feeding Period in the Southern Interior Plateau of British Columbia," *AAAG*, 44 (1954), p. 204; Arthur H. Bone, *Oregon Cattleman/Governor, Congressman: Memoirs and Times of Walter M. Pierce* (Portland, OR, 1981), p. 9; Bulman, *Kamloops*, pp. 23, 89; Gordon, "Report on Cattle," pp. 1090–91; Marriott, *Cariboo Cowboy*, pp. 31–33; Weir, *Ranching*, p. 124; Splawn, *Ka-Mi-Akin*, pp. 151–52; Drumheller, *Uncle Dan*, p. 65; Galbraith and Anderson, "Grazing History," pp. 9–10.

90. Meinig, *Great Columbia*, pp. 284–88; Thomas R. Weir, "Ranch Types and Range Uses within the Interior Plateau of British Columbia," *Canadian Geographer*, 2 (1952), p. 74; Gordon, "Report on Cattle," pp. 1090–91; Thomas, "British Columbia Ranching," p. 121; Marriott, *Cariboo Cowboy*, pp. 40, 41, 92.

91. Gossett, "Stock Grazing," p. 120; Bulman, *Kamloops*, pp. 98, 119; Meinig, *Great Columbia*, p. 236; Weir, *Ranching*, pp. 135–39; Gordon, "Report on Cattle," p. 1090; Marriott, *Cariboo Cowboy*, p. 40.

92. Oliphant, *On Cattle Ranges*, p. 186; Gordon, "Report on Cattle," pp. 1089–90.

93. Attebery, "Celts and Other," p. 21; Bulman, *Kamloops*, p. 83; Weir, *Ranching*, p. 89; Thomas, "British Columbia Ranching," pp. 45, 199–200; Lutz, "Interlude," p. 6.

94. Donovan Clemson, *Living with Logs: British Columbia's Log Buildings and Rail Fences* (Saanichton, BC, 1974); Terry G. Jordan and Matti Kaups, *The American Backwoods Frontier: An Ethnic and Ecological Interpretation* (Baltimore, MD, 1989), pp. 141–51; Terry G. Jordan, "The North American West: Continuity or Innovation?," in John Welsted and John Everitt, eds., *The Dauphin Papers: Research by Prairie Geographers*, Brandon Geographical Studies, no. 1 (Brandon, MB, 1991), pp. 1–17; Jon T. Kilpinen, "Material Folk Culture in the Adaptive Strategy of the Rocky Mountain Valley Ranching Frontier," M.A. thesis, University of Texas (Austin, TX, 1990), pp. 125–42, 155–75.

95. Simpson, "Studying the Cattleman," p. 7; Gordon, "Report on Cattle," p. 1071.

96. Edward J. Wood, "The Mormon Church and the Cochrane Ranch," *Canadian Cattlemen*, 8 (Sept. 1945), p. 84; Simon M. Evans, "The Origins of Ranching in Western Canada: American Diffusion or Victorian Transplant," *Great Plains Quarterly*, 3:2 (1983), p. 76; Don D. Walker, "The Cattle Industry of Utah, 1850–1900: An Historical Profile," *UTHQ*, 32 (1964), pp. 184–89; Gordon, "Report on Cattle," pp. 1071–72.

97. Robert M. Denhardt, "Driving Livestock East from California Prior to 1850," *CHSQ*, 20 (1941), pp. 344–46; Walker, "Cattle Industry," p. 183.

98. United States Senate, *Federal Census, Territory of New Mexico and*

the Territory of Arizona (Washington, DC, 1965); U.S. Census, 1880, manuscript population schedules for Arizona.

99. Bert Haskett, "Early History of the Cattle Industry in Arizona," *AZHR*, 6:4 (1935), pp. 37–39; Jane W. Brewster, "The San Rafael Cattle Company, a Pennsylvania Enterprise in Arizona," *Arizona and the West*, 8 (Summer, 1966), pp. 133–56; Gordon, "Report on Cattle," pp. 1047, 1051.

100. Fletcher, *Free Grass*, p. 5; Kilpinen, "Material Folk Culture," p. 20; Evans, "Origins of Ranching," p. 79; Saunderson, *Western Stock*, pp. 4, 32–35.

101. Nicholas Helburn, "Human Ecology of Western Montana Valleys," *Journal of Geography* 55 (1956), pp. 5–14; Rexford F. Daubenmire, "Vegetational Zonation in the Rocky Mountains," *Botanical Review*, 9 (1943), pp. 325–93; John M. Crowley, "Ranching in the Mountain Parks of Colorado," *GR*, 65 (1975), pp. 445, 447; Celia K. Atwood and Wallace W. Atwood, Jr., "Land Utilization in a Glaciated Mountain Range," *EG*, 13 (1937), pp. 371–72; André Fel, *Les hautes terres du Massif Central: Tradition paysanne et économie agricole* (Paris, France, 1962), pp. 30, 47; Clark I. Cross, "Geography of the Big Horn Basin of Wyoming," Ph.D. diss., University of Washington (Seattle, WA, 1951), pp. 20–96; Saunderson, *Western Stock*, pp. 4, 11–16; Tisdale, "Grasslands," p. 346; Burcham, *California Range*, p. 141.

102. M. H. Saunderson and D. W. Chittenden, *Cattle Ranching in Montana* (Bozeman, MT, 1937), pp. 4–5; Saunderson, *Western Stock*, pp. 32–33; Osgood, *Day of Cattleman*, p. 57; Evans, "Origins of Ranching," p. 88.

103. Dary, *Cowboy Culture*, p. 228; Kennedy, *Cowboys and Cattlemen* (1964), pp. 27–58; Kennedy, "Cowboys and Cattleman's Issue" (1961), pp. 24–28; Fletcher, *Free Grass*, pp. 13–15; Lusk, "Golden Cattle," p. 28; Osgood, *Day of Cattleman*, pp. 11–17.

104. Harold E. Briggs, "The Development and Decline of Open Range Ranching in the Northwest," *MVHR*, 20 (1934), pp. 523, 528; Matt J. Kelly, "Trail Herds of the Big Hole Basin," *Montana Magazine of History*, 2:3 (1952), pp. 57–58; Gordon, "Report on Cattle," p. 1021; Fletcher, *Free Grass*, pp. 22–24; Osgood, *Day of Cattleman*, pp. 23, 53.

105. Evans, "Origins of Ranching," p. 79.

106. Simon M. Evans, "American Cattlemen on the Canadian Range, 1874–1914," *Prairie Forum*, 4 (1979), pp. 123–24; D. MacMillan, "The Gilded Age and Montana's DHS Ranch," *Montana*, 20:2 (1970), pp. 50–57; Evans, "Origins of Ranching," pp. 87–88; Osgood, *Day of Cattleman*, pp. 57, 256; Fletcher, *Free Grass*, p. 52.

107. Kilpinen, "Material Folk Culture," p. 22; Fletcher, *Free Grass*, pp. 23–24, 29–30; Rollinson, *Wyoming Cattle*, p. 305.

108. James C. Dahlman, "Recollections of Cowboy Life in Western Nebraska," *NBH*, 10 (1927), pp. 335, 337; Briggs, "Development," p. 527; Dary, *Cowboy Culture*, p. 234; Meinig, *Great Columbia*, pp. 255–56; Gordon,

"Report on Cattle," p. 1025; Fletcher, *Free Grass*, p. 39; Rollinson, *Wyoming Cattle*, pp. 20–27, 114; Osgood, *Day of Cattleman*, p. 50; Grohman, "Cattle Ranches," p. 448; Splawn, *Ka-Mi-Akin*, pp. 226–28; Simpson, *Community*, p. 39; Pelzer, *Cattlemen's Frontier*, p. 201; Burns, "Newman Ranches," p. 31; Evans, "Origins of Ranching," p. 88.

109. William E. Jackson, "Diary of a Cattle Drive from La Grande, Oregon, to Cheyenne, Wyoming, in 1876," *AH*, 23 (1949), pp. 260–73; Rollinson, *Wyoming Cattle*.

110. David H. Breen, *The Canadian Prairie West and the Ranching Frontier, 1874–1924* (Toronto, ON, 1983), p. 9; John R. Craig, *Ranching with Lords and Commons, or, Twenty Years on the Range* (Toronto, ON, 1903), pp. 56, 241; Lutz, "Interlude," p. 8.

111. Grohman, "Cattle Ranches," p. 447.

112. Rollinson, *Wyoming Cattle*, pp. 183–86.

113. Wilson M. Rockwell, "Cow-Land Aristocrats of the North Fork," *Colorado Magazine*, 14 (Sept. 1937), pp. 161–63; Everett, *Cattle Cavalcade*, p. 367; Gordon, "Report on Cattle," p. 1003; Burns, "Newman Ranches," pp. 22–25.

114. H. S. Armitage, "Haying Operations in the Big Hole," in B. A. Francis, ed., *The Land of the Big Snows* (Caldwell, ID, 1955), p. 204; Eugene C. Mather, "The Production and Marketing of Wyoming Beef Cattle," *EG*, 26 (1950), pp. 83–85; Briggs, "Development," p. 523; Osgood, *Day of Cattleman*, pp. 56–57; Rollinson, *Wyoming Cattle*, p. 186; Everett, *Cattle Cavalcade*, p. 102; Saunderson, *Western Stock*, p. 16; Saunderson and Chittenden, *Cattle Ranching*, pp. 5, 17–18; Crowley, "Ranching in Mountain," p. 447; MacMillan, "Gilded Age," pp. 54, 55; Kelly, "Trail Herds," p. 58.

115. H. C. Cornwall, "Ranching on Ohio Creek, 1881–1886," *Colorado Magazine*, 32 (1955), p. 19.

116. Glen Barrett, "Stock Raising in the Shirley Basin, Wyoming," *JOW*, 14:3 (1975), p. 21.

117. Robert S. Fletcher, "The End of the Open Range in Eastern Montana," *MVHR*, 16 (1929), pp. 205–6; Armitage, "Haying," p. 204.

118. Everett, *Cattle Cavalcade*, pp. 26–27; Fletcher, "End of Open Range," pp. 205–6.

119. Charles M. Davis, "Land Utilization in North Park, Colorado," *EG*, 13 (1937), pp. 379–84; Crowley, "Ranching in Mountain," p. 451.

120. Robert H. Burns, Andrew S. Gillespie, and Willing G. Richardson, *Wyoming's Pioneer Ranches* (Laramie, WY, 1955), pp. 304, 681; Cornwall, "Ranching on Ohio," pp. 19–20.

121. John A. Alwin, "Montana's Beaverslide Hay Stacker," *Journal of Cultural Geography*, 3:2 (1982), pp. 42–50; Armitage, "Haying," pp. 204–12; Kilpinen, "Material Folk Culture," pp. 176–88; Young, "Hay Making," p. 326.

122. Everett, *Cattle Cavalcade*, p. 200; Osgood, *Day of Cattleman*, pp. 56–57.

123. Gisbert Rinschede, *Die Wanderviehwirtschaft im gebirgigen Westen des USA und ihre Auswirkungen im Naturraum* (Regensburg, Germany, 1984); Marion Clawson, *The Western Range Livestock Industry* (New York, NY, 1950), p. 97; Crowley, "Ranching in Mountain," pp. 455–58; Everett, *Cattle Cavalcade*, p. 237; Rockwell, "Cow-Land Aristocrats," pp. 164, 166; Cross, "Geography of Big Horn," pp. 162–64.

124. Osgood, *Day of Cattleman*, pp. 56–57.

125. Rollinson, *Wyoming Cattle*, p. 97; Saunderson, *Western Stock*, p. 13; Briggs, "Development," p. 523; on barns, see Kilpinen, "Material Folk Culture," pp. 125–42.

126. Frederick Simpich, "Missouri, Mother of the West," *National Geographic Magazine*, 43 (1923), p. 421.

Selected Annotated Bibliography

Journal Abbreviations

AAAG	Annals of the Association of American Geographers
AH	Agricultural History
AHR	American Historical Review
AZHR	Arizona Historical Review
CHSQ	California Historical Society Quarterly
CO	Chronicles of Oklahoma
EG	Economic Geography
FHQ	Florida Historical Quarterly
G	Geography
GHQ	Georgia Historical Quarterly
GJ	Geographical Journal
GR	Geographical Review
HAHR	Hispanic American Historical Review
HSSCQ	(Historical Society of) Southern California Quarterly
JAS	Journal of the African Society
JHG	Journal of Historical Geography
JOW	Journal of the West
JSH	Journal of Southern History
KHQ	Kansas Historical Quarterly
MR	McNeese Review
MVHR	Mississippi Valley Historical Review
NBH	Nebraska History
NMHR	New Mexico Historical Review
ORHQ	Oregon Historical Quarterly
PPHR	Panhandle-Plains Historical Review
QTSH	Quarterly of the Texas State Historical Association
SCHM	South Carolina Historical Magazine
SCQ	Southern California Quarterly
SGM	Scottish Geographical Magazine
SWHQ	Southwestern Historical Quarterly
UTHQ	Utah Historical Quarterly

WHQ   Western Historical Quarterly
WTYB   West Texas Historical Association Year Book

Chapter One
The Nature of Cattle Ranching

The secondary literature on cattle ranching in North America is simply vast. Few if any other topics have received so much attention, from scholars and amateurs alike. It would be quite impossible to list, much less annotate, this literature in its entirety in the available space, and I have by necessity been selective. Indeed so voluminous is this literature that I cannot claim to have perused it all, and for any major omissions I apologize. From the groaning library shelves I selected for inclusion only those sources I regard as essential to understanding the origin, regional differentiation, diffusion, adaptation, and repeated readaptations of this land-use strategy in the Old World and North America. To help remedy my deficiencies, see an even less complete listing: Henry E. Fritz, "The Cattleman's Frontier in the Trans-Mississippi West: An Annotated Bibliography," *Arizona and the West*, 14 (1972), pp. 45–70, 169–90.

On the place of ranching as a type in general classifications of agricultural land use, see the classic and much-copied work by Derwent S. Whittlesey, "Major Agricultural Regions of the Earth," *AAAG*, 26 (1936), pp. 199–240; the useful evaluation of such classifications provided by David Grigg, "The Agricultural Regions of the World: Review and Reflections," *EG*, 45 (1969), pp. 95–132; and the fuller development of this regional theme in Bernd Andreae, *Betriebsformen in der Landwirtschaft* (Stuttgart, Germany, 1964).

For the economic deterministic approach, in which ranching, as a commercial land use, competes for land space in the Thünenian open ranges of land rent, see Michael Chisholm, *Rural Settlement and Land Use* (London, UK, 1962); Edgar S. Dunn, Jr., *The Location of Agricultural Production* (Gainesville, FL, 1954); Richard Peet, "The Spatial Expansion of Commercial Agriculture in the Nineteenth Century: A von Thünen Explanation," *EG*, 45 (1969), pp. 283–301; Ernst Griffin, "Testing the von Thünen Theory in Uruguay," *GR*, 63 (1973), pp. 500–516; and my own youthful and in many ways lamentable excursion, Terry G. Jordan, "The Origin and Distribution of Open Range Cattle Ranching," *Social Science Quarterly*, 53 (1972), pp. 105–21.

In approaching ranching as an ecological adaptive strategy, the essential place of beginning is the stimulating and insightful piece by Arnold Strickon, "The Euro-American Ranching Complex," in Anthony Leeds and Andrew P. Vayda, eds., *Man, Culture, and Animals: The Role of Animals in Human Ecological Adjustments* (Washington, DC, 1965), pp. 229–58. Then sample relevant articles in David R. Harris, ed., *Human Ecology in Savanna Environments* (London, UK, 1980) and go directly to the outstanding study by

John W. Bennett, *Northern Plainsman: Adaptive Strategy and Agrarian Life* (Chicago, IL, 1969), chapter 6. But by far the best work on ranching that utilizes the ecological approach is James A. Young and B. Abbott Sparks, *Cattle in the Cold Desert* (Logan, UT, 1985), dealing with the intermontane region of the West; I only wish my present book could be as solid in range ecology and as sensitive to history and culture as this wonderful volume.

On the debate concerning the magnitude of ranching's impact on ecosystems and environmental deterioration, the literature is vast, but see representative works such as William C. Robison, "Grazing as a Natural Ecological Factor," *GR*, 51 (1961), pp. 308–10; Lincoln Ellison, "Influence of Grazing on Plant Succession of Rangelands," *Botanical Review*, 26 (1960), pp. 1–78, and P. J. Vickery, "Grazing and Net Primary Production of a Temperate Grassland," *Journal of Applied Ecology*, 9 (1972), pp. 307–14.

Comparing different ranching frontiers, always an instructive exercise, is preached in Jerome O. Steffen, *Comparative Frontiers: A Proposal for Studying the American West* (Norman, OK, 1980), pp. 51–70; and practiced in the tantalizing, brief piece by Otto Jessen, "Cosacos, cowboys, gauchos, Boers y otros pueblos a caballos propios de las estepas," *Runa*, 5:1/2 (1952), pp. 171–86; in the revealing work by H. C. Allen, *Bush and Backwoods: A Comparison of the Frontier in Australia and the United States* (East Lansing, MI, 1959); and in the suggestive effort by Paul F. Sharp, "Three Frontiers: Some Comparative Studies of Canadian, American, and Australian Settlement," *Pacific Historical Review*, 24 (1955), pp. 369–77. Some fine comparative regional vignettes, in the best tradition of the French school of human geography, are found in Pierre Deffontaines, *Contribution à la géographie pastorale de l'Amérique Latine* (Rio de Janeiro, Brazil, 1964).

Cowboy culture and mythology receive humanistic analysis in Marshall W. Fishwick, "The Cowboy: America's Contribution to the World's Mythology," *Western Folklore*, 11 (1952), pp. 77–92; in Mody C. Boatright, "The American Myth Rides the Range: Owen Wister's Man on Horseback," *Southwest Review*, 36 (Summer 1951), pp. 157–63; and in Lysius Gough, *Spur Jingles and Saddle Songs: Rhymes and Miscellany of Cow Camp and Cattle Trails in the Early Eighties* (Amarillo, TX, 1935). Philip A. Rollins, *The Cowboy: An Unconventional History of Civilization on the Old-Time Cattle Range*, 2nd ed. (New York, NY, 1936), is a popularized but reasonably accurate account of such aspects of cowboy life as garb, equipment, techniques, and recreation, lacking notes and bibliography. Joe B. Frantz and Julian E. Choate, Jr., *The American Cowboy: The Myth and the Reality* (Norman, OK, 1955), strikes me as too anecdotal but very readable. By far the best book on the subject is Richard W. Slatta's *Cowboys of the Americas* (New Haven, CT, 1990), an empirical, comparative analysis of the reality and myth of range life in the Western Hemisphere, nicely illustrated, fully annotated, and beautifully printed. Slatta could have given more attention to regional contrasts and to Old World roots, but his book is both scholarly and readable. Jack Weston's *The Real American Cowboy* (New York, NY, 1985), comes complete with movie stills of John Wayne, is laced with anecdotes, but still well worth reading, particularly chapter 7, "The Cowboy Myth." A real knee-slapper is Ramon F. Adams, *The Cowboy & His Humor* (Austin, TX, 1968).

Overviews of the essential subject of equestrian skills and equipment are presented by Harold B. Barclay, *The Role of the Horse in Man's Culture* (London, UK, and New York, NY, 1980), and Glenn R. Vernam, *Man on Horseback* (New York, NY, 1964). The more controversial and complicated issue of saddle evolution, while nicely handled by Barclay and Vernam, gets specialized scholarly treatment in Richard E. Ahlborn, ed., *Man Made Mobile: Early Saddles of Western North America* (Washington, DC, 1980); see also Arthur Woodward, "Saddles in the New World," *Quarterly of the Los Angeles County Museum*, 10 (Summer 1953), pp. 1–5.

Though spotty in coverage, amateurish in tone, and largely unreferenced, the treatment of drove trails in Garnet M. Brayer and Herbert O. Brayer, *American Cattle Trails, 1540–1900* (Bayside, NY, 1952), is useful and involves an effort to consider the subject in a wider national context. So does Harry S. Drago, in his *Great American Cattle Trails: The Story of the Old Cow Paths of the East and the Longhorn Highways of the Plains* (New York, NY, 1965), although coverage of the East is inexcusably thin and in dealing with the western trails, the author succumbs to the anecdotal model, providing rather minimal references.

The work by Manfred R. Wolfenstine, *The Manual of Brands and Marks* (Norman, OK, 1970), although billed as the "first complete and serious treatment" of this important subject, is quite disjointed (one chapter is titled "Miscellany") and too heavily focused on Anglo-America and Germany, but it still conveys the ancient, Old World origin of branding and earmarking.

Two studies that appeared at midcentury, describing livestock ranching in its present condition from the viewpoint of economics and range-resource management, are essential to any understanding of the subject, although both are largely ahistorical; these are Marion Clawson, *The Western Range Livestock Industry* (New York, NY, 1950), and Mont H. Saunderson, *Western Stock Ranching* (Minneapolis, MN, 1950). Both bear the mark of writers who knew the ranching business from firsthand experience.

Among the general studies, begin with Don D. Walker, *Clio's Cowboys: Studies in the Historiography of the Cattle Trade* (Lincoln, NE, 1981), covering everything from mythology to Marxism on the range, from classics of the secondary literature to poetry. Useful but far more prosaic is James W. Thompson, *A History of Livestock Raising in the United States, 1607–1860* (Washington, DC, 1942), a study not confined to ranching but far less inclusive than its ambitious title suggests, as well as rather spotty in coverage. Charles W. Towne and Edward N. Wentworth, *Cattle & Men* (Norman, OK, 1955), is readable but lacks focus—ranging from ancient Egypt to Viking

Greenland and colonial Carolina—and is quite disjointed, with chapters on stampedes and Mormons, but still it offers a good point of departure for any study of cattle ranching. By far the best and most comprehensive of the general histories is David Dary's *Cowboy Culture: A Saga of Five Centuries* (New York, NY, 1981); although environmentally naive and too Mexican-oriented in the search for origins, it otherwise presents a balanced view of the open-range era, providing, for example, a rare assessment of the California contribution, and offering a richness of quotations from firsthand accounts. The attention to the eastern seaboard missing in Dary's book can be partially corrected by reading David Wheeler, "The Beef Cattle Industry in the United States: Colonial Origins," *PPHR*, 46 (1973), pp. 54–67. Lewis Atherton, *The Cattle Kings* (Bloomington, IN, 1961), views ranching history from the perspective of the entrepreneurs who owned the operations—a useful analysis, given the status of ranching as a business.

Finally there is a classic contemporary work that portrays the cattle-ranching industry of the West at or near its apex. This remarkable report, accompanying the 1880 census of the United States, is Clarence Gordon, "Report on Cattle, Sheep, and Swine, Supplementary to Enumeration of Live Stock on Farms in 1880," *Report on the Productions of Agriculture as Returned at the Tenth Census (June 1, 1880)*, vol. 3 of the U.S. census (Washington, DC, 1883), pp. 951–1116. This contains a mixture of historical data, reminiscences, firsthand accounts by local officials and individual ranchers, and statistics. It is an essential document for the understanding of the cattle frontier of the West.

Chapter Two
Atlantic Fringe Source Regions

Iberian prototypes of American ranching received initial scholarly attention in two articles by C. Julian Bishko, "The Peninsular Background of Latin American Cattle Ranching," *HAHR*, 32 (1952), pp. 491–515, and "The Castilian as Plainsman: The Medieval Ranching Frontier in La Mancha and Extremadura," in Archibald R. Lewis and Thomas F. McGann, eds., *The New World Looks at its History* (Austin, TX, 1963), pp. 47–69, but both are flawed by geographical naivety and the assumption that "frontier" conditions still prevailed in central and western Iberia in 1500. Far more sophisticated is Bishko's later work, "The Andalusian Municipal Mestas in the 14th–16th Centuries: Administrative and Social Aspects," in *Andalucía medieval* (Córdoba, Spain, 1978), vol. I, pp. 347–74, an article that delivers far more than its modest title suggests, providing firm documentary evidence for the development, by 1500, of specialized, commercialized cattle ranching in the Sevilla area. Basically in agreement with Bishko's later work are ecologically based works by William E. Doolittle, "Las Marismas to Pánuco to Texas: The Transfer of Open Range Cattle Ranching from Iberia through Northeastern

Mexico," *Yearbook of the Conference of Latin Americanist Geographers,* 13 (1987), pp. 3–11, and Terry G. Jordan, "An Iberian Lowland/Highland Model for Latin American Cattle Ranching," *JHG,* 15 (1989), pp. 111–25. See also Sandra L. Myres, "The Ranching Frontier: Spanish Institutional Backgrounds of the Plains Cattle Industry," in Harold M. Hollingsworth and Sandra L. Myres, eds., *Essays on the American West* (Austin, TX, 1969), pp. 19–39; Karl W. Butzer, "Cattle and Sheep from Old to New Spain: Historical Antecedents," *AAAG,* 78 (1988), pp. 29–56; and Robert Aitken, "Routes of Transhumance on the Spanish Meseta," *GJ,* 106 (1945), pp. 59–69, each of which contains interesting ideas and information concerning the Iberian background and diffusion. A brief but suggestive discussion of Iberian herding heritage is included in George M. Foster, *Culture and Conquest: America's Spanish Heritage* (Chicago, IL, 1960), pp. 70–76. Essential for the understanding of traditional Castillan land systems is David E. Vassberg, *Land and Society in Golden Age Castille* (New York, NY, 1984).

Very useful for understanding the conditions in Andalucía and Extremadura about 1500 are *Huelva en la Andalucía del siglo XV* (Huelva and Sevilla, Spain, 1976); Manual González Jiménez, *La repoblación de la zona de Sevilla durante el siglo XIV* (Sevilla, Spain, 1975); Julius Klein, *The Mesta: A Study in Spanish Economic History, 1273–1836* (Cambridge, MA, 1920), the classic, definitive study of the privileged stock-raising guild; A. Gil Crespo, "La mesta de carreteros del reino," *Anales de la Asociación Española para el Progreso de las Ciencias,* 22 (1957), pp. 207–30, dealing with the unusual carter's guild of southern Avila and its attendant bovine transhumance; Miguel A. Ladero Quesada, "Algunos datos para la historia económica de las ordenes militares de Santiago y Calatrava en el siglo XV," *Hispania: Revista Española de Historia,* 30 (1970), pp. 637–62; and David E. Vassberg, "Concerning Pigs, the Pizarros, and the Agro-Pastoral Background of the Conquerors of Peru," *Latin American Research Review,* 13:3 (1978), pp. 47–62. The art of traditional horsemanship in Spain, from the standpoint of the gentleman, is described and displayed in the remarkable little book by Gregorio de Tapia y Salzedo, *Exercicios de la gineta al principe nuestro señor d. Baltasar Carlos* (Madrid, Spain, 1643), complete with colored drawings.

Valuable local studies covering the specific Andalusian source area include Mercedes Borrero Fernández, *El mundo rural sevillano en el siglo XV: Aljarafe y Ribera* (Sevilla, Spain, 1983); Alfonso Franco Silva, *El concejo de Alcalá de Guadaira a finales de la Edad Media* (Sevilla, Spain, 1974); Julio González, *Repartimiento de Sevilla,* 2 vols. (Madrid, Spain, 1951); Juan F. Ojeda Rivera, *Organización del territorio en Doñana y su entorno próximo (Almonte), siglos XVIII–XX* (Sevilla, Spain, 1987); Eduardo Camacho Rueda, *Propiedad y explotación agrarias en el Aljarafe sevillano: el caso de Pilas, 1760–1925* (Sevilla, Spain, 1984); and Ruth Pike, "Sevillian Society in the Sixteenth Century: Slaves and Freedmen," *HAHR,* 47 (1967), pp. 344–59.

Certain other studies and accounts, though based in the modern era, provide valuable insights concerning environments and traditional pastoral ways of life in Andalucía, Extremadura, and bordering provinces. These include, for the Andalusian lowlands, Michel Drain et al., *Le bas Guadalquivir: introduction géographique, le milieu physique* (Paris, France, 1971); Paul Gwynne, *The Guadalquivir: Its Personality, Its People and Its Associations* (London, UK, 1912); Otto Jessen, "Südwest-Andalusien," *Petermanns Mitteilungen Ergänzungsheft Nr. 186* (Gotha, Germany, 1924), pp. 1–84; Wilhelm Giese, "Von der Viehwirtschaft," in his *Nordost Cádiz: Ein Kulturwissenschaftlicher Beitrag zur Erforschung Andalusiens* (Halle/Saale, Germany, 1937) (*Beihefte zur Zeitschrift für romanische Philologie*, No. 89), pp. 138–42, part of a general ethnographical survey; and Javier Rubiales et al., *El Río: el bajo Guadalquivir* (Madrid, Spain, 1985).

For pastoralism in the mesetan interior of western Spain, see Miguel Muñoz de San Pedro, *Extremadura* (Madrid, Spain, 1961); James J. Parsons, "The Acorn-Hog Economy of the Oak Woodlands of Southwestern Spain," *GR*, 52 (1962), pp. 211–35; Luis L. Cortés y Vázquez, "Ganadería y pastoreo en Berrocil de Huebra (Salamanca)," *Revista de Dialectología y Tradiciónes Populares*, 8 (1952), pp. 425–64, 563–95, an indispensable and highly revealing ethnographic study of mesetan pastoral practices, with attention to material culture, traditional techniques, and vocabulary; Evelio Teijón Laso, "Los modos de vida en la dehesa salamantina," *Estudios Geográficos*, 9 (1948), 421–41; Nieves de Hoyos Sancho, "La vida pastoral en La Mancha," ibid., 623–36; Gisela Fiedler, *Kulturgeographische Untersuchungen in der Sierra de Gredos/Spanien* (Würzburg, Germany, 1970), a detailed and insightful account of life in the mountain region bordering Extremadura on the north; Adela Gil Crespo, "Hoyos del Espino," *Boletín de la Real Sociedad Geográfica*, 97 (1961), pp. 176–205, a case study revealing bovine transhumance between Avila province and Extremadura; Jan Hinderink, *The Sierra de Gata: A Geographical Study of a Rural Mountain Area in Spain* (Groningen, Netherlands, 1963), which demonstrates the relative unimportance of cattle in some Extremaduran districts; and Michael Weisser, *The Peasants of the Montes* (Chicago, IL, 1976), a revealing study of villages in the mountains near Toledo.

The definitive study of Spanish herding antecedents still remains to be written, and the existing body of material is local or superficial in nature. Similarly one finds no comprehensive study of the transfer of British highland herding practices to the New World. Indeed the only comprehensive statements of note on this subject are the seminal article by Grady McWhiney and Forrest McDonald, "Celtic Origins of Southern Herding Practices," *JSH*, 51 (1985), pp. 165–82, an imaginative, provocative, revealing, simplistic, and chauvinistic piece, which lapses into special pleading in places; and a somewhat revised version appearing as the chapter on "Herding" in Grady McWhiney, *Cracker Culture: Celtic Ways in the Old South* (Tuscaloosa, AL, 1988), pp. 51–79. The various works by McWhiney and McDonald should be read with the following antidote close at hand: Rowland Berthoff, "Celtic Mist over the South," *JSH*, 52 (1986), pp. 523–50. In a well-researched piece, Ray August, "Cowboys v. Rancheros: The Origins of Western American Livestock Law," *SWHQ* 96 (1993), forthcoming, demonstrates that the legal system of the ranch West came almost unaltered from English common law and ancient Germanic tradition, rather than from Hispanic Law.

Open-range cattle herding in the British Isles, aside of the diffusion issue, has received considerable scholarly attention. An excellent general study is Robert Trow-Smith, *A History of British Livestock Husbandry, 1700–1900* (London, UK, 1959), and another useful general treatment is E. H. Carrier, *The Pastoral Heritage of Britain: A Geographical Study* (London, UK, 1936). An excellent comparative analysis of four Celtic areas, with abundant attention to herding, is Pierre Flatrès, *Géographie rurale de quatre contrées celtiques: Irlande, Galles, Cornwall & Man* (Rennes, France, 1957). Books on individual British cattle breeds are numerous, but special mention should be made of G. Kenneth Whitehead, *The Ancient White Cattle of Britain and Their Descendants* (London, UK, 1953). The use of cattle-herding dogs, a diagnostic trait of the British system before about 1600, has scarcely been studied, but very useful information is provided in Xavier de Planhol, "Le chien de berger: développement et signification géographique d'une technique pastorale," *Bulletin de l'Association de Géographes Français*, 370 (1969), pp. 355–68. A wonderful general study of British cattle droving, lovingly researched, is K. J. Bonser, *The Drovers, Who They Were and How They Went: An Epic of the English Countryside* (London, UK, 1970). The enormous antiquity of open-range cattle raising and driving in Britain is suggested in Arthur J. and George Hubbard, *Neolithic Dew-Ponds and Cattle-Ways* (London, UK, 1905).

Cattle herding in the hilly shires of northern and western England receives scholarly attention in Edward Hughes, "A Pastoral Economy: Social Conditions," chapter 1 (pp. 1–27) in vol. 2 of his *North Country Life in the Eighteenth Century* (London, UK, 1965) and, for a representative section of the Pennines, see K. B. Cumberland, "Livestock Distribution in Craven," *SGM*, 54 (1938), pp. 75–93, a modern study but nevertheless revealing many traditions and ecological relationships. The problem of bovine epidemics in upland England is addressed in J. Bell, Jr., *The Cattle Plague in Westmorland and Cumberland, 1745–1754* (Kendal, UK, 1883). Seasonal livestock migration is briefly presented in T. H. Bainbridge, "A Note on Transhumance in Cumbria," *G*, 25 (1940), pp. 35–36, and Norman J. G. Pounds, "Note on Transhumance in Cornwall," *G*, 27 (1942), p. 34.

Scotland is nicely covered in Alexander Fenton, "The Traditional Pastoral Economy," in M. L. Parry and T. R. Slater, eds., *The Making of the Scottish Countryside* (London, UK, 1980), pp. 93–113, a very useful overview demonstrating the change from traditional subsistence, diversified highland herding to commercial beef-cattle raising in the 1500s and 1600s. Excellent local studies for Scotland include Malcolm D. MacSween, "Transhumance in North Skye," *SGM*, 75 (1959), pp. 75–88; Alexander Carmichael, "Grazing and Agrestic Customs of the Outer Hebrides," *Celtic Review*, 10 (1914–16), pp. 40–54, 144–48, 254–62, 358–75; and Alexander Fenton, "Cattle" (chapter 53), in his book *The Northern Isles: Orkney and Shetland* (Edinburgh, UK, 1978), pp. 428–43. In an award-winning article, Isobel M. L. Robertson, "The Head-Dyke: A Fundamental Line in Scottish Geography," *SGM*, 65 (1949), pp. 6–19, explains the "fence-out" concept of open-range grazing in northern Great Britain. A useful brief piece on range firing is A. J. Kayll, "Moor Burning in Scotland," *Tall Timbers Fire Ecology Conference, Proceedings*, 6 (1967), pp. 28–39. A good introduction to the traditional hamlet settlement of the Scots is Horace Fairhurst, "Scottish Clachans," *SGM*, 76 (1960), pp. 67–76. The definitive study of overland driving of Scottish cattle to English markets is A. R. B. Haldane, *The Drove Roads of Scotland* (London, UK, 1952); but see also the more anecdotal piece by William Thompson, "Cattle Droving between Scotland and England," *Journal of the British Archaeological Association*, n.s. 37 (1932), pp. 172–83.

For Wales the most abundant literature concerns cattle droving and the trade with England, although more general information on the herding system is also contained in these works. The most scholarly treatment is Richard J. Colyer, *The Welsh Cattle Drovers: Agriculture and the Welsh Cattle Trade before and during the Nineteenth Century* (Cardiff, UK, 1976). Far briefer, more anecdotal, and based heavily upon field interviews and observations, are Hugh R. Rankin, "Cattle Droving from Wales to England," *Agriculture*, 62 (1955), pp. 218–21, and John C. Lloyd, "The Black Cattle Droves," *Historical Memoranda of Breconshire* (1903), pp. 53–56. See also Caroline Skeel, "The Cattle Trade between England and Wales from the Fifteenth to the Nineteenth Centuries," *Transactions of the Royal Historical Society*, 4th s., 9 (1926), pp. 135–58. A more general and popularized treatment of Welsh cattle herding is Mary C. Harris, "Drovers and Hill Farms," in her book *Crafts, Customs and Legends of Wales* (Newton Abbot, UK, 1980), pp. 9–27.

The Irish literature is abundant and solidly researched. The best place to start is with three splendid, sensitive, and sympathetic works by the geographer E. Estyn Evans: *Irish Heritage: The Landscape, the People and Their Works* (Dundalk, Ireland, 1944); *Irish Folk Ways* (London, UK, 1957); and *The Personality of Ireland: Habitat, Heritage and History* (Cambridge, UK, 1973). Then turn to the excellent historical geography by Philip S. Robinson, *The Plantation of Ulster: British Settlement in an Irish Landscape, 1600–*

1670 (Dublin, Ireland, 1984), which reveals, among other things, the lack of an open-range cattle-herding tradition by the Scotch-Irish. John O'Donovan's *The Economic History of Live Stock in Ireland* (Dublin and Cork, Ireland, 1940), is acultural in approach and rather dry, but contains a very useful statistical base. The seasonal movement of cattle is described briefly in Mrs. J. M. Graham, "Transhumance in Ireland," *Advancement of Science,* 10 (1953), pp. 74–79; in F. H. A. Aalen, "Transhumance in the Wicklow Mountains," *Ulster Folklife,* 10 (1964), pp. 65–72; and in Seamus Ó Duilearga, "Mountain Shielings in Donegal," *Béaloideas,* 9 (1939), pp. 295–97. The seasonally occupied huts of herders are described in F. H. A. Aalen, "Clochans as Transhumance Dwellings in the Dingle Peninsula, Co. Kerry," *Journal of the Royal Society of Antiquaries of Ireland,* 94 (1964), pp. 39–45. Traditional clustered hamlet settlements are concisely treated in V. B. Proudfoot, "Clachans in Ireland," *Gwerin,* 2 (1959), pp. 110–22. For Manx cattle herding, see Peter S. Gelling, "Medieval Shielings in the Isle of Man," *Medieval Archaeology,* 6/7 (1962–63), pp. 156–72.

The possibility of African influence on North American cattle ranching has scarcely been studied at all. The first to suggest, briefly, such influence in the colonial Southeast was Peter H. Wood, *Black Majority: Negroes in Colonial South Carolina from 1670 through the Stono Rebellion* (New York, NY, 1975), but the possible connection between African and Hispanic-American herding remains unexamined. Beware of propagandistic overstatements such as "Africans Shaped American Cattle Industry," *UCLA African Studies Center Newsletter* (Fall 1986), p. 13.

The best place to begin is with several general cultural geographies of West Africa, the best of which are Augustin Bernard, *Afrique septentrionale et occidentale* (Paris, France, 1939); R. J. Harrison Church, *West Africa: A Study of the Environment and Man's Use of It,* 2nd ed. (London, UK, 1960); Reuben K. Udo, *The Human Geography of Tropical Africa* (Ibadan, Nigeria and London, UK, 1982); and, in briefer synthetic form, C. Daryll Forde, "The Cultural Map of West Africa: Successive Adaptations to Tropical Forests and Grasslands," in Simon and Phoebe Ottenberg, eds., *Cultures and Societies of Africa* (New York, NY, 1960), pp. 116–38. Geographical studies of regions and ecological zones are also useful, as for example, Hans Karl Barth, *Der Geokomplex Sahel* (Tübingen, Germany, 1977), which is good on local environment but focused on recent problems; Horst Mensching, "Der Sahel in Westafrika," *Hamburger Geographische Studien,* 24 (1971), pp. 61–73, a brief introduction to the Sahel as a transition zone; Reuben K. Udo, *Geographical Regions of Nigeria* (Berkeley and Los Angeles, CA, 1970), an insightful overview of one of the key states of the region; and Jean Gallais, *Le delta intérieur du Niger: étude de géographie régionale,* 2 vols. (Ifan-Dakar, Senegal, 1967), a splendid and comprehensive study of life in this unique part of the Sahel.

Comprehensive works on West African pastoralism include Douglas L.

Johnson, *The Nature of Nomadism: A Comparative Study of Pastoral Migrations in Southern Asia and Northern Africa* (Chicago, IL, 1969), a cursory cross-cultural approach; Walter Deschler, "Cattle in Africa: Distribution, Types, and Problems," *GR*, 53 (1963), pp. 52–58, valuable mainly because of the accompanying map; Théodore Monod, ed., *Pastoralism in Tropical Africa* (London, UK, 1975), a collection of scholarly articles, many of which deal with West Africa; Georges Doutressoulle, *L'élevage en Afrique Occidentale Français* (Paris, France, 1947), an excellent survey for the former French colonies; and Hans-Joachim Koloss, *Die Haustierhaltung in Westafrika* (München, Germany, 1968), a fine cross-cultural study focusing on folklore but containing, on pp. 26–38, useful data on herding. The important issue of horse usage is best approached by first reading Georges Doutressoulle, "Le cheval au Soudan Français et ses origines," *Bulletin de l'Institut Français d'Afrique Noire*, 2 (1940), pp. 342–46, who identified two breeds and traditions.

The Fulani (Fula, Fulbe, Peuls), the principal cattle-herding people, are well studied, especially those residing in Niger, Mali, and Nigeria, who best retain the traditional cattle focus. For Nigeria see C. Edward Hopen, *The Pastoral Fulbe Family in Gwandu* (London, UK, 1958), a work focusing upon societal matters; Derrick J. Stenning, *Savannah Nomads: A Study of the Wodaabe Pastoral Fulani of Western Bornu Province, Northern Region, Nigeria* (London, UK, 1959), one of the finest comprehensive studies of the Cattle Fulani; Derrick J. Stenning, "Transhumance, Migratory Drift, Migration: Patterns of Pastoral Fulani Nomadism," *Journal of the Royal Anthropological Institute of Great Britain and Ireland*, 87 (1957), pp. 57–73, the best analysis of Fulani seasonal movements; Werner Fricke, *Die Rinderhaltung in Nordnigeria und ihre natur- und sozialräumlichen Grundlagen* (Frankfurt, Germany, 1969), a good survey but dealing almost exclusively with the modern era; F. W. de St. Croix, "Some Aspects of the Cattle Husbandry of the Nomadic Fulani," *Farm and Forest*, 5 (1944), pp. 29–33, a sympathetic, firsthand observation by an expert in animal husbandry; E. A. Brackenbury, "Notes on the Bororo Fulbe or Nomad Cattle Fulani," *JAS*, 23 (1923), pp. 208–17, 271–77, containing a variety of firsthand ethnographical observations; and, of similar nature, J. R. Wilson-Haffenden, "Ethnological Notes on the Shuwalbe Group of the Bororo Fulani in the Kurafi District of Keffi Emirate, Northern Nigeria," *Journal of the Royal Anthropological Institute of Great Britain and Ireland*, 57 (1927), pp. 275–93. For the Fulani in former French territories, see Marguerite Dupire, *Peuls nomades: étude descriptive des Wodaabe du Sahel nigérien*, an excellent companion to the Stenning study cited earlier and containing an abundance of data on herd management; Z. Ligers, "Comment les Peul de Koa castrent leurs taureaux," *Bulletin de l'Institut Français d'Afrique Noire*, series B, 20 (1958), pp. 191–204, describing the most essential task assuring herd docility and controlled breeding; and Michael M. Horowitz, "Ethnic Boundary Maintenance among Pastoralists and Farmers in the Western Sudan (Niger)," *Journal of Asian and African Studies*, 7 (1972), pp. 105–14, describing the delicate territorial relationship between Fulani and sedentary agriculturists.

For the Taureg, see Johannes Nicolaisen, *Ecology and Culture of the Pastoral Taureg* (København, Denmark, 1963), which, in spite of its focus on the raising of livestock other than cattle, is richly illustrated and contains a splendid bibliography; Johannes Nicolaisen, "Some Aspects of the Problem of Nomadic Cattle Breeding among the Taureg of the Central Sahara," *Geografisk Tidsskrift*, 53 (1954), pp. 62–105, in which "cattle" refers to all horned livestock; Edmond Bernus, *Les Illabakan (Niger): une tribu touarèque sahélienne et son aire de nomadisation* (Paris, France, and 's Gravenhage, Netherlands, 1974), containing the best account of the cattle-raising southern Taureg; Edmond Bernus, "Les Touareg du Sahel nigérien," *Cahiers d'Outre-Mer*, 19 (1966), pp. 5–34, a briefer introduction to the topic; and Jean Bisson, "Eleveurs-caravaniers et vieux sédentaires de l'Air sud-oriental," *Travaux de l'Institut de Recherches Sahariennes*, 23 (1964), pp. 95–110, a cursory look at the more typical Saharan Taureg.

The Maures (Moors) are described in Paul Marty, *Etudes sur l'Islam et les tribus Maures* (Paris, France, 1921), a historical and genealogical study; Wolfgang Creyaufmüller, *Nomadenkultur in der Westsahara: Die materielle Kultur der Mauren, ihre handwerklichen Techniken und ornamentalen Grundstrukturen* (Stuttgart, Germany, 1983), largely unconcerned with pastoralism but still very useful; and P. Borricand, "La nomadisation en Mauritanie," *Travaux de l'Institut de Recherches Sahariennes*, 5 (1948), pp. 81–93, a brief complement to Creyaufmüller's work.

The slave trade and the important issue of the ethnic affiliation of Africans brought to the New World is best treated in Claude Meillassoux, "The Role of Slavery in the Economic and Social History of Sahelo-Sudanic Africa," in J. E. Inikori, ed., *Forced Migration: The Impact of the Export Slave Trade on African Societies* (London, UK, 1982), pp. 74–99; and Gonzalo Aguirre Beltran, "Tribal Origins of Slaves in Mexico," *Journal of Negro History*, 31 (1946), pp. 269–352. Promisingly titled but far less useful are Melville J. Herskovits, "On the Provenience of New World Negroes," *Social Forces*, 12 (1933–34), pp. 247–62; Montague F. Ashley-Montagu, "The African Origins of the American Negro and his Ethnic Composition," *Scientific Monthly*, 58 (1944), pp. 58–65; Philip D. Curtin, *The Atlantic Slave Trade: A Census* (Madison, WI, 1969); and Gabriel Debien et al., "Les origines des esclaves des Antilles," *Bulletin de l'Institut Français d'Afrique Noire*, series B, 23 (1961) through 27 (1967). The cultural linkages between Africa and the West Indies are explored in Sidney W. Mintz and Richard Price, *An Anthropological Approach to the Afro-American Past: A Caribbean Perspective* (Philadelphia, PA, 1976), and Margaret E. Crahan and Franklin W. Knight, eds., *Africa and the Caribbean: The Legacies of a Link* (Baltimore, MD, 1979), but both deal with the plantation-derived culture and do not consider herding.

Chapter Three
Implantments and Adaptations in the West Indies

The definitive study of colonial cattle ranching in the West Indies remains to be written. Indeed the subject has largely been neglected. Given this deficiency, the best place to begin is with two fine general historical geographies of the region: Carl O. Sauer, *The Early Spanish Main* (Berkeley and Los Angeles, CA, 1966) and David Watts, *The West Indies: Patterns of Development, Culture and Environmental Change since 1492* (Cambridge, UK, 1987). Then consult the insightful overview of the Americas by Alfred W. Crosby, Jr., *The Columbian Exchange: Biological and Cultural Consequences of 1492* (Westport, CT, 1972), which contains considerable information on the Indies. An exception to the general neglect of Antillean herding in John S. Otto and Nain E. Anderson, "The Origins of Southern Cattle Grazing: A Problem in West Indian History," *Journal of Caribbean History*, 21 (1988), pp. 138–53, a tantalizing and intriguing little article focusing upon the British Jamaicans and their role in the diffusion of herding to the mainland.

The floral resource base for cattle herding in the Antilles is cursorily portrayed in Peter R. Bacon, *Flora and Fauna of the Caribbean: An Introduction to the Ecology of the West Indies* (Port of Spain, Trinidad, 1978). For a more detailed and scholarly presentation, see J. S. Beard, "The Savanna Vegetation of Northern Tropical America," *Ecological Monographs*, 23 (1953), pp. 149–215. G. F. Asprey and R. G. Robbins, "The Vegetation of Jamaica," *Ecological Monographs*, 23 (1953), pp. 359–412, provide an overview of that important island. Detailed cataloging is available in A. S. Hitchcock and Agnes Chase, "Grasses of the West Indies," *Contributions from the United States National Herbarium*, 18 (1917), pp. 261–494, and Ovidio García-Molinari, *Grasslands and Grasses of Puerto Rico* (Río Piedras, PR, 1952).

Cultural adaptation and habitat modification are also poorly studied in the Indies, but suggestions of what this ecological approach might yield are found in David Harris, ed., *Human Ecology in Savanna Environments* (London, UK, 1980); and Carl O. Sauer, "Man in the Ecology of Tropical America," *Proceedings of the Ninth Pacific Science Congress* (Bangkok, Thailand, 1958), vol. 20, pp. 104–20. Promising more than it delivers is Dave D. Davis, "The Strategy of Early Spanish Ecosystem Management on Cuba," *Journal of Anthropological Research*, 30 (1974), pp. 294–314, which ought to be read in conjunction with Leo Waibel, "Place Names as an Aid in the Reconstruction of the Original Vegetation of Cuba," *GR*, 33 (1943), pp. 376–96, and the informal Raul F. Alonso Olivé, "Pastos y forrajes: una vista panorámica de su historia en Cuba," *Revista de Agricultura* (Cuba), 36:1 (1952–53), pp. 89–108, both of which contain valuable information on savannas.

Population origins, crucial to the evolution of ranching in the Indies, have received abundant attention. On Africans see the sources annotated at the end of the section on chapter 2. Peter Boyd-Bowman has produced the definitive works on Spanish migrants. For his overview see "Patterns of Spanish Emigration to the Indies until 1600," *HAHR*, 58 (1976), pp. 580–604, and "Spanish Emigrants to the Indies, 1595–98: A Profile," in Fredi Chiappelli, ed., *First Images of America: The Impact of the New World on the Old* (Berkeley, CA, 1976), vol. 2, pp. 723–35. For remarkable lists of individual migrants and their places of origin, see Boyd-Bowman's *Indice geobiográfico de cuarenta mil pobladores españoles de América en el siglo XVI*, 2 vols. (Bogotá, Colombia, and Mexico City, Mexico, 1964, 1968) and *Indice geobiográfico de más de 56 mil pobladores de la América hispánica* (Mexico City, Mexico, 1985).

The evolution of a mixed-blood work force is treated in Magnus Mörner, *Race Mixture in the History of Latin America* (Boston, MA, 1967). The deviant behavior and class struggle of the vaquero class in Latin America is nicely summarized in Silvio R. D. Baretta and John Markoff, "Civilization and Barbarism: Cattle Frontiers in Latin America," *Comparative Studies in Society and History*, 20 (1978), pp. 587–620, where a comparative approach is employed. In the same vein is Mario Góngora, "Vagabondage et société pastorale in Amérique latine," *Annales: Économies, Sociétés, Civilisations*, 21 (1966), pp. 159–77, although the article focuses on South America. Another highly significant population element, the buccaneers, are the subject of the remarkable firsthand account by A. O. Exquemelin, *The Buccaneers of America* (Baltimore, MD, 1969, originally published in 1678) and venerable secondary sources such as James Burney, *History of the Buccaneers of America* (London, UK, 1891) and Clarence H. Haring, *The Buccaneers in the West Indies in the XVII Century* (London, UK, 1910). Bovine immigration is treated in the very readable, popularized book by J. E. Rouse, *The Criollo: Spanish Cattle in the Americas* (Norman, OK, 1977).

The evolution of landed pastoral estates in the colonial Spanish Indies, no simple process, can be understood best by reading two splendid articles: Duvon C. Corbitt, "Mercedes and Realengos: A Survey of the Public Land System in Cuba," *HAHR*, 19 (1939), pp. 263–85, and James Lockhart, "Encomiendas and Haciendas: The Evolution of the Great Estate in the Spanish Indies," *HAHR*, 49 (1969), pp. 411–29. More detail on the early use of bonded Indian labor appears in Emilio R. Demorizi, *Los Dominicos y las encomiendas de indios de la isla Española* (Santo Domingo, Dominican Republic, 1971).

Local studies of the four Greater Antilles vary greatly in quality and quantity. For the mother island, Española, see the paired works by Médéric L. E. Moreau de Saint-Méry, *Descripción de la parte española de Santo Domingo* (Santo Domingo, Dominican Republic, 1944), and *Description topographique, physique, civile politique et historique de la partie française de l'isle Saint-Domingue*, 3 vols. (Paris, France, 1958), a firsthand account from the late 1700s rich in information on herding; and John M. Street, *Historical and Economic Geography of the Southwest Peninsula of Haiti* (Berkeley, CA, 1960). A glimpse of early Spanish Española as revealed through archaeology is Elizabeth J. Reitz, "Vertebrate Fauna from Locus 39, Puerto Real, Haiti," *Journal of Field Archaeology,* 13 (1986), pp. 317–28.

Jamaica, a crucially important island where Spanish, British, and African traditions met and mingled, has been well studied. For the Spanish era, see Francisco Morales Padrón, *Jamaica española* (Sevilla, Spain, 1952), a good general treatment, and Frank Cundall and Joseph L. Pietersz, *Jamaica under the Spaniards* (Kingston, Jamaica, 1919), an extremely valuable collection of translated Spanish documents. English Jamaica in the late seventeenth century is lavishly portrayed in the truly remarkable contemporary scholarly work by Hans Sloane, *A Voyage to the Islands Madera, Barbados, Nieves, S. Christophers and Jamaica,* 2 vols. (London, UK, 1707). Etymological evidence of Spanish-English-African cultural merging is nicely catalogued in Frederic G. Cassidy and R. B. LePage, *Dictionary of Jamaican English* (Cambridge, UK, 2nd ed., 1980). While focused on a later period, B. W. Higman's *Jamaica Surveyed: Plantation Maps and Plans of the Eighteenth and Nineteenth Centuries* (Kingston, Jamaica, 1988), contains valuable and revealing illustrations, especially chapter 7, which deals with cattle "pens." See also his later article, "The Internal Economy of Jamaican Pens, 1760–1890," *Social and Economic Studies* (Jamaica), 38 (1989), pp. 61–86.

For Cuba, the main ranching island of the Greater Antilles, we need go no further than Levi Marrero, *Cuba: economía y sociedad,* 14 vols. (Río Piedras, PR and Madrid, Spain, 1972–88), a simply incredible lifetime's work for which the term "definitive" is inadequate. Marrero's interdisciplinary approach is particularly appealing. See especially "La ganadería," in vol. 2 (Madrid, Spain, 1974), pp. 89–102.

Puerto Rico, a peripheral and rather unimportant island insofar as diffusion to the mainland is concerned, nevertheless long preserved a Spanish ranching system revealing Ibero-Antillean practices. Unfortunately we lack a definitive study of the subject. See the tantalizing, document-based piece by Dolores M. Nadal and Hugo W. Alberts, "The Early History of Livestock and Pastures in Puerto Rico," *AH,* 21 (1947), pp. 61–64.

Chapter Four
From the Indies to the Mainland

For Mexico, the most important continental implantment, begin with the excellent work by Richard J. Morrisey, "The Establishment and Northward Expansion of Cattle Ranching in New Spain," Ph.D. diss., University of California (Berkeley, CA, 1949), a portion of which was published as "Colonial Agriculture in New Spain," *AH,* 31 (1957), pp. 24–29. A competent, concise overview is offered in José Matesanz, "Introducción de la ganadería en Nueva España, 1521–1535," *Historia Mexicana,* 14 (1964), pp. 533–66, and useful background material is methodically provided in Peter Gerhard, *A Guide to the Historical Geography of New Spain* (Cambridge, UK, 1962), an uninspiring but essential encyclopedia. It should be read in conjunction with Lesley B. Simpson, *Exploitation of Land in Central Mexico in the Sixteenth Century* (Berkeley and Los Angeles, CA, 1952), which contains a summary in tabular form of early ranching land grants and allows regional comparisons to be made. Of far grander design, if now dated, is François Chevalier, *Land and Society in Colonial Mexico—The Great Hacienda*, trans. Alvin Eustis (Berkeley, CA, 1963), an analysis of the rise of the landed estates. The important role of Indian labor grants in early Mexico is treated in Lesley B. Simpson, *The Encomienda in New Spain: the Beginnings of Spanish Mexico* (Berkeley and Los Angeles, CA, 1982). See also the splendid firsthand account by Carl [Christian] Sartorius, *Mexico about 1850* (Stuttgart, Germany, 1961, reprint of 1858 edition), especially chapter 23—"The Cattle-Breeders and Herdsmen."

The local livestock-raisers' associations are the subject of William H. Dusenberry, *The Mexican Mesta: The Administration of Ranching in Colonial Mexico* (Urbana, IL, 1963), but the author pairs the Mexican mestas with the royal institution in Spain, rather than with the more comparable local mestas of Sevilla and other cities. An earlier article by the same author, "Ordinances of the Mesta in New Spain," *The Americas*, 4 (Jan. 1948), pp. 345–50, offers a translation of the 1530s regulations of the Mexico City mesta. The more comprehensive 1570s ordinances appear in the Morrisey dissertation cited earlier, pp. 233–54.

Mexican charro ways are nicely revealed in Carlos Rincón Gallardo, *El charro mexicano* (Mexico City, Mexico, 1939), written by an aristocratic member of this mythladen equestrian subculture. While reading it, keep at hand the beautifully illustrated volume by Leovigildo Islas Escárcega and Rodolfo García-Brava y Olivera, *Iconografía charra* (Mexico City, Mexico, 1969), containing color reproductions of paintings showing various aspects of charro cowboy life. Also useful is José Alvarez del Villar, *Historia de la charrería* (Mexico City, Mexico, 1941), by an author abundantly concerned with origins and culture history.

The development of ranching in early Mexico is best studied at the local or district level. Regrettably we have few such in-depth studies, but exceptions include Wolfgang Trautmann, *Der kolonialzeitliche Wandel der Kulturlandschaft in Tlaxcala* (Paderborn, Germany, 1983), and Jack A. Licate, *Creation of a Mexican Landscape: Territorial Organization and Settlement in the Eastern Puebla Basin, 1520–1605* (Chicago, IL, 1981), although neither deals with a major cattle-ranching area.

Market-influenced zonation is treated briefly in Ursula Ewald, "The von Thünen Principle and Agricultural Zonation in Colonial Mexico," *JHG*, 3 (1977), pp. 123–33, which delivers less than the title promises but suggests the viewpoint of an economic determinist. Approaching from the opposite, humanistic perspective, the interested scholar will find tantalizing etymological evidence strewn throughout Francisco J. Santamaría, *Diccionario de mejicanismos* (Mexico City, Mexico, 1959), although this work has annoying omissions. The same is true of Leovigildo Islas Escárcega, *Diccionario rural de México* (Mexico City, Mexico, 1961).

For data on specific population origins in Mexico, see the article by Aguirre cited for chapter 2 and those by Boyd-Bowman and Mörner for chapter 3. See also Peter Boyd-Bowman, *Patterns of Spanish Emigration to the New World (1493–1580)* (Buffalo, NY, 1973).

Spanish Florida, long and undeservedly neglected, received belated treatment in the excellent but all-too-brief article by Charles Arnade, "Cattle Raising in Spanish Florida, 1513–1763," *AH*, 35 (1961), pp. 116–24. Far more ecologically sensitive is the unpublished work by Brian G. Boniface, "A Historical Geography of Spanish Florida, circa 1700," M.A. thesis, University of Georgia (Athens, GA, 1971), but by focusing upon the early eighteenth century, he neglects the basic question of continuity. Quite minimal attention is devoted to the Spaniards in George H. Dacy, *Four Centuries of Florida Ranching* (St. Louis, MO, 1940), and Joe A. Akerman, Jr., *Florida Cowman: A History of Florida Cattle Ranching* (Kissimmee, FL, 1976). Transferral of the Spanish ranching complex to the Seminole Indians is treated peripherally in two articles by K. W. Porter, "The Founder of the Seminole Nation, Secoffee or Cowkeeper," *FHQ*, 27 (1949), pp. 362–84, and "The Cowkeeper Dynasty of the Seminole Nation," *FHQ*, 30 (1952), pp. 341–49. Far better on the subject of Indian ranching is Alan K. Craig and Christopher S. Peebles, "Ethnoecologic Change among the Seminoles, 1740–1840," *Geoscience and Man*, 5 (1974), pp. 83–96. But we still know all too little about Spanish Florida.

The South Carolina colonial core area of southern cattle ranching is far better studied. One begins with the seminal article by Gary S. Dunbar, "Colonial Carolina Cowpens," *AH*, 35 (1961), pp. 125–30, which first drew scholarly attention to the subject. A solid study, but unpublished, is James Maag, "Cattle Raising in Colonial South Carolina," M.A. thesis, University of Kansas (Lawrence, KS, 1964), the first comprehensive work on the subject. Another unpublished but very useful study, based on the archaeological excavation of one site, is Richard D. Brooks, "Cattle Ranching in Colonial South Carolina: A Case Study in History and Archaeology of the Lazarus/Catherina Brown Cowpen," M.A. thesis, University of South Carolina (Columbia, SC, 1988). John S. Otto has the annoying habit of self-plagiarism, but his several works, though greatly redundant, are essential to an understanding of early South Carolina cattle herding; see his "The Origins of Cattle-Ranching in Colonial South Carolina, 1670–1715," *SCHM*, 87 (1986), pp. 117–24; and "Livestock-Raising in Early South Carolina, 1670–1700: Prelude to the Rice Plantation Economy," *AH*, 61 (1987), pp. 13–24, among many others. Otto's knowledge fades rapidly with increasing diffusionary distance from Carolina, but his ideas seem, in general, sound. Of a more general nature, but still useful, is Converse D. Clowse, *Economic Beginnings in Colonial South Carolina, 1670–1730* (Columbia, SC, 1971). For a comparison with Virginia, where range cattle were far less important, see Wesley N. Laing, "Cattle in Seventeenth-Century Virginia," *Virginia Magazine of History and Biography*, 67 (1959), pp. 143–63.

The very important African slave majority in South Carolina receives attention in the work by Peter H. Wood listed in chapter 2, as well as in Elizabeth Donnan, "The Slave Trade into South Carolina before the Revolution," *AHR*, 33 (1928), pp. 804–28; W. Robert Higgins, "The Geographical Origins of Negro Slaves in Colonial South Carolina," *South Atlantic Quarterly* 70 (1971), pp. 34–47; and Albert H. Stoddard, "Origin, Dialect, Beliefs, and Characteristics of the Negroes of the South Carolina and Georgia Coasts," *GHQ*, 28 (1944), pp. 186–95. Special attention to the earliest South Carolina brand register, of crucial importance, is offered in Alexander S. Salley, Jr., "Stock Marks Recorded in South Carolina, 1695–1721," *SCHM*, 13 (1912), pp. 126–31, 224–28.

No study exists of pre-Cajun cattle ranching in Louisiana. We can only be tantalized by Lauren C. Post's, "The Domestic Animals and Plants of French Louisiana as Mentioned in the Literature, with Reference to Sources, Varieties and Uses," *Louisiana Historical Quarterly*, 16 (1933), pp. 554–86.

Chapter Five
Cattle Frontiers in Northern Mexico

The basic sources include Morrisey's dissertation, "The Establishment and Northward Expansion of Cattle Ranching in New Spain," cited for chapter 4; though ecologically naive and spatially weak, it provides a general overview. From it was derived Morrisey's "The Northward Expansion of Cattle Ranching in New Spain, 1550–1600," *AH*, 25 (1951), pp. 115–21, dealing with early diffusion in the highlands. A complementary article, Donald D. Brand's, "The Early History of the Range Cattle Industry in Northern Mexico," *AH*, 35 (1961), pp. 132–39, is a largely unsuccessful synthesis. The indispensable reference source is Peter Gerhard, *The North Frontier of New Spain* (Princeton, NJ, 1982), providing province-by-province background material and base maps. The essential documents for the northern frontier appear, translated, in Charles W. Hackett, *Historical Documents Relating to New Mexico, Nueva Vizcaya, and Approaches Thereto, to 1773*, 3 vols. (Washington, DC, 1923–37). A brief but informative overview of the northern frontier is Wolfgang Trautmann, "Geographical Aspects of Hispanic Colonization on the Northern Frontier of New Spain," *Erdkunde*, 40 (1986), pp. 241–50, and the makeup of the pioneer population is nicely presented in Oakah L. Jones, Jr., *Los Paisanos: Spanish Settlers on the Northern Frontier of New Spain* (Norman, OK, 1979). For an excellent collection containing numerous articles on the northern frontier of Mexico, including the present southwestern United States, see David H. Thomas, ed., *Columbian Consequences* (Washington, DC, 1989). The reader should also consult the general sources by Simpson, *Exploitation*, and Chevalier, *Land and Society*, listed for the previous chapter.

Nearly all of these general studies emphasize, or overemphasize, the cattle frontier of the interior plateau. Additional very useful studies of the altiplano ranching complex include Ramón M. Serrera Contreras, *Guadalajara ganadera: estudio regional novohispano, 1760–1805* (Sevilla, Spain, 1977), dealing with postfrontier times and economically focused, but still revealing, since it deals with the western-highland cattle focus, and Robert C. West, *The Mining Community in Northern New Spain: The Parral Mining District* (Berkeley, CA, 1949), an ecologically sensitive work that contains a wealth of information about cattle ranching in the northern reaches of the interior plateau. For additional essential environmental background concerning the floral resource base, see Jerzy Rzedowski and Rogers McVaugh, *La vegetación de Nueva Galicia*, University of Michigan Herbarium Contributions, 9:1 (Ann Arbor, MI, 1966) and Howard S. Gentry, *Los pastizales de Durango: estudio ecológico, fisiográfico y florístico* (Mexico City, Mexico, 1957).

The extension into New Mexico did not constitute a cattle-ranching frontier, but the sheep-raising complex there, with its subsequent diffusion, is nicely presented in John O. Baxter, *Las Carneradas: Sheep Trade in New Mexico, 1700–1860* (Albuquerque, NM, 1987), and Richard Nostrand, "The Century of Hispano Expansion," *NMHR*, 62 (1987), pp. 361–86. The spread of the plateau sheep-raising system into the interior parts of the Gulf coastal plain is clearly revealed in Eugenio del Hoyo, *Señores de ganado, Nuevo Reino de León, siglo XVII* (Monterrey, Mexico, 1987), and Valgene W. Lehmann, *Forgotten Legions: Sheep in the Rio Grande Plain of Texas* (El Paso, TX, 1969).

The spread of cattle ranching through the Gulf coastal plain of Tamaulipas, of crucial importance to Mexican Texas, remains essentially unstudied. William E. Doolittle, "Las Marismas to Pánuco to Texas: The Transfer of Open Range Cattle Ranching from Iberia through Northeastern Mexico," *Yearbook of the Conference of Latin Americanist Geographers*, 13 (1987), pp. 3–11, is merely suggestive, though an important idea piece. For an understanding of the floral resource base, see the excellent and insightful article by Marshall C. Johnston, "Past and Present Grasslands of South Texas and northeastern Mexico," *Ecology*, 44 (1963), pp. 456–65. For the climatological basis, see Richard J. Russell, "Climates of Texas," *AAAG*, 35 (1945), pp. 37–52. An instructive, ecologically sophisticated view of traditional ranching in Tamaulipas just prior to the American-inspired modernization is Basil M. Bensin, "Agroecological Exploration in the Soto la Marina Region, Mexico," *GR*, 25 (1935), pp. 285–97. Historian Dan E. Kilgore suggests the importance of Tamaulipas in the development of the feral Texas longhorn in the badly mistitled "The Spanish Missions and the Origins of the Cattle Industry in Texas," in Gilberto R. Cruz, ed., *Proceedings of the Second Annual Mission Research Conference* (San Antonio, TX, 1984), pp. 62–68.

Ranching in Hispanic Texas has received abundant attention. The best of the lot is Jack Jackson's *Los Mesteños: Spanish Ranching in Texas, 1721–1821* (College Station, TX, 1986), a work that displays both the strengths and weaknesses of local history—detail, thoroughness, poor organization, provincialism, and a complete lack of detectable editing. A welcome, far less detailed counterbalance is provided by Sandra L. S. Myres, "The Development of the Ranch as a Frontier Institution in the Spanish Province of Texas, 1691–1800," Ph.D. diss., Texas Christian University (Ft. Worth, TX, 1967), from which came "The Spanish Cattle Kingdom in the Province of Texas," *Texana*, 4 (1966), pp. 233–46, and *The Ranch in Spanish Texas, 1691–1800* (El Paso, TX, 1969), both of which are brief, modest in scope, but still useful. Both Jackson and Myres assume the existence of a monolithic Spanish ranching frontier. Odie B. Faulk's "Ranching in Spanish Texas," *HAHR*, 45 (1965), pp. 257–66, suffers from brevity, promising far more than it delivers. The largest mission ranch is the subject of Charles Ramsdell's "Espíritu Santo: An Early Texas Cattle Ranch," *Texas Geographic Magazine*, 13:2 (1949), pp. 21–25. Nor should one dismiss, on the grounds of its popular style, J. Frank Dobie's *The Longhorns* (Boston, MA, 1941), a book that, while absurdly colorful and overly anecdotal, was written by a man who knew ranching from firsthand experience. Dobie deals not just with the cattle, but with the whole ranching culture. For Dobie's insights on Hispanic cowhands, also based in his own observations on the Coastal Bend ranches, see "The Mexican Vaquero of the Texas Border," *Southwestern Political and Social Science Quarterly*, 8 (1927), pp. 15–26. Francis L. Fugate, "Origins of the Range Cattle Era in South Texas," *AH*, 35 (1961), pp. 155–58, is the weakest link in a special issue on cattle ranching. More worthwhile is C. Wayne Hanselka and Dan E. Kilgore, "The Nueces Valley: The Cradle of the Western Livestock Industry," *Rangelands*, 9:5 (1987), pp. 195–202. The only architectural analysis of pastoral Hispanic Texas is the fine article by Willard B. Robinson, "Colonial Ranch Architecture in the Spanish-Mexican Tradition," *SWHQ*, 83 (1979), pp. 123–50, a richly illustrated and insightful piece.

The vitally important connections between Spanish Texas and Louisiana, so crucial in the evolution of Anglo-American cattle ranching, remain largely unstudied. The underlying trade connection is treated by local historian Robert H. Thonhoff, *The Texas Connection with the American Revolution* (Burnet, TX, 1981). Only geographer Fred B. Kniffen, working from the perspective of material culture, has properly understood the connections, as was first revealed in his initial tentative inquiry, "A Spanish Spinner in Louisiana," *Southern Folklore Quarterly*, 13 (1949), pp. 192–99.

Pacific coastal Mexico lacks any sort of comprehensive ranching study.

One must piece the story together from such environmental studies as Forrest Shreve, "Lowland Vegetation in Sinaloa," *Bulletin of the Torrey Botanical Club*, 64 (1937), pp. 605–13, and Howard S. Gentry, *Río Mayo Plants: A Study of the Flora and Vegetation of the Río Mayo, Sonora* (Washington, DC, 1942). Read these in conjunction with such splendid original accounts as Juan Nentvig, *Rudo Ensayo: A Description of Sonora and Arizona in 1764* (Tucson, AZ, 1980), and Philipp Segesser, "Document: The Relation of Philipp Segesser," trans. Theodore E. Treutlein, *Mid-America: An Historical Review*, n.s. 27:3 (1945), pp. 139–87, among others. Nor should one overlook the fine regional portrait by Roger Dunbier, *The Sonoran Desert: Its Geography, Economy, and People* (Tucson, AZ, 1968), though its focus is not the ranching frontier.

Hispanic Arizona has been more adequately studied from the standpoint of cattle ranching. Begin with the survey study by J. J. Wagoner, *History of the Cattle Industry in Southern Arizona, 1540–1940* (Tucson, AZ, 1952), a weak work in the scholarly sense but a suitable introduction. Richard Morrisey's "The Early Range Cattle Industry in Arizona," *AH*, 24 (1950), pp. 151–56, is too brief to be of much use and deals only minimally with the Hispanic period. Far better, though still rather unscholarly, is Bert Haskett's "Early History of the Cattle Industry in Arizona," *AZHR*, 6:4 (1935), pp. 3–42. The excellent study by Ray H. Mattison, "Early Spanish and Mexican Settlements in Arizona," *NMHR*, 21 (1946), pp. 273–327, contains a definitive description of private pastoral grants. Baja California, of minimal importance to the Pacific ranching frontier, is best portrayed in Peveril Meigs III, *The Dominican Mission Frontier of Lower California* (Berkeley, CA, 1935).

Alta California has been well studied. The place to begin is the classic by Hubert H. Bancroft, *California Pastoral, 1769–1848* (San Francisco, CA, 1888), which in spite of its florid, anecdotal style and the brahmin author's arrogant, racist views, remains by far the best portrayal of Hispanic Californian herding culture. Then refer to Hazel A. Pulling's doctoral dissertation, listed in the chapter 8 sources. Balance these with the ecologically sound but historically weak study by geographer L. T. Burcham, *California Range Land: An Historico-Ecological Study of the Range Resources of California* (Sacramento, CA, 1957), which provides an essential environmental background and addresses the issue of range damage. An abbreviated version is his "Cattle and Range Forage in California, 1770–1880," *AH*, 35 (1961), pp. 140–49. Also dealing with environmental modification is the insightful article by Paul F. Starrs, "The Cultural Landscape of California Pastoralism: 200 Years of Changes," in W. James Clawson, ed., *Landscape Ecology: Study of Mediterranean Grazed Ecosystems* (Nice, France, 1989), pp. 49–61. A useful brief introduction to climate types is John E. Kesseli, "The Climates of California According to the Köppen Classification," *GR*, 32 (1942), pp. 476–80.

An overview of the mission system and its related herding activity is nicely presented in R. Louis Gentilcore, "Missions and Mission Lands of Alta California," *AAAG*, 51 (1961), pp. 46–72. The sequence of occupance in one representative coastal-range valley, including mission and private rancho periods, is revealed in Jan O. M. Broek, *The Santa Clara Valley, California: A Study in Landscape Changes* (Utrecht, Netherlands, 1932). From Richard L. Nostrand's "The Santa Ynez Valley: Hinterland of Coastal California," *HSSCQ*, 48 (1966), pp. 37–56, we get a tantalizing glimpse of the system of seasonal pasture shifting, suggesting a movement on the scale of transhumance.

The best brief study of the private rancho system, including elaborate cartographic display, is David Hornbeck, "Land Tenure and Rancho Expansion in Alta California, 1784–1846," *JHG*, 4 (1978), pp. 371–90. Much less scholarly or detailed, but still worth perusing, is Iris H. W. Engstrand, "California Ranchos: Their Hispanic Heritage," *SCQ*, 67 (1985), pp. 281–90. A useful list of pastoral land grants, together with endpaper maps, is Robert G. Cowan, *Ranchos of California: A List of Spanish Concessions, 1775–1822, and Mexican Grants, 1822–1846* (Fresno, CA, 1956). A number of instructive books and articles deal with individual Hispanic-era ranchos, including William R. Cameron, "Rancho Santa Margarita of San Luis Obispo," *CHSQ*, 36 (1957), pp. 1–20; Robert G. Cleland, *The Place Called Sespe: The History of a California Ranch* (San Marino, CA, 1957) and, by the same prolific author, *The Irvine Ranch of Orange County, 1810–1950* (San Marino, CA, 1952); Robert Gillingham, *The Rancho San Pedro* (Los Angeles, CA, 1961); Arthur B. Perkins, "Rancho San Francisco: A Story of a California Land Grant," *HSSCQ*, 39 (1957), pp. 99–126; H. F. Raup, "The Rancho Palos Verdes," *HSSCQ*, 19 (1937), pp. 7–21; Lois J. Roberts, "Rancho Jesús María, Santa Barbara County," *SCQ*, 68 (1986), pp. 1–35; and W. W. Robinson, "The Story of Rancho San Pasqual," *HSSCQ*, 38 (1955), pp. 347–53.

The Californian herder receives attention in Jo Mora's *Californios: The Saga of the Hard-Riding Vaqueros, America's First Cowboys* (Garden City, NY, 1949), written in a colorful, popular style without documentation, but valuable because of its basis in field observation. See also Arnold R. Rojas's book, listed for chapter 8. The classic contemporary account of the hide trade from Mexican California is Richard H. Dana, *Two Years before the Mast* (Boston, MA, 1884).

Chapter Six
Carolina's Children

Long largely ignored, the spread of open-range cattle herding across the eastern United States before 1850 has received increasing attention in recent decades. Some of these works, dealing with Old World antecedents and colonial implantations were discussed in the bibliographic section on chapters 2 and 4.

A useful overview of southern herding late in its lifespan can be gleaned from Sam B. Hilliard, *Hog Meat and Hoe Cake: Food Supply in the Old South, 1840–1860* (Carbondale and Edwardsville, IL, 1972), and the general thesis of southern diffusionary contributions to the West is treated in Terry G. Jordan, *Trails to Texas: Southern Roots of Western Cattle Ranching* (Lincoln, NE, 1981), chapters 2–5. The distinctive southern stock dog is the subject of Joseph W. LeBon, Jr., "The Catahoula Hog Dog: A Cultural Trait of the Upland South," M.A. thesis, Louisiana State University (Baton Rouge, LA, 1970), from which was distilled "The Catahoula Hog Dog: A Folk Breed," *Pioneer America*, 3:2 (1971), pp. 35–45. Challenging LeBon's ideas on the origin of this dog is Georges Lutz, "Catahoula Hog Dog ou Catahoula Cur: une opinion européenne ou le retour aux origines du chien," *Journal d'Agriculture Traditionnelle et de Botanique Appliquée*, 31 (1984), pp. 147–69. Fire, an essential southern forest range tool, receives useful attention in S. W. Greene, "Relation between Winter Grass Fires and Cattle Grazing in the Longleaf Pine Belt," *Journal of Forestry*, 33 (1935), pp. 338–41 and Merle C. Prunty, "Some Geographic Views of the Role of Fire in Settlement Processes in the South," *Tall Timbers Fire Ecology Conference, Proceedings*, 3 (1964), pp. 161–68. Pine barrens ecology in the broader context is nicely presented in W. G. Wahlenberg, *Longleaf Pine: Its Use, Ecology, Regeneration, Protection, Growth, and Management* (Washington, DC, 1946). The death throes of the open range tradition, as represented by Alabama and Mississippi, receives long overdue, if cursory, treatment in J. Crawford King, "The Closing of the Southern Range: An Explanatory Study," *JSH*, 48 (1982), pp. 53–70.

Regional studies of traditional southern cattle herding vary greatly in quality and coverage. The superbly documented, ecologically sensitive article by Mart A. Stewart, "Whether Wast, Deodand, or Stray: Cattle, Culture, and the Environment in Early Georgia," *AH*, 65 (1991), pp. 1–28, provides an excellent, if far from definitive study of that colony. See also the brief, tantalizing, and poorly documented articles by John H. Goff, "Cow Punching in Old Georgia," *Georgia Review*, 3 (1949), pp. 341–48, and James C. Bonner, "The Open Range Livestock Industry in Colonial Georgia," ibid., 17 (1963), pp. 85–92. The era of decline of the pine-barrens herding system in Georgia is nicely captured in Ann P. Malone, "Piney Woods Farmers of South Georgia, 1850–1900," *AH*, 60 (1986), pp. 51–84. The development of paired plantations and cowpens in Atlantic coastal Georgia is sketchily revealed in Alan Gallay's flawed case study, *The Formation of a Planter Elite: Jonathan Bryan and the Southern Colonial Frontier* (Athens, GA, 1989), a work that includes little about the specifics of cowpen significance or management.

Florida has received abundant attention. One should begin with the general works by Dacy and Akerman listed for chapter 4 and with the remarkable late contemporary account by Frederic Remington, "Cracker Cowboys of Florida," *Harper's New Monthly Magazine*, 91 (1895), pp. 339–45, complete with excellent drawings. Also worthy of attention is Jim B.

Tinsley, *Florida Cow Hunter: The Life and Times of Bone Mizell* (Orlando, FL, 1990). Far more insightful and scholarly is William T. Mealor, Jr., "The Open Range Ranch in South Florida and Its Contemporary Successors," Ph.D. diss., University of Georgia (Athens, GA, 1972), from which was abstracted the article, coauthored with Merle C. Prunty, "Open-Range Ranching in Southern Florida," *AAAG*, 66 (1976), pp. 360–76. The prolific and perceptive John S. Otto has given Florida abundant, and at times redundant, attention, carrying his annoying tendency to repeat himself in different journals to possibly unprecedented plateaus. Even so his works are well worth sifting through, for he always offers at least one new gem in each. Begin with his repetitive overviews, "Open-Range Cattle Herding in Antebellum South Florida (1842–1860)," *Southeastern Geographer*, 26 (1986), pp. 55–67, and "Open-Range Cattle Ranching in the Florida Pinewoods," *Proceedings of the American Philosophical Society*, 130 (1986), pp. 312–24, and then proceed to "Traditional Cattle-Herding Practices in Southern Florida," *Journal of American Folklore*, 97 (1984), pp. 291–309, which contains good detail and is based on the experiences of one specific family. Otto's useful county-level sample studies, based in part on the U.S. census manuscript schedules, appear as "Florida's Cattle Ranching Frontier," *FHQ*, 63 (1984–85), pp. 71–83, and 64 (1985–86), pp. 48–61. A twice-stated comparison with South America appears as "Open-Range Cattle-Ranching in Venezuela and Florida: A Problem in Comparative History," *Comparative Social Research*, 9 (1986), pp. 347–60, and (coauthored with Nain E. Anderson), "Cattle Ranching in the Venezuelan Llanos and the Florida Flatwoods: A Problem in Comparative History," *Comparative Studies in Society and History*, 28 (1986), pp. 672–83.

The Old Southwest, or at least Mississippi, is reasonably well covered. Begin with the summary by John D. W. Guice, "Cattle Raisers of the Old Southwest: A Reinterpretation," *WHQ*, 8 (1977), pp. 167–87; and its updated restatement in Thomas D. Clark and John D. W. Guice, *Frontiers in Conflict: The Old Southwest, 1795–1830* (Albuquerque, NM, 1989), chapter 6. Then proceed to the excellent, in-depth study by Kenneth D. Israel, "The Cattle Industry of Mississippi, Its Origin and Its Changes through Time up to 1850," Ph.D. diss., University of Southern Mississippi (Hattiesburg, MS, 1970). The early, Spanish-influenced period is treated in Jack D. L. Holmes, "Livestock in Spanish Natchez," *Journal of Mississippi History*, 23 (1961), pp. 15–37, a work solidly based on archival materials. Robert J. Baxter's "Cattle Raising in Early Mississippi," *Mississippi Folklore Register*, 10 (1976), pp. 1–23, is a remarkable narrative based in his own family's herding tradition in the pine barrens.

The development of prairie ranching in Louisiana and southeast Texas, an episode of major importance to the cattle frontier of the West, still lacks a definitive study, but see the popularized articles by Lauren C. Post, "The Old Cattle Industry of Southwestern Louisiana," *MR*, 9 (1957), pp. 43–55, and

"Cattle Branding in Southwest Louisiana," MR, 10 (1958), pp. 101–17, both of which have a Cajun focus, and Terry G. Jordan, "The Origin of Anglo-American Cattle Ranching in Texas: A Documentation of Diffusion from the Lower South," EG, 45 (1969), pp. 63–87, which deals with the Texas coastal prairie. Also useful is Jack D. L. Holmes, "Joseph Piernas and the Nascent Cattle Industry of Southwest Louisiana," MR, 17 (1966), pp. 13–26.

The study of open-range livestock herding in the southern Appalachians remains unwritten. We can only grasp at such straws as Philip J. Gersmehl, "Factors Leading to Mountaintop Grazing in the Southern Appalachians," Southeastern Geographer, 10 (1970), pp. 67–72; Eugene J. Wilhelm, Jr., "Animal Drives in the Southern Highlands," Mountain Life & Work, 42:2 (1966), pp. 6–11; and Edmund C. Burnett, "Hog Raising and Hog Driving in the Region of the French Broad River," AH, 20 (1946), pp. 86–103. Who will give us the much needed study of the Glades, South Branch, Big Levels, Asheville Basin, and other major centers of mountain herding?

The transmontane areas are far better understood, due largely to Paul C. Henlein's *Cattle Kingdom in the Ohio Valley, 1783–1860* (Lexington, KY, 1959), a splendid analysis abundantly based on source materials, though curiously acultural in approach. Component parts of this study can be seen in "Early Cattle Ranges of the Ohio Valley," AH, 35 (1961), pp. 150–54; "Cattle Driving from the Ohio Country, 1800–1850," AH, 28 (1954), pp. 83–95; and "Shifting Range-Feeder Patterns in the Ohio Valley before 1860," AH, 31 (1957), pp. 1–12. Another useful overview, though contributing no new insights, is David L. Wheeler's "The Beef Cattle Industry in the Old Northwest, 1803–1860," PPHR, 47 (1974), pp. 28–45. Other articles with promising titles prove, on inspection, to deal almost exclusively with cattle feeders rather than range-cattle producers, including R. L. Jones, "The Beef Cattle Industry in Ohio Prior to the Civil War," *Ohio Historical Quarterly*, 64 (1955), pp. 168–94; and the pair of articles by Paul W. Gates, "Cattle Kings in the Prairies," MVHR, 35 (1948), pp. 379–412 and "Hoosier Cattle Kings," *Indiana Magazine of History*, 44 (1948), pp. 1–24.

The early attention to improved cattle breeds, which eventually characterized the entire midwestern herding system, is the subject of two articles by George F. Lemmer: "The Spread of Improved Cattle through the Eastern United States to 1850," AH, 21 (1947), pp. 79–93, and "Early Leaders in Livestock Improvement in Missouri," *Missouri Historical Review*, 37 (1942–43), pp. 29–39. See also the venerable but unexcelled article by Charles T. Leavitt, "Attempts to Improve Cattle Breeds in the United States, 1790–1860," AH, 7 (1933), pp. 51–67.

The transition to the prairie environment can better be appreciated by reading Edgar N. Transeau, "The Prairie Peninsula," *Ecology*, 16 (1935), pp. 423–37, and Leslie Hewes, "The Northern Wet Prairie of the United States: Nature, Sources of Information, and Extent," AAAG, 41 (1951), pp. 307–23. Allan G. Bogue's *From Prairie to Corn Belt: Farming on the*

Illinois and Iowa Prairies in the Nineteenth Century (Chicago, IL, 1963) would seem to offer much, especially chapter 5, "The Passing of the Lean Kine," but in fact the book deals almost exclusively with the rise of the feeder industry.

The immediate trans-Mississippi states remain essentially unstudied. We must pick at pretty clean bones, such as Clifford D. Carpenter, "The Early Cattle Industry in Missouri," *Missouri Historical Review*, 47 (1953), pp. 201–15; Frank S. Popplewell, "St. Joseph, Missouri, as a Center of the Cattle Trade," *Missouri Historical Review*, 32 (1938), pp. 443–57, and J. H. Atkinson, "Cattle Drives from Arkansas to California Prior to the Civil War," *Arkansas Historical Quarterly*, 28 (1969), pp. 275–81. Upland southern portions of eastern Oklahoma and Texas are somewhat better covered, though still lacking in-depth studies. See Michael F. Doran, "Antebellum Cattle Herding in the Indian Territory," *GR*, 66 (1976), pp. 48–58; Norman A. Graebner, "History of Cattle Ranching in Eastern Oklahoma," *CO*, 21 (1943), pp. 300–311; and Terry G. Jordan, "Early Northeast Texas and the Evolution of Western Ranching," *AAAG*, 67 (1977), pp. 66–87.

Chapter Seven
The Anglo-Texan Ranching System

It would be impractical to list here most or even much of the secondary literature dealing with the rise and spread of the Anglo-Texas system of cattle ranching. The number of publications on this overemphasized, absurdly mythologized subject is simply mind-numbing. I list only those sources needed to gain an understanding of the subject.

Begin with the classic work by Walter P. Webb, *The Great Plains* (Boston, MA, 1931), chapter 6, in which the Texas system is, erroneously I feel, depicted as an environmentally determined, Mexican-derived livelihood. David Dary's "The Texian Culture," chapter 4 in his *Cowboy Culture: A Saga of Five Centuries* (New York, NY, 1981) is both highly readable and somewhat less dogmatic than Webb. Terry G. Jordan, *Trails to Texas: Southern Roots of Western Cattle Ranching* (Lincoln, NE, 1981), stresses the Anglo-Americans' heritage of cattle herding prior to the Texas experience and their input into the new system. James C. Malin, *The Grasslands of North America: Prolegomena to its History* (Lawrence, KS, 1947), is an indispensable guide to the literature about the Great Plains, including abundant attention to environment.

Among the general studies of Great Plains ranching, the best remains the venerable work by Edward E. Dale, *The Range Cattle Industry: Ranching on the Great Plains from 1865 to 1925* (Norman, OK, 1930), although he overemphasizes Texan influence on the Plains. Louis Pelzer's *The Cattleman's Frontier: A Record of the Trans-Mississippi Cattle Industry from Oxen Trains to Pooling Companies, 1850–1890* (Glendale, CA, 1936), is more descriptive and anecdotal, deals mainly with the northern plains, and stresses the economics rather than the culture or ecology of ranching. Mont H. Saunderson's *Western Stock Raising* (Minneapolis, MN, 1950), is a very useful study of contemporary ranching by a range economist with good ecological understanding. Lewis Atherton's *The Cattle Kings* (Lincoln, NE, 1972), deals in a Turnerian manner with the role of the large-scale Great Plains ranchers in American culture at large. Also focusing on the plains is Mari Sandoz, *The Cattlemen from the Rio Grande Across the Far Marais* (New York, NY, 1958), a beautifully written but overly anecdotal and poorly documented work. Also very useful but jarringly disjointed is Maurice Frink's section of the book *When Grass Was King: Contributions to the Western Range Cattle Industry Study* (Boulder, CO, 1956), pp. 3–132, which includes a year-by-year capsule of pastoral events on the Plains.

Works dealing with segments of the Texas-influenced Great Plains include Arrell M. Gibson, "Ranching on the Southern Great Plains," *JOW*, 6:1 (1967), pp. 135–53, an ecologically sensitive piece devoted mainly to the big operators in western Oklahoma, eastern New Mexico, and West Texas. A complementary article for the northern plains is Harold Briggs, "The Development and Decline of Open Range Ranching in the Northwest," *MVHR*, 20 (1934), pp. 521–36, a balanced and useful summary in which non-Texan influences also receive treatment. The Canadian plains have been thoroughly studied in a series of excellent works, unequalled south of the border, including David H. Breen, *The Canadian Prairie West and the Ranching Frontier, 1874–1924* (Toronto, ON, 1983), especially good on range leasing; John W. Bennett, "The Ranchers," chapter 6 in *Northern Plainsmen: Adaptive Strategy and Agrarian Life* (Chicago, IL, 1969), one of the most insightful pieces ever written about ranching.

Studies on the state and local level are legion, but some are particularly instructive. For Texas proper, see Daniel E. McArthur, "The Cattle Industry of Texas, 1865–1918," M.A. thesis, University of Texas (Austin, TX, 1918), a work that should have been published long ago; Louise S. O'Connor, *Cryin' For Daylight: A Ranching Culture in the Texas Coastal Bend* (Austin, TX, 1989), a wonderful book full of interviews and old photographs and dealing with a major zone of mixing of Anglo, African-American, and Mexican herding traditions on the coast; J. W. Williams, *The Big Ranch Country* (Wichita Falls, TX, 1954), a fine treatment of the Rolling Plains country west of the crucial north-central Texas region; Seymour V. Connor, "Early Ranching Operations in the Panhandle: A Report on the Agricultural Schedules of the 1880 Census," *PPHR*, 27 (1954), pp. 47–69, an all-too-rare use of one of the most basic source materials, implicitly admonishing us to cease our silly sweeping generalizations and anecdotal mythmaking and to deal with ranching as revealed in the documents; and Robert M. Utley, "The Range Cattle Industry in the Big Bend of Texas," *SWHQ*, 69 (1965–66), pp. 419–41, a rare look at ranching in montane Texas.

The best of the books on individual Anglo-Texan ranchers and cattle kings are two works by J. Evetts Haley, *The XIT Ranch of Texas and the Early Days of the Llano Estacado* (Norman, OK, 1953) and *Charles Goodnight: Cowman and Plainsman* (Boston, MA and New York, NY, 1936), both lovingly researched and splendidly presented; and William C. Holden, *A Ranching Saga: The Lives of William Electious Halsell and Ewing Halsell*, 2 vols. (San Antonio, TX, 1976), a model for such studies. David J. Murrah, *C. C. Slaughter: Rancher, Banker, Baptist* (Austin, TX, 1981), deals with a cattle king of the Texas High Plains whose family had earlier herded stock on the coastal pine barrens of the South.

A useful graphic collection of brand designs and histories is presented in Gus L. Ford, ed., *Texas Cattle Brands: A Catalog of the Texas Centennial Exposition Exhibit, 1936* (Dallas, TX, 1936). A more-than-usually enlightening view of the Texas cowboy is offered in Clifford P. Westermeier, "The Cowboy in His Home State," *SWHQ*, 58 (1954–55), pp. 218–34, based on contemporary newspaper accounts.

For Oklahoma the venerable works of Edward E. Dale remain unsurpassed, and the best summary is "History of the Ranch Cattle Industry in Oklahoma," *Annual Report of the American Historical Association for the Year 1920*, pp. 309–22. The best study of an individual Oklahoma ranch is Ellsworth Collings and Alma M. England, *The 101 Ranch* (Norman, OK, 1938).

The Texan influence in Colorado is well represented in Ora B. Peake, *The Colorado Range Cattle Industry* (Glendale, CA, 1937), still the best comprehensive treatment of ranching in that state. For the mountain parks of Colorado see George C. Everett, *Cattle Cavalcade in Central Colorado* (Denver, CO, 1966), an amateur history filled with useful information. On the most famous of the Colorado Texans, see Charles Kenner, "A Texas Rancher in Colorado: The Last Years of John Hittson," *WTYB*, 42 (1966), pp. 28–40.

Kansas and Nebraska both lack a definitive study of early ranching, but see Nellie S. Yost, *The Call of the Range: The Story of the Nebraska Stock Growers Association* (Denver, CO, 1960), which contains far more than its title suggests, and William D. Aeschbacher, "Development of Cattle Raising in the Sandhills," *NBH*, 28 (1947), pp. 41–64; and James C. Dahlman, "Recollections of Cowboy Life in Western Nebraska," *NBH*, 10 (1927), pp. 335–39, both of which reveal abundant Texas influences. For the Dakotas see Hazel A. Pulling's fine article, "History of the Range Cattle Industry of Dakota," *South Dakota Historical Collections*, 20 (1940), pp. 467–521, derived from her identically titled, unpublished 1931 Chicago master's thesis. The major focus of Texas-style ranching in the valley of the Little Missouri is treated in George F. Shafer, "Cattle Ranching in McKenzie County, North Dakota," *North Dakota Historical Quarterly*, 1 (1926), pp. 55–61; and the Texan presence in early South Dakota is revealed in August H. Schatz,

Longhorns Bring Culture (Boston, MA, 1961), the title of which alone demands attention and astonishment.

The Texan zone of Montana is expertly studied in Robert S. Fletcher, *History of the Range Cattle Business in Eastern Montana* (Washington, DC, 1928), written by an acknowledged expert. For a more recent treatment, see the collected articles in Michael S. Kennedy, ed., *Cowboys and Cattlemen: A Roundup from Montana* (New York, NY, 1964), and the special ranching issue of *Montana: The Magazine of Western History*, 11:4 (October 1961). The postfrontier industry is nicely presented in Mont H. Saunderson and D. W. Chittenden, *Cattle Ranching in Montana* (Bozeman, MT, 1937); and a useful account of a Texas cowboy in turn-of-the-century Montana is John Leakey, *The West that Was: From Texas to Montana* (Dallas, TX, 1958).

Alberta is particularly well studied. The best general work is Edward Brado, *Cattle Kingdom: Early Ranching in Alberta* (Vancouver, BC, and Toronto, ON, 1984), which contains some capsulized individual ranch histories and descriptions of cowboy life. Two articles by Simon M. Evans, "The Origins of Ranching in Western Canada: American Diffusion or Victorian Transplant," *Great Plains Quarterly*, 3:2 (1983), pp. 79–91, and "American Cattlemen on the Canadian Range, 1874–1914," *Prairie Forum*, 4 (1979), pp. 121–35, are useful antidotes for hyper-Texans such as Webb and Dobie and intelligently address the vital issue of cultural history in a revisionist way. Older and more personal in tone is Lewis G. Thomas, "The Ranching Period in Southern Alberta," M.A. thesis, University of Alberta (Calgary, AB, 1935), part of which was published, along with other assorted pieces, in *Ranchers' Legacy: Alberta Essays by Lewis G. Thomas*, ed. Patrick A. Dunae (Edmonton, AB, 1986). See also A. A. Lupton, "Cattle Ranching in Alberta, 1874–1910: Its Evolution and Migration," *Albertan Geographer*, 3 (1966–67), pp. 48–58, suggesting that ranching moved west-to-east across Alberta, rather than emanating from Texas. For Saskatchewan see W. J. Redmond, "The Texas Longhorn on Canadian Range," *Canadian Cattlemen*, 1 (Dec. 1938), pp. 112, 140; and Boyd M. Anderson, *Beyond the Range: A History of the Saskatchewan Stock Growers Association* (Saskatoon, SK, 1988), the latter fairly accurately described by its title, but containing a brief introduction concerning frontier ranching that stresses the Texas connection.

For Texans west of the continental divide, see the special issue of *UTHQ* on ranching, 32 (Summer 1964). Also about Texas cowboys in Utah is James H. Beckstead, *Cowboying: A Tough Job in a Hard Land* (Salt Lake City, UT, 1991). The Texan presence in Arizona is overstated in James A. Wilson, "West Texas Influence on the Early Cattle Industry of Arizona," *SWHQ*, 71 (1967–68), pp. 26–36, and should be read in conjunction with Terry G. Jordan, "Texan Influence in Nineteenth-Century Arizona Cattle Ranching," *JOW*, 14:3 (1975), pp. 15–17, a debunking work based upon manuscript census population schedules. For southern New Mexico, see Jack Parsons and Michael Earney, *Land and Cattle: Conversations with Joe Pankey, a*

New Mexico Rancher (Albuquerque, NM, 1978), a delightful modern study, anecdotal and photographic in emphasis, of a Texas-derived cattleman. The Texan presence in Idaho, while fleeting, is revealed in J. R. Keith, "When the Long-Horned Cattle of Texas Came to Idaho Territory," *Idaho State Historical Society, Biennial Report* 16 (1937–38), pp. 41–49.

On the great Texas cattle drives to the North and the West, begin with Joseph Nimmo, Jr., *Range and Ranch Cattle Traffic* (Washington, DC, 1885), a highly useful contemporary assessment, and J. Marvin Hunter, *The Trail Drivers of Texas* (Nashville, TN, 1925), a remarkable collection of reminiscences assembled while many who had participated were still living. The classic by Andy Adams, *The Log of a Cowboy* (Boston, MA, 1931), is a firsthand account of an 1880s drive of longhorns from South Texas to Montana. Other highly valuable trail diaries are contained in Martin F. Schmitt, ed., *The Cattle Drives of David Shirk: From Texas to the Idaho Mines, 1871 and 1873* (Portland, OR, 1956); James G. Bell, "A Log of the Texas-California Cattle Trail, 1854," *SWHQ*, 35 (1931–32), pp. 208–37, 290–316; 36 (1932–33), pp. 47–67, W. W. Baldwin, "Driving Cattle from Texas to Iowa, 1866," *Annals of Iowa*, 14 (1924), pp. 242–62. See also James H. Cook, "Trailing Texas Long-Horn Cattle through Nebraska," *NBH*, 10 (1927), pp. 339–43; Norbert R. Mahnken, "Early Nebraska Markets for Texas Cattle," *NBH*, 26 (1945), pp. 3–25, 91–103; and Don D. Walker, "Longhorns Come to Utah," *UTHQ*, 30 (1962), pp. 135–47. The best general source on the northern railroad markets of the Great Plains is Robert R. Dykstra, *The Cattle Towns* (New York, 1968), and the finest work on the major routes is Wayne Gard, *The Chisholm Trail* (Norman, OK, 1954), an authoritative study.

Cowboy life in the Texas system is revealed firsthand in Charles A. Siringo, *A Texas Cowboy* (Chicago, IL, 1885), perhaps the best of the reminiscences, and in James H. Cook, *Longhorn Cowboy* (Norman, OK, 1984), in which a Michigander becomes a Texan. See also the classic work by Philip A. Rollins, *The Cowboy: His Characteristics, His Equipment, and His Part in the Development of the West* (New York, NY, 1922). Dane Coolidge, author of *Texas Cowboys* (New York, NY, 1937), was a photographer living among cowboys in Arizona about 1910 who still practiced the old Texas ways in refuge, and his account is a valuable if unscholarly treatment. Far less useful, though beautifully illustrated, is Don Worcester, *The Texas Cowboy, by the Texas Cowboy Artists Association* (Fort Worth, TX, 1986).

Firsthand contemporary accounts describing the Texas system in its heyday abound. Among the best are "Stock-Raising," *Texas Almanac for 1861* (Galveston, TX, 1860), pp. 148–52, an anonymous description of the Texas system prior to the great northward expansion; Frank Wilkeson, "Cattle-Raising on the Plains," *Harper's New Monthly Magazine*, 72 (April 1886), pp. 788–95, a condemnation of the system on the eve of its collapse; Hiram Latham, *Trans-Missouri Stock Raising: The Pasture Lands of North America: Winter Grazing* (Omaha, NE, 1871), a promotional piece but remarkable for its early data; Reginald Aldridge, *Ranch Notes in Kansas, Colorado, the Indian Territory and Northern Texas*, a good account by a small-scale British investor who came to the Great Plains in the late 1870s, rich in detail of ranching techniques; Walter von Richthofen, *Cattle-Raising on the Plains of North America* (Norman, OK, 1964), a promotional tract originally published in 1885 but still containing much useful data; and Joseph G. McCoy, *Historic Sketches of the Cattle Trade of the West and Southwest* (Kansas City, MO, 1874), a self-serving piece by the founder of the Abilene, Kansas, trade but including a wealth of early information; John R. Craig, *Ranching with Lords and Commons, or, Twenty Years on the Range* (Toronto, ON, 1903), which deals with Alberta and the Texas influence on speculative, British-financed ranching; and W. Baillie Grohman, "Cattle Ranches in the Far West," *Fortnightly Review*, n.s. 28 (1880), pp. 438–57, written for potential investors and mistitled but still containing a useful description of Great Plains ranching. For a Texas cowboy in the north, see Michael Kennedy, ed., "Judith Basin Top Hand: Reminiscences of William O. Burnett, an Early Montana Cattleman," *Montana Magazine of History*, 3:2 (1953), pp. 18–23. An Illinoisan taking on Texan ways in Colorado is revealed in Rufus Phillips, "Early Cowboy Life in the Arkansas Valley," *Colorado Magazine*, 7 (1930), pp. 165–79. For the role of African-Americans on the Plains, see Philip C. Durham and Everett L. Jones, *The Negro Cowboys* (New York, NY, 1965), a much overstated but also overdue account, flawed mainly by shallowness of research.

On the environmentally induced collapse of the Texas ranching system, see Robert Fletcher, "That Hard Winter in Montana, 1886–1887," *AH*, 4 (1930), pp. 123–30, in which the author downplays the impact of winter dieoffs; T. Alfred Larson, "The Winter of 1886–87 in Wyoming," *Annals of Wyoming*, 14 (1942), pp. 5–17, which places the herd losses in that state at about 15 percent; Thadis W. Box, "Range Deterioration in West Texas," *SWHQ*, 71 (1967–68), pp. 37–45, which deals with overgrazing; and David L. Wheeler, "The Blizzard of 1886 and Its Effect on the Range Cattle Industry in the Southern Plains," *SWHQ*, 94 (1990–91), pp. 415–32, a well-written account of one of the many winter disasters.

Chapter Eight
Pastoral California Extended

Few writers have even recognized the existence of an Anglo-Californian ranching system, distinct from that of Texas, and no comprehensive study exists of pastoral California Extended. For its Hispanic roots, see sources cited for Alta California in the section for chapter 5. Some of these previously cited items, especially those by Burcham, Broek, Hornbeck, Mora, and Starrs, contain material on the Anglo period. A very useful if all-too-brief statement of the contrasts between the Texas and California ranching systems is Fred B. Kniffen, "The Western Cattle Complex: Notes on Differentiation and Diffusion," *Western Folklore*, 12 (1953), pp. 179–85.

The best book on the subject, though not written for a scholarly audience, is Arnold R. Rojas, *The Vaquero* (Charlotte, NC, and Santa Barbara, CA, 1964), an anecdotal treatment rich in lore gathered earlier in the century, especially on the Tejon Ranch. The major flaw in Rojas's work is his obvious anti-Texas bias. Dane Coolidge's equally popularized *Old California Cowboys* (New York, NY, 1939), though grossly mistitled, since it deals with Arizona and Sonora, has the virtue of being based on firsthand field experience among cowhands who worked in the Sonoran-Californian tradition. Its twin volume, *Arizona Cowboys* (New York, NY, 1938), is in a similar vein. Far more scholarly and one of the best works on Anglo-California, available in two editions differing slightly in title, is Robert G. Cleland, *The Cattle on a Thousand Hills: Southern California, 1850–1880* (San Marino, CA, 1951), but its focus is on the south rather than on the Anglo-dominated north; case studies enliven the book. Also consult the relevant chapters in Hazel A. Pulling, "A History of California's Range-Cattle Industry, 1770–1912," Ph.D. diss., University of Southern California (Los Angeles, CA, 1944), which contains a good treatment of the Anglo-Californian period, including changing range ecology and the retreat of ranching from the California nucleus. Pulling makes abundant use of the 1860 census manuscript schedules.

The Anglo period in California is also the focus of Clara M. Love's "History of the Cattle Industry in the Southwest," *SWHQ*, 19 (1915–16), pp. 370–99 and 20 (1916–17), pp. 1–18. The decline in the aftermath of the great drought of 1862–65 is treated in J. M. Guinn, "The Passing of the Cattle Barons of California," *Historical Society of Southern California, Publications* 8 (1909–10), pp. 51–60. A remarkable, if dry, early work in range ecology that also treats the expansion of ranching into northwestern reaches of the state, is Joseph B. Davy, *Stock Ranges of Northwestern California: Notes on the Grasses and Forage Plants and Range Conditions* (Washington, DC, 1902).

Two biographical or partially biographical works, in particular, are useful in understanding Anglo-Hispanic contact in California. A typical hispanicized Anglo entering the ranching business in Mexican California is the subject of Iris H. Wilson, *William Wolfskill, 1798–1866: Frontier Trapper to California Ranchero* (Glendale, CA, 1965), a well-written and fully documented study. By contrast, Esther B. Black's *Rancho Cucamonga and Doña Merced* (Redlands, CA, 1975) deals with the survival of Hispanic institutions in Anglo-ruled California.

The Anglo cattle kings of California Extended are well exemplified in the superb biography by Giles French, *Cattle Country of Peter French* (Portland, OR, 1964), a beautifully written work containing valuable detail and based upon a good knowledge of the land. Less impressive but still useful is Edward F. Treadwell, *The Cattle King: A Dramatized Biography* (New York, NY, 1931), which deals adoringly with Henry Miller.

Early cattle drives within and from Anglo-ruled California receive brief attention in James M. Jensen, "Cattle Drives from the Ranchos to the Gold Fields of California," *Arizona and the West*, 2 (1960), pp. 341–52, focusing upon the southern part of the state, and Robert M. Denhardt, "Driving Livestock East from California Prior to 1850," *CHSQ*, 20 (1941), pp. 341–47, a movement that included few cattle.

The definitive work on vertical cattle transhumance in the mountain West is Gisbert Rinschede, *Die Wanderviehwirtschaft im gebirgigen Westen der USA und ihre Auswirkungen im Naturraum* (Regensburg, Germany, 1984), a splendid geographical study based upon abundant field research. While its focus is contemporary, it includes the historical perspective.

For the environmental base of California Extended in the intermountain West, see James A. Young and B. Abbott Sparks, *Cattle in the Cold Desert* (Logan, UT, 1985), a sophisticated ecological analysis of ranching in northern Nevada and perhaps the finest interdisciplinary study ever done on the range-cattle industry. Also very useful is Thomas R. Vale, "Presettlement Vegetation in the Sagebrush-Grass Area of the Intermountain West," *Journal of Range Management*, 28, (1975), pp. 32–36, a reconstruction of the pristine ground cover based on early diaries and travel journals that leads the author to suggest that the extent of sagebrush has not markedly increased or grass diminished due to ranching activity. For a study of one of the most important bunchgrasses of the region, see A. L. Lesperance, James A. Young, Richard E. Eckert, Jr., and Raymond A. Evans, "Great Basin Wildrye," *Rangeman's Journal*, 5:4 (1978), pp. 125–27.

Few have considered the cultural imprint of Hispanic Californians in the intermontane region, but see the tantalizing piece by Louie W. Attebery, "Celts and Other Folk in the Regional Livestock Industry," *Idaho Yesterdays*, 28:2 (1984), pp. 20–29.

For Nevada, begin with Howard W. Marshall and Richard E. Ahlborn, *Buckaroos in Paradise: Cowboy Life in Northern Nevada* (Washington, DC, 1980), a brief monograph designed to accompany an exhibit at the National Museum of History and Technology. Oriented to material culture, it also contains an intelligent overview of contemporary ranching in a valley lying in the very heart of California Extended, addressing the perplexing question of Hispanic influence. Velma S. Truett, *On the Hoof in Nevada* (Los Angeles, CA, 1950), is little more than a listing of all livestock brands in the state. See also G. A. Carpenter, M. Clawson, and C. E. Fleming, *Ranch Organization and Operation in Northeastern Nevada* (Carson City, NV, 1941), a contemporary and pragmatic analysis of cattle raising on the border between California Extended and the Mormon culture area. A splendid local history, containing an abundance of information on ranching, is Edna B. Patterson, Louise A.

Ulph, and Victor Goodwin, *Nevada's Northeast Frontier* (Sparks, NV, 1969; reprinted 1991, University of Nevada Press).

For Oregon and Idaho, see the sources listed for chapter 9 and the book by French cited earlier for this chapter. In addition Idaho receives adequate, if uninspired, attention in C. W. Hodgson, *Idaho Range Cattle Industry* (Moscow, ID, 1948), dealing with the contemporary period, and Byron D. Lusk, "Golden Cattle Kingdoms of Idaho," M.S. thesis, Utah State University (Logan, UT, 1978), both acultural and maddeningly factual but the best comprehensive history of the state. Dana Yensen's *A Grazing History of Southwestern Idaho with Emphasis on the Birds of Prey Study Area*, revised ed. (Moscow and Boise, ID, 1982), focuses on the range alteration resulting from a century of ranching, as represented by an area in the Snake River Plain. For more on trans-Cascadian Oregon, see Margaret J. LoPiccolo, "Some Aspects of the Range Cattle Industry of Harney County, Oregon, 1870–1890," M.A. thesis, University of Oregon (Eugene, OR, 1962).

Studies of individual items of Californian-ranching material culture are very rare. We still await analysis of the various windlass devices. On corral construction, see J. Sanford Rikoon, "Traditional Fence Patterns in Owyhee County, Idaho," *Pioneer America Society, Transactions*, 7 (1984), pp. 59–69.

Chapter Nine
The Midwest Triumphant

Few writers on ranching have been sufficiently sensitive to matters of culture to recognize the midwestern influence on the cattle frontiers of the West. The most notable exceptions are Peter K. Simpson, *The Community of Cattlemen: A Social History of the Cattle Industry in Southeastern Oregon, 1869–1912* (Moscow, ID, 1987), a sophisticated, interdisciplinary study of Harney County that is a *big* cut above typical local history, and the venerable book by Ernest S. Osgood, *The Day of the Cattleman* (Chicago, IL, 1929), which deals mainly with Wyoming and Montana ranching in a scholarly, insightful manner.

Biographical studies of midwestern cattlemen on the Plains remain uncommon, but see C. W. McCampbell, "W. E. Campbell, Pioneer Kansas Livestockman," *KHQ*, 16 (1948), pp. 245–73; Walker D. Wyman, *Nothing but Prairie and Sky: Life on the Dakota Range in the Early Days* (Norman, OK, 1988); and Robert H. Burns, "The Newman Brothers: Forgotten Cattle Kings of the Northern Plains," *Montana*, 11 (Oct. 1961), pp. 28–36. The improvement of breeds, a basic midwestern concern on the Plains, is concisely treated in Lauren C. Post, "The Upgrading of Beef Cattle on the Great Plains," *California Geographer*, 2 (1961), pp. 23–30, and in T. R. Havins, "The Passing of the Longhorn," *SWHQ*, 56 (1952–53), pp. 51–58. Cattle raisers' associations, whose rise reflects Ohio Valley influences, appeared widely on the Plains and have been abundantly studied. See William W.

Savage, Jr., *The Cherokee Strip Live Stock Association* (Columbia, MO, 1973); and W. Turrentine Jackson, "The Wyoming Stock Growers' Association: Political Power in Wyoming Territory," *MVHR*, 33 (1947), pp. 571–94. For a misguided effort to link these associations genetically to the Spanish mesta instead of the American Midwest, see William Dusenberry, "Constitutions of Early and Modern Stock Growers' Associations," *SWHQ*, 53 (1949–50), pp. 255–75.

Midwestern entry into Oregon is revealed in William A. Bowen, *The Willamette Valley: Migration and Settlement on the Oregon Frontier* (Seattle, WA, 1978), which, however, demonstrates that this important valley never housed a notable cattle-ranching industry. J. Orin Oliphant's magnus opus, *On the Cattle Ranges of the Oregon Country* (Seattle, WA, 1968), covers the "Inland Empire" and is rich in facts and bibliography, to the point of being the definitive study, but flawed stylistically and essentially acultural in content. T. V. Boyce's "A History of the Beef Cattle Industry in the Inland Empire," M.A. thesis, Washington State University (Pullman, WA, 1937), offers little that is not in Oliphant's book. For midwestern cattle imports, see C. S. Kingston, "Introduction of Cattle into the Pacific Northwest," *Washington Historical Quarterly*, 14 (1923), pp. 163–85.

For the Columbia Basin, begin with Donald W. Meinig's massive work, *The Great Columbia Plain: A Historical Geography, 1805–1910* (Seattle, WA, 1968), from which a factual overview of the spread and decline of cattle ranching can be gleaned. The environmental base is concisely revealed in Otis W. Freeman, J. D. Forrester, and R. L. Lupher, "Physiographic Divisions of the Columbian Intermontane Province," *AAAG*, 35 (1945), pp. 53–75; and Rexford F. Daubenmire, "An Ecological Study of the Vegetation of Southeastern Washington and Adjacent Idaho," *Ecological Monographs*, 12 (1942), pp. 53–79. Roscoe Sheller's *Ben Snipes, Northwest Cattle King* (Portland, OR, 1957), is a badly dramatized biography of an Iowan who became a big rancher in the Yakima Valley. More useful are the grotesquely mistitled reminiscences of early Columbia Basin ranchers A. J. Splawn, *Ka-Mi-Akin: The Last Hero of the Yakimas* (Portland, OR, 1917), another Yakima Valley settler, and Daniel M. Drumheller, *Uncle Dan Drumheller Tells Thrills of Western Trails in 1854* (Spokane, WA, 1925). See also Gretta Gossett, "Stock Grazing in Washington's Nile Valley: Receding Ranges in the Cascades," *Pacific Northwest Quarterly*, 55 (1964), pp. 118–27.

For midwesterners in California, see the chapters on the Anglo-American era in the Pulling dissertation, cited in chapter 8, as well as the biographies by French and Wilson. The best account of a trail drive from the heartland is Cyrus C. Loveland, *California Trail Herd: The 1850 Missouri-to-California Journal of Cyrus C. Loveland*, ed. Richard H. Dillon (Los Gatos, CA, 1961).

Additional sources on Anglo-American ranchers in southern Idaho include the excellent local history by Adelaide Hawes, *The Valley of Tall Grass* (Bruneau, ID, 1950) and the useful companion book, about the same county,

Mildretta Adams' *Owyhee Cattlemen, 1878–1978: 100 Years in the Saddle,* 2nd ed. (Homedale, ID, 1979). See also the special ranching issue of *Idaho Yesterdays,* 28:2 (1984).

As was true of the Great Plains, ranching on the Canadian side of the intermountain region has been studied far more thoroughly than the area south of the border. The definitive study is Thomas R. Weir, *Ranching in the Southern Interior Plateau of British Columbia,* revised ed., (Ottawa, ON, 1964), a historical and geographical study based in intimate knowledge of the region. Less impressive is Gregory E. G. Thomas, "The British Columbia Ranching Frontier, 1858–1896," M.A. thesis, University of British Columbia (Vancouver, BC, 1976), which suffers from an acultural, ecologically naive methodology. A very useful, if disjointed, firsthand account, written by the son of an early rancher and depicting the industry in the early twentieth century, is T. Alex Bulman's *Kamloops Cattlemen: One Hundred Years of Trail Dust* (Sidney, BC, 1972). Henry Marriott, who worked on the early Gang Ranch and later became a rancher in his own right, wrote the interesting and valuable *Cariboo Cowboy* (Sidney, BC, 1966). See, in addition, the anecdotal, popularized book by Ed Gould, *Ranching in Western Canada* (Saanichton, BC, and Seattle, WA, 1978), which covers both British Columbia and Alberta.

The floral basis of ranching in interior British Columbia is described in E. W. Tisdale, "The Grasslands of the Southern Interior of British Columbia," *Ecology,* 28 (1947), pp. 346–82. Also useful are the biographical sketches in F. W. Laing, "Some Pioneers of the Cattle Industry," *British Columbia Historical Quarterly,* 6 (1942), pp. 257–75.

For the Rocky Mountains, begin with the overview of the very complicated environmental pattern by Rexford F. Daubenmire, "Vegetational Zonation in the Rocky Mountains, *Botanical Review,* 9 (1943), pp. 325–93. Then refer to the excellent German study of western transhumance by Rinschede, cited for chapter 8. Jon T. Kilpinen, in "Material Folk Culture in the Adaptive Strategy of the Rocky Mountain Valley Ranching Frontier," M.A. thesis, University of Texas (Austin, TX, 1990), uses the relict built environment as an index to decipher the modified midwestern ranching system that developed in the high valleys.

For the Colorado Rockies, see the book by Everett cited for chapter 7, as well as John M. Crowley, "Ranching in the Mountain Parks of Colorado," *GR,* 65 (1975), pp. 445–60, a useful abstract of his dissertation, which was also listed for chapter 7. An interesting reminiscence about Gunnison County, west of the Continental Divide, is H. C. Cornwall, "Ranching on Ohio Creek, 1881–1886," *Colorado Magazine,* 32 (1955), pp. 16–27.

For Wyoming see Robert H. Burns, Andrew S. Gillespie, and Willing G. Richardson, *Wyoming's Pioneer Ranches* (Laramie, WY, 1955), which contains a wealth of information on the local level that facilitates the analysis of culture history, and John K. Rollinson, *Wyoming Cattle Trails: History of the Migration of Oregon-Raised Herds to Mid-Western Markets* (Caldwell, ID, 1948), which documents the influence of the Pacific Northwest upon montane Wyoming, including original trail diaries. A useful local-level study is Glen Barrett, "Stock Raising in the Shirley Basin, Wyoming," *JOW*, 14:3 (1975), pp. 18–24.

The Montana Rockies, where the mountain cattle industry began, and the adjacent foothill zone receive abundant attention in Robert Fletcher, *Free Grass to Fences: The Montana Range Cattle Story* (New York, NY, 1960), the best work on the subject, and in Michael S. Kennedy, ed., *Cowboys and Cattlemen: A Roundup from Montana* (New York, NY, 1964), a collection of brief articles that appeared in the journal *Montana*. Much valuable information concerning postfrontier range conditions and operating methods is available in Mont H. Saunderson and D. W. Chittenden, *Cattle Ranching in Montana* (Bozeman, MT, 1937), a work sensitive to the environmental contrasts within the state. The evolution of one element of haying material culture, a crucially important development centered in western Montana, is the subject of John A. Alwin's article, "Montana's Beaverslide Hay Stacker," *Journal of Cultural Geography*, 3:2 (1982), pp. 42–50.

INDEX

Aberdeen-Angus, 16, 46
Africa, 14, 15; emigration from, 64; as hearth area, 310–12; herding in, 55–64; maps of, 16, 56; plants from, 71, 84; slave trade from, 67, 85
African-Americans. *See* Africans
Africans: as herders, 58–64, 83, 92–3, 112–13, 117, 132, 174, 180, 188–9, 213–15, 220, 311; influence of, 82–5, 93, 112–13, 311–12; in Mexico, 92–3, 132; as ranchers, 188–9; in South, 112–13, 117, 121, 177, 180, 188; in Spain, 34; in Texas, 188–9, 213, 215, 220–1; in West, 215; in West Indies, 66–7, 72, 83, 85
Agriculture: British, 45, 48; in California, 163; Mexican, 99–100; in New Mexico, 146; in South, 116; Spanish, 29–30, 40, 41. *See also* Corn Belt; plantations.
Alabama, 108, 120, 121, 178, 179, 189, 191
Alberta, 221, 226–7, 238, 300, 301, 311
Amerindians: agencies of, 221–2; Apache, 143, 254; Arawak, 66, 68; Aztec, 126; and brands, 94; in California, 159, 162, 165, 166; Cherokee, 183, 190, 191, 225; Chichimecan, 126–8, 130, 132–3, 135, 156; Chickasaw, 199; Choctaw, 183; as cowboys, 73, 93, 107, 112, 117, 132, 141, 167, 180, 234, 256, 289, 294, 311; Creek, 107, 108, 178, 183, 225; demography of, 70, 88, 98, 126; firing by, 70; in Great Plains, 222; Huastecan, 87, 88, 136; Karankawa, 156; in Mexico, 87–8, 93, 98, 126–35, 140–1; Navajo, 147, 228, 298; in New Mexico, 146; in Oklahoma, 225; Paiute, 281; Pima, 141, 143; Pueblo, 146; raiding by, 110, 121, 127, 130, 143–4, 148, 171, 254, 290; as ranchers, 98, 107, 108, 127, 142, 176, 182–3, 289; rustling by, 9, 128, 148; Seminole, 108, 178, 183; as sheep raisers, 147; in South, 106, 108, 110; in Texas, 156; Ute, 147, 228; in West Indies, 66, 68, 70, 72; Yamassee, 110, 171
Andalucía, 15; cattle from, 67; as hearth area, 18–35, 85, 309–11, 312; maps of, 16, 19; as source of emigrants, 21, 35, 65, 66, 67, 91, 104
Anglo-Saxons: in Britain, 42, 50; in California, 241–9; in Midwest, 201–5; in South, 110–13, 170–201; in Texas, 205–7, 214, 269; in West, 214–15, 220–1, 234, 245, 249–66, 267–307; in West Indies, 78–84, 109–10; *see also* Crackers
Antilles. *See* West Indies
Appalachians, 190, 192–9, 205, 311
Arabs, 18, 71; in Africa, 58, 59, 60,

429

Arabs (*cont.*)
　61, 63; as herders, 34, 58, 59; in Spain, 34, 312
Architecture. *See* House types
Arizona: cattle from, 162; cowboys in, 214–5; drought in, 238; maps of, 209, 242; ranching in, 142, 143, 228, 230, 238, 254–5, 297; Spaniards in, 142, 143, 144
Arkansas, 190, 203, 204
Associations, stockraiser, 201, 212, 267, 274–5, 313. *See also* Mesta
Attebery, Louie W., 4
Australia, 49, 118, 194

Bahamas, 119
Baja California, 124, 144, 162, 165
Bajío, El, 123–8
Baltimore, MD, 201
Barbados, 78, 79, 80, 109–10, 119
Barbecue, 78, 112
Barns, 293, 296, 306, 313
Barrens, 197, 199. *See also* Pine barrens
Beef, as food: in Africa, 64; in British Isles, 51, 310; in California, 162, 247, 265; in Canada, 293; in Mexico, 97, 140; in South, 106, 119, 177; in Spain, 34, 41, 310; in Texas, 187, 213, 218; in West Indies, 70; *see also* Jerky; trade in: 51–2, 119, 191; *see also* Markets
Berbers, 15, 18, 58–61, 63, 67
Bermuda, 110
Bishko, C. Julian, 37, 42
Bits, 168
Blacks. *See* Africans
Blood, as food, 50–51, 64, 310
Boots, 214–15, 257, 312, 313
Branding: in Africa, 61, 310; in British Isles, 50, 310; in California, 168; illustrated, 28; in Mexico, 94, 103, 126, 130, 142; in South, 118, 180, 186, 188, 196; in Spain, 26,
41, 310; in Texas, 206, 211, 212, 213, 217; in West, 234, 255, 259; in West Indies, 72, 80, 81
British Colombia: cattle sent to, 243, 245, 247, 249; map of, 242; ranching in, 248, 250, 254, 257, 259–60, 291–6, 300; section on, 291–5; spread from, 300, 301
British Isles: emigrants from, 44, 45, 54, 80, 85, 113, 115, 193; as hearth area, 310, 311; map of, 43; section on, 42–55; topography of, 42–43
Buccaneers, 77–78, 79, 82, 84, 85
Bulldogging, 119

Cajuns. *See* French
California, 312; cattle sent from, 243, 245, 299; cattle sent to, 143, 144, 205, 218, 228, 279–80; drought in, 248, 265; environment of, 159–61; lasso in, 94; maps of, 160, 164, 242, 244, 252, 276; Mexican rule of, 165, 241, 243, 245–6, 279, 311; sections on, 159–69, 241–66, 279–82; Spaniards in, 159–69, 243, 245; spread from, 242, 243, 245–6, 247–55, 313; Trail, 280, 283. *See also* Baja California
Canada, 232, 314. *See also* Alberta, British Columbia, Saskatchewan
Canary Islands, 67
Canebrakes, 69, 88, 105, 114, 170, 171, 178, 185, 192, 198, 204, 210. *See also* Glades
Castilla (Castile), 19. *See also* Andalucía
Castration: in Africa, 63; in California, 168; in Mexico, 93, 99, 100, 137; in Midwest, 202; in South, 118, 180, 194, 196; in Spain, 23, 39, 41, 42; in Texas, 149, 156, 187, 213, 217; in West, 232, 234, 295; in West Indies, 73, 80, 81

Cattle: in Africa, 58–64; black, 46, 109, 203, 243; breeds of, 29, 30, 37, 46, 59, 67, 78, 108, 110, 184, 187, 194, 201, 205, 206, 231–2, 247, 264, 266, 267, 269, 270, 274, 280, 283, 310; in British Isles, 45–54; in California, 162–9, 243–9; diseases of, 171, 191; domestication of, 14; fattening of, 52, 194, 197–202, 221, 269; feral, 76–7, 79, 139, 143; in Great Basin, 230, 249–66, 282–96; in Great Plains, 221–27, 231–40; introduction of, 67; longhorn, 29, 30, 46, 67, 108, 120, 153, 162, 184, 202, 210, 229, 230–2, 270, 274; in Mexico, 86–104, 123–45; in Midwest, 199–207; in Rocky Mountains, 227–8, 230–1, 298–307; shorthorn, 46, 270, 274, 275, 283, 306–7; in South, 106–22, 170–201; in Spain, 18–42; in Texas, 148–56, 208–24, 269–70; veneration of, 14, 59–60; in West, 227–31; in West Indies, 70–85

Celts, 15, 37, 42, 44–6, 113. *See also* Irish, Scotch-Irish, Scots, Welsh

Charleston, SC, 110, 111, 119, 177, 189

Charro, 40, 100–2, 128, 132, 167, 243

Chicago, IL, 221

Chile, 94

Chisholm Trail, 222

Chisum, John, 220, 223

Climate, 9; of Africa, 55–6; of British Isles, 44; of California, 159, 248, 265; of Great Basin, 250–1, 288–9; of Great Plains, 222–3, 237–8; of Iberia, 20, 22, 36; of Mexico, 87–8, 97, 125, 128–9, 133, 138–9; of South, 105, 113–4, 192; of Texas, 205; of West Indies, 67–8

Clothing, 26, 100–1, 132, 167, 211, 212, 243, 257–8

Colombia, 74

Colorado, 209, 214–5, 225, 230–1, 237, 271, 272, 301–2

Córdoba, 19, 34

Core/periphery models, 11–15

Corn Belt, 198, 202, 205, 221, 269

Corrals: in Africa, 61, 62; in British Isles, 48; in California, 168; illustrations of, 101, 154, 183; in Mexico, 100, 101, 137, 142; in South, 107, 116, 118, 119, 177, 182; in Spain, 25, 39, 40; in Texas, 149, 151, 152, 154, 183, 206; in West, 234, 260, 265, 267, 275, 286, 288; in West Indies, 72, 76–77, 81, 82

Cortés, Herman, 37, 100

Costa Rica, 68, 139, 256

Cowboys: in Africa, 61–64; in British Isles, 50; in California, 167, 168, 247, 265; in Canada, 294; in Great Basin, 255–6, 265; in Great Plains, 233–4, 272; in Mexico, 92, 93, 94, 100–102, 127, 132; in South, 107, 117; in Southwest, 214–15, 297; in Spain, 23, 25, 26, 34, 39–40; in Texas, 156, 211, 214–15, 217–18, 220, 269; traits of, 8–9; in West Indies, 73, 82; as word, 50, 180, 213, 232

Cowhunts. *See* Roundups

Cowpens, 112, 115–6, 170–1, 173–8, 180–1, 191, 195; Battle of the, 191. *See also* Corrals

Crackers, 117, 170, 178, 179, 182

Cuba, 65–6, 68, 70, 72, 78, 88, 106, 120

Dairying. *See* Milking complex

Dally roping, 167, 262

Dary, David, 95

Devon: breed, 201, 274, 275, 280; province, 16, 113, 182

Dogs, 55, 312; in British Isles, 49, 52, 310; in Canada, 296; illustrations of, 83, 181; map of usage, 16; in Midwest, 202; in South, 119, 121–2, 180, 193, 194; in Texas, 181, 206, 213, 269; in West, 232; in West Indies, 70, 77, 79, 82, 83, 85
Doolittle, William E., 4, 20, 27, 90, 153
Drives, cattle: in Britain, 44, 49, 50, 52–53, 310; to California, 205, 218, 228, 280; in Canada, 295; in Mexico, 102, 103, 151; in Midwest, 201, 221; to northern markets, 191–2, 194, 196, 201, 206, 221–2, 225; in South, 157, 180, 187, 191–2, 204; in Spain, 23, 25, 310; in Texas, 213, 218; in West, 221–2, 245, 301, 307. *See also* Transhumance, Trails
Drumheller, Dan, 290, 292
Dunbar, Gary, 171
Durham breed, 16, 201, 206, 274, 275, 280, 283, 291, 300

Earmarking: in Africa, 61, 310; in British Isles, 50, 310; in Mexico, 94, 103; in South, 118; in Spain, 29, 310; in Texas, 212; in West Indies, 72, 81
Egypt, 14, 15
Encomienda, 73, 92, 107
England. *See* Anglo-Saxons, British Isles, Devon, Somerset
Escandón, José de, 135–6, 150, 151, 153, 155
Española, 65; cattle from, 88; dogs in, 70; labor in, 73; population of, 72; ranches in, 72–8; saddles in, 27, 94; settlement of, 65–67; vegetation of, 69, 71. *See also* Haiti, Vega Real
Estates, pastoral: in Mexico, 96, 127, 128, 132, 137, 140, 143, 312; in Spain, 30, 33; in Texas, 150; in West Indies, 74–6, 312. *See also* Land grants, Land tenure
Etymology, 312; in Britain, 47; in South, 108, 112, 180–2, 187, 197; in Spain, 30–3, 43; in Texas, 153, 155, 313; in West, 231–3, 256–7, 298, 313; in West Indies, 68, 75–6, 78, 80, 83–4
Europe, 15, 16, 18, 55. *See also* British Isles, France, Greece, Iberia, Ireland, Scotland, Spain, Wales
Evans, Simon, 226, 227
Extremadura, 15, 19; emigrants from, 37, 41, 100, 104; as hearth area, 35, 42, 74, 85, 99, 309–10; maps of, 16, 19

Fencing: in British Isles, 48–49; in Great Plains, 274–5; illustrations of, 183; in Mexico, 100–102, 137; in Midwest, 313; in South, 116, 117, 175, 182, 183; in Spain, 39; in Texas, 183, 206, 275; in West, 234, 251, 265, 267, 286–8, 295, 296, 302, 303, 305–7. *See also* Corrals
Fire, use of, 10, 11; in Africa, 57, 310; in Britain, 44, 47, 310; in Great Plains, 223, 273; in Mexico, 96, 139; in South, 115, 157, 180, 185, 193; in Spain, 25, 41, 310; in Texas, 206, 213; in West Indies, 70, 71, 78–80
Florida: breeds in, 108, 184; cattle from, 110, 120; culture of, 184, English in, 108, 178, 179, 180; map of, 66; Spaniards in, 66, 105–8, 178, 184, 309
France, 16, 20, 25, 54–5, 84, 298
Franciscans, 106, 143–4, 147–8, 151, 153, 161–2
French: Cajun, 188, 189; as cattle raisers, 54–5, 84–5, 120–2, 176;

432   INDEX

in Haiti, 84; in Midwest, 120, 202–4; in South, 120–22, 176, 178, 181, 183, 188, 203–4, 210; in Texas, 188, 189
French, Peter, 253, 255, 261, 282, 285
Frontier: in California, 159–69, 241–9, 279–82; in Canada, 226–7, 291–5; defined, 7; in Great Basin, 249–55, 282–91; in Great Plains, 221–7, 269–75; in Mexico, 86–104, 123–45; in Midwest, 201–7; in mountain West, 227–31, 298–307; in Pacific Northwest, 275–9, 282–91; in South, 105–22, 170–201; in Southwest, 145–7, 228–30, 296–8; in Texas, 147–58, 179, 186–9, 205–7, 208–21; in West Indies, 65–85
Fulani, 58–64, 67, 85, 113, 310

Georgia, 108, 173–7, 179–80, 191
Germans: Goths, 15, 18; Pennsylvania-, 194, 197, 198, 200; Russian-, 239; Texas-, 219
Glades, 192, 197, 199, 205, 294; The, 195–6, 198, 201. *See also* Canebrakes
Glenn, Hugh, 253, 280, 282
Goats, 41, 45, 47, 59–60, 70, 74, 96, 99, 100, 131, 148, 151, 297
Goodnight, Charles, 220, 223, 270
Grasslands: in Africa, 57; in California, 160–61; ecology of, 10–11, in Great Basin, 251, 288–9, 292; in Great Plains, 222–3; maps of, 160, 172, 190, 209; in Mexico, 88, 99, 125, 129, 133; midlatitude, 133; in Midwest, 201–2, 204; in South, 181, 184, 193; in Spain, 20–22; in Texas, 150, 184, 205–6, 215, 220; in West Indies, 68–70. *See also* Barrens, Glades, Marshes, Savannas

Great Basin, 230, 249–57, 259–61, 263, 282–97
Great Plains, 209, 221–7, 231–9, 269–75, 301, 311
Guadalajara, Mexico, 123, 126, 138, 140
Guinea, 55, 56

Hackamore, 256, 262
Hardin, James, 254
Harper, Jerome, 250, 294, 295
Haiti, 70, 83–5, 120, 121, 311
Hats, 26, 100, 101, 132, 167, 211
Haymaking: in British Isles, 45, 47, 49, 310; in Mexico, 130; in Midwest, 203, 205, 313; in West, 234, 265–6, 272–4, 281, 285–6, 287, 295, 302–5, 313
Henlein, Paul, 267
Herefords, 46, 201, 270, 274, 283
Hides, 7, 9; in Africa, 64; in British Isles, 51, 52; in California, 167, 243, 247; curing of, 156, 167; in Mexico, 96, 137; in South, 106, 107, 119, 120, 177, 188; in Spain, 34; in Texas, 156, 213, 218; in West Indies, 73, 74, 77, 79
Hispanic-Americans: in Arizona, 254–5; in California, 241–3, 246–7, 262, 313; as cowboys, 212, 215, 218, 243, 255–6, 265, 285, 313; in Great Plains, 212, 215; influence of, 255–65, 311; in New Mexico, 231; as ranchers, 254–5; as sheepherders, 219; in Texas, 212, 215, 218; in West, 215, 285. *See also* Spaniards
Hispaniola. *See* Española
Hittson, John, 220, 225
Honduras, 66
Hooker, H. C., 228, 298
Horses, 8, 312; in Africa, 59, 63; breaking, 262; in British Isles, 45, 49, 52; in California, 163, 166,

Horses (*cont.*)
245; in Canada, 301; feral, 155; map of usage, 16; in Mexico, 93, 102, 137, 145; in Midwest, 200, 202; in South, 107, 108, 118, 119, 121, 180, 184, 185, 187, 188, 196; in Spain, 25, 29; in Texas, 155, 156, 187, 211, 218; theft of, 191; in West, 232, 262, 297, 307; in West Indies, 70, 71, 79, 84. *See also* Bits, Saddles

House types: in Mexico, 87, 129, 136–7, 151; in South, 116; in Spain, 23; in Texas, 151, 218, 275; in West, 235, 287, 306

Hudson's Bay Company, 245, 289, 291, 292, 294

Iberia, 15, 16, 18–42. *See also* Spain

Idaho: cattle sent to, 247, 249–50; cowboys in, 214–15, 221, 311; map of, 242; Mormons in, 296; ranching in, 230, 251, 282–8, 290, 299, 313

Ikard, W. S., 270

Iliff, John W., 271

Illinois, 120, 203, 204, 218, 271, 282, 283, 297, 301, 313

Indiana, 202, 203, 204, 313

Indians. *See* Amerindians

Iowa, 205, 207, 218, 271, 297, 298, 300, 313

Ireland, 43–5, 48, 50–2, 54

Irish, 15, 42, 47, 51, 54; Maol, 46; in North America, 113, 283, 296; in West Indies, 80

Irrigation, 265–6, 273, 281, 286, 291, 302–4, 313

Jamaica, 65; British in, 78–84, 311; exports from, 74; feral cattle in, 77; imports to, 119; influence of, 109–12, 114–6, 118; land grants in, 75; maps of, 66, 81; population of, 72; ranches in, 72, 73, 78–84; saddles in, 94; Spaniards in, 66–75, 78, 79, 309; vegetation of, 69

Jerky, 74, 78, 85, 107, 112, 140, 168, 187–8

Jesuits, 140–3, 144

Kansas, 209, 214–15, 221–2, 225, 237, 270–3

Kentucky, 108, 196–201, 204, 206, 270–1, 279, 299, 300

King Ranch, 150, 155, 213

Kino, Eusebio Francisco, 141–2

Kniffen, Fred, 254

Lance, cattle, 25–6, 73, 77, 79, 93–5, 107, 132, 143, 167, 169, 186, 233. *See also* Staff

Land grants: in California, 164, 165–6, 241–2, 265; in Louisiana, 152, 157, 188; maps of, 91, 152, 164; in Mexico, 91, 127, 135, 143; in Missouri, 203; in Texas, 149, 150, 152, 157; in West Indies, 75. *See also* Land tenure

Land tenure, 30–4, 36, 42, 45–6, 50, 54, 74–6, 82, 96, 104, 127, 132, 166. *See also* Estates; Land grants

Lariat. *See* Lasso

Lasso, 107, 310; in Africa, 63–4; in California, 167, 262; illustrated, 95, 211; map of usage, 16, 152; in Mexico, 93–5, 102, 132, 143, 312; in South, 152, 157, 185–6, 188; in Spain, 25, 41, 310; in Texas, 149, 152, 187, 210–12, 217–18; in West, 232, 246; in West Indies, 27, 73, 85. *See also* Dally roping

Laws: in Anglo-America, 113, 116–7, 118, 180, 191, 213, 251, 302; in Britain, 44, 50, 310; in Mexico, 103–4; in Spain, 26, 33, 34

Leather, 34, 52, 64, 73

434 INDEX

León, Martín de, 150, 155–6, 158, 210
Line riding, 186, 187, 213, 235, 313
Longhorns. *See* Cattle, longhorn
Los Angeles, CA, 161, 163, 165, 166, 262
Louisiana: Anglos in, 178, 179, 180, 187; cattle from, 203; emigration from, 189; French in, 120–22, 156, 188, 210; as hearth area, 187–88, 208, 211; herding in, 120–22, 156–158, 188, 309; land grants in, 152, 157; maps of, 66, 152; sections on, 120–22, 156–58; Spaniards in, 121, 156–58, 184
Loving, Oliver, 220, 223
Lux, Charles, 252–5, 281

McCoy, Joseph, 271
Marismas, Las, 19–26, 28–35, 309
Markets: in Africa, 64; in British Isles, 51–2, 54; in California, 167, 246–7, 254, 280; in Canada, 231, 292, 294–5; influence of, 11–2, 310; in Mexico, 96–7, 126–27, 138, 142; in North, 192, 194, 197, 218, 221–2, 227; in South, 119–20, 157, 177, 178, 197, 218; in Spain, 25, 34; in West, 222, 239, 247, 251, 283, 287, 299; in West Indies, 73, 74, 79, 178
Marshes: in British Isles, 47; in California, 161, 281; in Great Basin, 251, 292; maps of, 19, 21, 24, 32, 90; in Mexico, 88, 129, 139; salt, 20–22, 69, 114, 139, 150, 185, 189; in South, 114, 172, 185; in Spain, 20–22; in Texas, 150, 185, 189; tule, 161, 251, 265, 281; in West Indies, 69. *See also* Glades; Marismas
Maryland, 195, 197
Meinig, D. W., 245, 255

Mesta, 33, 39, 103–4, 143
Mexico: cattle from, 120; charros in, 100–2; coastal plain of, 86–97, 133–8, 309; environment of, 86–8, 123–5, 133, 138–9; influence of, 184; maps of, 66, 87, 90–1, 124, 134; markets in, 74, 96–7, 126–7, 129–30; Pacific coastal, 138–45; Plateau of, 97–104, 124, 128–32, 309; ranching hearths in, 66, 88–9, 97–8, 123–8; sections on, 86–104, 123–45; Texans in, 238. *See also* Baja California; El Bajío; Guadalajara; Sonora; Tamaulipas; Veracruz
Midwest: cattle feeding in, 200–205, 218; drives to, 218–9; herding in, 199–205, 207; influence of, 205–7, 267–307, 313
Milking complex: in Africa, 64, 310; in Barbados, 109; in British Isles, 49, 51, 310; in South, 117, 118, 177, 180, 186, 194, 196; in Spain, 41, 310; in Texas, 213; in West, 280, 307
Miller, Henry, 252–5, 281
Mining: in Canada, 231, 292–4; in Mexico, 126–7, 129, 130, 142; in West, 222, 228, 230–1, 246–7, 287, 299–300
Miscegenation, 311; in California, 167; in Mexico, 92, 93, 127, 135, 136, 137, 163, 311; in South, 107, 117, 180, 188, 311; in Spain, 311; in Texas, 189, 311; in West, 234; in West Indies, 73, 311
Missions: in California, 161–6, 312; in Florida, 106–7, 312; in Mexico, 127–8, 132, 136, 140–3, 144, 145, 312; in Pacific Northwest, 289, 291; in Texas, 147–9, 151, 158
Mississippi, 120–1, 178–80, 184, 189

Missouri: Anglos in, 204–5; cattle from, 200, 271, 299; drives to, 218, 221; emigration from, 206, 270–2, 279, 280, 282, 283, 286, 290, 297, 300, 301; French in, 203, 204; influence of, 307

Mobile, AL, 120, 121

Montana: cattle sent to, 249–50, 264, 301; cowboys in, 214–15; dieoffs in, 238; fences in, 303, 305, 307; haying in, 304–5; maps of, 209, 284; markets in, 222, 231; ranching in, 226, 227, 238, 254, 272, 299–300, 306–7, 314; saddles in, 257; terrain of, 298

Moors. *See* Arabs

Mormons, 229–30, 231, 245, 251, 254, 296–7

Mules, 137, 155, 166, 188, 218, 245

Native Americans. *See* Amerindians

Nebraska, 209, 214–5, 221, 225–6, 237, 239, 271, 301

Nevada, 230, 242, 247–9, 251–6, 282–6, 300

New Mexico, 144–7, 209, 214–5, 221–3, 228–9, 231, 264

New Orleans, LA, 138, 155–7, 187, 218, 221

New York, NY, 110, 192, 201, 218

Nomadism, 47, 60–61, 310

North Carolina, 171, 173, 179, 182, 191, 196, 197

Ohio, 199–200, 271, 300, 301

Oklahoma, 108, 183, 209, 221, 225, 239, 271

Open Range. *See* Range, open

Oregon: beef wheels in, 259–61; cattle sent to, 205, 243, 245, 247–9, 275, 277; cowboys in, 214–5, 255–6; maps of, 242, 252, 276, 278, 284; ranching in, 251–4, 265, 275–9, 282–8, 290; saddles in, 257; spread from, 254, 264, 298–301; Trail, 268, 270–1, 275, 277, 280, 283, 289, 297, 299–301, 307. *See also* Willamette Valley

Osgood, Ernest, 286

Overgrazing, 10, 52; in California, 248, 265; in Mexico, 128, 130; in Texas, 217, 220; in West, 239; in West Indies, 71

Ozarks, 204–5

Panama, 66, 74

Pánuco: delta, 87–90, 92, 93, 104, 133, 135; town, 96; river, 134

Parsons, James J., 4

Pens. *See* Corrals

Philadelphia, PA, 192, 194, 201, 218

Pickett, Bill, 119, 221

Pigs. *See* Swine

Pilgrim cattle, 271, 274, 275, 283, 291, 299

Pine barrens, 105, 114, 170–84

Plantations, 76, 79, 81, 116, 120, 180, 312

Portugal, 19, 20

Prairies. *See* Grasslands

Predators, 9, 22, 48, 61, 70, 115, 143, 148, 238

Puerto Rico, 65, 66, 68, 69, 72, 75

Ranching: in California, 161–9, 228, 241–9, 252, 256–65; in Canada, 226–7, 291–6; defined, 7–8; ecology of, 9–11; economics of, 11–3; in Great Basin, 230, 249–57, 259–61, 263, 282–96; in Great Plains, 221–7, 231–9, 269–75; in Mexico, 86–104, 123–45; in Midwest, 199–205; Old World origin of, 14–64; in Rockies, 227–8, 230–1, 298–307; in South, 106–22, 170–201; in Southwest, 228, 296–8; in Texas, 148–56, 186–9, 205–7, 208–40, 269–70; in West,

227–31, 249–66, 267–307; in West Indies, 72–85
Range: carrying capacity of, 57, 114, 129, 138, 170, 239; defined, 9; ecology, 10–11, 20–2, 47, 57, 71, 99, 103, 114–5, 128, 138, 171, 191, 217; leasing of, 236, 238; open, 7–8, 23, 44, 48, 100, 107, 116–7, 180, 195, 196–7, 217, 230
Road ranches, 268, 270, 283, 289, 291, 292–3, 297, 299, 301
Rocky Mountains, 227–8, 230–1, 298–307
Rodeo, 297. *See also* Roundups
Ropes, 41, 79; fiber, 93, 153, 210, 232, 256, 262, 313; rawhide, 41, 102, 167, 262, 313. *See also* Dally; Lasso
Roundups: in British Isles, 49, 50; in California, 168; in Mexico, 94, 100, 142; in South, 107, 118, 121, 175, 180, 186, 195–6; in Spain, 25; in Texas, 149, 156, 206, 213, 217; in West, 234, 260, 265, 272, 275; in West Indies, 72–3
Rustling: in Africa, 61; in British Isles, 45, 49; in California, 246; in Mexico, 92, 128; in South, 108, 191

Saddles, 312; in California, 167, 257–8, 313; illustrations of, 27, 211, 258; in Mexico, 93–5; in South, 188; in Texas, 187, 188, 211, 212, 258, 312, 313; Visalia, 257, 313; in West, 257; in West Indies, 27
Sahel, 16, 55–9
Salamanca: charro culture in, 40, 101; emigrants from, 104; as hearth area, 99, 100, 167; herding in, 35, 36, 37, 40; Mexico, 101, 126
Salt: in food preservation, 51, 119, 188; in herd management, 23, 119, 142–3, 180, 193–4, 202–3, 206, 214, 235, 269, 307; marshes, 20–2, 69, 185, 210
San Antonio, TX, 147–9, 151; River, 150
San Diego, CA, 162–5
San Francisco, CA, 241, 254, 295
Santa Fe Trail, 268, 270, 297
Saskatchewan, 226
Savannah, GA, 173, 175–7
Savannas: in Africa, 57; map of, 81; in Mexico, 88, 133; in South, 105–6, 112, 115, 170, 171, 174, 180, 181, 192; in West Indies, 68–71, 81
Scotch-Irish, 196, 197
Scotland, 43–4, 48, 50–4
Scots, 15, 45, 51, 80, 113, 182, 283. *See also* Scotch-Irish
Scythians, 15
Semites, 14, 18. *See also* Arabs
Sevilla: cattle raisers in, 33, 35; diet in, 34; emigrants from, 21, 35, 65, 66, 309; hinterland of, 22; map showing, 21; as market, 25, 34, 73, 96; population of, 34; province of, 29; site of, 20; slaves in, 67
Sheep, 313; in Africa, 59, 60; in British Isles, 45, 47, 54; in California, 163, 166, 265; in Mexico, 96, 99–100, 102, 103, 128, 131, 134, 136, 142, 145; in Midwest, 202; in New Mexico, 146–7, 264; in South, 115; in Spain, 36–8, 39, 41; in Texas, 148, 151, 156, 211, 213, 219; in West, 142, 228, 239, 307; in West Indies, 70, 74
Shorthorns. *See* Cattle, shorthorn
Siemens, Alfred, 4
Simpson, Peter, 267, 283, 285
Slaughter, C. C. ("Lum"), 220, 270
Slavery: in Africa, 61, 67; in Mexico, 92; in South, 117, 175, 177, 180,

Slavery (*cont.*)
  188–9; in Spain, 34, 67; in West Indies, 66–7, 82–3, 85
Smith, J. S., 271, 273
Snipes, Ben, 289, 290, 291, 292
Soils, 20, 22–3, 44–5, 68, 128
Somerset, 43, 47, 49, 80, 113, 115
Sonora, 140–5, 162–3, 167, 228, 254–5, 298
South Carolina: breeds in, 108, 110; climate of, 113–4; emigrants from, 179, 189, 191; as hearth area, 170–207; herding in, 109–20, 171, 189–91, 193, 309; map of, 111; section on, 109–20; settlement of, 110–13; vegetation of, 114–5
South Dakota, 214–5, 226, 237–8, 271, 273
Spain, 18–42. *See also* Andalucía; Extremadura; Salamanca; Sevilla
Spaniards: in California, 159–69; in Iberia, 18–42; in Louisiana, 156–8; in Mexico, 86–104; in Missouri, 203; in New Mexico, 146–7; in South, 105–8, 156–8, 178, 183–4; in Texas, 147–57; in West Indies, 65–78
Spanish windlass, 259–60
Spinners (whirligigs), 25, 93, 152, 153, 157, 210
Splawn, A. J., 289, 290
Spurs, 132, 167, 211, 257–9
Staff, cattle, 40, 63, 94, 95, 102, 310. *See also* Lance
Steers. *See* Castration
Stewart, Mart, 177
Sudan, 16, 55–8
Swine: in British Isles, 45; in Iberia, 26, 36–7, 41; in Mexico, 96, 99, 100; in Midwest, 202; in South, 106–7, 110, 112, 114–5, 117, 120, 178, 180, 185, 191, 192, 197; in Texas, 187, 213, 218; in West, 161;
  in West Indies, 70, 72, 74, 76–8, 82, 84

Tallow, 8, 9, 51, 52, 74, 77, 107, 119, 120, 126, 177, 188, 191, 213, 218, 243, 247
Tamaulipas, 133–8, 150–5
Tanning, 73, 119, 120, 167, 188
Tejon Ranch, 262
Tennessee, 108, 197, 199, 200
Terrain: of Africa, 57–8; of Britain, 42–3; of California, 159; of Great Basin, 250, 291; of Great Plains, 223; influence of, 9; of Mexico, 86–7, 97, 123, 128, 138; of Spain, 20–3, 36; of Texas, 219; of West, 298–9; of West Indies, 67
Texas: Amerindians in, 183; Anglos in, 179, 180, 186–7, 204, 205–7, 208–40, 269; cattle from, 120, 200; chapter on, 208–40; dogs in, 181; environment of, 148, 150, 205–6; maps of, 124, 134, 209, 216, 224, 229; Midwestern influence in, 205–7, 269–70; ranching in, 205–7, 208–24; Spaniards in, 147–56, 179; spread from, 223–40, 254, 291, 300, 313
Thünen, J. H. von, 11–13, 98, 251, 308–9
Todhunter, W. B., 253, 285
Toponyms, 22, 32–3, 68–9, 78, 92, 112, 137
Trails, cattle: in British Isles, 51–3; in Canada, 292; in East, 180, 190, 192, 195, 196, 201, 218; in Great Plains, 209, 218, 221–2; maps of, 24, 53, 190, 209; in Mexico, 102, 126; photograph of, 102; in Spain, 23–4; in West, 228, 262–4, 280, 300. *See also* Chisholm Trail; Oregon Trail; Santa Fe Trail; Transhumance
Transhumance: in Appalachians,

438   INDEX

193, 195, 198–9; in British Isles, 47–8, 310; in California, 168, 262, 281; in Great Plains, 273–4; maps of routes, 24, 263; in Mexico, 95, 102–4, 128, 131, 145; in Midwest, 200, 202–3, 267; in New Mexico, 147; in Ozarks, 204; in Spain, 35–6, 39, 41, 310; in West, 251, 262–4, 286–7, 289. *See also* Nomadism

Turner, Frederick J., 13, 86, 110–1, 191, 198–9, 249, 290, 297, 308–9

Utah, 229–30, 237, 249, 276, 284, 296–7

Vega Real, 66, 69, 72, 75
Vegetation: of Africa, 57; of British Isles, 44; of California, 159–61; of Great Basin, 251, 292; of Great Plains, 222–3; maps of, 21, 90, 160, 172; of Mexico, 87–8, 125, 129, 133; of Midwest, 201–2; of Oregon, 277, 279; of Rockies, 299; of South, 105–6, 114–5, 170, 178, 184–5, 192–3; of Spain, 20–3, 35–6; of Texas, 148, 150, 215, 220, 227–8; of West, 228; of West Indies, 68–70. *See also* Canebrakes; Grasslands; Glades; Marshes; Pine barrens; Savannas
Veracruz: city, 88, 91, 92, 104; state, 86–97
Verdugo, José M., 165–6
Vikings, 42
Virginia, 110, 196, 197
Vocabulary. *See* Etymology

Waggoner, Dan, 212, 220, 269–70
Wales, 44, 45, 50, 52
Washington state: cattle sent to, 243, 248–50; cowboys in, 214–5; map of, 242; ranching in, 254, 277, 287–91, 295–6; spread from, 292, 300–1
Weaning, calf, 61, 118, 186, 233, 272, 313
Webb, Walter P., 9, 187, 308–9
Welsh, 15, 42, 54, 113, 283
West Indies: Africans in, 66–7, 72, 83; British in, 78–84; chapter on, 65–85; climate of, 67–8; French in, 84–5; as hearth area, 85; land tenure in, 74–6; livestock in, 70–85; map of, 66; as market, 178, 191; population of, 72; ranching in, 72–85; Spaniards in, 66–79; vegetation of, 68–70
West Virginia, 196, 197, 198, 294
West, Robert C., 4
Wetlands. *See* Canebrakes; Glades; Marshes
Whips: British, 52, 79, 310; illustration of, 233; in Midwest, 201, 202; in South, 118, 180, 193, 194; in Texas, 213, 269; in West, 233, 292; in West Indies, 79, 80
Whirligigs. *See* Spinners
Wichita, KS, 225, 271
Willamette Valley, 242, 245–6, 275–9, 282, 289
Windlasses, 259–61, 313
Wyoming: cowboys in, 214–5, 231, 272; haying in, 272, 304; irrigation in, 303–4; map of, 209; ranching in, 226, 230, 271, 301; terrain of, 298; transhumance in, 264; weather in, 237

XIT Ranch, 236

Yukon, 295